T0044621

Parasitology

Volume 123   *Supplement 2001*

Flukes and snails re-visited

EDITED BY

D. ROLLINSON

CO-ORDINATING EDITOR

L. H. CHAPPELL

CAMBRIDGE
UNIVERSITY PRESS

CAMBRIDGE UNIVERSITY PRESS
Cambridge, New York, Melbourne, Madrid, Cape Town, Singapore, São Paulo

Cambridge University Press
The Edinburgh Building, Cambridge CB2 8RU, UK

Published in the United States of America by Cambridge University Press, New York

www.cambridge.org
Information on this title: www.cambridge.org/9780521891066

© Cambridge University Press 2001

This publication is in copyright. Subject to statutory exception
and to the provisions of relevant collective licensing agreements,
no reproduction of any part may take place without the written
permission of Cambridge University Press.

First published 2001

A catalogue record for this publication is available from the British Library

ISBN 978-0-521-89106-6 paperback

Transferred to digital printing 2008

Front Cover Illustration: A. Scanning electron microscope image of male and female *Schistosoma nasale* (courtesy of V. R. Southgate). B. Tentacle of *Biomphalaria glabrata* enlarged by the presence of a mother sporocyst of *Schistosoma mansoni,* that is releasing daughter sporocysts (courtesy of E. S. Loker and A. Woodwards). C. Shell of *Bulinus forskalii*. D. *Dicrocoelium dendriticum* cercariae (courtesy of M. Y. Manga-Gonzalez, this volume).

Background Illustration: Transmission electron micrograph of a section of a trypanosome-infected tsetse fly salivary gland showing the attachment junctions between parasite flagellum and the microvillar border of the salivary gland epithelium. Original micrograph, Dr L. Tetley, University of Glasgow.

Contents

Contents   v

Contents

List of contributions

Flukes and snails re-visited

EDITED BY D. ROLLINSON

CO-ORDINATING EDITOR L. H. CHAPPELL

Preface

It has been a pleasure and a privilege to edit this special supplement on trematodes and their snail hosts. The task has been made much easier by enthusiastic authors, who have contributed an excellent mix of authoritative reviews and research papers and made exceptional efforts to keep to deadlines. I must also acknowledge Dr Les Chappell, the co-ordinating editor, who has been the bedrock, providing the gentle prompts and professional advice when required.

Those with longer memories will recall the book from which we take our title 'Flukes and Snails' by Chris Wright published by George Allen and Unwin Ltd in 1971. This rather special volume set out many new ideas and interpretations concerning the relationships between flukes and snails. The book found its way onto the shelves of malacologists and parasitologists of the day and indeed made a significant impact on thinking and future research. Many of the chapters in this supplement find supportive quotes and citations from the original, illustrating the pioneering value of Wright's endeavours. Fortunately, the publishers must have over-estimated sales, as for many years students and others could find the paperback version in the bargain basements of book shops all around London. Could it be that Chris, to his credit, used his persuasive powers to convince the publishers of the importance of the subject that was so dear to his heart?

This supplement to volume 123 *Parasitology* will testify to the progress that has been made over recent years in unravelling the complexities of the snail-parasite relationship. The papers draw on the considerable expertise of leading research scientists from around the world and cover many different aspects including ecology, parasite transmission, parasite interactions, evolutionary biology, molecular systematics and snail defence systems. By necessity the subjects reflect the main research interests of the contributors but, taken as whole, I hope the supplement will provide a useful overview of snail-parasite interactions. More importantly, I hope it will also provide a stimulus for future research. For snail–parasite biologists there is much to do and much to be learnt by incorporating new techniques to study the intricacies of host parasite relationships. Genomic and post-genomic studies are set to revolutionise much of biology; the schistosome genome project is making good progress (http://www.nhm.ac.uk/hosted_sites/schisto/index.html) but greater efforts are needed to unravel the complexities of the molluscan genome.

In the following pages will be found fascinating accounts and details of the complex interactions between trematodes and their intermediate hosts. Those that read carefully may well find small differences of opinion and interpretation and identify areas where further study is required. One thing however is quite clear 'the final laugh is still with the flukes and their snail hosts' (Chris Wright 1971, *Flukes and Snails*).

DAVID ROLLINSON
May 2001

Parasitology (2001), **123**, S1.   © 2001 Cambridge University Press
DOI: 10.1017/S0031182001008447

Snail–trematode life history interactions: past trends and future directions

R. E. SORENSEN* and D. J. MINCHELLA

Department of Biological Sciences, Purdue University, West Lafayette, IN 47907, USA

SUMMARY

Life history traits expressed by organisms vary due to ecological and evolutionary constraints imposed by their current environmental conditions and genetic heritage. Trematodes often alter the life history of their host snails by inducing parasitic castration. Our understanding of the variables that influence the resulting changes in host growth, fecundity and survivorship is insufficient to confidently predict specific outcomes of novel snail–trematode combinations. In a literature review of the last 30 years, we found 41 publications examining various life history characteristics of trematode-infected snails. These publications reported 113 different field and laboratory experiments involving 30 snail species and 39 trematode species and provided a data set for assessing factors that potentially affect life history outcomes. Analysis of the diverse responses across various snail–trematode systems and experimental conditions teased out general patterns for the expression of host growth, fecundity and survival. These were used to address existing hypotheses and develop several new ones relating the response of snail-trematode interactions to environmental and genetic factors. Finally, we propose directions for future experiments that will better assess the ecological and evolutionary factors influencing snail life history responses to trematode parasitism.

Key words: Castration, gigantism, stunting, trematode, snail life history, fecundity compensation.

INTRODUCTION

The life history of an organism is shaped by energetic trade-offs between current opportunities and future fitness gains (Williams, 1966; Stearns, 1976). Life history traits are typically separated into three interrelated categories: those associated with the growth of the individual over the short term, those yielding progeny either sexually or asexually, and those influencing health and long-term survival (Stearns, 1992). The relative investment that an organism puts into growth, reproduction and survival can depend heavily on prevailing environmental conditions; as a result, life history traits tend to diverge across populations subjected to distinct environmental influences (Law, 1979).

Parasites and parasitoids are unique among organisms in that a host is their habitat, and fitness gains for the inhabitant are accrued at a direct expense to the habitat (Anderson & May, 1979; Holmes, 1983). Hosts, on the other hand, gain fitness benefits when they employ adaptations preventing or counteracting detrimental effects of the parasite (Price, 1980; Bayne & Loker, 1987). The selective forces working within host and parasite populations turn hosts into biotic battlefields where parasites strive to convert the host's energy and resources into parasite tissue, while at the same time hosts try to avert these

reallocation efforts. Thus, two distinct genotypes interact to shape the host's phenotype, that of the parasite diverts host assets from their intended use and that of the host retaliates to prevent further losses (Dawkins, 1982). This conflict plays itself out at two levels. At the proximate level, this conflict involves the direct day-to-day competitive interactions between the two agonists as they attempt to use the same resources. The ultimate level of antagonism expressed by parasites and their hosts depends on microevolutionary processes which establish a dynamic equilibrium between the two participants that yields optimal fitness values for both parties in light of the inherent discord (Poulin, 1998). Parasites are generally given the upper hand, evolutionarily speaking, because their generation times tend to be shorter than that of their hosts (May & Anderson, 1979). However, the degree of damage a parasite ultimately inflicts upon its host is kept in check by the requirement that the host survive long enough to allow the parasite to mature and spread its offspring among subsequent hosts (Ewald, 1994). The key point here is that a host's traits must be interpreted in terms of the life history strategies of both the parasite and the host, remembering that each participant strives to maximize its own fitness both proximately and ultimately.

Some of the best evidence of antagonistic interactions between parasites and hosts involve alteration of a host's fecundity and somatic growth schedules (Sousa, 1983; Minchella, 1985). Terms commonly used to describe these phenomena include parasitic

* Corresponding author: R. E. Sorensen, Department of Biological Sciences, Purdue University, West Lafayette, IN 47907. Tel: (765) 494-9103. Fax: (765) 494-0876. E-mail: sorensen@purdue.edu

Parasitology (2001), **123**, S3–S18. © 2001 Cambridge University Press
DOI: 10.1017/S0031182001007843

Table 1. Summary of field studies examining life history traits of infected marine prosobranch snails. Parasite species are described according to the predominant intramolluscan larvae (Lrv). Associations between increased prevalence and host life-history traits are identified in the 'Prevalence Bias' column. 'Host Traits' refers to the growth (Grwth), fecundity (Fec), and survival (Surv) patterns of hosts are relative to uninfected snails. An '=' in this column means no trends were noted distinguishing infected and uninfected individuals; conversely, '+' or '–' symbols indicate that hosts showed either increases or decreases in a trait relative to uninfected snails, respectively. Types of experiments (Exp. Type) were identified as either surveys or mark-recaptures (Mk-Recap). The abbreviation n/a means information on this feature was not available in the original publication

Host	Parasite Species	Lrv*	Prevalence Bias	Host Traits Grwth	Fec	Surv	Exp. Type	Reference
Littorina littorea	Cryptocotyle lingua	R	size	=	n/a	n/a	Mk-Recap	Hughes & Answer (1982)
L. littorea	C. lingua	R	n/a	=	–	n/a	Survey	Mouritsen et al. (1999)
L. littorea	Renicola roscovita	S	size	=	n/a	n/a	Mk-Recap	Hughes & Answer (1982)
L. littorea (Rønbjerg)	R. roscovita	S	size	–	–	n/a	Survey	Mouritsen et al. (1999)
L. littorea (Grenå)	R. roscovita	S	n/a	=	–	n/a	Survey	Mouritsen et al. (1999)
L. littorea	Himasthla leptosoma	R	size	=	–	n/a	Survey	Mouritsen et al. (1999)
Littorina saxatilis	Various spp.	n/a	age	n/a	n/a	n/a	Survey	Sokolova (1995)
L. saxatilis	Microphallus piriformes	S	size	=	–	n/a	Survey	Gorbushin & Levakin (1999)
Littorina obtusa	M. piriformes	S	size	=	–	n/a	Survey	Gorbushin & Levakin (1999)
Onoba aculeus	Microphallus pseudopygmaeus	S	size	+	–	n/a	Survey	Gorbushin & Levakin (1999)
Cerithidea californica	Echinoparyphium sp.	R	smaller†	n/a	n/a	n/a	Survey	Sousa (1983)
C. californica	Acanthoparyphium spinulosum	R	smaller†	n/a	n/a	n/a	Survey	Sousa (1983)
C. californica	Parorchis acanthus	R	larger†	n/a	n/a	n/a	Survey	Sousa (1983)
C. californica	Himasthla rhigedana	R	larger†	n/a	n/a	n/a	Survey	Sousa (1983)
C. californica	Euhaplorchis californiensis	R	intermediate†	n/a	n/a	n/a	Survey	Sousa (1983)
C. californica	Catatropis johnstoni	R	intermediate†	n/a	n/a	n/a	Survey	Sousa (1983)
C. californica	Microphallid sp.	S	intermediate†	n/a	n/a	n/a	Survey	Sousa (1983)
C. californica (Juvenile)	Various spp.	R	n/a	–	n/a	n/a	Mk-Recap	Sousa (1983)
C. californica (Juvenile)	Microphallid sp.	S	n/a	=	n/a	n/a	Mk-Recap	Sousa (1983)
C. californica (Adult)	Catatropis or Echinoparyphium	R	n/a	–	n/a	n/a	Mk-Recap	Sousa (1983)
C. californica (Adult)	Various spp.	n/a	n/a	=	n/a	n/a	Mk-Recap	Sousa (1983)
C. californica	Various spp.	n/a	size	–	n/a	n/a	Survey	Lafferty (1993)
C. californica	Various spp.	n/a	n/a	–	n/a	–	Mk-Recap	Lafferty (1993)
Ilyanassa obsoleta	Zoogonus lasius	S	shell weight	n/a	n/a	n/a	Survey	Cheng et al. (1983)
I. obsoleta	Various spp.	n/a	none	n/a	n/a	n/a	Survey	Cheng et al. (1983)
I. obsoleta	Various spp.	n/a	size, age	–	n/a	n/a	Mk-Recap	Curtis (1995)
Buccinum undatum	Neophasis sp.	R	size, age, sex	n/a	–	n/a	Survey	Tetreault et al. (2000)

* Predominant intramolluscan larvae present: S = sporocysts, R = rediae.

† Different parasite species were more prevalent among different-sized hosts (e.g. "smaller" means prevalence of this trematode was highest among smaller hosts).

Table 2. Summary of field studies examining life history traits of infected freshwater snails. Parasite species are described according to the predominant intramolluscan larvae (Lrv). Associations between increased prevalence and host life-history traits are identified in the 'Prevalence Bias' column. 'Host Traits' refers to the growth (Grwth), fecundity (Fec), and survival (Surv) patterns of hosts are relative to uninfected snails. An '=' in this column means no trends were noted distinguishing infected and uninfected individuals; conversely, '+' or '−' symbols indicate that hosts showed either increases or decreases in a trait relative to uninfected snails, respectively. Types of experiments (Exp. Type) were identified as either surveys or mark-recaptures (Mk-Recap). The abbreviation n/a means information on this feature was not available in the original publication

Host	Parasite Species	Lrv*	Prevalence Bias	Host Traits Grwth	Fec	Surv	Exp. Type	Reference
Subclass Prosobranchia								
Hydrobia ulvae	Various spp.	n/a	size	+	n/a	n/a	Survey	Huxham et al. (1995)
H. ulvae (Juvenile)	Cryptocotyle sp.	R	n/a	=	−	n/a	Survey	Gorbushin (1997)
H. ulvae (Adult)	Cryptocotyle sp.	R	n/a	+	−	n/a	Survey	Gorbushin (1997)
H. ulvae (Dangast)	Himasthla sp.	R	n/a	=	−	n/a	Survey	Gorbushin (1997)
H. ulvae (Lendrup)	Himasthla sp.	R	n/a	+	−	n/a	Survey	Gorbushin (1997)
H. ulvae	Cryptocotyle sp.	R	n/a	+	−	n/a	Survey	Gorbushin (1997)
H. ulvae	Bunocotyle progenetica	R	n/a	+	−	n/a	Survey	Gorbushin (1997)
H. ulvae	Maritrema subdolum	S	n/a	=	−	n/a	Survey	Gorbushin (1997)
H. ulvae	Microphallus claviformis	S	n/a	+	−	n/a	Survey	Gorbushin (1997)
H. ulvae	Microphallus pirum	S	n/a	+	−	n/a	Survey	Gorbushin (1997)
Hydrobia ventrosa	Notocotylus sp.	R	n/a	=	−	n/a	Survey	Gorbushin (1997)
H. ventrosa	Cryptocotyle sp.	R	n/a	+	−	n/a	Survey	Gorbushin (1997)
H. ventrosa	Bunocotyle progenetica	R	n/a	+	−	n/a	Survey	Gorbushin (1997)
H. ventrosa (low density)	B. progenetica	R	n/a	+	n/a	n/a	Mk-Recap	Gorbushin (1997)
H. ventrosa (high density)	B. progenetica	R	n/a	=	n/a	n/a	Mk-Recap	Gorbushin (1997)
Subclass Pulmonata								
Lymnaea elodes	Various spp.	n/a	size	−	−	+	Both	Minchella et al. (1985)
L. elodes	Various spp.	n/a	size	n/a	n/a	n/a	Survey	Brown et al. (1988)
L. elodes (5 study sites)	Various spp.	n/a	size	n/a	n/a	n/a	Survey	Sorensen & Minchella (1998)
L. elodes (Shock Lake)	Various spp.	n/a	none	n/a	n/a	n/a	Survey	Sorensen & Minchella (1998)
Helisoma anceps	Various spp.	n/a	size	n/a	−	n/a	Both	Crews & Esch (1986)
H. anceps	Halipegus occidualis (mature)	R	n/a	n/a	−	n/a	Mk-Recap	Crews & Esch (1986)
H. anceps	H. occidualis (immature)	R	n/a	n/a	=	n/a	Mk-Recap	Crews & Esch (1986)
H. anceps	H. occidualis	R	size	=	n/a	=	Mk-Recap	Goater et al. (1986)
H. anceps	Megalodiscus temperatus	R	none	n/a	n/a	n/a	Mk-Recap	Goater et al. (1986)
H. anceps	Diplostomulum scheuringi	S	none	n/a	n/a	n/a	Mk-Recap	Goater et al. (1986)
H. anceps (Juvenile)	H. occidualis	R	n/a	−	n/a	n/a	Mk-Recap	Fernandez & Esch (1991)
H. anceps (Adult)	H. occidualis	R	n/a	=	n/a	n/a	Mk-Recap	Fernandez & Esch (1991)
Helisoma trivolvis	Echinostoma trivolvis	R	age, size	n/a	n/a	n/a	Survey	Sorensen (unpubl.)
H. trivolvis	Plagiorchis sp.	S	age, size	n/a	n/a	n/a	Survey	Sorensen (unpubl.)

* Predominant intramolluscan larvae present: S = sporocysts, R = rediae.

castration, gigantism and stunting (Rothschild, 1936; Wright, 1971; Baudoin, 1975). Although definitions for these phenomena are not strict and vary from study to study, the general effects are well recognized. Castration involves a reduction or complete cessation in the production of host offspring coincident with the development and maturation of parasites. Gigantism and stunting, as the names imply, refer to hosts that are atypically large or small, respectively, when compared to other members of their population or cohort. For the sake of this paper, we view gigantism and stunting as time-specific phenomena and not as ultimate endpoints. We chose this perspective because both host growth (in terms of a size increase) and parasite growth (in terms of number of offspring produced) are dynamic processes. In other words, hosts may demonstrate enhanced growth during one period of an infection and stunting later as the parasite exacts a greater toll on host resources. Although gigantism and stunting appear in a number of invertebrate taxa, the most cited examples are from molluscs, especially snails infected with trematodes (Poulin, 1998). In this paper, we focus on these life history traits as they relate to flukes and snails; however, several of our arguments may also have implications for other parasite-host systems.

Parasite-induced snail life history changes, like castration and gigantism, are readily observable and well described, but attempts to discern the ultimate causes for these modifications tend to provide more conjecture than hard evidence. The following two paragraphs briefly review the hypotheses that have been offered to explain host life history alterations. Explanations for castration, stunting and gigantism often posit parasite control over host reproductive activities with the energy and resources diverted from host reproduction primarily serving the needs of the intruder (Baudoin, 1975; Dawkins, 1982). Therefore, a host only has access to the energy and resources remaining after the parasite takes its share and stunting occurs when the parasite leaves little or nothing for the host to apply to its own growth. Synergistic interactions between growth and reproduction provide a means to associate gigantism with castration. In this case, increased growth results from energy made available to the host once parasitic castration impedes the host's reproductive activities (Sousa, 1983). Thus, when castration frees more host resources than the parasite currently requires, the host channels some of the remaining energy and resources into its growth rather than their intended use in reproductive activities.

As an alternative to total control by the parasite, some hosts may alter either their reproductive activities or growth, presumably to enhance survival, which could increase the probability of outliving or sequestering the parasite. Variations on this theme suggest that the parasite eventually causes castration but the host gains short-term fitness advantages by increasing either its fecundity or its growth early in the infection before the castration is complete (Minchella, 1985). Accordingly, individuals reproducing at a greater rate prior to castration maximize their fitness before their reproductive death. Under the 'increased growth before castration' scenario, enhanced host growth rate early in an infection yields future fitness rewards by at least two mechanisms (Sorensen & Minchella, 1998). Since host fecundity typically is a function of size, a growth spurt early in an infection may provide future fecundity rewards in the few weeks before castration ensues and warrant the increased investment in growth. Similarly, in cases where mechanical destruction of the ovotestes causes castration, larger hosts might maintain viable reproductive tissues longer than smaller ones. Overall, these explanations are reasonable and they undoubtedly describe the host-parasite systems for which they were developed. However, a better understanding of the interaction would result if researchers could associate the expected outcomes with natural life history characteristics of the snail and the trematode.

Wright's (1971) seminal book, *Flukes and Snails*, contains a concise discussion of trematode influences on snail life history traits and forms the groundwork for developing just such a systematic interpretation of this interaction. His discussion of the topic reviewed existing literature as it related to the various proximate mechanisms inducing castration and gigantism. He also acknowledged that the life history consequences of infection likely varied across parasite and host species, environmental conditions and host age. Finally, Wright addressed the strengths of certain experimental methods and prescribed studies to bolster our understanding of the interaction between flukes and snails. Many studies of the host life history consequences associated with trematode infection have been performed during the three decades since publication of Wright's book. These studies, in combination with those of earlier researchers, provide an empirical database that allows tests of predictions about the interaction between the life histories of parasites and their hosts.

Our goal is to meld ideas from life history theory with empirical studies of snail trematode systems in order to develop a more comprehensive view of the host-parasite interaction. The remainder of this paper is divided into three sections. The first explains how information was assembled from a variety of studies assessing parasite influences on host life history in order to create a summary data set (Tables 1–4). The second section uses this data set to identify associations between experimental variables, parasite characteristics and host life history patterns. In the end, we synthesize ideas from life history theory with the empirical outcomes of the second section and suggest directions for future research efforts,

Table 3. Summary of laboratory studies examining life history traits for infected prosobranch and planorbid snails. Maturity of snails is identified in the 'Age' column (J = juvenile, A = adolescent, M = mature adult). The predominant intramolluscan stage in the parasite life-cycle is given in the 'Lrv' column (R = rediae, S = sporocysts). Life history characteristics for the hosts are distinguished based upon the stage of parasite development (Pre = prepatent stage of infection before cercariae are produced, Patnt = patent stage of infection during which cercariae are produced, End = size difference between hosts and uninfected snails when the data was last collected). Symbols (=, +, −) used here refer to the extent a trait is expressed in hosts relative to uninfected individuals. The 'Conditions' column tells how many snails were housed together in experimental containers (Hsg; S = housed singly, G = housed in groups) and how many miracidia were used per snail during the exposure phase of the experiment (Mr/Snl). The abbreviation n/a designates information not available from the original publication

Host	Age	Parasite	Host Size				Fecundity		Survival		Conditions		References
			Lrv	Pre	Patnt	End	Pre	Patnt	Pre	Patnt	Hsg	Mr/Snl	
Subclass Prosobranchia													
Cerithidea californica	A	Various species	n/a	n/a	n/a	n/a	n/a	n/a	n/a	+*	G	n/a	Sousa & Gleason (1989)
Potamopyrgus antipodarum	A	Microphallus sp.	S	−*	−*	−*	−	−	n/a	n/a	G	n/a	Krist & Lively (1998)
Subclass Pulmonata Family Planorbidae													
Biomphalaria glabrata	J	Schistosoma mansoni	S	−	+*	−*	−†	−†	n/a	n/a	S	1	Théron et al. (1992a)
B. glabrata	J	S. mansoni	S	+	−*	−*	−†	−†	n/a	n/a	S	1	Gérard et al. (1993)
B. glabrata	J	S. mansoni	S	+*‡	−*	−*	−†‡	−†‡	n/a	n/a	S	8	Gérard et al. (1993)
B. glabrata	J	S. mansoni	S	+*†	−*	−*†	n/a	−*	n/a	n/a	S	1	Gérard & Théron (1995)
B. glabrata	J	S. mansoni	S	+*	−*	−*	=	−*	n/a	n/a	S	1	Gérard & Théron (1997)
B. glabrata	J	Echinostoma liei	R	+	+*	−*	+*	−*	−*	−*	G	n/a	Kuris (1980)
B. glabrata	A	S. mansoni	S	+	+*	+*	+*	−*	n/a	n/a	S	7	Thornhill et al. (1986)
B. glabrata	A	S. mansoni	S	−	+*	+*	n/a	n/a	n/a	n/a	S	1	Théron et al. (1992b)
B. glabrata	A	S. mansoni	S	−	+*	+*	−	−*	n/a	−*	S	8	Théron et al. (1992b)
B. glabrata	M	S. mansoni	S	n/a	n/a	n/a	+*	−*	n/a	n/a	S	10	Minchella & LoVerde (1981)
B. glabrata	M	S. mansoni	S	+*	=	=	+*	−*	n/a	n/a	G	12	Crews & Yoshino (1989)
B. glabrata	M	S. mansoni	S	=	=	=	n/a	n/a	n/a	n/a	S	1	Gérard et al. (1993)
B. glabrata	M	S. mansoni	S	=	=	=	n/a	n/a	n/a	n/a	S	8	Gérard et al. (1993)
B. glabrata	M	S. mansoni	S	+*†	−*†	−*†	+*†	−*	n/a	n/a	S	1	Gérard & Théron (1995)
B. glabrata	M	S. mansoni	S	+*	+*	−*	+*	−*	n/a	n/a	S	1	Gérard & Théron (1997)
Biomphalaria pfeifferi	A	S. mansoni	S	n/a	n/a	n/a	−	+	=	=	G	1	Makanga (1981)
B. pfeifferi	A	S. mansoni	S	n/a	n/a	n/a	+	−	=	−	G	2	Makanga (1981)
B. pfeifferi	A	S. mansoni	S	n/a	n/a	n/a	−	−	−	−	G	3	Makanga (1981)
Bulinus natalensis	J	Schistosoma margrebowei	S	−*	−*	−*	n/a	n/a	n/a	n/a	S	5	Raymond & Probert (1993)
B. natalensis	A	S. margrebowei	S	=	−	−	=	−	n/a	n/a	S	5	Raymond & Probert (1993)
B. natalensis	M	S. margrebowei	S	=	−	−	=	−	n/a	n/a	S	5	Raymond & Probert (1993)
Bulinus truncatus	M	Schistosoma haematobium	S	n/a	n/a	n/a	−*	−*	=	=	G/S	5	Schrag & Rollinson (1994)
B. truncatus	J	S. haematobium "truncatus"	S	n/a	+*	+	n/a	+	n/a	+	G/S	5	Fryer et al. (1990)
Bulinus senegalensis	J	S. haematobium "globosus"	S	n/a	+	+	n/a	+	n/a	+	G/S	5	Fryer et al. (1990)
B. senegalensis	J	S. haematobium "truncatus"	S	n/a	+*	+	n/a	+*	n/a	n/a	G/S	5	Fryer et al. (1990)
Bulinus globosus	J	S. haematobium "globosus"	S	n/a	−	−	n/a	+*	n/a	−*	G/S	5	Fryer et al. (1990)
Helisoma anceps (Low)	A	Halipegus occidualis	R	−	+*	+*	=	−	n/a	+	G	3	Keas & Esch (1997)
H. anceps (High)	A	H. occidualis	R	=	+*	+*	−	−	n/a	+	G	3	Keas & Esch (1997)
H. anceps (Low)	M	H. occidualis	R	+	=	=	+	−	n/a	−	G	3	Keas & Esch (1997)
H. anceps (High)	M	H. occidualis	R	−	−	=	−	−	n/a	+	G	3	Keas & Esch (1997)

* Original authors report statistically significant differences at $P < 0.05$ at some point during the described time period.
† Growth of digestive gland area (mm²) used as measure of investment in growth.
‡ Growth of gonad gland area (mm²) used as measure of investment in fecundity.

Table 4. Summary of laboratory studies examining life history traits for infected lymnaeid snails. Maturity of snails is identified in the 'Age' column (J = juvenile, A = adolescent, M = mature adult). The predominant intramolluscan stage in the parasite life-cycle is given in the 'Lrv' column (R = rediae, S = sporocysts). Life history characteristics for the hosts are distinguished based upon the stage of parasite development (Pre = prepatent stage of infection before cercariae are produced, Patnt = patent stage of infection during which cercariae are produced, End = size difference between hosts and uninfected snails when the data was last collected). Symbols (=, +, −) used here refer to the extent a trait is expressed in hosts relative to uninfected individuals. The 'Conditions' column tells how many snails were housed together in experimental containers (Hsg; S = housed singly, G = housed in groups) and how many miracidia were used per snail during the exposure phase of the experiment (Mr/Snl). The abbreviation n/a designates information not available from the original publication

Host	Age	Parasite		Host Size			Fecundity		Survival		Conditions		References
			Lrv	Pre	Patnt	End	Pre	Patnt	Pre	Patnt	Hsg	Mr/Snl	
Subclass Pulmonata Family Lymnaeidae													
Lymnaea catascopium	J	*Schistosomatium douthitti*	S	n/a	n/a	n/a	=	−*	−	+	G	3	Loker (1979)
L. catascopium	M	*S. douthitti*	S	+*	−*	−*	−*	−*	=	−*	S	3	Loker (1979)
L. catascopium	M	*S. douthitti*	S	n/a	n/a	n/a	n/a	n/a	−	−*	S	10	Loker (1979)
Lymnaea stagnalis	A	*Trichobilharzia ocellata*	S	−	+	+	+*	−*	n/a	n/a	S	40	Schallig et al. (1991)
L. stagnalis	M	*T. ocellata*	S	=	+	+	+	−*	n/a	n/a	S	40	Schallig et al. (1991)
Lymnaea elodes	J	*Plagiorchis elegans*	S	+*	+*	+*	−*	−*	=	−*	S	1	Zakikhani & Rau (1999)
L. elodes	J	*P. elegans*	S	+*	+*	+*	−*	−*	n/a	n/a	S	4	Zakikhani & Rau (1999)
L. elodes	J	*P. elegans*	S	+*	+*	+*	−*	−*	−*	−*	S	8	Zakikhani & Rau (1999)
L. elodes	J	*P. elegans*	S	=	=	=	−*	−*	−*	−*	S	16	Zakikhani & Rau (1999)
L. elodes	A	*P. elegans*	S	+*	+*	+*	−*	−*	=	−	S	1	Zakikhani & Rau (1999)
L. elodes	A	*P. elegans*	S	+*	+*	+*	−*	−*	n/a	n/a	S	4	Zakikhani & Rau (1999)
L. elodes	A.	*P. elegans*	S	=	=	=	−*	−*	=	−*	S	8	Zakikhani & Rau (1999)
L. elodes	A	*P. elegans*	S	+*	+*	+*	−*	−*	−*	−*	S	16	Zakikhani & Rau (1999)
L. elodes	M	*P. elegans*	S	+*	+*	+*	−*	−*	=	−	S	1	Zakikhani & Rau (1999)
L. elodes	M	*P. elegans*	S	=	+*	+*	−*	−*	=	−*	S	4	Zakikhani & Rau (1999)
L. elodes	M	*P. elegans*	S	+*	+*	=	−*	−*	n/a	−	S	8	Zakikhani & Rau (1999)
L. elodes	M	*P. elegans*	S	+*	+*	+*	−*	−*	=	−*	S	16	Zakikhani & Rau (1999)
L. elodes	M	*Echinostoma revolutum*	R	+*	+*	+*	−*	−*	−*	−*	S	10	Sorensen & Minchella (1998)
Lymnaea truncatula	J	*Fasciola hepatica*	R	=	+	+*	−	−*	−	+*	G	5	Hodasi (1972)
L. truncatula	A	*F. hepatica*	R	=	+*	+*	−	−*	n/a	n/a	G	5	Hodasi (1972)
L. truncatula	A	*F. hepatica*	R	=	+*	+*	−	−*	−	=	G	5	Wilson & Denison (1980)
L. truncatula	M	*F. hepatica*	R	+	+	+	−*	−*	n/a	n/a	G	5	Hodasi (1972)
L. truncatula	M	*F. hepatica*	R	n/a	n/a	−*	−*	−*	−*	−*	S	1–3	Dreyfuss et al. (1999)
Subclass Pulmonata Family Physidae													
Physa sayii	n/a	Echinostome sp.	R	n/a	+*	+*	n/a	n/a	n/a	n/a	n/a	20	Cheng (1971)

* Original authors report statistically significant differences at $P < 0.05$ at some point during the described time period.

which offer the most promise for evaluating the concepts presented in this review.

THE DATA SET − USES AND LIMITATIONS

To compile the studies reviewed in this paper, we searched journals containing primary literature on the topic written in English since 1970. We found 41 publications examining various life history characteristics of snails infected with trematode parasites. Sixteen of them contained only laboratory data while another 25 contained either field data exclusively or a combination of field and laboratory data. Furthermore, the publications contained 113 different studies of various experimental treatments or conditions (56 field studies and 57 laboratory studies) using 30 snail species and 39 trematode species. Very few of the hosts considered in these studies were concurrently infected by multiple trematode species; therefore, we limited our review and discussion to hosts infected by a single trematode species. Although we tried to be thorough in our search, it is likely that we overlooked some works. We hope that excluded researchers understand the difficulty in scouring 30 years of literature.

Tables 1–4 compare the expression of growth, fecundity and survival traits for infected snails relative to uninfected snails. Symbols in these tables depict a comparison of the infected snail's phenotype relative to that of uninfected individuals. We chose this approach in our scoring criteria because the original papers often lacked sufficient statistical analysis to allow strict consideration of the comparisons we sought. Rather than omitting or attempting to reanalyze the original data, relative differences based on figures or tables served as the basis for a number of our comparisons.

These tables are organized according to features of the host and the parasite as well as experimental considerations. Tables 1 and 2 summarize the results of field studies whereas Tables 3 and 4 describe experiments conducted under laboratory conditions. The choice to separate the studies in this manner rather than another organizing factor is somewhat arbitrary, but it is useful in that the degree of resolution in field and laboratory studies differs substantially. For the field studies, we identify whether parasite prevalence correlates with host characteristics, such as age, size or gender, and describe how parasitism influences expression of host life history traits. For laboratory studies, we document expression of the host traits during two stages of the parasite's development: the prepatent period (before cercariae are released) and the patent period (when cercariae are present). Cercariae release was presumed to begin at five weeks unless noted otherwise by the original authors. An additional category was required to describe growth during patency under laboratory conditions because patterns of size among infected and uninfected snails periodically reversed during this time. Therefore, we report growth differences between infected and uninfected individuals during patency's early stages and again at the end of each experiment.

Because of the inherent conflict between trematodes and their hosts we assume that parasites always exact a cost on host fitness and in response, hosts always strive to minimize this cost. A host's ability to counter the negative aspects of parasitism could depend upon a multitude of variables. The variables we postulate as being important are reported in each of the tables and these include: the predominant intramolluscan parasite stage, the phylogenetic association of the trematode and snail, the host's age, the density of the host population and the number of miracidia used to infect hosts.

The information from the large number of studies we reviewed provides a useful data set for evaluating the various hypotheses used to explain the effects of intramolluscan trematode larvae on host life history traits; however, there may be limitations in this data worth addressing at the outset. We postulate that the factors identified in the tables (i.e. parasite species, host maturity, housing conditions, etc.) exert the principal forces generating the observed life history traits, recognizing that many other factors also influence phenotypic expression. When comparing across studies, failure to identify the most influential components of any one study promotes variation across the compiled works, impeding our ability to draw generalizations. Likewise, variation in the methodology of different studies sometimes necessitates comparison of dissimilar metrics. This type of variation may be introduced when comparing the growth and fecundity characteristics of snails across studies where these traits were assessed in different manners. For instance, most studies measured growth based on changes in shell size over time; in other cases, tissue mass or volume was measured. For fecundity, most researchers counted the number of eggs that were released, but other means of assessing reproductive activity were also used. For instance, a few investigators examined the number of eggs and/or the amount of sperm within the ovotestes, while others measured the actual volume of the reproductive organs in dissected individuals to assess reproductive investment. Each of the various estimates of growth or reproductive ability were treated equally in our analysis.

At this point, the value of any comparisons drawn from these tables may seem questionable when so many factors potentially hinder recognition of patterns in the data. To the contrary, recognizing the limitations associated with a data set strengthens claims drawn from observed patterns in that data. In other words, robust patterns may still emerge in spite of weaknesses associated with a data set. Furthermore, when robust patterns do not appear, comparisons remain valuable because they potentially demonstrate differences in the design and conduct of experiments that are important. Thus, they become worthwhile candidates as experimental variables for future studies.

RESULTS – IDENTIFYING PATTERNS

Growth

Laboratory studies and field mark-recapture techniques facilitate the assessment of size as a function of time allowing a direct determination of host growth rates. Surveys, on the other hand, typically yield a single measure of host size at the time of the snail's collection, hence they typically lack an estimate of growth. A technique recently developed by Gorbushin (1997) vastly improves the information gained from field surveys, because it quantifies the amount of shell material added beyond the last annual growth line. Nonetheless, it is important to remember that all field studies using naturally infected snails cannot adequately assess host consequences during prepatency since the timing of the infection onset remains unknown.

One method that has been used frequently to determine whether trematodes alter their host's life history is to assess whether prevalence varies as a function of host size among field collected snails. An association between prevalence and large size might be due to enhanced growth of hosts following infection; other possibilities, such as increased survival without a concomitant reduction in growth, or higher infection rates for large snails would produce the same effect (Baudoin, 1975). For the field studies reviewed (Tables 1 and 2), a positive association between snail size and parasite presence (prevalence) occurred in 21 of 30 cases. This trend was detected in both marine and freshwater snail species with 68 % and 73 %, respectively, showing increased rates of parasitism among larger individuals. The type of intramolluscan larvae infecting the snails did not seem to influence strongly the relationship between prevalence and host size, since no patterns emerged to distinguish the effects of sporocysts and rediae at this level of resolution.

Much stronger evidence for parasite-induced changes in host somatic growth are provided when the growth rates of infected and uninfected individuals are compared instead of comparing size differences at a particular time. Thirty-four of the reviewed field comparisons provided this type of information based on the addition of shell material. A noteworthy outcome was that freshwater snails seemed much more likely to display gigantism than marine snails. Of 19 comparisons using freshwater snails, 10 (53 %) indicated that hosts grew faster than uninfected individuals, while only 1 of 15 studies involving marine snails shared the same result. On the other hand, stunting appeared more commonly among marine snails (40 % of 15 comparisons) than among freshwater snails (11 % of 19 comparisons). As with the prevalence data discussed earlier, we examined whether relative growth differences existed among studies involving either sporocysts or rediae and, again, no definitive patterns emerged. We also attempted to discern age-specific growth differences following parasitism but too few field studies existed to allow confident consideration of this question.

Although laboratory studies offered a more refined view of tendencies over time, there were limitations in the diversity of snails used. Only two of the laboratory studies examined the life history traits of infected prosobranchs, so consideration of laboratory studies was restricted to snails in subclass Pulmonata. As in earlier analyses, comparisons were made between experiments using hosts infected with either rediae or sporocysts. In studies involving rediae, the frequency of gigantism increases with time (8 % during prepatency, 60 % at patency, 58 % at the end of the experiment). Sporocyst infections display less variation in the frequency of gigantism over time (57 % during prepatent period, 58 % during patency and 42 % when the experiment ended). Thus, it appears that enhanced growth is relatively common in pulmonate–trematode systems studied under laboratory conditions. Moreover, host growth seems to be enhanced during prepatency primarily when sporocysts inhabit host tissues rather than rediae.

Most of the freshwater hosts examined were from two families, Lymnaeidae and Planorbidae, enabling comparisons between them. Striking growth differences were apparent between these groups with lymnaeids displaying gigantism more often than planorbids. This was most pronounced during patency when 90 % of the experiments demonstrated gigantism among infected *Lymnaea* snails; whereas only 36 % of infected planorbids exhibited gigantism. This proportion decreased over time for both types of snails, indicative of a growth cost associated with parasite maturity; even at the end of the experiments, gigantism existed in over 75 % of the *Lymnaea* cases. Rather than exhibiting gigantism, the planorbid genera *Biomphalaria*, *Bulinus* and *Helisoma* showed stunting to be the more common outcome and the degree to which stunting was shown increased as the infection progressed (prepatency, 35 %; patency, 48 %, end of experiment, 64 %). In contrast, the proportion of studies showing stunting among *Lymnaea* snails never exceeded 10 % during the three time periods. In combination, these results suggest that lymnaeid and planorbid snails show altered growth patterns in response to trematode infections. However, there seem to be differences in the predominant outcome for the two snail types and in the time during which this effect is most noticeable.

Organisms normally alter their energetic investments in life history traits as they mature; therefore, we examined the influence of snail age on growth during the three stages of infection. The evidence for gigantism among juvenile and adult snails was very similar over all the time frames we examined; for both ages, the frequency of gigantism among these studies ranged from 35–50%, with 50% being the modal value. For adolescent snails, the pattern seemed different, since only 29% of the studies showed gigantism during the prepatent period but this proportion increased dramatically during early patency (71%) and then declined as the infection progressed (57%). Overall, this seems to show that the proportion of studies displaying stunting increased over time for juvenile and adult snails but not so for adolescent snails. This suggests that juvenile and adult snails suffer more in terms of growth as the parasite matures than adolescents do. It is interesting that juveniles and adults have such similar growth responses given the purported trade-offs between growth and reproduction. Juveniles have no opportunity to reproduce, so the cost of parasitism should be more strongly expressed in their somatic growth patterns; adults can spread the burden of parasitism across their growth and reproduction allocation efforts. Perhaps adult snails shunt energy from growth towards early reproduction when parasitized; consequently adults hosts grow less than uninfected individuals.

Density-dependent phenomena, in terms of either parasite or host density, may affect snail growth; therefore, housing conditions and infection dose were considered in the analysis. During the first two infection stages, studies using hosts living in isolation tended to exhibit gigantism (prepatent, 55%; patent, 61%) to a greater extent than those living in groups of two or more (prepatent, 9%; patent, 33%). Stunting, on the other hand was somewhat more evident among hosts living in groups (prepatent 18%; patent, 53%; end, 53%) relative to their counterparts housed singly (prepatent, 23%; patent, 29%; end, 44%). The fact that stunting becomes more common during the course of the infections under both types of housing conditions suggests that the costs of parasitism increase as the intramolluscan larvae mature. To compare the effects of parasite dose, we separated the studies into two groups: those using ≤ 5 miracidia, and those using > 5 miracidia. Comparisons across infection doses showed that snails exposed to the most miracidia also demonstrated the highest incidence of gigantism among hosts. Furthermore, this pattern existed during all periods but became most notable during the patent period (≤ 5 miracidia: prepatent, 36%; patent, 53%; end, 36% vs. > 5 miracidia: prepatent, 67%; patent, 85%; end, 77%). Combining the effects of housing conditions and miracidial dose, it appears that gigantism occurs more frequently among snails housed in isolation and those exposed to more than five miracidia, while stunting is predominantly found among snails living in groups. Although this outcome could be explained in a number of ways, one possibility is that hosts with heavier parasite burdens feed at a higher rate than other snails. In the absence of negative influences on feeding patterns due to other snails, extra host resources become incorporated into somatic tissue and/or shell material more effectively when infected snails are housed singly.

Fecundity

Reproductive tissues are not essential for the survival of individual hosts, so it is not surprising that parasites become adapted to using these resource-rich tissues to support their growth and development within hosts. This effect was well demonstrated in the studies we reviewed, since trematode infection led to reduced or completely inhibited reproductive activity in all studies reviewed. Furthermore, castration was nearly complete in most cases by the time cercariae were released. Thus, field studies were of no use in evaluating early effects of parasitism on host fecundity, since they determine infection status based on cercariae release. Laboratory studies, on the other hand, proved useful for investigating parasite effects on host fecundity as the infection progressed. Although a steady reduction in reproductive activity appeared as the norm in these studies, several cases presented an alternate outcome with hosts producing more eggs than uninfected individuals in the prepatent period. Several cases showed increased egg production during prepatency and this seems to happen more frequently when sporocysts were the primary intramolluscan stage (23%) rather than rediae (10%). The single case of increased host fecundity involving rediae was *Halipegus occidualis* infections among mature *Helisoma anceps* snails on a low quality diet. Schistosome infections accounted for all of the instances where sporocyst infections induced hosts to increase their fecundity relative to controls. The parasite–host combinations yielding this result were *Schistosoma mansoni* infecting either *Biomphalaria glabrata* or *B. pfeifferi* and *Trichobilharzia ocellata* parasitizing *Lymnaea stagnalis*. None of the other factors affecting growth appeared to influence this compensatory response, since it occurred irrespective of host age, housing conditions (isolation vs. grouped), or infection dose.

Rarely, trematodes did not completely castrate their hosts and in two studies, hosts laid more eggs than uninfected snails during patency. This increase in host fecundity occurred among adolescent *B. pfeifferi* snails monomiracidially infected with *S. mansoni* and among infected juvenile *Bulinus senegalensis* snails exposed to 5 *S. haematobium* miracidia.

In both of these studies, the snails were grouped together. These two studies also utilized strains of the host and the parasite that did not have a long coevolutionary history. In the first study, the *B. pfeifferi* snails originated in Tanzania while the *S. mansoni* miracidia came from a West Nile population (Makanga, 1981), making it unlikely that this snail–trematode combination shared a recent history of microevolution. In the second combination, the hosts and parasites originated from the same natural populations, but the *S. haematobium* strain used in the *B. senegalensis* infections also parasitized *Bulinus truncatus* snails from the same location (Fryer *et al.* 1990). *B. senegalensis* displayed increased egg production when infected, while *B. truncatus* showed a marked reduction, suggesting that this parasite strain is more closely associated to the latter host species.

Survivorship

Parasites may alter host survival by different means during an infection. Mortality that occurs soon after exposure to the parasite probably reflects compatibility issues, whereas death observed later presumably indicates excessive energy demands or tissue damage imposed by the parasite (Bayne & Loker, 1987). All the field studies that surveyed snails at a single point in time necessarily used naturally infected snails, so the extent of mortality occurring earlier, during the prepatent period, is unknown. Two of the three field studies using mark-recapture techniques assessed host mortality and showed that host survival decreased as a consequence of parasitism. Twenty-two laboratory studies reported snail survival data during the prepatent period; unfortunately, only six involved planorbid snails and just five used redial parasites. Our interpretation of mortality data from these studies suggests that rediae impose their strongest negative effect on host survival during the prepatent period whereas sporocyst infections decrease host survival more later. This conclusion emerged because all of the experiments using rediae showed reduced host survival during the prepatent stage, whereas merely 47 % of the sporocyst infections showed the same outcome at that time. In contrast, sporocysts' negative impact seemed more pronounced during patency as host survivorship decreased in 82 % of these studies compared to just 44 % for those with rediae infections. Interestingly, snails that survived prepatency with rediae infections apparently outlived uninfected individuals more frequently than those with sporocysts (rediae, 44 %; sporocysts, 9 %). When considering the different types of hosts, we found that lymnaeids and planorbids showed similar patterns of reduced host survival during prepatency (lymnaeids, 50 %; planorbids, 62 %), but *Lymnaea* snails seemed to suffer greater mortality as the infection matured (prepatency, 50 %; patency,

81 %). Planorbids, on the other hand, showed relatively constant levels of mortality over time (prepatency, 62 %; patency, 60 %).

Snail age, host density and infection dose also appeared to affect host survival. Mortality during the prepatent stage of the infection was most likely in juvenile snails (juveniles, 83 %; adolescents, 50 %; adults, 50 %). Although the incidence of increased mortality was above 50 % during prepatency for all host ages, both mature and juvenile hosts appeared to die later in the infection more frequently than adolescent snails (juveniles, 70 %; adolescents, 54 %; adults, 82 %). When examining the effect of host density on mortality, we found that snail density tends to affect host survival differently during early and late stages of the infection. Mortality was most common in studies of grouped snails (groups, 75 %; singles, 50 %) during the parasite's early development. In contrast, isolated snails showed increased host mortality 93 % of the time compared to 50 % for grouped individuals during patency. Infection dose effects were somewhat equivocal but a general trend of increased mortality coincided with increased miracidia doses during all periods.

DISCUSSION – A MORE COMPREHENSIVE VIEW

Our analysis of over 100 experiments investigating trematode influences on host life history traits identified a number of salient trends. Although these results are substantially informative in their own right, we can increase their value by using them to test hypotheses. In so doing, the robustness and applicability of earlier hypotheses can be measured against this broad data set to better define the experimental, ecological and evolutionary conditions most compatible with particular scientific questions. As we develop a more comprehensive view of snail–trematode interactions, the new perspective should guide the direction of future research efforts.

Evaluation of existing hypotheses

Field and laboratory studies of gigantism are not equivalent. The parity of field and laboratory studies on host life history responses to trematodes has been challenged primarily due to the relative abundance of food resources provided for snails under typical laboratory conditions compared to natural conditions (Fernandez & Esch, 1991). These investigators felt that this difference might explain the paucity of field studies demonstrating gigantism. Fernandez & Esch (1991) are justified in their thesis, given the paradox that results when field and laboratory studies on a particular host–parasite combination yield startlingly different outcomes in terms of snail size. However, we found little evidence among the studies we reviewed to support their claim. In fact, among all field-collected snails we found a strong bias (87 % of comparisons examined) relating parasitism to large size based on the positive correlation between size and prevalence. Our finding that 53 % of field-collected freshwater snail comparisons show evidence of enhanced growth rates compares well with the 50 % value obtained when averaging across all the laboratory experiments and time frames using freshwater snails. Thus, it appears that gigantism may be as evident in the field as it is in the laboratory for trematode-infected freshwater snails.

In defence of Fernandez & Esch (1991), prior to their work few investigators had assessed the growth of infected snails under natural conditions. Aside from two studies using *Helisoma anceps* (Crews & Esch, 1986; Goater *et al.* 1989) most used marine snails (Rothschild, 1941; Hughes & Answer, 1982; Sousa, 1983). We found that only 7 % of the experiments on field-collected marine snails demonstrate evidence of increased growth rates, as expected by their hypothesis. The limited ability to demonstrate gigantism within natural, marine snail populations is surprising given that 65 % of the experiments we examined involving marine snails showed a positive relationship between prevalence and host size. If enhanced growth has not led to this positive correlation, then one of the other five mechanisms proposed by Baudoin (1975) may be important. It is also noteworthy that planorbid snails, like *H. anceps*, are less likely to show gigantism under laboratory conditions compared to *Lymnaea* snails. Nonetheless, when Keas & Esch (1997) investigated *H. anceps* growth in the laboratory, they found that adult hosts on a high quality diet do show faster growth than uninfected individuals. Perhaps food resources are limiting for this snail population in Charlie's Pond, but that does not appear to be the case for the other freshwater snails that have been studied.

The type of intramolluscan stage determines the host life history outcome. Because of their different feeding mechanisms, sporocysts and rediae cause host castration via distinct histopathologies (Wright, 1966; Malek & Cheng, 1974). Different costs associated with each larval form have been proposed to influence the growth and reproduction of hosts (Wright, 1971; Sousa & Gleason, 1989; Gorbushin, 1997). There are clearly differences in the timing of both mortality and growth. Rediae infections tend to reduce host survival earlier in the infection than sporocyst infections do. Our analysis of host growth patterns in the laboratory also support this position, since we found that sporocysts tend to promote enhanced growth earlier in an infection than do rediae. However, as patency continues, it appears that sporocysts affect host growth differently than rediae do; during patency, the incidence of gigantism increases among rediae-infected snails and remains relatively constant for those with sporocysts. Interestingly, data collected from field studies also show gigantism occurring more frequently among rediae-infected snails than among those bearing sporocysts; however, this difference is slight. Thus, it appears that among the studied trematode–snail combinations, the two types of larvae alter host growth differentially over time.

Both rediae and sporocysts seem to promote complete castration of their hosts during patency. Most of the differences in host fecundity we were able to detect between infections with either of these two larvae occurred during the prepatent period and involved a 'fecundity compensation' response as described by Minchella & LoVerde (1981). Our survey indicated that this response, although relatively rare, happens most frequently in schistosome infections, which rely on sporocysts. We found that 'fecundity compensation' appeared in 4 snail species from 3 genera, *Biomphalaria*, *Bulinus* and *Lymnaea* following parasitism by schistosome sporocysts (Makanga, 1981; Minchella & LoVerde, 1981; Thornhill, Jones & Kusel, 1986; Fryer *et al.* 1990). The extent to which other sporocyst infections might elicit this response remains unknown because so few of the necessary experiments have been performed. However, a comprehensive laboratory study recently conducted by Zakikhani & Rau (1999) showed that *Lymnaea elodes* snails do not increase their egg production following infection with *Plagiorchis elegans* sporocysts. Of the studies we reviewed, only one demonstrated an association between rediae and increased host reproduction. In this case, mature *Helisoma anceps* snails appeared to increase their reproductive output in response to the early stages of a *Halipegus occidualis* infection when fed a low quality diet (Keas & Esch, 1997). Thus, for now it appears that a 'fecundity compensation' response is reserved to schistosome infections and infections involving hosts on restricted diets.

What, if anything, do schistosome sporocysts and dietary constraints have in common? Schistosome mother sporocysts typically inhabit the head-foot region of the host putting the parasite near to neurosecretory cells in the cerebral ganglion that are influential in host reproductive activities (Bayne & Loker, 1987; De Jong-Brink, 1995). An infection close to the centres regulating reproduction may be important in explaining the association between schistosomes and increased host reproduction. It would be interesting to know if other parasites residing adjacent to cerebral ganglia evoke a similar host response and how dietary constraints influence the release of hormones produced in that region. Such studies might also elucidate the biochemical and physiological mechanisms promoting increased host growth prior to castration since neurosecretory cells regulating growth are also located in the cerebral ganglion (De Jong-Brink, 1995).

Long-lived snails show different responses to infection than short-lived snails

Researchers have formed vastly different hypotheses regarding the likelihood of gigantism based on snail longevity (Sousa, 1983; Minchella, 1985; Gorbushin & Levakin, 1999). Sousa (1983) considered increased host growth rates to be a selectively neutral side effect of parasitic castration. As such, he expected gigantism to be most commonly observed among short-lived (< 1·5 years) semelparous hosts due to their high investments in growth and reproduction relative to maintenance. In other words, because short-lived snails have limited reparative capacities they should allot energy freed by parasitic castration to growth. In contrast, Minchella (1985) proposed that long-lived (> 1·5 years) iteroparous species would be more likely to show gigantism than short-lived species are. Under this hypothesis, gigantism is viewed as a host adaptation that improves host survival beyond the longevity of the parasite. Since short-lived hosts have little opportunity to outlive the parasite, only long-lived hosts benefit from this strategy. Gorbushin & Levakin (1999) recognized a wider range of snail life-spans beyond the annual vs. perennial distinctions used earlier. Based upon mathematical models and empirical evidence, they posited that snails with intermediate life spans (2–3 years) were most likely to show gigantism. This conclusion relies upon long-lived species paying a high energetic cost for repair mechanisms and short-lived hosts suffering severe parasite-induced pathogenesis.

The data in Tables 1 and 2 provide sufficient information to evaluate the hypothesis of Gorbushin & Levakin (1999) for prosobranch snails. According to their definitions, all of the snails in Table 1, except *Onoba aculeus*, should be considered long-lived, while the 2 *Hydrobia* species listed in Table 2 and *O. aculeus* all have intermediate longevity. In separating the comparisons in this manner, we found that no long-lived snails exhibited gigantism, whereas 69 % of the shorter-lived prosobranch-trematode combinations exhibited gigantism. These results agree with Gorbushin & Levakin (1999). Sousa's (1983) hypothesis would also be supported by the analysis if the age he used to distinguish short-lived and long-lived snails were increased to 4 years.

From a broader perspective, it is interesting to note that the distinction we made above based on snail age is essentially a distinction between freshwater and marine species. Therefore, we should ask what characteristics of freshwater and marine environments favour the alternative host responses to trematode parasitism. In general, freshwater habitats are less stable and have a larger net primary productivity than marine habitats, resulting in different life history traits among inhabitants of the two environments (Begon, Harper & Townsend, 1990). Population turn-over rates tend to increase as a function of increasing primary productivity and decreasing stability; therefore, life history traits that maximize reproductive ability early in an organism's life, like early maturity and semelparity, are favoured in these environments (Stearns, 1976). At the other end of the life history continuum, competition plays a greater role in shaping an organism's life history in less productive and more stable habitats, so increased longevity with iteroparous reproduction are favoured under these conditions. In light of these habitat-specific influences on snail life histories, it may be more appropriate to associate the potential for host gigantism to environmental criteria, like productivity and permanence, rather than demographic features of the host, like longevity.

Development of new hypotheses

Environmental conditions ultimately dictate host life history alternatives following parasitism. Ecological theory and empirical evidence show that environmental conditions influence the evolution of life history traits in a population specific manner (Stearns, 1976; Brown, Devries & Leathers, 1985). Furthermore, experimental evidence indicates that demographic traits of the host influence the life history consequences associated with trematode parasitism (Lafferty, 1993; Gorbushin, 1997). Therefore, an attempt to connect the environmental causes of life history variation, in the absence of parasitism, to the host life history effects we observe in the presence of parasites seems worthwhile.

Evolutionarily successful individuals living in unpredictable habitats require life history traits that allow for rapid population growth since environmental conditions typically prevent these populations from reaching stable sizes. Therefore, early maturity, semelparous reproduction and short life-spans will be favoured in growing populations. In contrast, stable habitats with populations existing under equilibrium conditions, favour individuals with delayed maturity, iteroparous reproduction and long life-spans. Because different environmental conditions favour different longevity patterns, the amount of energy an organism devotes to maintenance (i.e. maintaining its health) should vary. Thus, long-lived individuals residing in stable habitats should devote more resources to reparative processes than short-lived individuals in unstable environments.

If we consider gigantism and 'fecundity compensation' to be phenotypically-plastic host responses to parasitism in which hosts reallocate resources, they should be most commonly observed among individuals living in less stable habitats. In these populations, little energy is allocated to maintenance compared to the resources invested in reproduction. Trematode-induced cessation of reproduction in these habitats leads to reallocation of those sizable resources to growth. Conversely, stunting should be considered evidence of parasite-induced limitations on either the energy or resources available to the host. In this case, trematode-induced stunting would be expected more frequently under environmental conditions favouring stable population sizes. In these populations, host growth rates are typically low and much of the energy is already allocated to maintenance, so trematode-induced changes in reproduction do not result in an increase in energy available for growth, especially if the host's internal defences mount a costly attack on the intruder. These hypotheses allow for population-specific analyses, which may help to explain different growth responses from the same host species under different environmental conditions. For instance, Mouritsen & Jensen (1994) showed that differences in the food availability in two populations affected the growth rates of infected *Hydrobia* snails in those populations differently. Similarly, Gorbushin (1997) found that competitive interactions in dense populations of *Hydrobia* limited the snails' ability to increase their growth rate in response to trematode parasitism. Aside from the influence of primary productivity and population density, it would also be useful to consider the influence of habitat permanence, because some snail species switch from semelparity to iteroparity as permanence decreases (Hunter, 1975; Brown *et al.* 1985).

Parasite transmission rates, not rediae or sporocysts, ultimately determine the life history costs trematodes impose on their hosts. As mentioned earlier, rediae are presumed to inflict greater damage to their hosts than sporocysts; however, the amount of damage a parasite inflicts upon its hosts ultimately depends upon parasite transmission factors (Ewald, 1987). The first intermediate hosts in trematode life cycles serve as a means to asexually increase the number of clones available for infecting subsequent hosts. The degree of damage that trematodes cause their hosts, ultimately depends upon the relative success cercariae have in finding and infecting their next host and the extent to which this damage jeopardizes host survival. Therefore, the optimal virulence level that a trematode invokes on members of a first intermediate host population must be sensitive to ecological factors influencing host mortality and parasite transmissibility.

From an evolutionary perspective, the extent that subsequent hosts overlap temporally and spatially dictates the optimal virulence level that a parasite can express (Williams & Nesse, 1991; Ewald, 1994). For trematodes having two hosts and non-motile cercariae, such as *Microphallus*, both hosts must provide a window of opportunity for infection by being present simultaneously for the parasite population to persist. Furthermore, a vertebrate host must come in direct contact with the patently infected first intermediate host and ingest it for the parasite to reach maturity. Under these conditions, we predict that the parasite's virulence level will be limited, because increased virulence is likely to lead to a reduced probability that the host survives long enough to encounter a definitive host. The presence of motile cercariae in a two-host life cycle, like *Schistosoma*, may remove some of the constraint on virulence because cercariae can disperse over a wide area, increasing the probability that definitive hosts become infected. Residual resting stages (e.g. metacercariae) in a two-host system, such as *Fasciola*, may further reduce constraints on virulence because both hosts need not overlap temporally for the life cycle to continue. Virulence within first intermediate host snails should be least constrained when resting stages reside in or on second intermediate hosts, as with *Echinostoma*, because the additional hosts further expand the time and space that first intermediate and definitive hosts can be separated from each other.

A number of factors influence how sequential hosts overlap temporally and spatially, such as seasonal effects and parasite-altered intermediate host behaviours; therefore, exceptions to our scheme for ranking virulence patterns across host combinations are likely to be found. Nonetheless, the virulence differences it proposes may help explain the variation detected in host responses to trematode parasitism. The patterns we offered for the evolution of virulence levels support two similar yet distinct hypotheses. The first hypothesis proposes that the host life history variation observed across different trematode species which use the same type of intramolluscan larvae can be explained by life cycle characteristics. In the studies we reviewed, an increased incidence of both gigantism and host mortality was observed among studies using ten or more miracidia. If we can equate the host response to a virulent parasite with the response to increasing numbers of miracidia (since both expand the size of the infrapopulation) then we predict that gigantism and decreased survival will be observed more frequently among hosts harbouring virulent parasites relative to those harbouring less virulent ones. This hypothesis could be tested by evaluating the response of a single host species to various trematodes with different life cycle characteristics which we predict to be associated with different virulence characteristics. For example, in an experiment considering three trematodes that all utilize sporocysts, such as *Microphallus*, *Schistosoma* and *Plagiorchis*, we would expect to see systematic differences in the degree to which host growth and survival are altered following parasitism across these systems. Therefore, *Plagiorchis* should induce gigantism and decrease host survival to a greater extent than the other species, while *Microphallus*-infected snails should be least affected by the parasite.

The second hypothesis considers distinct populations of the same host–parasite combination when different levels of spatial and temporal overlap exist between subsequent hosts in the life cycle. It seems likely that the average prevalence values among first intermediate hosts over several seasons signify the degree to which intermediate and definitive hosts coexist; therefore, prevalence can be used as an indicator of the amount of host overlap at that site. Thus, in populations where prevalence is higher, trematode infections will alter host life history more so than in the same parasite–host combination at lower prevalence. According to the arguments given for the first hypothesis, gigantism and increased host mortality should be more predominant in host–parasite populations with higher prevalence values when compared to hosts at sites with lower prevalence.

Directions for future research

Of the 113 different experiments we reviewed, less than one-third (35) presented survival data for any of their experiments, only three were from field studies. It is not surprising that so few field studies quantify mortality since field surveys offer no opportunity to assess mortality directly, and mark-recapture studies typically show low recapture rates, making parasite-induced mortality difficult to appraise. However, the fact that only 56 % (32) of the studies conducted under controlled laboratory conditions attempted to measure the effect of parasitism on host survival seems more disconcerting. Theoretical arguments and empirical evidence consistently confirm the importance of trade-offs between growth, fecundity and survival when explaining an organism's life history. Future laboratory studies of host life history responses need to simultaneously assess all three of these life history traits. The next generation of field studies should employ mark-recapture techniques within enclosures using experimentally infected hosts in order to gain a clearer insight into the relationship between results derived from field and laboratory studies. This type of experimental design will allow investigators the opportunity to assess the host response as the parasite infrapopulation matures and multiplies.

It is becoming increasingly clear that a myriad of influences impact the interaction between intra-molluscan trematode larvae and their hosts. As we demonstrated earlier, host maturity, parasite dose and housing conditions all appear to influence the snail's response to trematodes. In future studies, these experimental variables need to be held in common to allow valid comparisons among studies or systematically manipulated within a study to assess their individual effects. In addition, future

studies must consider the origin of the host and parasite. Results based on the use of laboratory parasite strain 'A' and hosts from pond 'B' tell nothing about how evolutionary forces shape the outcome of host–parasite interactions in either laboratory 'A' or pond 'B'. In other words, if the host–parasite interactions are to be interpreted as the result of coevolution, it is best that the host and parasite share a recent history of coevolution under natural conditions.

In conclusion, it appears that sufficient studies have been conducted to say confidently that flukes alter snail life history traits by influencing their growth rates and fecundity patterns under many conditions, but the ability to predict confidently the direction of this outcome escapes us. Generalizations can be drawn across studies as we did during our review but in the end it would be difficult to predict the actual outcome for a novel host–parasite combination. Too little is known about which factors are most important in shaping the way that flukes affect their host snails. Thirty years ago, Wright (1971) wrote, "The extent of the parasite-induced modifications of host growth, reproduction, and survivorship are not only delimited by the genotypes of the participants, but also environmental conditions". Results from our review reiterate the importance of the 'Wright' stuff in shaping the variety of host life history responses we observe. In order to both refine and expand our understanding of trematode influences on snail life history traits we need to design experiments that will systematically assess specific hypotheses. These hypotheses would benefit greatly from serious consideration of the effects of both genotypic and environmental influences, which could be more readily assessed by applying the following approaches: (1) Utilize a more diverse set of snail–trematode combinations; (2) Compare the host-specific infection response with several trematodes both singly and in combination; (3) Design experiments so that growth, fecundity and survival can each be assessed over the appropriate time scales to view the process rather than the outcome; (4) Make greater use of field experiments that involve experimentally infected snails; (5) Restrict host–parasite combinations to naturally coevolving strains, or clearly identify the source of both host and parasite.

By exploring the energetic and evolutionary trade-offs in a variety of snail–trematode systems under an array of semi-natural environmental conditions, we will extend our view of parasite-induced host life history variation, thus advancing our understanding of host-parasite evolutionary biology.

ACKNOWLEDGEMENTS

We gratefully acknowledge A. Bieberich, J. Curtis and G. Sandland for their constructive criticism, helpful comments and editorial advice.

REFERENCES

ANDERSON, R. M. & MAY, R. M. (1979). Population biology of infectious diseases: Part I. *Nature* **280**, 361–367.

BAUDOIN, M. (1975). Host castration as a parasitic strategy. *Evolution* **29**, 335–52.

BAYNE, C. J. & LOKER, E. S. (1987). Survival within the snail host. In *The Biology of Schistosomes : From Genes to Latrines* (ed. Rollinson, D. & Simpson A. J. G.), pp. 321–346. San Diego, Academic Press Ltd.

BEGON, M., HARPER, J. L. & TOWNSEND, C. R. (1990). *Ecology : Individuals, Populations and Communities*, 2nd edn., Cambridge, MA, Blackwell Scientific Publications, Inc.

BROWN, K. M., DEVRIES, D. R. & LEATHERS, B. K. (1985). Causes of life history variation in the freshwater snail *Lymnaea elodes. Malacologia* **26**, 191–200.

CHENG, T. C. (1971). Enhanced growth as a manifestation of parasitism and shell deposition in parasitized mollusks. In *Aspects of the Biology of Symbiosis* (ed. Cheng, T. C.), pp. 103–137. Baltimore, University Park Press.

CHENG, T. C., SULLIVAN, J. T., HOWLAND, K. H., JONES, T. F. & MORAN, H. J. (1983). Studies on parasitic castration: soft tissue and shell weights of *Ilyanassa obsoleta* (Mollusca) parasitized by larval trematodes. *Journal of Invertebrate Pathology* **42**, 143–150.

CREWS, A. E. & ESCH, G. W. (1986). Seasonal dynamics of *Halipegus occidualis* (Trematoda: Hemiuridae) in *Helisoma anceps* and its impact on fecundity of the snail host. *Journal of Parasitology* **72**, 645–651.

CREWS, A. E. & YOSHINO, T. P. (1989). *Schistosoma mansoni*: effect of infection on reproduction and gonadal growth in *Biomphalaria glabrata. Parasitology* **68**, 326–334.

CURTIS, L. A. (1995). Growth, trematode parasitism, and longevity of a long-lived marine gastropod (*Ilyanassa obsoleta*). *Journal of the Marine Biological Association of the United Kingdom* **75**, 913–925.

DAWKINS, R. (1982). *The Extended Phenotype*, Oxford, Freeman.

DE JONG-BRINK, M. (1995). How schistosomes profit from the stress responses they elicit in their hosts. *Advances in Parasitology* **35**, 177–256.

DREYFUSS, G., RONDELAUD, D. & VAREILLE-MOREL, C. (1999). Oviposition of *Lymnaea truncatula* infected by *Fasciola hepatica* under experimental conditions. *Parasitology Research* **85**, 589–593.

EWALD, P. W. (1987). Transition modes and evolution of the parasitism-mutualism continuum. *Annals of the New York Academy of Science* **503**, 175–306.

EWALD, P. W. (1994). *Evolution of Infectious Disease*, New York, Oxford University Press.

FERNANDEZ, J. & ESCH, G. W. (1991). Effect of parasitism on the growth rate of the pulmonate snail *Helisoma anceps. Journal of Parasitology* **77**, 937–944.

FRYER, S. E., OSWALD, R. C., PROBERT, A. J. & RUNHAM, N. W. (1990). The effect of *Schistosoma haematobium* infection on the growth and fecundity of three sympatric species of bulinid snails. *Parasitology* **76**, 557–563.

GERARD, C., MONE, H. & THÉRON, A. (1993). *Schistosoma mansoni–Biomphalaria glabrata*: dynamics of the sporocyst population in relation to the miracidial dose and the host size. *Canadian Journal of Zoology* **71**, 1880–885.

GERARD, C. & THÉRON, A. (1995). Spatial interaction between parasite and host within the *Biomphalaria glabrata/Schistosoma mansoni* system: influence of host size at infection time. *Parasite* **2**, 345–350.

GERARD, C. & THÉRON, A. (1997). Age/size- and time-specific effects of *Schistosoma mansoni* on energy allocation patterns of its snail host *Biomphalaria glabrata. Oecologia* **112**, 447–452.

GOATER, T. M., SHOSTAK, A. W., WILLIAMS, J. A. & ESCH, G. W. (1989). A mark-recapture study of trematode parasitism in overwintered *Helisoma anceps* (Pulmonata) with special reference to *Halipegus occidualis* (Hemiuridae). *Journal of Parasitology* **75**, 553–560.

GORBUSHIN, A. M. (1997). Field evidence of trematode-induced gigantism in *Hydrobia* spp. (Gastropoda: Prosobranchia). *Journal of the Marine Biological Association of the United Kingdom* **77**, 785–800.

GORBUSHIN, A. M. & LEVAKIN, I. A. (1999). The effect of trematode parthenitae on the growth of *Onoba aculeus*, *Littorina saxatilis* and *L. obtusata* (Gastropoda: Prosobranchia). *Journal of the Marine Biological Association of the United Kingdom* **79**, 273–279.

HODASI, J. K. M. (1972). The effects of *Fasciola heptica* on *Lymnaea truncatula. Parasitology* **65**, 359–369.

HOLMES, J. C. (1983). Evolutionary relationships between parasitic helminths and their hosts. In *Coevolution* (ed. Futumya, D. J. & Slatkin, M.), pp. 161–185. Sunderland, MA, Sinauer Associates Inc.

HUGHES, R. N. & ANSWER, P. (1982). Growth, spawning and trematode infection of *Littorina littorea* (L.) from an exposed shore in North Wales. *Journal of Molluscan Studies* **48**, 321–330.

HUNTER, R. D. (1975). Growth, fecundity, and bioenergetics in three populations of *Lymnaea palustris* in upstate New York. *Ecology* **56**, 50–63.

KEAS, B. E. & ESCH, G. E. (1997). The effect of diet and repoductive maturity on the growth and reproduction of *Helisoma anceps* (Pulmonata) infected by *Halipegus occidualis* (Trematoda). *Journal of Parasitology* **83**, 96–104.

KRIST, A. C. & LIVELY, C. M. (1998). Experimental exposure of juvenile snails (*Potamopyrgus antipodarum*) to infection by trematode larvae (*Microphallus* sp.): infectivity, fecundity compensation and growth. *Oecologia* **116**, 575–582.

KURIS, A. M. (1980). Effect of exposure to *Echinostoma liei* on growth and survival of young *Biomphalaria glabrata* snails. *International Journal for Parasitology* **10**, 303–308.

LAFFERTY, K. D. (1993). The marine snail, *Cerithidea californica*, matures at smaller sizes where parasitism is high. *Oikos* **68**, 3–11.

LAW, R. (1979). Ecological determinants in the evolution of life histories. In *Population Dynamics* (ed. Anderson, R. M., Turner, B. D. & Taylor, L. R.), pp. 81–103. Oxford, Blackwell Scientific Publications.

LOKER, E. S. (1979). Effects of *Schisomatium douthitti* infection on the growth, survival, and reproduction of *Lymnaea catascopium. Journal of Invertebrate Pathology* **34**, 138–144.

MAKANGA, B. (1981). The effect of varying the number of *Schistosoma mansoni* miracidia on the reproduction and survival of *Biomphalaria pfeifferi*. *Journal of Invertebrate Pathology* **37**, 7–10.

MALEK, E. A. & CHENG, T. C. (1974). *Medical and Economic Malacology*, New York, Academic Press.

MAY, R. M. & ANDERSON, R. M. (1979). Population biology of infectious diseases: Part II. *Nature* **280**, 455–461.

MINCHELLA, D. J. (1985). Host life history variation in response to parasitism. *Parasitology* **90**, 205–216.

MINCHELLA, D. J., LEATHERS, B. K., BROWN, K. M. & MCNAIR, J. N. (1985). Host and parasite counteradaptations: an example from a freshwater snail. *American Naturalist* **126**, 843–854.

MINCHELLA, D. J. & LOVERDE, P. T. (1981). A cost of increased early reproductive effort in the snail *Biomphalaria glabrata*. *American Naturalist* **118**, 876–881.

MOURITSEN, K. N., GORBUSHIN, A. & JENSEN, K. T. (1999). Influence of trematode infections on *in situ* growth rates of *Littorina littorea*. *Journal of the Marine Biological Association of the United Kingdom* **79**, 425–430.

MOURITSEN, K. N. & JENSEN, K. T. (1994). The enigma of gigantism: effect of larval trematodes on growth, fecundity, egestion and locomotion in *Hydrobia ulvae* (Pennant) (Gastropoda: Prosobranchia). *Journal of Experimental Marine Biology and Ecology* **181**, 53–66.

POULIN, R. (1998). *Evolutionary Ecology of Parasites*, London, Chapman Hall.

PRICE, P. (1980). *Evolutionary Biology of Parasites*, Princeton, NJ, Princeton University Press.

RAYMOND, K. & PROBERT, A. J. (1993). The effect of infection with *Schistosoma margrebowiei* on the growth of *Bulinus natalensis*. *Journal of Helminthology* **67**, 10–16.

ROTHSCHILD, M. (1936). Gigantism and variation in *Peringia ulvae* Pennant 1777, caused by infection with larval trematodes. *Journal of the Marine Biological Association of the United Kingdom* **20**, 537–546.

ROTHSCHILD, M. (1941). Observations on the growth and trematode infections of *Peringia ulvae* (Pennant) 1777 in a pool in the Tamar Saltings, Plymouth. *Parasitology* **33**, 406–415.

SCHALLIG, H. D. F. H., SASSEN, M. J. M., HORDIJK, P. L. & DE JONG-BRINK, M. (1991). *Trichobilharzia ocellata*: influence of infection on the fecundity of its intermediate snail host *Lymnaea stagnalis* and cercarial induction of the release of schistosomin, a snail neuropeptide antagonizing female gonadotropic hormones. *Parasitology* **102**, 85–91.

SCHRAG, S. J. & ROLLINSON, D. (1994). Effects of *Schistosoma haematobium* infection on reproductive success and male outcrossing ability in the simultaneous hermaphrodite, *Bulinus truncatus* (Gastropoda: Planorbidae). *Parasitology* **108**, 27–34.

SOKOLOVA, I. M. (1995). Influence of trematodes on the demography of *Littorina saxatilis* (Gastropoda: Prosobranchia: Littorinidae) in the White Sea. *Diseases of Aquatic Organisms* **21**, 91–101.

SORENSEN, R. E. & MINCHELLA, D. J. (1998). Parasite influences on host life history: *Echinostoma revolutum* parasitism of *Lymnaea elodes* snails. *Oecologia* **115**, 188–195.

SOUSA, W. P. (1983). Host life history and the effect of parasitic castration on growth: a field study of *Cerithidea californica* Haldeman (Gastropoda: Prosobranchia) and its trematode parasites. *Journal of Experimental Marine Biology and Ecology* **73**, 273–296.

SOUSA, W. P. & GLEASON, M. (1989). Does parasitic infection compromise host survival under extreme environmental conditions? The case for *Cerithidea californica* (Gastropoda: Prosobranchia). *Oecologia* **80**, 456–464.

STEARNS, S. C. (1976). Life history tactics: a review of the ideas. *Quarterly Review of Biology* **51**, 3–47.

STEARNS, S. C. (1992). *The Evolution of Life Histories*, New York, Oxford University Press.

TETREAULT, F., HIMMELMAN, J. H. & MEASURES, L. (2000). Impact of a castrating trematode, *Neophasis* sp., on the common whelk, *Buccinum undatum*, in the northern gulf of St. Lawrence. *Biological Bulletin* **198**, 261–271.

THÉRON, A., GERARD, C. & MONE, H. (1992 a). Early enhanced growth of the digestive gland of *Biomphalaria glabrata* infected with *Schistosoma mansoni*: side effect or parasite manipulation? *Parasitology Research* **78**, 445–450.

THÉRON, A., MONE, H. & GERARD, C. (1992 b). Spatial and energy compromise between host and parasite: the *Biomphalaria glabrata–Schistosoma mansoni* system. *International Journal for Parasitology* **22**, 91–94.

THORNHILL, J. A., JONES, J. T. & KUSEL, J. R. (1986). Increased oviposition and growth in immature *Biomphalaria glabrata* after exposure to *Schistosoma mansoni*. *Parasitology* **93**, 443–450.

WILLIAMS, G. C. (1966). *Adaptation and Natural Selection*, Princeton, NJ, Princeton University Press.

WILLIAMS, G. C. & NESSE, R. M. (1991). The dawn of Darwinian medicine. *Quarterly Review of Biology* **66**, 1–22.

WILSON, R. A. & DENISON, J. (1980). The parasitic castration and gigantism of *Lymnaea truncatula* infected with the larval stages of *Fasciola hepatica*. *Zeitschrift für Parasitenkunde* **61**, 109–119.

WRIGHT, C. A. (1966). The pathogenesis of helminths in the mollusca. *Helminthological Abstracts* **35**, 207–224.

WRIGHT, C. A. (1971). *Flukes and Snails*, London, Allen and Unwin, Ltd.

ZAKIKHANI, M. & RAU, M. E. (1999). *Plagiorchis elegans* (Digenea: Plagiorchiidae) infections in *Stagnicola elodes* (Pulmonata: Lymnaeidae): Host susceptibility, growth, reproduction, mortality, and cercarial production. *Journal of Parasitology* **85**, 454–463.

Trematode infection and the distribution and dynamics of parthenogenetic snail populations

C. M. LIVELY

Department of Biology, Indiana University, Bloomington, IN 47405-3700, USA.

SUMMARY

According to the Red Queen hypothesis for sex, cross-fertilization should be positively associated with the probability of exposure (*risk*) to virulent parasites. Unfortunately, risk is difficult to measure in the wild, and *prevalence* of infection is often substituted for risk. Here I suggest that prevalence of infection may not generally suffice as a surrogate for risk, since the Red Queen model can make opposite predictions depending on the distribution of risk in the wild. Specifically, the results of a matching-alleles model suggest that asexual populations should be more infected than sexual populations, when (1) the variance in risk among populations is small, and (2) the mean risk of exposure to parasites is near the point where selection switches to favouring sex over asex. If, however, the variance in risk among populations is large, sexual reproduction should be positively associated with the prevalence of infection. In addition, the coefficient of variation for reproductive mode should increase sharply at the switch point. In light of these results, I re-evaluated data from two studies on the distribution of males in 95 populations of a freshwater snail (*Potamopyrgus antipodarum*). Populations of these snails are often mixtures of sexual and asexual individuals, and the frequency of males is correlated with the frequency of sexual females in the population. The results show a large, highly skewed variance among populations for prevalence of infection by larval trematodes. The results also show a positive, significant relationship between prevalence of infection and the frequency of males, with a sharp increase in the coefficient of variation at intermediate prevalence. In addition, experimental studies suggest that some of the necessary conditions of the Red Queen hypothesis are also met in this system. Specifically, the most common trematode infecting these snails is (1) adapted to infecting local host populations of the snail, and is (2) more infective to clones that were common in the recent past. It is too early to know if the parasite theory is sufficient to explain the widespread distribution of sex. I suggest that the theory is not sufficient, but that parasites in combination with mutation accumulation in clonal lines may explain the maintenance of sex in species that occasionally produce apomictic mutants.

Key words: Asexual reproduction, parasitic castration, *Potamopyrgus antipodarum*, Red Queen hypothesis, sexual reproduction, trematodes.

INTRODUCTION

Ecologists and evolutionary biologists are increasingly interested in parasites. This interest stems in part from: (1) models showing that parasites can control the density of host populations (May & Anderson, 1978); (2) models showing that host–parasite coevolution could lead to the maintenance of genetic diversity and, possibly, the maintenance of sexual reproduction (e.g. Jaenike, 1978; Bremermann, 1980; Hamilton, 1980, 1982; Bell, 1982; Judson, 1995), and (3) phylogenetic studies showing the co-speciation of host and parasites (e.g. Brooks, Thorson & Mayes, 1981). In addition, parasites were suggested as influential in the choice of mates, and hence the divergence of sexually selected traits in natural populations (Hamilton & Zuk, 1982). Thus parasite–host interactions, which were almost totally ignored in the ecological and evolutionary literature 20 years ago, were suddenly suggested as

Address for correspondence: Department of Biology, Indiana University, Bloomington, IN 47405-3700, USA (Fax: 812-855-6705; email: clively@indiana.edu)

potentially involved in regulating the density and genetic diversity of populations, the maintenance of sex, mate choice, and the generation of new species.

One challenge for any theory is to make unambiguous and, ideally, unique predictions. The Red Queen theory of sex is based on the idea that parasites should be under strong selection to be able to infect the most common local host genotypes (Haldane, 1949; Jaenike, 1978; Bremermann, 1980; Hamilton, 1980, 1982). Hence, as a clonal mutant spreads in a sexual population, it should be selected against in direct proportion to its frequency. If infections are sufficiently virulent to compensate for the reproductive advantage of clonal females, then sexual females should be preserved in the short term. This leads to the expectation that asexual populations, where they exist, should be associated with a low risk of infection. However, the individual hosts from asexual populations should be easier to infect by local coevolving parasites than individuals from a sexual population. So, on one hand, sex should be associated with the relevant selection pressure (parasites); but, on the other hand, asexual popu-

Parasitology (2001), **123**, S19–S26. © 2001 Cambridge University Press
DOI: 10.1017/S0031182001008113

lations are expected to have higher levels of infection (assuming all else equal). This kind of reasoning has lead to the criticism that the parasite theory can predict any kind of association between parasites and host sex in the wild, thereby making the theory unfalsifiable (e.g. Kondrashov, 1993).

I suspect that part of the solution to this apparent paradox may lie in cleanly discriminating between *risk* of infection and *prevalence* of infection. I use risk here to mean the probability of exposure to infective parasite propagules, and prevalence to mean the frequency of infected individuals in a population. A simple model is first presented to estimate risk and prevalence in sexual and asexual populations. I then compare the results of the model to results gained from a freshwater snail (*Potamopyrgus antipodarum*), which has populations that are either completely asexual or contain mixtures of sexual and asexual individuals.

A MODEL

In what follows, I present a simple model to generate the expected relationship between risk and prevalence of infection in sexual and asexual populations. Two haploid loci are assumed, each with two alleles. Contact between host and parasite genotypes occurs at random, and infection takes place if the parasite matches the host at both loci. The parasite is killed if it does not match the host at both loci. This is the matching-alleles model of host–parasite coevolution (see Otto & Michalakis, 1998), which is based on the idea that hosts possess self/non-self recognition systems for the detection and elimination of parasites.

The recursion equation for the frequency of the ijth host type (H_{ij}) in the next generation is given by

$$H'_{ij} = (1-r)\frac{H_{ij}(1-TVP_{ij})}{\bar{W}_H} + rh_ih_j,$$

where i and j give the alleles at the first and second locus respectively (e.g. H_{11} has allele 1 at both loci); T gives the probability of host contact with a parasite propagule (i.e. risk); V gives the reduction in fitness for infected hosts (i.e. infected hosts have a fitness of $1-V$); and $\bar{W}_H$ gives the mean fitness of hosts after selection. In addition, h_i gives the normalized frequency of the ith allele at the first locus, and h_j gives the normalized frequency of the jth allele at the second locus. Finally, r is the frequency of recombination between loci (see Gillespie, 1998 for a derivation of r).

Similarly, the recursion equation for the frequency of the ijth parasite genotype (P_{ij}) in the next generation is given by

$$P'_{ij} = (1-r)\frac{TP_{ij}H_{ij}}{\bar{W}_P} + rp_ip_j,$$

where i and j give the alleles at the first and second parasite locus respectively (e.g. P_{11} has allele 1 at both loci), and $\bar{W}_P$ gives the mean fitness of parasites after selection. p_i gives the normalized frequency of the ith allele at the first parasite locus, and p_j gives the normalized frequency of the jth allele at the second parasite locus. These last two variables were modified in the simulation to incorporate a mutation rate of 0·003. The high mutation rate is a proxy for both mutation and migration, and is set above zero to prevent parasite genotypes from going to fixation, especially under the high virulence that leads to high amplitude oscillations (e.g. Lively, 1999).

The simulations were initiated in linkage equilibria for both host and parasite and allowed to run for 900 generations; this period was enough to allow the cycles to reach stable values for their periods and amplitudes. I then calculated the prevalence of infection as:

$$Prev = \sum_{i=1}^{2} \sum_{j=1}^{2} TP_{ij}H_{ij}.$$

I also calculated the mean of this value for generations 901–1000 for both sexual and asexual hosts.

For sexual host populations, I set the recombination rate (r) to 0·5 to simulate the situation where the interaction loci are on different chromosomes for both host and parasite. To simulate asexual host populations, I fixed the host for a single genotype; parasites remained sexual in these runs with a recombination rate of one half ($r = 0·5$). In all simulations, I set virulence (V) to 0·8, to mimic the situation for the highly virulent larvae of castrating trematodes. These simulations assume that, even if the entire host population is infected, the fitness of individuals is sufficient to replace the population in the next generation (in other words, uninfected intermediate hosts are producing greater than 5 offspring each).

RESULTS

The results show that (all else equal) an asexual population should have a higher prevalence of infection for any given risk of infection (Fig. 1A/). For example, if each individual from a single asexual clone was exposed to one propagule from a population of coevolving parasites, we would expect all of them to become infected under this model. However, the same rate of exposure in a sexual population would produce only about one third of this number of infected individuals. The genetic diversity of the sexual population has reduced the prevalence of infection by about two thirds.

This result, by itself, would suggest that we should see populations composed of a single clone associated with higher rates of infection. But the

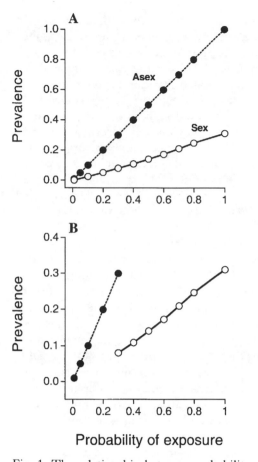

Fig. 1. The relationship between probability of exposure to a single parasite propagule and the prevalence of infection in sexual and asexual hosts. (A) The results for sexual and asexual populations assuming that parasites are responding to reproductive mode, but not selecting for it. (B) The results for sexual and asexual populations assuming that sexuals are favoured by selection if the probability of exposure is greater that 0·3. I refer to this value as the "switch point". Here the switch point was determined from computer simulations that considered host–parasite coevolution in concert with the accumulation of deleterious mutations (Howard & Lively, 1994). The mutation rate (U) per genome per generation was 0·5 in a population of 1000 hosts, and selection against individual mutations was 0·025; the mutations were assumed to have independent effects. Parasite virulence (V) was 0·8. The switch point is sensitive to the parameter values chosen, but the main ideas of the figure should hold as long as sexuals show prevalences greater than zero at the switch point.

argument under the Red Queen hypothesis is that parasites are selecting for the mode of reproduction, not simply responding to it. Thus at some risk of infection, the selection imposed by parasites selects for sexual reproduction in the host population. In a recent simulation study, we asked: what is the probability of infection that would shift the selection on hosts to reproduce sexually versus asexually (Howard & Lively, 1994)? The results of that model suggest that for highly virulent parasites, such as castrating species of trematodes ($V \geqslant 0·8$), a 30%

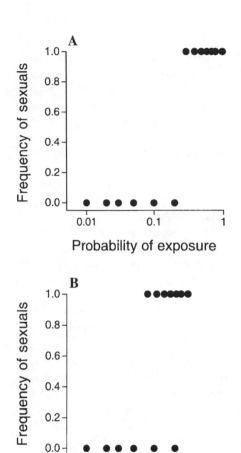

Fig 2. The expected relationship between (A) risk of infection and the frequency of sexual individuals in the host population (see also Sá Martins, 2000), and (B) prevalence of infection and the frequency of sexual individuals in the host population. Note the overlap for prevalences in B. This overlap is due to a drop in prevalence in sexual populations at the switch point (see Fig. 1B).

chance of exposure to miracidia would shift selection on the host to cross-fertilization, even for a two-fold cost of sex (details given in legend to Fig. 1).

Fig. 1B gives the expected results for infection in sexual and asexual populations given that a risk of exposure of 0·3, or greater, shifts the population to sexual reproduction. The results show that, if the risk of exposures is great enough, the most infected populations should also be sexual. But the figure also shows that there should be a large overlap in prevalence in sexual and asexual populations, especially in the segment of risk parameter space in which the shift in host reproductive modes occurs (here at 0·3).

This overlap in prevalence in sexual and asexual populations is more clearly illustrated in Fig. 2, which gives the expected frequency of sexual individuals in the simple model against risk of exposure (Fig. 2A) and parasite prevalence (Fig. 2B). It is especially obvious here that risk does not translate directly into prevalence, and that there is a large overlap zone for expected prevalence in sexual and asexual populations. Nonetheless, if there is a wide distribution of risk and prevalence in the wild, we should expect, if the Red Queen hypothesis is correct, that the frequency of sexual individuals should be positively correlated with the prevalence of infection (Fig. 2B). But we should also expect the data to show substantial scatter at intermediate levels of prevalence.

If, however, the variance in risk is small and the mean risk is positioned very near the transition point, then sexual and asexual populations would not be expected to differ in prevalence of infection in the wild or sexuals may be even less infected than asexuals. Hence, the general criticism of the Red Queen hypothesis appears to be correct because the existence and direction of any significant correlation depends on the position of the mean and the variance in risk. However, it is possible to reformulate the predictions for the biogeographic distribution of sexual host populations: sexual host populations should be positively associated with the prevalence of infection by virulent parasites but only if the range of prevalences is high and includes populations with very low and high prevalences. In addition, the variance in reproductive mode should increase dramatically with the prevalence of infection.

THE DISTRIBUTION OF ASEXUAL SNAIL POPULATIONS

Based on these predictions, I reanalysed data for the frequency of males and the prevalence of castrating digeneans for a common freshwater snail (*Potamopyrgus antipodarum*) in New Zealand lakes and streams (Lively, 1987, 1992). The frequency of males in this gonochoric gastropod is significantly correlated with the frequency of sexual females in populations that contain both diploid sexual individuals and triploid asexual individuals (Fox *et al.* 1996) and it serves as a measure of the frequency of sexual individuals in the population. The frequency of males is highly variable among populations and many populations that contain no males appear to be composed completely of asexual females. The data set includes 65 lakes that were sampled between 1985 and 1991, and 30 streams that were sampled between 1985 and 1986. The snails (usually about 100) from all 95 locations were dissected to determine their gender and the presence/absence of trematode larvae; these larvae from about a dozen different species invariably destroy the reproductive organs of both males and females. Most of the lakes were sampled in multiple locations and/or multiple years; means were calculated from these samples to provide a single datum for each location.

The distribution of prevalences is given in Fig. 3. It is apparent from this distribution that there is a large variance in prevalence and that the distribution is highly skewed. Specifically, there are many populations with low prevalence, and a few populations with high prevalence.

If the simple model given above can be generalized to the more complicated situation for the snails, then we would expect to find that prevalence is positively associated with sex and that there are very few highly sexual populations having low prevalences. The results are consistent with these expectations (Fig. 4). The product-moment correlation between prevalence of infection and male frequency is highly significant ($r = 0.363$; $df = 94$; $P < 0.001$ for log transformed data; identical results were obtained for the Spearman rank correlation). In addition, there are very few highly sexual populations ($> 10\%$ males) where the prevalence of parasites is low ($< 5\%$). Finally, the coefficient of variation for male frequency increases sharply at intermediate values of prevalence (Fig. 5), as predicted by the model.

Although these results are consistent with expectations under the Red Queen, there are alternative explanations that might also fit the data. One sensible alternative is that males are simply more susceptible to infection than females. Such a gender asymmetry would produce a positive relationship between parasite prevalence and male frequency without any underlying coevolution. In laboratory infection experiments, however, males were not more susceptible than females (Lively, 1989). This set of experiments involved fully reciprocal cross-exposure treatments between one set of two lakes, and another set of three lakes. Hence, I could estimate relative susceptibility of males and females to both local and foreign sources of parasites. There were no significant differences between males and females in experimental infections for any of the five comparisons using the local parasite source or any of the eight comparisons involving remote sources of parasites. There is therefore no indication that males are more susceptible to parasites from either potentially coevolving sources (the local sources) or non-coevolving sources (the remote sources). To determine whether there might have been a small, difficult-to-detect bias towards greater male susceptibility, I collapsed all thirteen comparisons into a single 2×2 matrix (infected females: 286; infected males 225; uninfected females: 871; uninfected males: 338). The difference in infection in this more powerful test was also non-significant ($\chi^2 = 0.01$; $df = 1$; $P = 0.916$).

These latter results suggest that the observed correlation between male frequency and parasite prevalence (Fig. 4) is not likely to be an artifact of greater male susceptibility. I nonetheless reexamined the data by calculating the correlation between male frequency and the prevalence of infection in females. Thus any bias in male susceptibility under field conditions could not affect the relationship between male frequency and prevalence of infection. I am less

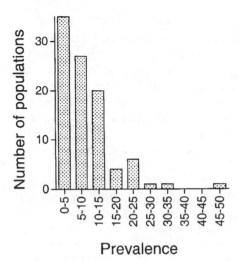

Fig. 3. Distribution of prevalence for infection by digenetic trematodes in *P. antipodarum.*

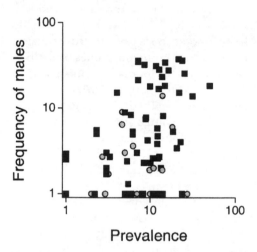

Fig. 4. Relationship between prevalence of infection by trematodes and the frequency of males for 95 populations of the freshwater snail, *P. antipodarum.* Squares represent lake populations (from Lively, 1992); circles represent stream populations (from Lively, 1987).

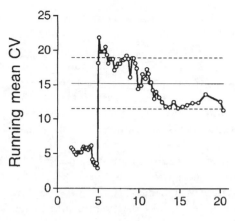

Running mean prevalence

Fig. 5. Relationship between the running mean for prevalence of infection and the running mean for the coefficient of variation (CV) for male frequency. Each circle indicates a "sliding window" of 20 data points taken from data ranked from lowest to highest with respect to prevalence of infection. For example, the first circle gives the mean prevalence and CV for male frequency calculated for the 20 sites with the lowest prevalence of infection; the second circle gives the same calculations for the next 20 sites; and the last circle gives the calculations for the 20 sites having the highest prevalence of infection. Note the steep increase in the CV, which is associated with the inclusion of the first highly sexual populations in the running mean. The solid horizontal line is the mean CV from randomization runs. In these runs, male frequencies were randomized 20 times and the CV calculated as above. The dashed horizontal lines are one standard deviation on either side of this mean.

confident in these estimates for prevalence, because all data were excluded for which gender and infection were not determined for the same set of individuals (Note: I often dissected many extra snails in an effort to detect rare parasites where gender was not examined). The results of this reanalysis, however, were virtually identical to those gained from the fuller data set ($r = 0.393$; $df = 94$; $P < 0.001$). Thus there is a positive, significant relationship between male frequency and prevalence of trematode infection in the wild, independent of whether males are included in the estimates of prevalence.

I wish to emphasize that male *P. antipodarum* are not environmentally induced by cues that might be associated with risk of infection but rather are genetically determined (chromosomal sex determination). Hence the above results suggest the possibility that the sexual females (which produce the males) have not been replaced by asexual females in those populations were parasites are abundant enough to generate strong, frequency-dependent selection against asexual lineages; but there is no indication that either sex or gender is a facultative response to the presence of parasites.

DISCUSSION

Here I have considered expectations for the biogeography of sex if parasites are responsible for the persistence of sex in species that occasionally produce ameiotic mutants. Based on the results of a simulation model, I argue that sexual populations may be either more infected or less infected than asexual host populations depending on the distribution of risk of infection and parasite prevalence in natural populations. However, the model predicts that a negative association should only be observed between host sex and parasite prevalence under very narrow conditions which can be specified. Hence, the parasite model of sex is subject to test based on biogeographic data.

Specifically, assuming that parasites contribute to selection for sex in local host populations, a negative association between host sex and parasite prevalence would require that the variance in risk is small and that the mean risk of infection is positioned very near the "switch point". (The switch point is the minimum risk of infection that selects for sexual reproduction.) For the model present above, this combination should show asexual populations to have higher prevalences of infection than sexual populations. It should also show a narrow range for the distribution of prevalences among populations, with few populations that have low prevalences.

If however there is a wide variance for risk, the variance in prevalence should also be high and sexual populations should show a tendency to have higher prevalences of infection than asexual populations. This association is largely driven by selection in low-risk, low-prevalence populations to favour asexual reproduction. But, because of the expected overlap in prevalence between sexual and asexual populations around the "switch point", the data should be very messy. In particular, the variance in sex should increase dramatically in the region of the switch point (which would have the side effect of making any positive association more difficult to detect).

Data for the frequency of snails (*P. antipodarum*) infected by castrating trematode larvae show a large range (0–50 %) and highly skewed distribution for 95 populations in New Zealand (Fig. 3). Many populations show a very low prevalence and a few populations show a high prevalence. If the arguments given above are correct, this is the kind of situation where sexual reproduction in hosts should be associated with higher levels of infection than asexual host populations. Such an association is in fact observed (Fig. 4). In addition, the coefficient of variation spikes at an intermediate level of prevalence, as expected under the model (Fig. 5).

These results are interesting in their own right, but I have a few reservations about the model. Firstly, the model is much simpler than the complicated situation observed in the snails. Most pressing is that the genetics underlying resistance in the snails (and in general) will be much more complicated and interesting than the two-locus, two-allele haploid model presented here. Secondly, the snails seem to have either asexual populations or mixed populations (sexuals and asexuals together) instead of either sexual or asexual populations as assumed by the model. It nonetheless seems that the basic conclusions of the model are likely to hold, provided mixed populations have greater genetic diversity for resistance and that this diversity reduces, but does not eliminate, infection at the switch point.

Another way to attack the ecological hypotheses for sex is to determine if the necessary conditions of the theories are met. One of these conditions for the Red Queen hypothesis is that the most common host genotypes become disproportionately infected by parasites and driven down in frequency. If this is not the case in the wild, then the Red Queen theory is surely incorrect. We have found both to be true for the New Zealand snail system. Specifically, we have shown that among lakes, the most common clone is sometimes significantly overinfected and sometimes significantly underinfected (Dybdahl & Lively, 1995). This result is expected if parasites are responding to common host genotypes with a time lag (Dybdahl & Lively, 1995; Morand, Manning & Woolhouse, 1996; Gandon *et al.* 1996). In addition, we have found that change in the frequency of host clones is correlated with the lagged change in infection of these same clones, which meets a clear prediction of the Red Queen model (Dybdahl & Lively, 1998). Finally, we have found in two separate experiments that recently common clones are more susceptible to infection by local parasites than clonal genotypes that were rare during the same period of time (Dybdahl & Lively, 1998; Lively & Dybdahl, 2000*a*). In the second of these experiments, we also found that the recently common clones were not more susceptible than rare clones to a remote source of parasites (Lively & Dybdahl, 2000*a*). This later finding suggests that local coevolutionary interactions with parasites accounts for the advantage of possessing a rare genotype, rather than greater resistance *per se* (see also Lively & Dybdahl, 2000*b*); and it explains previous experimental results showing strong local adaptation by parasites in this system (Lively, 1989). None of these experimental studies confirm that sex in the snails is maintained by parasites; but the data could have put the Red Queen in an extremely embarrassing position.

The Red Queen theory is, in fact, put into a rather embarrassing position by a recent study of parasitoids. Kraaijeveld & Godfray (2001) exposed individuals from 20 populations of *Drosophila melanogaster* to both sympatric parasitoids and to a reference strain of parasitoids (the braconid wasp, *Asobara tabida*). In addition, they exposed a reference strain of hosts to each of the 20 populations of parasitoids. The idea is that if local coevolution is working in this interaction, the parasitoids should generally be best able to infect hosts drawn from the same population (or *vice versa*, depending on the genetic basis for resistance). Instead they found that success of parasitoids was significantly and positively correlated with success on the host reference strain, and that the residuals were significantly and negatively correlated with host resistance to the parasitoid reference strain. They concluded that directional selection for resistance to attack by the parasitoid, and not any kind of rare advantage, is responsible for their results. In addition, in experimental studies, they found that replicate host lines selected for resistance to one strain of parasitoid in the lab were also more resistant to other strains of the parasitoid

(Kraaijeveld & Godfray, 1999). Hence, a general resistance, and not one depending on a specific genetic match, was observed. These results are contrary to expectation under the Red Queen theory.

These results points to the most frustrating aspect of the Red Queen theory of sex. One could rightly ask: "Is there another parasite of these flies that attacks in a frequency-dependent way? Perhaps they simply got the wrong biological antagonist." The problem with this kind of reasoning is that it could render the Red Queen model unfalsifiable. It also leads to the question of which parasites are the most likely to generate a rare advantage that is sufficient to overcome the reproductive advantages of asexuals when common. The trematode parasites of molluscs seem to be a likely choice, because of the sterilizing effect they have on their hosts, and the evidence for local adaptation to host populations (Lively, 1989; Lively & Dybdahl, 2000 *a*). But what of other invertebrate groups which also have the capacity to produce asexual mutants? Do they also have parasites that have such debilitating effects? Do these parasites attack in a frequency-dependent manner?

A final way to attack the Red Queen involves theory. In simulation studies, we challenged a sexual population of haploid hosts with competition from one or more clones (Lively & Howard, 1994). The sexual population had two segregating alleles at each of three loci, making for eight possible genotypes. We found that the introduction of two clones with different genotypes was sufficient to drive the sexual host population extinct even when the effects of infection were severe. Thus, in these simulations at least, parasites were selecting for clonal diversity and not sex *per se* (see also Judson, 1997; Lythgoe, 2000). This result makes sense, because, under the Red Queen hypothesis, parasites are a source of frequency-dependent selection. They cannot respond to reproductive mode of their hosts *per se*.

If this result is generally true then, strictly speaking, the Red Queen theory for genetic diversity is correct; but the extension of the theory to the maintenance of sex does not hold. We also found, however, that if we allowed mildly deleterious mutations to accumulate in the clones, that parasites aided in the stochastic accumulation of these mutations by periodically driving clones through bottlenecks (Howard & Lively, 1994, 1998; Lively & Howard, 1994). In other words, parasites facilitated the action of Muller's ratchet and the ratchet drove the clones to extinction. The combination of mutation accumulation and parasites then could favour sexual reproduction even when clones were periodically sampled from the sexual population, since clones were driven to extinction as fast as they were generated in the simulation. More generally, any process that keeps clones from fixing in the short term and also aids the accumulation of mutations would accomplish the same thing (Lively & Howard,

1994). The advantage of antagonistic coevolution over other forms of selection is that the clonal bottlenecks may be more predictable.

Finally, it may be premature to conclude that highly virulent parasites are required to protect sexual populations from replacement by asexual females. In our simulations that included spontaneous deleterious mutations, we found that parasites that reduce host fitness by as little as 30 % could be sufficient to select for sex. In addition, the models of Hamilton, Axelrod & Tanese (1990) show that sexual reproduction is favoured if there are synergistic effects of combining different species of parasites. So, even if each parasite has a low virulence, the combination of many different parasite species in the same host may reduce the ability of that host to successfully compete for limited resources. We have found support for this idea in an annual plant infected by rust (Lively *et al.* 1995).

Unfortunately, the combinations of ideas (e.g. parasites and mutations, or parasites and resource competition) make tests of models even more difficult. The effort, nonetheless, seems worthwhile. And, fortuitously, the consideration of combinations of models does not preclude the finding that sex has one simple and elegant solution (West, Lively & Read, 1999).

ACKNOWLEDGEMENTS

I thank Lynda Delph, Steve Howard, Jukka Jokela, Tom Little, Maurine Neiman, Eric Osnas, Dick Repasky, David Rollinson, Michele Tseng and Andrew Read for helpful suggestions on the manuscript. I also thank the US National Science Foundation (DEB 9904840) for research funding. Special thanks to Jukka Jokela for suggesting the time-series-like analysis associated with Fig. 5.

REFERENCES

BELL, G. (1982). *The Masterpiece of Nature: The Evolution and Genetics of Sexuality*. Berkeley, CA, University of California Press.

BREMERMANN, H. J. (1980). Sex and polymorphism as strategies in host–pathogen interactions. *Journal of Theoretical Biology* **87**, 671–702.

BROOKS, D. R., THORSON, T. B. & MAYES, M. A. (1981). Freshwater stingrays (Potamotrygonidae) and their helminth parasites: testing hypotheses of evolution and coevolution. In *Advances in Cladistics: Proceedings of the First Meeting of the Willi Hennig Society* (ed. Funk, V. A. & Brooks, D. R.), pp. 147–175. New York, New York Botanical Gardens.

DYBDAHL, M. M. & LIVELY, C. M. (1995). Host–parasite interactions: infection of common clones in natural populations of a freshwater snail (*Potamopyrgus antipodarum*). *Proceedings of the Royal Society of London, B* **260**, 99–103.

DYBDAHL, M. F. & LIVELY, C. M. (1998). Host–parasite coevolution: evidence for rare advantage and time-lagged selection in a natural population. *Evolution* **52**, 1057–1066.

FOX, J. A., DYBDAHL, M. F., JOKELA, J. & LIVELY, C. M. (1996). Genetic structure of coexisting sexual and clonal subpopulations in a freshwater snail (*Potamopyrgus antipodarum*). *Evolution* **50**, 1541–1548.

GANDON, S., CAPOWIEZ, Y., DUBOIS, Y., MICHALAKIS, Y. & OLIVIERI, I. (1996). Local adaptation and gene-for-gene coevolution in a metapopulation model. *Proceedings of the Royal Society of London, B* **263**, 1003–1009.

GILLESPIE, J. H. (1998). *Population Genetics: A Concise Guide*. Baltimore, Johns Hopkins University Press.

HALDANE, J. B. S. (1949). Disease and evolution. *La Ricerca Scientifica* (Suppl.) **19**, 68–76.

HAMILTON, W. D. (1980). Sex versus non-sex versus parasite. *Oikos* **35**, 282–290.

HAMILTON, W. D. (1982). Pathogens as causes of genetic diversity in their host populations. In *Population Biology of Infectious Diseases* (ed. Anderson, R. M. & May, R. M.), pp. 269–296. New York, Springer-Verlag.

HAMILTON, W. D., AXELROD, R. & TANESE, R. (1990). Sexual reproduction as an adaptation to resist parasites (A review). *Proceedings of the National Academy of Sciences, USA* **87**, 3566–3573.

HAMILTON, W. D. & ZUK, M. (1982). Heritable true fitness and bright birds: a role for parasites? *Science* **218**, 384–387.

HOWARD, R. S. & LIVELY, C. M. (1994). Parasitism, mutation accumulation and the maintenance of sex. *Nature* **367**, 554–557. (Reprinted figures in *Nature* **368**, 358.)

JAENIKE, J. (1978). An hypothesis to account for the maintenance of sex within populations. *Evolutionary Theory* **3**, 191–194.

JUDSON, O. P. (1995). Preserving genes: a model of the maintenance of genetic variation in a metapopulation under frequency-dependent selection. *Genetical Research, Cambridge* **65**, 175–191.

JUDSON, O. P. (1997). A model of asexuality and clonal diversity: cloning the Red Queen. *Journal of Theoretical Biology* **186**, 33–40.

KONDRASHOV, A. S. (1993). Classification of hypotheses on the advantage of amphimixis. *Journal of Heredity* **84**, 372–387.

KRAAIJEVELD, A. R. & GODFRAY, H. C. J. (1999). Geographic patterns in the evolution of resistance and virulence in *Drosophila* and its parasitoids. *American Naturalist* **153**, S61–S74.

KRAAIJEVELD, A. R. & GODFRAY, H. C. J. (2001). Is there local adaptation in *Drosophila*-parasitoid interactions? *Evolutionary Ecology Research* **3**, 107–116.

LIVELY, C. M. (1987). Evidence from a New Zealand snail for the maintenance of sex by parasitism. *Nature* **328**, 519–521.

LIVELY, C. M. (1989). Adaptation by a parasitic trematode to local populations of its snail host. *Evolution* **43**, 1663–1671.

LIVELY, C. M. (1992). Parthenogenesis in a freshwater snail: reproductive assurance versus parasitic release. *Evolution* **46**, 907–913.

LIVELY, C. M. (1999). Migration, virulence, and the geographic mosaic of adaptation by parasites. *American Naturalist* **153**, S34–S47.

LIVELY, C. M. & DYBDAHL, M. F. (2000*a*). Parasite adaptation to locally common host genotypes. *Nature* **405**, 679–681.

LIVELY, C. M. & DYBDAHL, M. F. (2000*b*). In search of the Red Queen: a response. *Parasitology Today* **16**, 508.

LIVELY, C. M. & HOWARD, R. S. (1994). Selection by parasites for clonal diversity and mixed mating. *Philosophical Transactions of the Royal Society of London, B* **346**, 271–281.

LIVELY, C. M., JOHNSON, S. G., DELPH, L. F. & CLAY, K. (1995). Thinning reduces the effect of rust infection on jewelweed (*Impatiens capensis*). *Ecology* **76**, 1851–1854.

LYTHGOE, K. A. (2000). The coevolution of parasites with host-acquired immunity and the evolution of sex. *Evolution* **54**, 1142–1156.

MAY, R. M. & ANDERSON, R. M. (1978). Regulation of stability of host–parasite population interactions. II. Destabilizing processes. *Journal of Animal Ecology* **47**, 249–267.

MORAND, S., MANNING, S. D. & WOOLHOUSE, M. E. J. (1996). Parasite–host coevolution and geographic patterns of parasite infectivity and host susceptibility. *Proceedings of the Royal Society of London, B* **263**, 119–128.

OTTO, S. P. & MICHALAKIS, Y. (1998). The evolution of recombination in changing environments. *Trends in Ecology and Evolution* **13**, 145–151.

SÁ MARTINS, J. S. (2000). Simulated coevolution in a mutating ecology. *Physical Review E* **61**, R2212–R2215.

WEST, S. A., LIVELY, C. M. & READ, A. F. (1999). A pluralist approach to sex and recombination. *Journal of Evolutionary Biology* **12**, 1003–1012.

Genetic structure in natural populations of flukes and snails: a practical approach and review

P. JARNE[1]* *and* A. THÉRON[2]

[1] *Centre d'Ecologie Fonctionnelle et Evolutive, UPR 9056 du CNRS, 1919 route de Mende, 34295 Montpellier cedex 5, France*
[2] *Centre de Biologie et Ecologie Tropicale et Méditerranéenne, UMR 5555 du CNRS, Université de Perpignan, Avenue de Villeneuve, 66860 Perpignan cedex, France*

SUMMARY

Several aspects of the coevolutionary dynamics in host-parasite systems may be better quantified based on analyses of population structure using neutral genetic markers. This includes, for example, the migration rates of hosts and parasites. In this respect, the current situation, especially in fluke-snail systems is unsatisfactory, since basic population genetics data are lacking and the appropriate methodology has rarely been used. After reviewing the forces acting on population structure (e.g. genetic drift or the mating system) and how they can be analysed in models of structured populations, we propose a simplified, indicative framework for conducting analyses of population structure in hosts and parasites. This includes consideration of markers, sampling, data analysis, comparison of structure in hosts and parasites and use of external data (e.g. from population dynamics). We then focus on flukes and snails, highlighting important biological traits with regard to population structure. The few available studies indicate that asexual amplification of flukes within snails strongly influences adult flukes populations. They also show that the genetic structure among populations in strongly affected by traits in other than snails (e.g. definitive host dispersal behaviour), as snails populations have limited migration. Finally more studies would allow us to deepen our current understanding of selective interference between flukes and snails (e.g. manipulation of host mating system by parasites), and evaluate how this affect population structure at neutral markers.

Key words: Population structure, flukes, snails, genetic markers, coevolution.

INTRODUCTION

The general question addressed in this paper is why is it important to analyse the population genetic structure of hosts and parasites, especially in flukes and snails, based on neutral genetic markers? There is a series of possible answers at both the intra- and inter-population level, revolving around the idea that the coevolutionary dynamics, including that of resistance and virulence genes, cannot be understood without a clear view on the genetic structure of populations as described using neutral markers. A few examples will clarify this point.

(1) Recent coevolutionary models are based on the idea that hosts and parasites are distributed in metapopulations. Migration and population size of both partners then become instrumental features of the coevolutionary dynamics (Gandon et al. 1996, 1998). For example, the relative rate of migration of hosts and parasites conditions the evolution of local adaptation of hosts to parasites. This is also a crucial issue in the debate on the role of parasites in the evolution of sex (the Red Queen hypothesis; Johnson, Lively & Schrag, 1997). This may also explain the patchy distribution of host susceptibility to parasite infections, especially in snail–digenean

pairs (Michelson & DuBois, 1978). (2) Several studies have shown that individual fitness partially depends on individual heterozygosity, or in other words on individual inbreeding (review in David, 1998). By extension, it might be hypothesized that the resistance of hosts towards parasites is also a function of individual inbreeding. This hypothesis has been substantiated by recent work conducted in Soay sheep populations parasitized by intestinal nematodes (Coltman et al. 1999). Similarly the virulence of parasites might be related to individual inbreeding, although no data are currently supporting this idea. (3) The dynamics of parasite populations and the epidemiology of disease transmission have been widely studied from a theoretical point of view, mostly using differential equations. Initial models focused on single populations, but metapopulation models have recently been developed (see e.g. Rohani & Ruxton, 1999 and references therein). These studies highlighted the influence of both intrapopulation processes, such as density-dependent effects on fitness, and dispersal. Quite clearly, empirical studies are far behind, especially in flukes, since it is extremely difficult to access to key parameters of their population dynamics. Even important aspects of snail population dynamics, such as dispersal rates, are basically unknown, despite the large number of studies that have monitored snail populations (e.g. Woolhouse, 1992).

* Corresponding author. Tel: 33 (0)4 67 61 32 27. Fax: 33 (0)4 67 41 21 38. E-mail: jarne@cefe.cnrs-mop.fr

Parasitology (2001), **123**, S27–S40. © 2001 Cambridge University Press
DOI: 10.1017/S0031182001007715

Extensive, thorough studies of the genetic structure within populations of both hosts and parasites would allow us to clarify these three aspects of host–parasite interactions. However such data remain scarce, especially in fluke–snail interactions. Even worse, parasite population structures have hardly been studied, and almost no data have been gathered in flukes (Mulvey *et al.* 1991; Dybdahl & Lively, 1996). A similar situation prevails in definitive hosts, and it is only slightly more favourable in snails (Jarne, 1995; Städler & Jarne, 1997). Genetic data derived from neutral markers are therefore crucially required. Moreover, these data should be gathered based on an appropriate theoretical and statistical framework. This is the impetus for this paper. More specifically, three aspects will be considered: (1) we will first present an overview of forces involved in shaping population genetic structure. (2) A framework will be proposed for analysing the population structure of intermediate and definitive hosts and flukes. Some suggestions will also be made on how they might be compared. Note that although this paper is mostly devoted to flukes and snails, the methodology can be used with any parasite-host system. (3). We will analyse life-history traits and aspects of life-cycles that make flukes and snails special (or ordinary) in a population genetics perspective, highlighting aspects for which genetic (neutral) markers should bring a bonus. The available data will be reviewed, highlighting studies analyzing both host and parasite population structure.

A VERY SHORT PRIMER OF POPULATION STRUCTURE ANALYSIS

The forces acting on population structure

Forces acting on the within- and among-population structure are detailed in excellent textbooks (e.g. Hartl & Clark, 1997) and reviews (Slatkin, 1985; Rousset, 2001). They will be reviewed only briefly here.

What is a population? Populations are ideally defined by population geneticists as 'a group of organisms of the same species living within a sufficiently restricted geographical area that any member can potentially mate with any other member' (Hartl & Clark, 1997, p. 71). This definition is of course hardly operational in the wild, since individuals are typically distributed following a non-random pattern. In free-living organisms, analyses of population dynamics, aiming at describing the distribution of individuals in space and time, may be used to specify the operational definition of populations. Parasites, on the other hand, are restricted to their hosts. This led to the definition of infrapopulations, that is individuals of a given species inhabiting a given host individual at some point in time (see Margolis *et al.* 1982; Combes 1995, p. 18). Sampling an infrapopulation of parasites is therefore straightforward. However, an infrapopulation is not necessarily a population, as it may either uncover several populations if the habitat within hosts is structured (probably not a classical situation in flukes), or part of a population if individuals are freely exchanging genes with individuals from other infrapopulations. Genetic analyses may therefore be used to define populations *a posteriori*, once migration among infrapopulations or *a priori* defined populations has been evaluated.

Mutation. Two aspects of the mutation process are relevant to the study of population structure, namely the mutation model and the mutation rate. Several mutation models have been proposed. Two extreme forms are the infinite alleles model (IAM) under which each mutation introduces a distinct new allele, and the stepwise mutation model (SMM) under which each mutation introduces an allele differing from the progenitor allele by one 'step' (see details in Jarne & Lagoda, 1996; Estoup & Angers, 1998). Depending on the genetic marker considered, one model may be more appropriate. For example, the recent rise of microsatellites was followed by a revival of interest for the SMM. The mutation rate has long been considered of little interest because it was assumed to be negligible with regard to other forces, such as migration. Although this was certainly an appropriate position with allozymes which mutation rate is of the order of 10^{-6} per locus and generation, this cannot systematically be held with microsatellites which mutation rate may reach 10^{-2}.

Genetic drift. It results from the random sampling of gametes for zygote formation in finite populations (see e.g. Hartl & Clark, 1997, chapter 7). Given enough time, genetic drift results in the fixation or loss of alleles at particular loci. An important concept defining its strength is the effective population size (N): the smallest this size, the strongest the effect of drift. N is affected by factors such as the sex ratio or the variance in individual fitness. How it can be calculated in metapopulations is exposed in Barton & Whitlock (1997).

The mating system. It defines the way gametes are united and plays an important role in the transmission of genetic information across generations. Random-mating is often assumed in models of population genetics. However, as some vector snails and most flukes are hermaphroditic, self-fertilization becomes an open option. Parthenogenesis is also possible in some vector snails (see Lively, this supplement). Other forms of non-random mating should also be considered, especially in vertebrate hosts, such as homogamy or biparental inbreeding.

Migration. This introduced the idea of subdivided populations (for reviews, see Slatkin, 1985; Rousset, 2001). As for mutation, several models have been proposed, the most famous of which being the island model. It assumes an infinite number of populations exchanging genes drawn from a unique pool, to which each population contributes equally at a proportion m. Finite versions of the model have since been derived, in which migrants may originate from only one source population (see references in Pannell & Charlesworth, 1999). Lattice models have also been proposed in which populations are distributed on a n-dimension grid. This formally introduced a geographic structure, since migration may take the form of a random variable depending on the geographic distance between populations. For example, in isolation-by-distance models, the probability of exchanging genes between two populations decreases with the geographic distance. Note that all these models are not intended to describe a geographic reality. Rather, they represent a convenient, powerful framework for analysing the distribution of variability (Slatkin, 1985; Rousset, 2001).

Equilibrium and non-equilibrium situations in structured populations

Introducing Wright's F-statistics. S. Wright introduced the F-statistics to describe the distribution of variation within and among populations (reviewed in Slatkin, 1985; Excoffier, 2001; Rousset, 2001). They can conveniently be defined by reference to the notion of gene identity (see e.g. Rousset, 2001):

$$F \equiv \frac{Q_w - Q_b}{1 - Q_b},\qquad(1)$$

where Q_w and Q_b are the probabilities of identity within and between classes of genes. For example, if this class is a deme (an individual), we have the definition of the famous F_{st} (F_{is}). Prominent features of the F-statistics are that (1) they can be related, sometimes simply, to basic parameters of population structure (see below); (2) probabilities of gene identity can be related to coalescence times of genes, bridging the gap between two views on population genetics; (3) they can be estimated from genetic data, that is from genes sampled in natural populations. Certainly the most widely used estimates are those of Weir & Cockerham (1984).

The mutation-drift-mating system equilibrium within populations. The classical approach has been to analyse the mutation-drift equilibrium in a random-mating population, meaning that the two parameters of interest are N and u. Theoretical expectations have been derived for two widely used, as well as easily estimated in natural populations, parameters, namely the number of alleles and gene diversity per locus. For example, gene diversity – defined as $1 - Q_w$ – is approximately equal to $4Nu$, when $Nu \leqslant 1$ under both the IAM (Hartl & Clark, 1997, chapter 7) and the SMM. The interesting aspect is that gene diversity will strongly depend upon genetic drift. Populations always maintaining a small size or going through regular bottlenecks are therefore expected to maintain less diversity than larger populations for a given mutation rate. This may hold for mollusc hosts, especially in tropical and mediterranean areas because water availability strongly varies in time and space (Brown, 1994; Städler & Jarne, 1997; Viard, Justy & Jarne, 1997*b*). Similarly fluke populations may markedly fluctuate in size.

As mentioned above, selfing may not be less common than random-mating in flukes and snails. Selfing erodes heterozygous genotypes, compared to Hardy-Weinberg expectations, and therefore increases the probability of gene identity within individuals. Assuming a combination of random outcrossing and selfing – the mixed-mating model – a remarkable result at equilibrium is:

$$F_{is} = \frac{S}{2 - S},\qquad(2)$$

with S the proportion of selfed offspring; this holds whatever the mutation model (Rousset, 1996). Another important consequence of partial selfing is the very slow erosion of multilocus genotypic associations, i.e. gametic disequilibria. Completely homozygous genotypes are transmitted intact across generations, except for mutations. A last consequence is the loss of gametic variability which has several sources (review in Charlesworth, Morgan & Charlesworth, 1993; Jarne, 1995). First, the effective size N_{in} of an inbreeding population is

$$N_{in} = \frac{N_{out}}{1 + F}\qquad(3)$$

with F the inbreeding coefficient and N_{out} the effective size of the corresponding random-mating population. The genetic variability is therefore expected to be halved in populations with very high frequencies of self-fertilization. A second reason is that indirect selection may be more efficient, whether it acts on favored alleles (genetic hitchhiking) or on deleterious mutations (background selection). Milder forms of inbreeding than selfing have milder consequences on variability.

The migration-drift-mating system equilibrium among populations. Here the classic approach is to consider the equilibrium between the size of populations (N) and the migration rate (m) in island or stepping-stone models, based on identity of genes within and among populations. This has giving rise to the quasi-mythical relationship:

$$F_{st} \approx \frac{1}{1+4Nm},\qquad(4)$$

which holds only under restricted conditions. The analysis of a two-dimensional lattice model suggests a relationship for which the parameter space of pertinence is more firmly established (Rousset, 1997, 2001):

$$\frac{F_{st}}{1-F_{st}} \approx \frac{\ln\dfrac{r}{\sigma}}{4D\pi\sigma^2}+b,\qquad(5)$$

where r is the distance between populations, D the population density per surface unit, σ^2 the average squared distance between parent and offspring and b depends on D, σ^2 and the surface unit. This relationship which holds in the limit of small r has practical implications since the slope of the regression of $F_{st}/(1-F_{st})$ on distance (or its logarithm) provides an estimate of $D\sigma^2$.

The limits of validity for equations (4) and (5) are set by, among other forces, the mutation process. Analogues of F_{st} have for example been devised for the SMM (Slatkin, 1995; Rousset, 1996). The question then arises whether it is more appropriate to use F_{st} or its analogues for estimating population differentiation, when one assumes a specific mutation model (e.g. the SMM if one is concerned with microsatellites). Theoretical analyses have shown that it is generally better to use F_{st}, meaning that the mutation process does not matter. However the mutation rate should not be too elevated. The mating system also modifies the interpretation of these equations. For example, selfing acts on both the effective size of populations and the effective migration rate, if migration occurs at the diploid stage (Maruyama & Tachida, 1992; Jarne, 1995). In purely selfing populations, $4Nm$ should be changed for Nm in equation (4). This means that for the same number of individuals per population and the same number of migrants, inbreeding populations should on average exhibit more differentiation than outcrossing populations.

Non-equilibrium situations. Two sources of non-equilibrium have been theoretically analysed. First, the variation in size of single populations (bottleneck or expansion) has a marked influence on the allelic spectrum (numbers and frequencies of alleles). For example, bottlenecks, especially when recurrent decrease the available variability in a population going through a bottleneck, and then recovering its initial size. Gene diversity and the number of alleles go back to their initial values with different dynamics, in a fashion depending on the mutation model assumed (see Cornuet & Luikart, 1996). This has been a basis for testing for bottlenecks (Cornuet & Luikart, 1996; see also Beaumont, 1999; Galtier, Depaulis & Barton, 2000). Metapopulation dynamics including extinction and recolonization is another source of non-equilibrium (see references in Pannell & Charlesworth, 1999). The mode of recolonization (e.g. the probability that colonizers are from a single population) and the relative values of the extinction and migration rates greatly affect the conclusions drawn from models. However one can retain that the amount of variability is lower and differentiation is stronger than in a situation without extinctions for quite a large window of parameters. The good news though is that F-statistics recover their stationary values quite quickly after a single perturbation (Rousset, 2001), which pleads for studies at short geographic scale for which perturbations can be ignored or monitored.

Selection

Selection has been almost excluded from discussions above, since we assumed neutral genetic markers. A note of caution should be introduced though. Selection may act indirectly through statistical or physical associations of neutral markers with selected genes, such as genes directly involved in resistance or virulence. In the latter case, allelic frequencies would be affected which would bias the analysis. However strong selection should result in peculiar patterns of allelic frequencies (see Ross et al. 1999 for a recent example). On the other hand, the probability that randomly-cloned markers are physically associated with genes involved in coevolution is probably low. We also noted above that selection may strongly affect patterns of variation when recombination is limited, as for example in inbred organisms.

A FRAMEWORK FOR ANALYSING POPULATION STRUCTURE

Molecular markers

Molecular markers have been widely used in a population genetics and/or phylogeographic context (see Avise, 1994; Hillis, Moritz & Mable, 1996; Carvalho, 1998; Goldstein & Schlötterer, 1999). They have primarily been used in flukes and snails for distinguishing phenotypic variants or geographic isolates and for systematics (e.g. Jelnes, 1986; Després et al. 1992; Dias Neto et al. 1993; review in Johnston et al. 1993), and few studies bear on population structure. Note also that no reference will here be made to mtDNA for length constraints (see Avise, 2000 for general references and Anderson, Blouin & Beech, 1998 for examples in parasites).

A central question is how to choose markers. They should indeed exhibit several qualities, including codominant expression (all genotypes can be distinguished), allelism (alleles can be attributed to loci), Mendelian transmission, neutrality of alleles and genotypes (alleles or genotypes are equivalent with regard to natural selection at all stages of the life-cycle), and variability. It is also worthwhile to be able to analyse several loci in a single study (ideally 10 at least), and to have *a priori* knowledge on the location of markers within genomes (e.g. to avoid selected areas) or on their mutation process. This qualifies markers such as allozymes, microsatellites, intron length polymorphism and single-nucleotide polymorphisms (SNP) although the influence of selection on thee markers is debatable. On the other hand, markers such as randomly amplified polymorphic DNAs (RAPDs) or amplified fragment length polymorphisms (AFLPs) are of lesser interest, especially because of dominance which prevents any clear analysis of the within-population structure, but also because of lack of allelism.

Technical aspects are also extremely important and are generally understated in the literature. Assuming that readable patterns are obtained, at least three aspects deserve consideration. The first, of which parasitologists are well aware, is the small size of the organisms studied and the often-associated difficulty of collecting large samples. Here PCR-based techniques should certainly be favoured. Difficulties of sampling are certainly no excuse to conduct analyses of population structure based on small samples. The second aspect is the repeatability of results, within or among laboratories. RAPDs are notoriously weak on this side. The third aspect is null alleles at codominant markers, such as allozymic alleles which are not expressed (Voelker *et al.* 1980) or microsatellite alleles which are not amplified (Pemberton *et al.* 1995). They quite likely explain a good part of the heterozygote deficiencies reported in the literature, especially in outbred species.

We mentioned above that variability was an important quality of markers. Their discriminatory power indeed increases with increasing variability, and the power of numerous statistical techniques (Rousset & Raymond, 1997) as well. This has to be scaled with the questions addressed, the geographical scale of studies and the temporal scale of sampling, if repeated sampling is included. For example, if the focus is on local adaptation of parasites and hosts, a microgeographic approach would certainly be required, scaling from e.g. a few km to a few hundreds of km. The level of variation offered by microsatellites would then be appropriate. However one should note that the efficiency of the methods used to analyse data may also depend on the geographic scale and on the number of markers used. For example, Rousset's (1997) method should not in principle be used at large geographic distances (below a few tens of σ^2; see equation (5)). Another important aspect is that the markers used should not be too polymorphic, the standard approximations failing above a certain mutation rate (which can prudently be evaluated at about 10^{-3} mutations per locus and generation).

No single marker fulfils all requirements indicated above. Allozymes would not rank so badly, but microsatellites and other PCR-based markers (e.g. intron length) would certainly be *primi intra pares*. However, if the technology of DNA chips democratize to a point at which any species could be studied, SNPs will certainly become strong competitors in this market.

Sampling

This is one of the weakest aspects of empirical studies concerned with the coevolutionary process in natural populations. Objective reasons at the source of such a situation include taxonomical ambiguities, the minute size of parasites, their low prevalence and intensity, ethical problems when the definitive host is man or endangered species. There are no absolute recommendations for sampling intensity, as they will be related to the biological questions addressed. However, one should realize that studying too few individuals in species with very large natural ranges could lead to inappropriate inferences about population biology.

Parasite populations. We are still waiting for studies of parasite populations in definitive hosts which would have the following minimum features: 20 parasites per individual host (if biologically possible), five hosts per site and ten sites separated by a few units of dispersal. This would allow some statistical confidence in the results, and provide some idea on both within- and among-population structure. These values are of course indicative and should be adapted to specific situations with possible trade-offs between the numbers of parasites, hosts and sites. Ideally a similar design should be constructed for intermediate and definitive hosts, although this would likely be hindered by low prevalence and intensity, especially in snail–schistosome pairs.

Host populations. The sampling schemes presented above for parasitized hosts should be completed with non-infected hosts, such as to analyse enough individuals. Sampling of snails causes in general little problem, especially on the ethical side. Several studies have included several hundreds of individuals or even more, with 20–30 individuals per population and populations separated by a few km to a few thousands of km (Viard *et al.* 1997*b, c*). Sampling

Table 1. *A simplified, indicative framework for analysing population genetic structure of both hosts and parasites. More details are given in text. 'Mating system' refers to inbreeding, homogamy and asexual amplification, and 'technical problems' to null alleles and weak repeatability. Estimation of D (linkage disequilibrium) is rarely possible. (4) assumes temporal sampling; (5) could more efficiently be conducted if external demographic data are available. See text for references. (7) can be analysed on both populations and individuals. (1) to (4) should be compared across loci.*

Level	Analysis/hypothesis	Tests/parameters	Aspects to be discussed
Within populations	(1) Variation of basic parameters	Number of alleles and genotypes Gene diversity	Technical problems, comparisons across sexes, populations and species
	(2) Hardy–Weinberg equilibrium	Exact test per locus Estimation of F_{is}	Null alleles, mating system, Wahlund effect, sibling species
	(3) Genotypic disequilibrium	Exact test per locus Estimation of D	Physical and statistical linkage, mating system
	(4) Temporal variation	See (1), (2), (3) and (6) Waples' test Estimation of N_e	Sampling effect, genetic drift, selection, migration
Among populations	(5) Demographic variation	Various tests	Bottleneck and expansion
	(6) Population structure	Exact test per pair of populations	Genetic drift, migration (rate and models; comparison with direct estimates), mating system
		Estimation of F_{st}	
	(7) Isolation by distance	Regression of $F_{st}/(1 - F_{st})$ on geographic distance Mantel test on F_{st} and geographic distance	
	(8) Phylogenetic approach	Various distances Tree building	Genetic drift, phylogeography, phylogenetic relationships

definitive hosts is often more difficult (Mulvey *et al.* 1991; Sire *et al.* in press) and therefore deserve special attention.

Data analysis

The population genetic literature is replete with methods devised for analysing population structure. Their statistical bases are not always settled and their connections to theoretical models often unexplored. This certainly leaves much room for incorrect intuitions when analysing data. We propose in Table 1 a basic framework that can be straightforwardly implemented with markers such as allozymes or microsatellites, and has been routinely used in the context of host-parasite interactions (Viard *et al.* 1997*b*; Delmotte, Bucheli & Shykoff, 1999; Martinez *et al.* 1999). Its statistical basis is sound, it is formally related to models of population structure and specific hypotheses are formally tested using exact tests. Finally both the spatial and temporal variation is taken into account. Of course this is not the only possible framework, and other methods could be used.

Within populations. Analysing the number of alleles and gene diversity per population and loci is a first step, allowing to avoid major technical problems. Testing for Hardy-Weinberg proportions provides information about population sub-structure, whether it arises because of inbreeding, clonal propagation, homogamy or admixture of populations (Wahlund effect). It is also possible to test for demographic processes, such as recent increase in size of bottlenecks (Cornuet & Luikart, 1996; Beaumont, 1999; Galtier *et al.* 2000). However the power of these techniques, when known, is maximal under very specified windows of parameters. For example, the method of Cornuet & Luikart (1996) detects only those bottlenecks that occurred about a few N_e generations before the present time. Moreover the influence of migration on such tests has still to be evaluated. Sampling on a short temporal scale is a clear bonus in population structure analysis, as lucidly exposed by Waples (1989). Temporal variation in allelic frequencies may be explained by genetic drift if one assumes closed populations. However, extremely large variation may not be consistent with such an hypothesis. For example, Viard *et al.* (1997*c*) found variation in Niger populations of *Bulinus truncatus* that may be explained by drift only if N_e is lower than about 50, an unlikely biological picture. They therefore resorted to the idea of an open population, with migration as the source of variation in allelic frequencies.

Among populations. Examining the matrix of pairwise F_{st} and the corresponding matrix of P values from exact tests constitutes the basic analysis. Isolation by distance can also routinely be tested using a regression method (Rousset, 1997) which significance is evaluated using Mantel-like procedures. This analysis can also be performed between pairs of individuals (Rousset, 2000). On the other hand, we do not recommend estimating gene flow from equation (4) (Whitlock & McCauley, 1999; Rousset, 2001). As mentioned above, the mutation model and rate may affect the values of F-statistics which led to the development of analogues of F_{st} for various models of mutation (review in Excoffier, 2001). However it is unlikely that these analogues perform better than the classical parameters, as shown by theoretical analyses (Rousset, 1996) and simulation results (Gaggiotti *et al.* 1999). More problematic is the mutation rate and we can only suggest to choose markers with intermediate rate.

The approach described above can be completed by a phylogenetic approach. The general idea is to calculate genetic distances between entities (e.g. populations or even individuals) and then to build trees or networks (Hillis *et al.* 1996; Li, 1997). When the goal is to compare populations, these methods should in principle not be used when migration remains strong enough to blur the influence of genetic drift and mutation. In other words populations should evolve under genetic drift or drift-mutation alone. As this is hardly a temporal scale of interest when analysing ongoing coevolutionary interactions, the approach described above seems more appropriate.

Practical aspects. Most of the tests and analyses included in the framework exposed above can be performed using e.g. GENEPOP (ftp://ftp.cefe. cnrs−mop.fr/genepop/) or ARLEQUIN (http:// anthropologie.unige.ch/arlequin). A list of software is provided by Luikart & England, 1999). One should also consult the Web site maintained by J. Felsenstein (http://evolution.genetics.washington. edu).

Other methods. New methods are regularly feeding the specialized literature (Luikart & England, 1999 for an overview and references), although their statistical validity and biological pertinence has not systematically been evaluated. Let us cite a few of the most promising. On the statistical side, the Bayesian approach brings a serious alternative to the more classical hypothesis testing approach. It is based on the incorporation of *a priori* information. The Bayesian approach is extremely powerful in formal genetics where explicit information can generally easily be incorporated, such as expectations about segregation ratios among offspring of crosses. However, more vague information can be incorporated as well. For example, Rannala & Moutain (1997) showed that a Bayesian approach provides more consistent results than current alternative methods when evaluating Nm in an island model. Maximum-likelihood approaches are also increasingly used.

The idea is to build a theoretical scenario based on several variables to be estimated. Data are then compared to the model such as to infer the variables maximizing their likelihood. Nested models can also be tested. For example, Galtier *et al.* (2000) proposed a framework for testing for bottlenecks and selective events when data are available at several loci. Three nested models are considered. The first one assumes mutation-drift equilibrium, the second one a bottleneck in the past of given strength and duration, and the third one a bottleneck and selective effects at each locus. Models are fitted to the data using a maximum-likelihood procedure, and comparing the likelihoods of the three models allows to keep one model. For example, if the second model explains the data better than the first one, but that the third one is not better than the second, then this suggests that the population studied experienced a bottleneck and that selection can be rejected.

On the genetical side, autocorrelation methods have been developed and used by several authors (e.g. Slatkin & Arter, 1991) without any *a priori* connection with theoretical population genetics models. It was therefore difficult to interpret any negative relationship between genetic and geographical distances. However the recent demonstration that classical index of autocorrelation such as Moran's I can be related to F-statistics give more weight to these methods (Hardy & Vekemans, 1999). Assignment methods are also likely to be increasingly used in the near future, in conjunction with the above-mentioned statistical methods. The idea is to evaluate the probability that a given individual (genotype) belongs to a given group (e.g. population) (Paetkau *et al.* 1997; Cornuet *et al.* 1999). This can be useful for evaluating a recent migration event, as an alternative to classical methods which are more concerned with long-term migration events. An example can be found in Paetkau *et al.* (1997).

Comparing population structures

Once population structures have been thoroughly analyzed in hosts and parasites, we are back to the central question addressed in this paper, that is how to compare these structures. Several key-aspects of comparisons are: (1) sampling bears on about the same number of individuals and populations in both hosts and parasites. The sampling area is also not easily defined *a priori*, since we have shown that the validity of some techniques presented above depends on the spatial scale of migration. (2) The markers used have about the same mutation rate. This invalidates analyses using allozymes for example in hosts and hypervariable microsatellites in parasites. (3) Differences in mating system and ploidy levels, which strongly affect within- and among-population variability, should be accounted for. In such a situation, the only unknown parameter governing intra-population variation is genetic drift. Its respective influence can be evaluated by conducting comparisons of parameters describing intra-population variation using non-parametric tests (Estoup et al. 1998). Parameters describing among-population variation can be compared as well, especially F_{st}. Rousset's (1997) technique will provide estimates of $D\sigma^2$ for hosts and parasites. Once again, external information would be useful, here on D, for evaluating the dispersal component. Of course it is not exactly clear how to estimate D in highly agregated species, as are often parasite species (e.g. Théron et al. 1992). Mantel-like procedures may then be used for testing for a correlation between pairwise F_{st} estimates in hosts and parasites (Martinez et al. 1999). A positive correlation would indicate that the pattern of differenciation has the same form, even if the absolute values of differentiation are different. However it is unclear what correlation is expected in the general situation, that is when hosts and parasites have their own population size and dispersal rates.

External data

Analyses of population structure cannot rely on genetic markers only. In the light of the models presented above, it clearly appears that external information, either at the individual, or at the population level, can be extremely useful for interpreting data. A few aspects only are presented here. At the individual level, the mating system (selfing rate, paternity) of both flukes and snails can be analysed based on progeny-arrays and genetic markers (Ritland, 1990). Social structure (e.g. biparental inbreeding) within definitive hosts can be approached using similar techniques, as well as direct observations. Dispersal distribution of hosts can be evaluated directly using capture-mark-recapture methods (Lebreton, Pradel & Clobert, 1993; Hanski, Alho & Moilanen, 2000). At the population level, classical parasitological parameters such as prevalence and intensity are useful for interpreting dispersal patterns of parasites. Data on extinction and recolonization dynamics in hosts and parasites are also welcome. Does recolonization occur from a limited number of infrapopulations in definitive hosts? What is the life-expectancy of infrapopulations? Are extinctions frequent in snail populations? These questions may receive answers from classical, though long-term, population dynamics surveys, coupled with methods using genetic markers.

FLUKE AND SNAIL POPULATION STRUCTURE

We highlight here specific traits of flukes and snails having a significant influence on their population structures. However one is engaged to consider more general aspects (see above), as well as to thoroughly consider traits of parasite cycles, in order to figure out the most relevant aspects of specific host-parasite pairs.

Important traits of snail hosts

Two groups of snail traits may be considered here, namely those (1) affecting snail population genetic structure irrespective of parasites and (2) directly affecting the population structure of parasites. These are overlapping groups, and group (2) will be deferred to below.

Classical parameters influencing population structure have been considered above. Of particular interest in snails are the mating system, especially in highly selfing species (Viard et al. 1997a; Charbonnel et al. 2000; Trouvé et al. 2000), and the metapopulation structure, that is the role of subdivision, extinction and recolonization processes. Readers are referred to a recent review on this topic (Städler & Jarne, 1997). Note also that both can affect indirectly the genetic structure of parasite populations. Inbreeding, distributing the variability more among than within individuals and populations, will certainly affect the host–parasite relationship. For example, one can assume that parasite virulence is positively related to host inbreeding, in which case high selfing rates in molluscs may entail higher prevalence. Similarly, snails from populations regularly experiencing bottlenecks may be more sensitive to parasites. On the other hand, parasites from such host populations may also be less variable, because they themselves experience genetic drift, and therefore less virulent.

The framework developed above for molluscan hosts could also be used for non-molluscan intermediate hosts and definitive host. Important traits of flukes will only be alluded below because of space constraints.

Important traits of flukes

The interpretation of population genetic structure in flukes depends not only on parasite-specific characters, but also on host-specific characters and on host-parasite relationships.

Important factors for appreciating the extent of genetic drift in parasite populations are, among others, the size, number, sex ratio (in *Schistosoma*) and life expectancy of infrapopulations, and the migration rate connecting them. Infrapopulation size is primarily the product of recruitment of new individuals and local mortality. Based on prevalence and intensity data, both the size and number of infrapopulations probably varies greatly within and among fluke species (Théron et al. 1992). The life expectancy of parasite infrapopulations results from individual fluke and host survival, and from the possibility of continuous recruitment within a given host which is precluded by concomitant immunity. Note that (1) parasite distribution among hosts may be particularly important for infrapopulation life-expectancy through parasite-induced host mortality processes (Anderson & Gordon, 1982), and (2) the extinction of infrapopulations is a trivial consequence of host death which should be balanced by colonization of new hosts for a local population to be maintained (Bush & Kennedy, 1994).

As previously mentioned, digeneans are simultaneous hermaphrodites reproducing by mixed mating (Nollen, 1983; Trouvé *et al.* 1996). Schistosomes are gonochoric outcrossing exceptions to this pattern with marked sexual dimorphism. Mate pairs have long been thought to be maintained life-long, but divorce has recently been proven (Pica-Mattocia *et al.* 2000). Adult parthenogenesis is possible (Jourdane, Imbert-Establet & Tchuem Tchuenté, 1995). Interestingly, overdispersion of parasites among hosts may open possibilities of partial selfing depending on local density (Combes & Théron, 2000). Related to the mating system is parasite asexual amplification within the intermediate snail host (10^4–10^5). This fluke-specific demographic process has been interpreted as an adaptation compensating the high mortality of free-living cercariae in the external environment during the transmission phase (Combes, 1991). It may also have genetic consequences akin to that of adult parthenogenesis as genetically identical infective cercariae are massively produced (Mulvey *et al.* 1991; Sire *et al.* 2001).

Parasite dispersal depends on (1) hosts dispersal and (2) dissemination of larval free-living stages (miracidium and cercariae). (1) *A priori* parasite dispersal should closely mimic that of its host with higher dispersal rate. For example, gene flow in schistosomes probably depends to a small extent only on mollusc gene flow, since mostly adult snails are parasitized, a stage at which gene flow is limited. Thus gene flow in schistosomes will be more dependent on that of vertebrate definitive hosts. However, the relationship between parasite dispersal and host dispersal is also a function of infection patterns (prevalence and intensity). An efficient parasite disperser is therefore not only a species with substantial dispersion, but also one in which either all dispersing individuals are infected, or a few of them harbour large parasite infrapopulations. These aspects would certainly deserve closer quantitative examination. Another important aspect is that the genetic structure of parasite populations, especially that of adult infrapopulations, is also shaped at a local scale by host foraging pattern. This indeed allows random recruitment of parasite larvae originating from a large number, non-aggregated intermediate hosts. (2) Free-living stages are not *a priori* considered as contributing significantly to

dispersion. However this has to be evaluated more thoroughly. Cercariae and miracidia may indeed be dispersed by water currents, and even limited amounts of long-distance dispersal have a significant influence on gene flow. Similarly long-lived stages, such as eggs or metacercariae, may contribute to dispersal in some trematode species.

An aspect that has not been considered up to now is host specificity, that is the number of definitive host species exploited by a given parasite species. By definition, specialists exploit one host species, while generalists exploit more than one. If more than one species of the host spectrum offers different environments (e.g. in immune reactions or behaviour), host selection may occur during parasite recruitment or development (LoVerde *et al.* 1985). Diversifying selection can then maintain a stable polymorphism as demonstrated in sympatric populations of *S. mansoni* exploiting different (human and murine) host resources (Théron & Combes, 1995; Combes & Théron, 2000). This would certainly affect parasite population structure through non-random mating between host-adapted parasites. Rousset's (1999) generalization of structured single-state population models to multistate models would constitute an appropriate framework for analysing such a situation.

A review of data

Forces acting on snail population structure have been recently reviewed (Städler & Jarne, 1997; see also Viard *et al.* 1997 *b, c*). We will therefore focus on fluke population structure and analyses of co-structures.

Fluke populations. Price (1980) suggested that populations of parasites are characterized by high levels of inbreeding, low intrapopulation genetic variability and high level of interpopulation differentiation due to genetic drift, founder effect and patch dynamics. This has since then been the dominant paradigm, though the (very slow) build-up of genetic data suggests that this is likely to be incorrect (Nadler, 1995; Anderson *et al.* 1998). For example, most nematode species studied up to now show little evidence of limited gene flow (Anderson *et al.* 1998). Whether this is typical of parasitic nematodes is an open question since the few species analysed have been actively dispersed by human activities. Studies of genetic population structure in digeneans are still in their enfancy and very few data are available. We distinguish between two kinds of studies bearing on (1) within-host genetic diversity (infrapopulation level) and (2) population genetic structure (metapopulation level).

(1) A critical factor for interpreting adult population structure is the genetic diversity within the intermediate host (e.g. the fraction of adults with

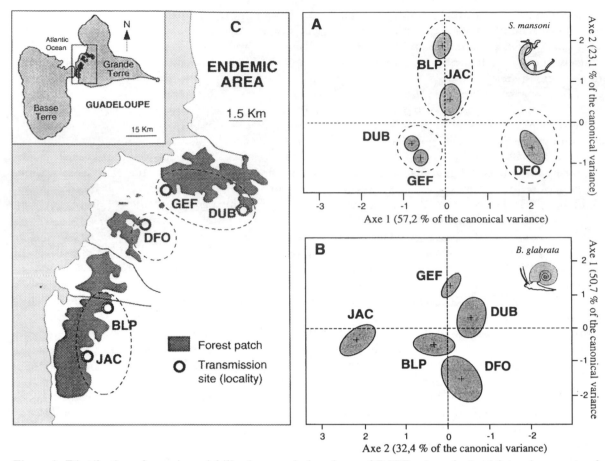

Figure 1. Distribution of genetic variability in canonical analyses of RAPDs data obtained from the swampy forest focus of Guadeloupe. (A) Adult *Schistosoma mansoni* populations sampled from *Rattus rattus*. (B) The corresponding *Biomphalaria glabrata* populations. C. Snails and rats were sampled in five populations representing three forest patches (BLP and JAC; DFO; DUB and GEF) delineated by dotted lines. The percentage of variation represented by the two first canonical axes is given in (A) and (B). See text for comments (modified from Sire *et al.* 2001).

identical genotypes; Curtis & Minchella, 2000; Sire *et al.* 2001). It has been evaluated using various markers, though for intramolluscan larval stages of schistosomes exclusively. Minchella, Sollenberger & Pereira de Sousa (1995) analysed naturally infected *Biomphalaria glabrata* individuals from human foci in Brazil using repetitive DNA elements, and found 3·3 genotypes (miracidia; maximum = 9) of *Schistosoma mansoni* per host. Different was the pattern described in Guadeloupe for the same host–parasite system (Sire *et al.* 1999). Within that area, definitive hosts (*Rattus rattus*) are highly infected (100%) with heavy schistosome load (150 worms per host on average). In contrast only 1·1 parasite genotypes (maximum = 3) were found per snail. This might be explained by a regulatory process limiting the number of developing parasites during simultaneous recruitment (Théron, Pagès & Rognon, 1997) or by concomitant immunity preventing delayed re-infection of snails (Sire, Rognon & Théron, 1998). For the *S. haematobium/Bulinus* pairs in African, Dabo *et al.* (1997) and Davies *et al.* (1999) found 1·6 and 6·2 genotypes per snail, respectively.

Genetic variation of adult flukes within definitive hosts has also been little investigated. Mulvey *et al.*

(1991) showed that white-tailed deer infected with the liver fluke *Fascioloides magna* tended to harbour both few genotypes and several copies of the same genotype. The latter result is probably a consequence of the localized distributions of metacercariae of the same clone. Similarly schistosome infrapopulations in rats from Guadeloupe are made of both a large number of genotypes based on RAPDs markers and a few largely replicated genotypes (Barral *et al.* 1996; Sire *et al.* in press). As an example, 105 multilocus genotypes were detected among 193 schistosomes sampled from an individual rat.

(2) Adult fluke population genetic structure has been seriously analysed in two studies only. As is clear from the definition of parasite population and infrapopulation, patterns of variation are advantageously analysed using a hierarchical sampling design. This is well illustrated by the above-mentioned study on *F. magna* (Lydeard *et al.* 1989; Mulvey *et al.* 1991). Genetic divergence among flukes was analysed within and between hosts and among geographic areas using enzyme polymorphism. Populations of *F. magna* representing hunt units of deer showed low average differentiation ($\hat{F}_{st} = 0.016$), a part of which is due to the occurrence

of multiple-copy genotypes (see above). Genetic distances between fluke populations were related neither to geographic distance between hunt units, nor to definitive host genetic distances. At this scale, both the large population size of the parasite and movements of the definitive host tend to counteract parasite spatial differentiation. In contrast, populations of *F. magna* collected from different states (South Carolina and Tennessee) exhibited markedly greater levels of differenciation ($\hat{F}_{st} = 0.176$) and isolation by distance. The second study (Sire *et al.* in press) shows a clear genetic differentiation between populations of adult *S. mansoni* that were sampled from naturally infected rats trapped in a small endemic area in Guadeloupe. A significant part of the total genetic variation (19%) occurred among the five local populations studied, separated by a few kilometers only. Additionally there was significant differentiation at a higher hierarchical level of sampling, that is forest patches, but no pattern of isolation by distance (Fig. 1). Such a pattern is suggestive of low gene flow among local populations. In absence of human infection, a likely explanation is the restricted movement and dispersal of both rodents and snails.

Co-structures between flukes and snails. There is only a handful of studies analysing population genetic structure in both hosts and parasites (Michalakis *et al.* 1993; Delmotte *et al.* 1999; Martinez *et al.* 1999), and a few are concerned with flukes. One bears on definitive hosts (Mulvey *et al.* 1991; results presented above). The three others analysed flukes and snails. Davies *et al.* (1999) compared the population structures of *S. haematobium* intramolluscan larval stages and adult *Bulinus globosus*. Results suggest more gene flow in the parasite than in the host. Similarly Dybdahl & Lively (1996) using allozymic data found a stronger differentiation among populations of *Potamopyrgus antipodarum* than among populations from the same New Zealand lakes of its trematode parasite *Microphallus* sp. This is due to the higher dispersal of definitive hosts, that is waterbirds. Genetic distances between pairs of host/parasite populations were however correlated. On the other hand, similar levels of differentiation were detected among populations of adult *S. mansoni* and *B. glabrata* in Guadeloupe (Sire *et al.* 2001), and there was no significant correlation between parasite and host genetic distances. No general rule therefore emerges from these four studies. Finally we would like to highlight Dybdahl & Lively's (1996) studies, since it was coupled with an analysis of parasite infectivity indicating local adaptation of parasites to molluscan hosts (Johnson *et al.* 1997). This last result may be viewed as a consequence, or a side effect, of low gene flow in molluscs and high gene flow in parasites (see Gandon *et al.* 1996).

CONCLUSION

Scientific papers generally end with a call for more work and data in their relevant fields. Unfortunately we can only echo this conclusion: basic data, arising from well-designed studies, on the population genetic structure of hosts and parasites (or, even better, of both) are critically lacking. This seriously hinders our understanding of the coevolutionary dynamics. One conclusion though from the few available studies is that there is little reason to think that digenean parasites present specific population structure (Price, 1980). This is hardly surprising given the range of ecological conditions they experience. Future studies willing to fill this gap might try to fulfill the following conditions: (1) use well-behaved markers, such as microsatellites, which are now available in several parasite and host species; (2) conduct small- and large-scale spatial and temporal studies that may allow to contrast parasites with different life-cycles (e.g. two *versus* three hosts) or experiencing different environmental conditions (e.g. more or less open environments), (3) combine population dynamics and population genetics into single studies, something that has not been performed at this point in studies of flukes and snails, and more generally in host-parasite relationships. Theoretical works aiming at improving the analysis of co-structures would also be welcome.

Such studies should be helpful in clarifying several aspects. For example, we still do not know whether a parasite population as defined above corresponds to an infrapopulation or a local population as defined by parasitologists (Margolis *et al.* 1982). There is also much to be learned about recruitment of parasites by their various hosts, inbreeding within local populations and dispersal patterns. Coupled with studies on influence on parasites on host fitness and local adaptation, they should allow to deepen our understanding of selective interference between hosts and parasites and their consequences on population structure. For example, several studies in which flukes and snails feature prominently suggest that parasite are able to manipulate their host mating system (Johnson *et al.* 1997). On the other hand, nothing indicates that parasites manipulate their host dispersal to modify their own dispersal (Boulinier, McCoy & Sorci 2001). Fascinating avenues of research are therefore open to those brave empiricists willing to simultaneously investigate host and parasite population genetic structures.

ACKNOWLEDGEMENTS

The authors thank F. Rousset, Y. Michalakis and F. Viard for discussions on population structure, and L. Excoffier, F. Rousset and G. Sorci for access to unpublished manuscripts. Their work is partly supported by the Ministère de l'Education Nationale de la Recherche et de la Technologie (PRFMMIP).

REFERENCES

ANDERSON, R. M. & GORDON, D. M. (1982). Processes influencing the distribution of parasite numbers within host populations with special emphasis on parasite-induced host mortalities. *Parasitology* **85**, 373–398.

ANDERSON, T. J. C., BLOUIN, M. S. & BEECH, R. N. (1998). Population biology of parasitic nematodes: applications of genetic markers. *Advances in Parasitology* **41**, 219–283.

AVISE, J. C. (1994). *Molecular markers, Natural History and Evolution*. London, Chapman & Hall.

AVISE, J. C. (2000). *Phylogeography. The History and Formation of Species*. Harvard, Harvard University Press.

BARRAL, V., MORAND, S., POINTIER, J.-P. & THÉRON, A. (1996). Distribution of schistosome genetic diversity within naturally infected *Rattus rattus* detected by RAPD markers. *Parasitology* **113**, 511–517.

BARTON, N. H. & WHITLOCK, M. C. (1997). The evolution of metapopulations. In *Metapopulation biology: Ecology, Genetics and Evolution* (ed. Hanski, I. & Gilpin, M.), pp. 183–210. London & New York, Academic Press.

BEAUMONT, M. (1999). Detecting population expansion and decline using microsatellites. *Genetics* **153**, 2013–2029.

BOULINIER, T., MCCOY, K. D. & SORCI, G. (2001). Dispersal and parasitism. In *Dispersal: Individual, Population and Community* (ed. Clobert, J., Danchin, E., Dhondt, A. A. & Nichols, J. D.), pp. 169–179. Oxford, Oxford University Press.

BROWN, D. S. (1994). *Freshwater Snails of Africa and their Medical Importance*. London, Taylor & Francis Ltd.

BUSH, A. O. & KENNEDY, C. R. ((1994). Host fragmentation and helminth parasites: hedging your bets against extinction. *International Journal for Parasitology* **24**, 1333–1343.

CARVALHO, G. (Editor) (1998). *Advances in Molecular Ecology*. Amsterdam, IOS press.

CHARBONNEL, N., ANGERS, B., RAZATAVONJIZAY, R., BREMOND, P. & JARNE, P. (2000). Microsatellite variation in the freshwater snail *Biomphalaria glabrata*. *Molecular Ecology* **9**, 1006–1007.

CHARLESWORTH, B., MORGAN, M. T. & CHARLESWORTH, D. (1993). The effect of deleterious mutations on neutral molecular variation. *Genetics* **134**, 1289–1303.

COLTMAN, D. W., PILKINGTON, J. G., SMITH, J. A. & PEMBERTON, J. M. (1999). Parasite-mediated selection against inbred soay sheep in a free-living, island population. *Evolution* **53**, 1259–1267.

COMBES, C. (1991). Ethological aspects of parasite transmission. *American Naturalist* **138**, 866–880.

COMBES, C. (1995). *Interactions Durables. Ecologie et Évolution du Parasitisme*. Paris, Masson.

COMBES, C. & THÉRON, A. (2000). Metazoan parasites and resource heterogeneity: constraints and benefits. *International Journal for Parasitology* **30**, 299–304.

CORNUET, J.-M. & LUIKART, G. (1996). Description and power analysis of two tests for detecting recent population bottlenecks from allele frequency data. *Genetics* **144**, 2001–2014.

CORNUET, J.-M., PIRY, S., LUIKART, G., ESTOUP, A. & SOLIGNAC, M. (1999). New methods employing multilocus genotypes to select or exclude populations as origins of individuals. *Genetics* **153**, 1989–2000.

CURTIS, J. & MINCHELLA, D. J. (2000). Schistosome population genetic structure: when clumping worms is not just splitting hairs. *Parasitology Today* **16**, 68–71.

DABO, A., DURAND, P., MORAND, S., LANGAND, J., IMBERT-ESTABLET, D., DOUMBO, O. & JOURDANE, J. (1997). Dispersion and genetic diversity of *Schistosoma haematobium* within its Bulinid intermediate hosts in Mali. *Acta Tropica* **66**, 15–26.

DAVID, P. (1998). Heterozygosity–fitness correlations: new perspectives on old problems. *Heredity* **80**, 531–537.

DAVIES, C. M., WEBSTER, J. P., KRÜGER, O., MUNATSI, A., NDAMBA, J. & WOOLHOUSE, M. E. J. (1999). Host-parasite population genetics: a cross-sectional comparison of *Bulinus globosus* and *Schistosoma haematobium*. *Parasitology* **119**, 295–302.

DELMOTTE, F., BUCHELI, E. & SHYKOFF, J. A. (1999). Host and parasite population structure in a natural plant–pathogen system. *Heredity* **82**, 300–308.

DESPRES, L., IMBERT-ESTABLET, D., COMBES, C. & BONHOMME, F. (1992). Molecular evidence linking Hominid evolution to recent radiation of schistosomes (Platyhelminthes: Trematoda). *Molecular Phylogenetics and Evolution* **4**, 295–304.

DIAS NETO, E., PEREIRA DE SOUZA, C., ROLLINSON, D., KATZ, N., PENA, S. D. J. & SIMPSON, A. J. G. (1993). The random amplification of polymorphic DNA allows the identification of strains and species of schistosomes. *Molecular and Biochemical Parasitology* **57**, 83–88.

DYBDAHL, M. F. & LIVELY, C. M. (1996). The geography of coevolution: comparative population structures for a snail and its trematode parasite. *Evolution* **50**, 2264–2275.

ESTOUP, A. & ANGERS, B. (1998). Microsatellites and minisatellites for molecular ecology: theoretical and empirical considerations. In *Advances in Molecular Ecology* (ed. Carvalho, G.), pp. 55–86. Amsterdam, IOS press.

ESTOUP, A., ROUSSET, F., MICHALAKIS, Y., CORNUET, J.-M., ADRIAMANGA, M. & GUYOMARD, R. (1998). Comparative analysis of microsatellite and allozyme markers: a case study investigating microgeographic differentiation in brown trout (*Salmo trutta*). *Molecular Ecology* **7**, 339–353.

EXCOFFIER, L. (2001). Analysis of population subdivision. In *Handbook of Statistical Genetics* (ed. Balding, D., Bishop, M. & Cannings, C.), in press. New York, Wiley & Sons.

GAGGIOTTI, O. E., LANGE, O., RASSMANN, K. & GLIDDON, C. (1999). A comparison of two indirect methods for estimating average levels of gene flow using microsatellite data. *Molecular Ecology* **8**, 1513–1520.

GALTIER, N., DEPAULIS, F. & BARTON, N. H. (2000). Detecting bottlenecks and selective sweeps from DNA sequence polymorphism. *Genetics* **155**, 981–987.

GANDON, S., CAPOWIEZ, Y., DUBOIS, Y., MICHALAKIS, Y. & OLIVIERI, I. (1996). Local adaptation end gene-for-gene coevolution in a metapopulation model.

Proceedings of the Royal Society of London B **263**, 1003–1009.

GANDON, S., EBERT, D., OLIVIERI, I. & MICHALAKIS, Y. (1998). Differential adaptation in spatially heterogeneous environments and host-parasite coevolution. In *Genetic Structure and Local Adaptation in Natural Insect Populations. Effects of Ecology, Life History, and Behavior* (ed. Mopper, S. & Strauss, S. Y.), pp. 325–342. London & NY, Chapman & Hall.

GOLDSTEIN, D. B. & SCHLÖTTERER, C. (Editors) (1999). *Microsatellites. Evolution and Applications.* Oxford, Oxford University Press.

HANSKI, I., ALHO, J. & MOILANEN, A. (2000). Estimating the parameters of survival and migration of individuals in metapopulations. *Ecology* **81**, 239–251.

HARDY, O. J. & VEKEMANS, X. (1999). Isolation by distance in a continuous population: reconciliation between spatial autocorrelation and population genetic models. *Heredity* **83**, 145–154.

HARTL, D. L. & CLARK, A. G. (1997). *Principles of Population Genetics.* Sunderland, Sinauer Associates.

HILLIS, D. M., MORITZ, G. & MABLE, B. K. (1996). *Molecular Systematics.* Sunderland, Sinauer Associates.

JARNE, P. (1995). Mating system, bottlenecks and genetic polymorphism in hermaphroditic animals. *Genetical Research* **65**, 193–207.

JARNE, P. & LAGODA, P. J. L. (1996). Microsatellites, from molecules to populations and back. *Trends in Ecology and Evolution* **11**, 424–429.

JELNES, J. E. (1986). Experimental taxonomy of *Bulinus* (Gastropoda: Planorbidae): the West and North African species reconsidered, based upon an electrophoretic study of several enzymes per individual. *Zoological Journal of the Linnean Society* **87**, 1–26.

JOHNSON, S. G., LIVELY, C. M. & SCHRAG, S. J. (1997). Evolution and ecological correlates of uniparental and biparental reproduction in freshwater snails. In *Evolutionary Ecology of Freshwater Animals* (ed. Streit, B., Städler, T. & Lively, C.), pp. 263–291. Basel, Birkhäuser Verlag.

JOHNSTON, D. A., DIA NETO, E., SIMPSON, A. J. G. & ROLLINSON, D. (1993). Opening the can of worms: molecular analysis of schistosome populations. *Parasitology Today* **9**, 286–291.

JOURDANE, J., IMBERT-ESTABLET, D. & TCHUEM TCHUENTE, L.-A. (1995). Parthenogenesis in Schistosomatidae. *Parasitology Today* **11**, 427–430.

LEBRETON, J. D., PRADEL, R. & CLOBERT, J. (1993). The statistical analysis of survival in animal populations. *Trends in Ecology and Evolution* **8**, 91–95.

LI, W. H. (1997) *Molecular Evolution.* Sunderland, Sinauer Associates.

LOVERDE, P. T., DEWALD, J., MINCHELLA, D. J., BOSSHARDT, S. C. & DAMIAN, R. T. (1985). Evidence for host-induced selection in *Schistosoma mansoni. Journal of Parasitology* **71**, 297–301.

LUIKART, G. & ENGLAND, P. R. (1999). Statistical analysis of microsatellite DNA data. *Trends in Ecology and Evolution* **14**, 253–256.

LYDEARD, C., MULVEY, M., AHO, J. M. & KENNEDY, P. K. (1989). Genetic variability among natural populations of the liver fluke *Fascioloides magna* in white-tailed deer, *Odocoileus virginianus. Canadian Journal of Zoology* **67**, 2021–2025.

MARGOLIS, L., ESCH, G. W., HOLMES, J. C., KURIS, A. M. & SCHAD, G. A. (1982). The use of ecological terms in parasitology (report of an ad hoc committee of the American society of parasitologists). *Journal of Parasitology* **68**, 131–133.

MARTINEZ, J. G., SOLER, J. J., SOLER, M., MOLLER, A. P. & BURKE, T. (1999). Comparative population structure and gene flow of a brood parasite, the great spotted cuckoo (*Clamator glandarius*), and its primary host, the magpie (*Pica pica*). *Evolution* **53**, 269–278.

MARUYAMA, K. & TACHIDA, H. (1992). Genetic variability and geographical structure in partially selfing populations. *Japanese Journal of Genetics* **67**, 39–51.

MICHALAKIS, Y., SHEPPARD, A. W., NOEL, V. & OLIVIERI, I. (1993). Population structure of a herbivorous insects and its host plant on a microgeographic scale. *Evolution* **47**, 1611–1615.

MICHELSON, E. A. & DUBOIS, L. (1978). Susceptibility of Bahian populations of *Biomphalaria glabrata* to an allopatric strain of *Schistosoma mansoni. American Journal of Tropical Medicine and Hygiene* **27**, 782–786.

MINCHELLA, D. J., SOLLENBERGER, K. M. & PEREIRA DE SOUSA, C. (1995). Distribution of schistosome genetic diversity within molluscan intermediate hosts. *Parasitology* **111**, 217–220.

MULVEY, M., AHO, J. M., LYDEARD, C., LEBERG, P. L. & SMITH, M. H. (1991). Comparative population genetic structure of a parasite (*Fascioloides magna*) and its definitive host. *Evolution* **45**, 1628–1640.

NADLER, S. A. (1995). Microevolution and the genetic structure of parasite populations. *Journal of Parasitology* **81**, 395–403.

NOLLEN, P. M. (1983). Patterns of sexual reproduction among parasitic platyhelminths. *Parasitology* **86**, 99–120.

PAETKAU, D. L., WAITS, L. P., CLARKSON, P. L., CRAIGHEAD, L. & STROBECK, C. (1997). An empirical evaluation of genetic distance statistics using microsatellite data from bear (Ursidae) populations. *Genetics* **147**, 1943–1957.

PANNELL, J. R. & CHARLESWORTH, B. (1999). Neutral genetic diversity in a metapopulation with recurrent local extinction and recolonization. *Evolution* **53**, 664–676.

PEMBERTON, J. M., SLATE, J., BANCROFT, D. R. & BARRETT, J. A. (1995). Nonamplifying alleles at microsatellite loci: a caution for parentage and population studies. *Molecular Ecology* **4**, 249–252.

PICA-MATTOCIA, L., MORONI, R., TCHUEM TCHUENTE, L.-A., SOUTHGATE, V. R. & CIOLI, D. (2000). Changes of mate occur in *Schistosoma mansoni. Parasitology* **120**, 495–500.

PRICE, P. W. (1980). *Evolutionary Biology of Parasites.* Princeton, Princeton University Press.

RANNALA, B. & MOUTAIN, J. L. (1997). Detecting immigration by using multilocus genotypes. *Proceedings of the National Academy of Sciences, USA* **94**, 9197–2001.

RITLAND, K. (1990). A series of FORTRAN computer programs for estimating plant mating systems. *Journal of Heredity* **81**, 235–237.

ROHANI, P. & RUXTON, G. D. (1999). Dispersal-induced

instabilities in host-parasitoid metapopulations. *Theoretical Population Biology* **55**, 23–36.

ROSS, K. G., SHOEMAKER, D. D., KRIEGER, M. J. B., DEHEER, C. J. & KELLER, L. (1999). Assessing genetic structure with multiple classes of molecular markers: a case study involving the introduced fire ant *Solenopsis invicta*. *Molecular Biology and Evolution* **16**, 525–543.

ROUSSET, F. (1996). Equilibrium values of measures of population subdivision for stepwise mutation processes. *Genetics* **142**, 1357–1362.

ROUSSET, F. (1997). Genetic differentiation and estimation of gene flow from F-statistics under isolation by distance. *Genetics* **145**, 1219–1228.

ROUSSET, F. (1999). Genetic differentiation within and between two habitats. *Genetics* **151**, 397–407.

ROUSSET, F. (2000). Genetic differentiation between individuals. *Journal of Evolutionary Biology* **13**, 58–62.

ROUSSET, F. (2001). Inferences from spatial population genetics. In *Handbook of Statistical Genetics* (ed. Balding, D., Bishop, M. & Cannings, C.), in press. New York, Wiley & Sons.

ROUSSET, F. & RAYMOND, M. (1997). Statistical analyses of population genetic data: new tools, old concepts. *Trends in Ecology and Evolution* **12**, 313–317.

SIRE, C., DURAND, P., POINTIER, J.-P. & THÉRON, A. (1999). Genetic diversity and recruitment pattern of *Schistosoma mansoni* in a *Biomphalaria glabrata* snail population: a field study using random-amplified polymorphic DNA markers. *Journal of Parasitology* **85**, 436–441.

SIRE, C., LANGAND, J., BARRAL, V. & THÉRON, A. (2001). Parasite (*Schistosoma mansoni*) and host (*Biomphalaria glabrata*) genetic diversity: population structure in a fragmented landscape. *Parasitology*, **122**, 545–554.

SIRE, C., ROGNON, A. & THERON, A. (1998). Failure of *Schistosoma mansoni* to reinfect *Biomphalaria glabrata* snails: acquired humoral resistance or intra-specific larval antagonism. *Parasitology* **117**, 117–122.

SLATKIN, M. (1985). Gene flow in natural populations. *Annual Review of Ecology and Systematics* **16**, 393–430.

SLATKIN, M. (1995). A measure of population subdivision based on microsatellite allele frequencies. *Genetics* **139**, 457–462.

SLATKIN, M. & ARTER, H. E. (1991). Spatial autocorrelation methods in population genetics. *American Naturalist* **138**, 499–517.

STÄDLER, T. & JARNE, P. (1997). Population biology, genetic structure, and mating system parameters in freswater snails. In *Ecology and Evolution of Freshwater Organisms* (ed. Streit, B., Städler, T. & Lively, C. M.), pp. 231–262. Basel, Birkhäuser Verlag.

THÉRON, A. & COMBES, C. (1995). Asynchrony of infection timing, habitat preference, and sympatric speciation of schistosome parasites. *Evolution* **49**, 372–375.

THÉRON, A., PAGES, J. R. & ROGNON, A. (1997). *Schistosoma mansoni*: distribution patterns of miracidia among *Biomphalaria glabrata* snail hosts as related to host susceptibility and sporocyst regulation process. *Experimental Parasitology* **85**, 1–9.

THÉRON, A., POINTIER, J.-P., MORAND, S., IMBERT-ESTABLET, D. & BOREL, G. (1992). Long-term dynamics of natural populations of *Schistosoma mansoni* among *Rattus rattus* in patchy environment. *Parasitology* **104**, 291–298.

TROUVE, S., DEGEN, L., MEUNIER, C., TIRARD, C., HURTREZ-BOUSSES, S., DURAND, P., GUEGAN, J.-F., GOUDET, J. & RENAUD, F. (2000). Microsatellites in the hermaphroditic snail, *Lymnaea truncatula*, intermediate host of the liver fluke, *Fasciola hepatica*. *Molecular Ecology* **9**, 1662–1664.

TROUVE, S., RENAUD, F., DURAND, P. & JOURDANE, J. (1996). Selfing and outcrossing in a parasitic hermaphrodite. *Heredity* **177**, 1–8.

VIARD, F., DOUMS, C. & JARNE, P. (1997*a*). Selfing, sexual polymorphism and microsatellites in the hermaphroditic freshwater snail *Bulinus truncatus*. *Proceedings of the Royal Society of London B* **264**, 39–44.

VIARD, F., JUSTY, F. & JARNE, P. (1997*b*). The influence of self-fertilization and bottlenecks on the genetic structure of subdivided populations: a case study using microsatellite markers in the freshwater snail *Bulinus truncatus*. *Evolution* **51**, 1518–1528.

VIARD, F., JUSTY, F. & JARNE, P. (1997*c*). Population dynamics inferred from temporal variation at microsatellite loci in the selfing snail *Bulinus truncatus*. *Genetics* **146**, 973–982.

VOELKER, R. A., LANGLEY, C. H., LEIGH BROWN, A. J., OHNISHI, S., DICKSON, B., MONTGOMERY, E. & SMITH, S. C. (1980). Enzyme null alleles in natural populations of *Drosophila melanogaster*: Frequencies in a North Carolina population. *Proceedings of the National Academy of Sciences, USA* **75**, 1091–1095.

WAPLES, R. S. (1989). Temporal variation in alleles frequencies: testing the right hypothesis. *Evolution* **43**, 1236–1251.

WEIR, B. S. & COCKERHAM, C. C. (1984). Estimating F-statistics for the analysis of population structure. *Evolution* **38**, 1358–1370.

WHITLOCK, M. C. & MCCAULEY, D. E. (1999). Indirect measures of gene flow and migration: Fst ≠ $1/(4Nm+1)$. *Heredity* **82**, 117–125.

WOOLHOUSE, M. E. J. (1992). Population biology of the freshwater snail *Biomphalaria pfeifferi* in the Zimbabwe highveld. *Journal of Applied Ecology* **29**, 687–694.

Coevolution and compatibility in the snail–schistosome system

J. P. WEBSTER* *and* C. M. DAVIES

Wellcome Trust Centre for the Epidemiology of Infectious Disease (WTCEID), University of Oxford, South Parks Road, Oxford, OX1 3FY

SUMMARY

In stark contrast to the huge body of theoretical work on the importance of hosts and parasites as selective agents acting on each other, until recently, little systematic empirical investigation of this issue has been attempted. Research on snail–schistosome interactions have, therefore, the potential for making an important contribution to the study of coevolution or reciprocal adaptation. This may be particularly pertinent since snail–schistosomes represent an indirectly transmitted macroparasite system, so often overlooked amongst both theoretical and empirical studies. Here we review ideas and experiments on snail–schistosome interactions, with particular emphasis on those that may have relevance to the potential coevolution between host resistance and parasite infectivity and virulence. We commence with an introduction and definition of the general concepts, before going into detail of some specific studies to illustrate these: evidence of snail–schistosome coevolutionary process in the field; evidence of coevolutionary processes in the laboratory; a general assessment of the applicability of coevolutionary models in snail–schistosome interactions; and finishing with a section on conclusions and areas for further study.

Key words: Coevolution, compatibility, resistance, virulence, infectivity, snail, schistosome.

INTRODUCTION

There is a huge body of theoretical work on the importance of hosts and parasites as selective agents acting on each other. However, there has until recently been little systematic empirical investigation of this topic. One area in which research on snail–schistosome interactions has the potential for making an important contribution is, therefore, the study of coevolution or reciprocal adaptation. In particular, the evolution of snail resistance and schistosome infectivity and virulence may offer the prospect of an insight into the genetics of adaptation.

One approach from which to infer snail–schistosome coevolution is to investigate the current 'end points' of coevolutionary interactions in the field. An alternative approach is to demonstrate coevolution in action through controlled laboratory experiments. The ultimate demonstration may be longitudinal field studies that incorporate both these factors. There are several ways to achieve these aims. The first way is to document additive genetic variation in host resistance, parasite infectivity and/or virulence using quantitative genetic techniques or artificial selection experiments. This may help elucidate the genetic architecture underlying these traits. One must also show that host resistance affects parasite fitness and conversely that parasite infection and virulence affect host fitness

(Kraaijeveld *et al.* 1998). If genetic variability is demonstrated, one must then identify how such variability is maintained in natural populations, and whether it varies at different times or places. Likewise, one should also consider whether evolutionary changes in resistance, infectivity or virulence influence host-parasite population dynamics.

Our aims here are to review ideas and experiments on snail–schistosome interactions, with particular emphasis on those that may have relevance to the potential coevolution between host resistance and parasite infectivity and/or virulence. Much of this paper concerns examples from *Biomphalaria glabrata–Schistosoma mansoni* interactions, as this is the system that has been most intensively investigated. Nevertheless, an attempt is made to put this work within a broader snail–trematode framework.

The review will be split into five sections. We commence with an introduction and definition of the general concepts, before going into detail of some specific studies to illustrate these: evidence of snail–schistosome coevolutionary process in the field; evidence of coevolutionary processes in the laboratory; a general assessment of the applicability of coevolutionary models in snail–schistosome interactions; and finally finishing with a section on conclusions and areas for further study. Several other issues, such as the physiological basis of resistance and/or infectivity, are of undoubted relevance to understanding this system, but are outside the scope of our account. Moreover, these aspects will addressed in detail in other parts of this issue (see e.g. de Jong-Brink; Lewis, Patterson & Richards, both this supplement).

* Corresponding author: Tel: 01865 271288. Fax: 01865 281245.
E-mail: joanne.webster@wellcome-epidemiology.oxford.ac.uk

Parasitology (2001), **123**, S41–S56.   © 2001 Cambridge University Press
DOI: 10.1017/S0031182001008071

COEVOLUTION: DEFINITIONS AND MODELS

Before progressing further, it is appropriate to define the terms we will use in this account, given that there are inconsistencies across the literature.

Co-evolution

Co-evolution is evolution in one species in response to selection imposed by a second species, accompanied by evolution in the second species in response to reciprocal selection imposed by the first species. Host–parasite coevolution is driven by the reciprocal evolution of host resistance and parasite infectivity and/or virulence (Sorci, Moller & Boulinier 1997). However, reciprocal selection has seldom been measured in natural, particularly animal, populations (Clayton *et al.* 1999).

Resistance

The term 'resistance' (sometimes referred to in the literature as 'non-susceptibility' or 'refractoriness') specified here refers to the genetic, biochemical and/or physiological profiles that inhibit parasite establishment, survival and/or development within the host (Coustau, Chevillon & Ffrench-Constant, 2000).

Within this definition two forms of host resistance may be considered (Gandon & Michalakis, 2000). First, the host could adopt a quantitative form of resistance, which can be used to limit the deleterious effects induced by the parasites. In this case, hosts may be infected by the parasites but more resistant ones are harmed less. Alternatively, the host could adopt a qualitative form of resistance that would prevent any infection by the parasite (i.e. resistant hosts cannot be infected at all). The majority of models of host–parasite coevolution regard 'resistance' to be the qualitative type referred to here, and will be the major focus of this article.

Infectivity

Infectivity is defined as the infective capacity of the pathogen when applied to suitable host tissues. Infectivity can variably be thought of as the number or type of host genotypes a parasite is capable of infecting (Read, 1994) or, at the population level, the proportion of a particular host strain the parasite will infect. In snail–schistosome interactions, an infected host is usually considered to be one that produces the next infective stage, the cercariae (Wright, 1974).

Virulence

Virulence may be defined as the reduction in host fitness (lifetime reproductive success) attributable to parasitic infection (Read, 1994). Parasite-induced host mortality, a commonly used measure of virulence in evolutionary models, may be a direct or indirect result of parasitic infection. Examples of the latter include increased predation risk, susceptibility to other pathogens, or decreased competitive ability (Poulin, Hecker & Thomas, 1998). Virulence, by definition, also represents a fitness cost to most parasites, since parasites are intimately dependent on their hosts for reproduction and survival (Combes, 1997). For example, killing the host presents an obvious evolutionary cost to the parasite, at least for those parasites where host death prevents further parasite propagation and transmission.

Definitions of virulence, however, do differ across the literature. Although we will use that presented above, it is worth considering a commonly used alternative: the infective capacity of pathogens (here termed 'infectivity'). This is the usage of virulence in many models of host–pathogen coevolution that have been developed primarily for plant–pathogen systems (e.g. Frank, 1993). It is interesting to note that such models implicitly assume a positive association between 'infectivity' and 'virulence' (the severity of the infection). However, this is not always the case (e.g. Barbosa, 1975; Ebert, 1994). Moreover, these factors are not equivalent *vis à vis* cost-benefit evolution (see below). It is therefore important to distinguish between the two in this review.

Compatibility

Virulence, infectivity and resistance are complex features of host–parasite interactions. Though they may appear to be attributes of the parasite or of the host, they are in fact the net effect of the physiological, morphological and behavioural interactions between parasite and host (Toft & Karter, 1990). Thus it should always be noted that definition of a host as resistant or a parasite as having high infectivity or virulence, may be specific to a particular host–parasite species or strain combination. The relationship may thus be better described as 'compatibility'.

As regards snail–schistosome interactions, in compatible interactions, the parasite recognizes, penetrates and develops within the snail, giving rise to the parasites next infective stage, the cercariae. Alternatively, in incompatible interactions, the larval trematode either fails to recognize, penetrate or develop in the snail, or penetrates and is recognized as non-self, and is destroyed by the mollusc's internal defence system (van der Knapp & Loker, 1990).

Coevolutionary models

Coevolution may take different forms and several models have been proposed. Some of the most important ones, described below, may have relevance to snail–schistosome co-evolution, but even these are not mutually exclusive for a coevolving system.

Frequency-dependent or 'Red Queen' coevolution.
To understand the kind of selection pressure acting on resistance, infectivity and virulence, it is important to understand the specificity of the interaction. This question is important, as specificity will give rise to adaptation to the most common genotype and hence frequency-dependent advantage for rare genotypes. The result is locally dynamic or 'Red Queen' coevolution with a constant flux of genotype frequencies and high heritability of infectivity, virulence and resistance (Kraaijeveld & Godfray, 1999; see also Lively, this supplement). In models of dynamic coevolution of host resistance and parasite infectivity (Morand, Manning & Woolhouse, 1996), parasite genotypes track common host genotypes, promoting a frequency-dependent advantage to rare host genotypes, since they escape the deleterious consequences of parasitic infection. The 'rare' advantage of a host genotype will depend not only on its likelihood of being infected (parasite infectivity), but also on the severity of any such infection (virulence) (Lively, 1999). Just as genotype-specific resistance, infectivity and virulence mechanisms lead to a constant turnover of gene frequencies, so species-specific resistance, infectivity and virulence may lead to changes in species frequency (Kraaijeveld & Godfray, 1999).

Density-dependent coevolution. Non-specific resistance and infectivity involves increased resistance improving survival against all genotypes of a parasite, and increased infectivity improving performance against all genotypes of host, such that the rank order of resistance of different host strains exposed to different parasite strains (or of infectivity to different hosts) will be constant. Frequency-dependent dynamics and Red Queen coevolution are thus not expected. However, where parasites have a major effect on host population dynamics, coupled coevolution could still lead to spatio-temporal variation in resistance, infectivity and virulence. In this situation, higher parasite densities may lead to selection for enhanced resistance, which might then cause either a reduction in parasite numbers (and hence a relaxation of selection for resistance) or a corresponding increase in the level of infectivity. Virulence is again important as it determines the strength of selection on hosts (Lively, 1999). Models of this type of interaction in host–parasitoid interactions that assume density-dependent rather than frequency-dependent selection suggest that the population and genetic dynamics may reach an equilibrium or show persistent cycles, depending on the initial conditions (Kraaijeveld & Godfray, 1999).

Cost-benefit trade-offs and locally fixed optima.
Where frequency-dependent or density-dependent selection does not occur, host-coevolution might lead to locally fixed optima for the levels of resistance, infectivity and virulence, which may vary across time or space as costs and benefits change. Genetic constraints are considered to be fundamental in life-history evolution (Messenger, Molineux & Bull, 1999). While increased resistance or infectivity is clearly beneficial in the context of host–parasite interactions, there may be trade-offs involving other components of fitness that might cause the optimum level of either trait to be lower than the maximum achievable. Thus, for example, host resistance may bear metabolic costs that are higher than the increases in fitness achieved through avoiding parasite infection (Frank, 1994). This might be particularly true where parasite prevalence is low. Virulence represents a fitness cost to most parasites, since parasites are intimately dependent on their hosts for their reproduction and survival (Combes, 1997). Thus the fitness cost of being virulent may influence the direction of parasite evolution (see below).

The geographic mosaic theory of coevolution. The geographic mosaic theory of coevolution differs slightly from that of the models described above in that it is a general hypothesis about how the raw materials of coevolution are organized (Thompson, 1994, 1999). It suggests that there is a selection mosaic among populations, favouring different evolutionary directions to interactions in different populations. Thus for example, variations in opportunities for transmission or parasite-independent mortality (Ebert & Herre, 1996), and environmental effects on the expression of host resistance (Abdullah, 1997) are expected to alter the balance of cost-benefit trade-offs governing the evolution of parasite virulence (Kraaijeveld & Godfray, 1999). The geographic mosaic theory further assumes that there are 'coevolutionary hotspots', such that reciprocal selection need not occur in all populations. Finally, the hypothesis suggests that there is a continuous remixing of the range of coevolving traits, resulting from the mosaic, gene flow, random genetic drift and the local extinction of populations (Thompson, 1994, 1999).

EVIDENCE OF SNAIL–SCHISTOSOME COEVOLUTION IN THE FIELD

Variability in host–parasite compatibility may be taken as evidence of coadaptation and, in some cases, potential coevolution in natural populations. Snail–trematode compatibility is a highly specific relationship, often at the population or strain levels for both participants (Lo & Lee, 1995; Webster & Woolhouse, 1998). This specificity has the important practical effect of limiting medically important trematodes such as schistosomes to geographic areas occupied by compatible snails (van der Knaap & Loker, 1990).

Whilst variations in snail–schistosome compatibility was first reported by Files & Cram (1949), perhaps the clearest example of the presence of compatibility factors in both snail and schistosome is provided by Paraense & Correa (1963), who showed that a *S. mansoni* strain adapted to *Bi. tenagophila* will not infect *Bi. glabrata*, and vice versa. Within species, Manning, Woolhouse & Ndamba (1995) used a reciprocal cross-infection design with *Bulinus globosus* snails and *Schistosoma haematobium/ mattheei* from Zimbabwe and found that sympatric parasite–host combinations were more compatible than allopatric combinations across two sites 60 km apart. Similar findings have also been suggested by Lively (1989) and Lively & Dybdahl (2000) for *Microphallus* spp. infections of *Potamopygrus antipodarum* in New Zealand. However, not all studies demonstrate such local adaptation (see Morand *et al.* 1996). For example, Vera *et al.* (1990) also used reciprocal cross infection experiments of wild *Bu. truncatus* snails and *S. haematobium* parasites, and did not find any difference in compatibility across three sites up to 800 km apart. Several other studies which tested the infectivity of a single trematode population to two or more snail populations also found exceptions to this rule (reviewed by Richards & Shade, 1987; Morand *et al.* 1996). Morand *et al.* (1996) therefore used field data taken from the literature to develop a mathematical model based on the dynamics of the host–parasite interaction. In the model, parasite infectivity and host susceptibility were defined by the matching of genotypes in a diploid system, and it was shown that frequency-dependent coevolution could explain such local adaptation. They further demonstrated that whilst there is a tendency for sympatric combinations to be more compatible than allopatric combinations, instances of the reverse pattern also occur. This may be explained by the fact that frequency-dependent models of host–parasite coevolution, such as that of the Red Queen hypothesis, predict that changes in sympatric parasite allele frequencies tend to lag behind host allele frequencies (Lively & Apanius, 1995; Morand *et al.* 1996). Thus allopatric parasite allele frequencies may therefore, by virtue of being in a different phase of the cycle, chance to correspond more closely to host allele frequencies (Woolhouse & Webster, 2000).

EVIDENCE OF SNAIL–SCHISTOSOME COEVOLUTION IN THE LABORATORY

The results of the compatibility studies described above may suggest snail–schistosome coevolution. Further substantiation, nevertheless, requires complimentary investigation in the laboratory, where phenotypic and other variables can be tightly controlled.

Heritability of host resistance

Models of host–parasite coevolution require that variation in host resistance to parasite infection and of parasite infectivity and/or virulence is, at least partially, genetically determined (Anderson & May, 1982), since heritable genetic variation is a prerequisite for natural selection. Richards and colleagues have made extensive studies of the genetics of resistance and susceptibility of *Bi. glabrata* to *S. mansoni*. As these are reviewed in a separate chapter of this issue (Lewis *et al.* this supplement), we will not duplicate information here. Instead we will focus on other studies aimed to elucidate the heritability of host resistance.

Webster & Woolhouse (1998) used artificial selection experiments to determine the heritability of snail–schistosome compatibility. Two unselected populations of *Bi. glabrata* snails, and two unselected *S. mansoni* parasite populations, were chosen for artificial selection (see Fig. 1). For each snail–schistosome strain combination, adult P_1 snails were individually exposed to five *S. mansoni* miracidia and subsequently divided up into groups containing either uninfected 'resistant-selected' or infected 'susceptible-selected' individuals. F_1 progeny were then exposed to the same strain of *S. mansoni* as their parents and only snails consistent with their selection group were maintained. This breeding and selection protocol was continued until the F_3 generation. A matched number of unselected control snails were also exposed to both parasite strains at each generation.

The results suggested that compatibility in this system is heritable. Snails, of either strain, selected for resistance were significantly more resistant than controls to *S. mansoni* by the F_1 generation. Likewise, snails selected for susceptibility were significantly more susceptible than controls to *S. mansoni*, although increased susceptibility did not develop before two generations of selection. By the F_3 generation, infection prevalence was approximately 25 % among resistant-selected snail lines and 75 % among susceptible-selected snail lines. Unselected control snail lines remained at approximately 50 % infection rate when exposed to five miracidia per snail throughout each generation (Fig. 2).

A subsequent study on these same host–parasite lines (Webster, 2001) then investigated whether such resistance is dominant over susceptibility, following simple Mendelian inheritance, as is common for many plant–pathogen (Fritz & Simms, 1992) and other animal–helminth (Richards, 1975 *a*,*b*; Behnke *et al.* 2000) interactions. Individual adult snails from each artificially-selected replicate snail line described above were paired in small pots with a selected partner. In the first two groups (of $n \geqslant 14$ pairs each) the compatibility status of each member of the pair

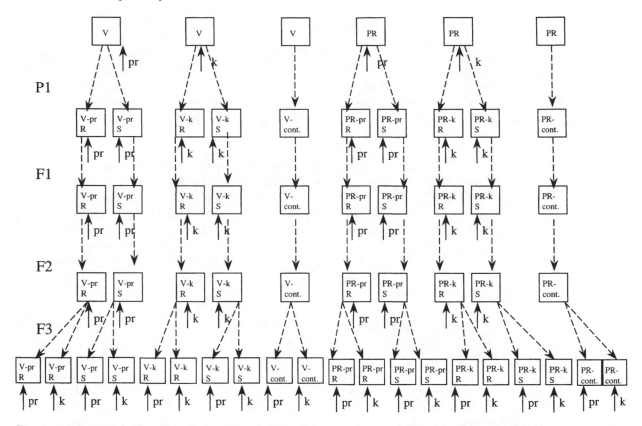

Fig. 1. Artificial selection methods for *Biomphalaria glabrata* resistance and susceptibility to *Schistosoma mansoni*. V = Vespiano snail strain; PR = Puerto Rican snail strain; pr = Puerto Rican parasite strain; k = Kenyan parasite strain; R = Resistant-selected snail lines; S = Susceptible-selected snail lines. Each box represents a tank/snail line (new tank for each generation, split into two tanks per line in P_1 and F_3 generations). Each solid arrow represents exposure of all individuals to named parasite strain. Each broken arrow represents the division of snail lines into those to be resistant-selected or susceptible-selected. Each snail tank/line was maintained at matched population sizes within each generation. Heritability of compatibility was investigated following exposure to the same parasite strain across all generations. Strain-specificity of compatibility was investigated following exposure to a novel parasite within the F_3 generation only (see text for further details).

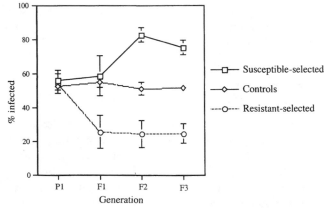

Fig. 2. Heritability of resistance and susceptibility in a *Biomphalaria glabrata–Schistosoma mansoni* system. % infection rate across generations P_1 to F_3 amongst resistant-selected (R), susceptible-selected (S), and unselected control (C) snail lines following exposure to 5 miracidia. Bars represent S.E.M.s across pooled snail–parasite combinations ($n = 4$ replicate combinations per group per generation). Data adapted from Webster & Woolhouse (1998).

was matched, where both were either resistant- or susceptible-selected. In contrast, in the third groups one member of each pair was from a resistant-selected line and the other from a susceptible-selected line. Finally, a group of unselected snails were included to serve as controls. Following an initial period to accommodate for potential sperm storage, snails were left to reproduce, and cross-breeding was identified by Randomly Amplified Polymorphic DNA–PCR (RAPD–PCR; Williams *et al.* 1990). The resulting compatibility phenotype of all offspring was determined.

As would be predicted if resistance were a dominant trait, only the resistance phenotype (75 % infection rate) was displayed amongst progeny from either matched resistant-selected pairs and from non-matched pairs where one parent was from a resistant-selected line and the other from a susceptible-selected. The susceptibility phenotype (25 % infection rate) was only displayed amongst crosses from matched susceptible-selected pairs. Infection rates, as for the previous Webster &

Woolhouse (1998) study, remained at approximately 50 % amongst unselected control snails exposed to the same parasites (Fig. 1).

These two studies suggest that both resistance and susceptibility to schistosome infection, at least for the two *Bi. glabrata–S. mansoni* strain combinations used, are heritable traits, with resistance dominant over susceptibility (Webster & Woolhouse, 1998; Webster, 2001).

Heritability of parasite infectivity and virulence

As for the host resistance described above, one of the best ways to ascertain heritability of virulence and/or infectivity is to document genetic differences in the laboratory.

Hybridisation between closely related schistosome species has been an important tool in demonstrations of the genetic basis of infectivity to a particular snail species or strain. For example, Wright (1974) reported that progeny of a cross between *S. mattheei* and *S. intercalatum* were equally infective to *Bu. globosus* and *Bu. scalarias*, whereas the parental forms were restricted in the case of *S. mattheei* to *Bu. globosus* and in the case of *S. intercalatum* to *Bu. scalaris*. This duel infectivity persisted to the F_3 generation, even though the parasite was passaged solely through *Bu. globosus*; infectivity to *Bu. scalaris* was apparently lost in the F_4 generation. A similar situation occurs with *S. haematobium* and *S. intercalatum* from Cameroon, where parental lines develop only in *Bu. rohlfsi* and *Bu. forskalii* respectively, but their F_1 hybrids can develop in both snail hosts (Rollinson & Southgate, 1987).

Another powerful way of demonstrating the presence of genetic variation for a trait is through isofemale lines, in which a laboratory strain is bred from a single mated individual (Parsons, 1980). Arrays of lines are scored for a trait under identical laboratory conditions, and the presence of significant between-line variation is attributed to genetic differences. Several such studies using inbred parasite lines have found evidence of the heritability of infectivity in snail–schistosome systems. For example, Cohen & Eveland (1988) reported consistent differences between clones of *S. mansoni* derived from monomiracidial infections of an inbred laboratory strain and maintained by serial microsurgical transplantation of sporocysts from infected to uninfected *Bi. glabrata*. The infectivity of individual clones in snails ranged from 44 to 100 % and were highly consistent within each clone, irrespective of time or subpassage frequency, thereby suggesting that the differences had a genetic basis. Likewise, McManus & Hope (1993) conducted a series of experiments with inbred *S. mansoni* strains originating from different areas and revealed a wide range of infectivities. Moreover, they determined that *S. mansoni* derived from a single geographic

population compromise a diverse population with respect to infectivity to snails.

Whilst the aforementioned studies investigated the heritability of infectivity, a recent study by Davies, Webster & Woolhouse (2001) aimed to determine the heritability of both infectivity and, for the first time in the snail host, virulence. Five substrains of *S. mansoni* from a laboratory strain originally from Puerto Rico were developed. In the F_0 generation snails were exposed to a single miracidium and infected snails were randomly paired and used to infect a single mouse host. Since single miracidial infections had been used, cercariae arising from a single snail represented a single clonal population, derived by asexual reproduction. Five egg-producing lines were recovered, arising from crosses between a male and female cercarial clone infection. In two subsequent generations, groups of sexually mature *Bi. glabrata* were individually exposed to *S. mansoni* miracidia from each substrain and cercariae harvested used to infect groups of mice. As shown in Fig. 3a, b, there were significant differences in both infectivity (measured as the frequency of patent infections) and virulence (measured as a reduction in host survival) between substrains. Moreover, such patterns were stable across two generations. This indicated that both these parasite traits have a genetic basis, and that these were heritable over two generations.

In a subsequent study, Davies & Webster (in press) used artificial selection to produce lines differing in virulence to the snail host. Selection was conducted on replicate laboratory strains from two widely differing geographic regions on the intensity of infection, which has been shown to be a correlated factor of virulence (Barbosa, 1975, Davies *et al.* 2001). As shown in Fig. 4, virulence was significantly reduced by artificial selection. This thus provided further support for the heritability of virulence in the snail–schistosome system.

Selective pressures for the evolution of host resistance and parasite infectivity and virulence

The aforementioned studies demonstrate genetic variation and/or heritability in host resistance and susceptibility, as well as parasite infectivity and virulence. However, in order to infer coevolution it still remains necessary to demonstrate that parasite infection and virulence affect host fitness and, conversely, that host resistance affects parasite fitness. Once again there has been some progress in this area using examples from snail–schistosome interactions.

Parasites, by definition, have fitness-reducing effects on their hosts (Ebert & Herre, 1996). Accordingly, schistosomes can be extremely virulent parasites of their intermediate hosts and infection has been reported to cause a number of host fitness-

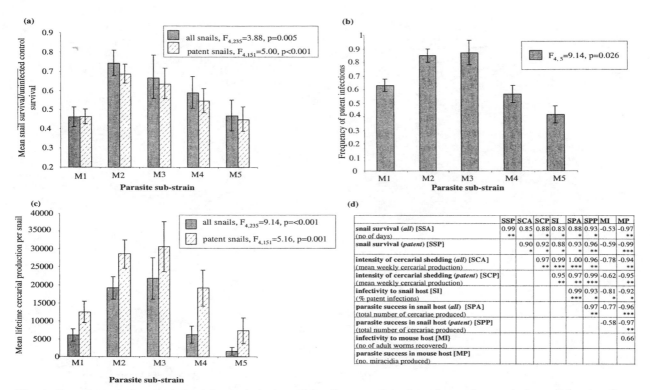

Fig. 3. Genetic variation for infectivity and virulence in a *S. mansoni* strain and genetic correlations with other fitness traits. Five sub-strains were developed (M1–M5) from a laboratory *S. mansoni* strain originally from Puerto Rico in order to investigate evidence of parasite genetic variation for virulence and infectivity. Sub-strains were developed from the mating of single clones of female schistosomes to single male clones. Groups of snails were exposed to each sub-strain and life-history parameters recorded. Mean values of each parasite sub-strain over two generations were compared by analysis of variance and are shown for (a) virulence (the opposite of host survival), (b) infectivity (the frequency of patent infections) and (c) lifetime cercarial production in the snail host (the number of cercariae produced was recorded weekly for all snails until snail death). Genetic correlations of virulence and infectivity with other fitness traits including parasite reproduction in snail hosts, and infectivity and parasite reproduction in the mouse definitive host, were demonstrated by comparing mean values per parasite sub-strain using Pearson's correlation coefficients. Fitness traits are shown for all snails (*all*), the subset of patently infected snails (*patent*), and mouse definitive hosts (d). Significant correlations are highlighted ***$P < 0.001$, **$P < 0.01$, *$P < 0.05$. Data adapted from Davies, Webster & Woolhouse (in press).

reducing effects. For example, Woolhouse (1989), through capture-recapture techniques, reported reductions in survival in naturally infected populations of *Bi. pfeifferi* and *Bu. globosus* in the field, as did Sturrock (1973) for *Bi. glabrata*. Likewise, Pan (1965) and Webster & Woolhouse (1999) showed significantly higher mortality rates amongst infected than uninfected lines of *Bi. glabrata*, and Woolhouse (1989) for *Bi. pfeifferi*, in the laboratory. However, increased mortality rates need not always be direct. For example, schistosome infection has also been shown to reduce the tolerance of infected snails to elevated temperature, molluscicidal chemicals and to heavy metals such as zinc (Bayne & Loker, 1987).

Schistosomes can also affect host fitness through reducing the reproductive success of infected individuals. Reductions of fecundity in molluscan hosts due to infection with schistosomes are observed where the parasite is thought to 'castrate' its host in order to divert resources towards its own development. For example, reductions in the number of egg masses laid by infected snails and/or the number

of embryos hatched have been reported in the *S. mansoni–Bi. pfeifferi* (Sturrock. 1966), *S. mansoni–Bi. glabrata* (Sturrock & Sturrock, 1970) and *S. haematobium–Bu. globosus/truncatus/senegalensis* systems (Fryer *et al.* 1990). However, the extent of the inhibition may be linked to the stage of reproductive maturity of the host (Fryer *et al.* 1990), and in several cases fecundity inhibition may be preceded by a short-term burst in egg output (Minchella & LoVerde, 1981; Minchella *et al.* 1985). Other ways in which schistosome infection may affect host reproductive success is through influencing host behaviour. Rupp (1996) studied the mating behaviour of *S. mansoni*-infected and uninfected lines of *Bi. glabrata* and *Bi. alexandrina* in the laboratory. The mating frequencies of patently infected snails were lower than those of controls, which was concluded to result from stress induced by the pathology of infection.

There is less evidence specifically documenting the effects of hosts on parasite fitness. Nevertheless, of the limited data available, *Bi. glabrata* snails

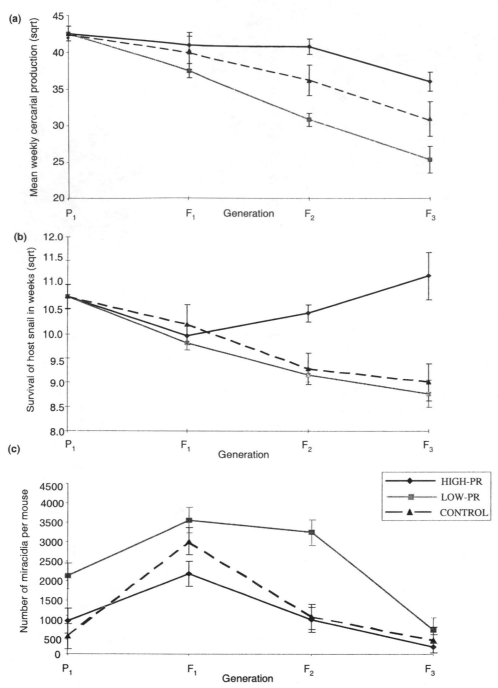

Fig. 4. Heritability of parasite reproduction and virulence in two replicate *S. mansoni* strains. (a) Artificial selection of two laboratory *S. mansoni* strains, one originally from Puerto Rico, and one from Kenya for high and low parasite reproduction (PR) (measured as the number of cercariae produced at week 7 post infection). The mean of two replicates and two unselected control lines are shown. The average weekly cercarial production was significantly increased in HIGH PR-selected and reduced in LOW-PR selected lines ($P < 0.001$). There was no overall difference in reproductive rate between the two replicates ($P = 0.09$) or in the response of the replicates to artificial selection ($P = 0.48$). (b) Snail survival was measured daily. Survival was significantly higher (i.e. virulence was lower) in snails infected with parasites selected for HIGH reproduction than those infected with parasites with a LOW intensity of infection ($P < 0.001$). (c) Transmission and success in the definitive host was measured as the number of miracidia present in the livers of mice infected with 220 cercariae at 7 weeks post snail infection. Miracidial production was significantly increased in mice infected with LOW PR-selected parasites and significantly reduced in HIGH-PR selected parasites ($P < 0.001$). Data from Davies (2000).

differing slightly in their susceptibility to *S. mansoni* infection have shown dramatic differences in cercarial output per snail (Ward *et al.* 1988), although this was not found by Manning *et al.* (1995) for their

Bu. globosus–S. haematobium/mattheei combinations. Preliminary work in our laboratory has also shown differences in parasite reproductive success and/or transmissibility depending on the resistance status of

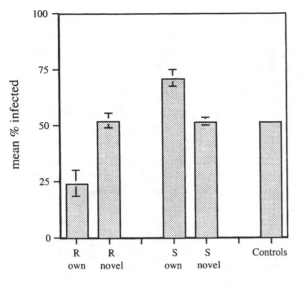

Fig. 5. Strain-specificity of resistance and susceptibility in a *Biomphalaria glabrata–Schistosoma mansoni* system. Following three generations of artificial selection, mean (and SEM) infection rates averaged 75 % amongst resistant-selected snail lines (R) and 25 % amongst susceptible-selected lines (S) following exposure to their own familiar parasite strain (i.e. the parasite strain to which their selection had been based). In contrast, infection rates averaged 50 % amongst unselected control snail lines and also amongst selected snail lines following exposure to a novel parasite strain. Pooled results from four replicate groups (of $n = 60$ snails) per snail line. Data adapted from Webster & Woolhouse (1998).

their intermediate hosts (Blair *et al.* in press; Webster, J. P. unpublished).

ASSESSMENT OF THE APPLICABILITY OF COEVOLUTIONARY MODELS IN SNAIL–SCHISTOSOME INTERACTIONS

The results of the field and laboratory studies described above may suggest snail–schistosome co-evolution. What now remains is a consideration of how applicable the currently available data on snail–schistosome interactions are to that of the major coevolutionary models and theories proposed within the first section.

Frequency-dependent or 'Red Queen' coevolution

The specificity of host–parasite interactions is a fundamental assumption of frequency dependent or Red-Queen coevolution (Hamilton, 1980; Hamilton, Axelrod & Tanese, 1990; Lively, 1999). Moreover, the greater compatibility amongst many sympatric host–parasite interactions revealed by the reciprocal cross-infection studies described previously in the second section (Manning *et al.* 1995; Morand *et al.* 1996; Woolhouse & Webster, 2000), suggests such

specificity in snail–schistosome interactions. Webster & Woolhouse (1998) thus investigated the potential strain-specificity of snail–schistosome compatibility in the laboratory. The same initial artificial-selection protocol towards resistant-selected and susceptible-selected snail lines, as described above (Fig. 1), was performed on two populations of *Bi. glabrata* snails and two *S. mansoni* parasite populations until the F_2 generation. At the F_3 generation, only half the selected snails from each line were exposed to the same (their 'own') parasite strain to which they had been selected. The other half were exposed to a novel *S. mansoni* strain to which they had not been selected. A matched number of control snails were exposed to both parasite strains. The results suggested that compatibility in this system is strain specific, as the infection rate amongst F_3 snails exposed to a novel parasite was approximately 50 % amongst both resistant- and susceptible-selected snails, and hence matched to that of unselected controls. The resistance (75 % infection rate) and susceptible (25 % infection rates) phenotypes were only observed in snails following exposure to the same single parasite strain to which their artificial selection was focused (Fig. 5).

Webster (2001) then investigated whether it is possible to cross-breed such single strain-specific *Bi. glabrata* snails to be resistant and/or susceptible to more than one *S. mansoni* strain. As for the dominance study described above, individual adult snails from each artificially-selected replicate snail line were paired with a selected partner of matched or non-matched compatibility status. In this case, the first two groups (of $n \geqslant 14$ pairs each) consisted of matched snails in each pair, where each were selected towards the same parasite strain. In contrast, in the third group one member of each pair was of a compatibility status selected towards one (a Kenyan) parasite and the other was selected towards a different (a Puerto Rican) parasite strain. Cross-breeding was identified by RAPD–PCR, and the resulting compatibility phenotype of all offspring was determined. The resistance phenotype (75 % infection rate) was simultaneously displayed against both parasite strains amongst resistant-selected crosses arising from non-matched pairs. The converse was also suggested amongst some susceptible-selected crosses (25 % infection rate). Likewise, as for the previous study (Webster & Woolhouse, 1998), single strain-specificity of compatibility was demonstrated, as all effects were lost if snails were exposed to a novel parasite strain (50 % infection rate, as for unselected controls).

The results from both these studies (Webster & Woolhouse, 1998; Webster, 2001) thus suggest that *Bi. glabrata–S. mansoni* resistance and susceptibility is, as predicted by coevolutionary models such as that of the Red Queen hypothesis, strain-specific. Such strain-specificity may also provide empirical support for genetic theories such as the matching allele at multiple loci model of coevolution, which states that each host allele confers resistance to one parasite allele, and parasites may successfully infect a host only when there is an exact match between host and parasite alleles (Frank, 1996). This contrasts with, for example, the 'gene-for-gene' hypothesis, where two alleles at a single locus in both host and parasite are involved (Frank, 1994). Even a system based on a single locus with multiple alleles responsive to selection for graded increases in recognition and affinity to certain parasite-strain characteristics and not others, is unlikely to reflect this snail–schistosome data. Any change in allele frequencies for one effect, such as selection for resistance to one parasite strain, would naturally result in a change in allele frequencies in the remainder, such as for resistance to another strain. No such associations were observed. The results thus support the existence of a multi-locus trait, at least a two-locus, two allele model with the resistance, susceptible, and control lines differing in allele frequencies. Although it is unlikely that there is an allele determining compatibility for every parasite strain, at least some, such as the two examined in the studies above, appear to involve different alleles.

Unfortunately, the potential strain-specificity of parasite infectivity and virulence has not yet been so intensely investigated as has host resistance. However, Richards (1975b,c) did expose well-characterized strains of *Bi. glabrata* snails to miracidia from strains of *S. mansoni* differing in infectivity, and showed that infectivity was determined by a number of factors with at least one such genetic factor being sex-linked. Similarly, the ability to breed schistosome hybrids infective to the snail strain or species of both parental forms suggests a form of complementation and involvement of multiple infectivity factors. Finally Davies (2000), in an extension of the artificial selection study for reduced *S. mansoni* virulence described above (Fig. 4), demonstrated that the effect was specific to the host strain used in selection. In the F_0 generation, each of two parasite strains were used to infect two groups of inbred laboratory strains of *Bi. glabrata*, one originally from Brazil and one from the Caribbean. One parasite strain was subjected to two-way artificial selection in the Brazilian snail strain, and the other parasite strain in the Caribbean snail strain. Exposure of all selected parasite lines to both snail strains in the F_3 generation demonstrated that the reduced virulence of 'HIGH–PR' selected parasites in the F_3 generation compared to the original F_0 population, and to the F_3 control and 'LOW–PR' lines, was only seen for each replicate in the snail host strain which had been used in selection (i.e. in the Brazilian snail strain for one and the Caribbean strain for the other parasite replicate). There was no evidence of a difference in virulence of the F_3 generation selected parasites and the original F_0 parasites in the snail strain not used for selection (Davies, 2000). Such apparent strain-specificity may suggest that schistosome virulence may also involve a multi-locus trait (Webster & Woolhouse, 1998; Webster, 2001).

Thus strain-specificity has been demonstrated on the sides of both snail and schistosome. This not only suggests that different gene combinations may be involved with each local host–parasite interaction (Rollinson & Southgate, 1987), but it also suggests the potential for Red-Queen coevolution in this system.

However, evidence of Red-Queen dynamics in the field also requires documentation of frequency-dependent selection of host genotypes and a 'rare advantage' (reduced susceptibility of rare hosts). This is a difficult question to answer in sexual populations such as that of *Biomphalaria* and *Bulinus* snails because there are currently no existing markers for the relevant genotypes – i.e. those that are directly involved in the host–parasite interaction. Nevertheless, support is available from another snail–trematode system: *Microphallus* spp.–*P. antipodarum* (Dybdahl & Lively, 1995, 1998, and see Lively, this supplement). Dybdahl & Lively (1995) selected a lake in New Zealand in which all the snails are asexual. In these clonal lines, the genotypes for resistance in the snails are inextricably linked with their multilocal allozyme genotypes. In an initial survey of this lake they discovered four relatively common clones (defined by their allozyme genotypes) and found that the most common clone was significantly over infected (Dybdahl & Lively, 1995). The authors argued that if parasites are driving coevolutionary cycles, they would expect this most-common clone to be driven down in frequency and a different clone would become the most common in the population. Furthermore, they predicted that any changes between years in clone frequency would be correlated with changes in the frequency of infections in that clone at some future point. Support for both of these predictions were provided across a five-year study (Dybadhl & Lively, 1998), and thus the parasites did appear to be driving oscillatory dynamics in the host population. Laboratory experiments confirmed that clones that had been rare in the population for the previous four years were significantly less susceptible than common clones. The combination of results from this study therefore show strong evidence for a rare advantage, as well as evidence for time-lagged selection in the field. Both of these results are again consistent with the Red Queen hypothesis in this system (Lively, 1999).

Cost-benefit trade-offs and locally fixed optima

Cost-benefit trade-offs will be important in host–parasite coevolution, whether locally fixed optima or dynamic fluctuating coevolution is expected, as they may have implications regarding, for example, optimal levels observed or the speed of evolution. Trade-offs may also be an important mechanism for the maintenance of genetic variation for resistance, infectivity and virulence in natural populations. Trade-offs may take many forms, although empirical evidence is often lacking. Nevertheless, once again, some of the few animal host–parasite examples come from snail–schistosome interactions.

Costs of resistance. It has frequently been argued that for there to be a stable genetic polymorphism in disease resistance the fitness of the resistant geno-types should be less than that of the susceptible genotypes in the absence of the disease. The argument is intuitively obvious: without such a cost, an allele for resistance should continue to increase in frequency as long as some disease is present (Antonovics & Thrall, 1994). Accordingly, there is no evidence for a fixation of resistance amongst wild snail populations (Manning *et al.* 1995). Webster & Woolhouse (1999) investigated potential costs of resistance in a *Bi. glabrata–S. mansoni* host–parasite system. Once again, using artificial selection to breed snails that are resistant or susceptible to schistosome infection, their study investigated whether com-patibility has any associated cost in terms of snail fertility (defined as actual reproductive performance, measured as the number of offspring produced) and/or fecundity (defined as potential reproductive capacity, measured as number of eggs and embryos formed). Indeed, susceptible-selected snail lines showed significantly higher fertility than resistant-selected or unselected control snail lines, irrespective of current infection status. In contrast, there were no significant differences between snail lines in fec-undity, proportion of abnormal egg masses pro-duced, or mean number of eggs per egg mass. These results are consistent with snails incurring costs of resistance to schistosome infection in the absence of the parasite.

Similar results were also found by Cooper *et al.* (1994), where again fertility rates were significantly lower amongst resistance-selected snails lines. How-ever, Cousin *et al.* (1995) showed that significant abnormalities exist in the snail strain used by Cooper *et al.* (1994) which were suspected to have resulted from the intense inbreeding of this stock. Thus the direct link between fertility and resistance proposed by Cooper *et al.* (1994) remains to be ascertained (Cousin *et al.* 1995).

Finally, working on the *Bi. glabrata–Echinostoma caproni* system, Langand *et al.* (1998) also found an apparent cost of resistance, manifested as a delay in reproductive maturity. Again utilising snail lines artificially selected towards resistance or suscep-tibility, they found that (although only analysing offspring from a single snail pair) that resistant-selected individuals reached maturity approximately four days later than did susceptible snails.

Costs of infectivity. It is also frequently assumed that infectivity alleles must have a negative effect on parasite fitness that offset the benefit of wider host range. This assumption is again necessary because without a fitness cost, the infectivity allele would spread to fixation (Frank, 1993), and polymorphisms would not be maintained. Costs will also be im-portant in determining the optimal level of infection in certain situations. Davies *et al.* (2001) determined that though there was an overall cost of infection in the snail–schistosome system utilized, manifested as a reduction in host, and hence parasite, survival, higher levels of infectivity were not apparently associated with increasing fitness costs to the host (and hence indirectly to the parasite). In fact, as shown in Fig. 3d, infectivity was positively associated with both snail survival and cercarial shedding intensity. This resulted in significant differences in parasite fitness (measured as the lifetime production of cercarial stages) between inbred lines (Fig. 3c). Similar patterns of positive associations of cercarial shedding intensity, infec-tivity and survival have been previously reported by Barbosa (1975), who examined 16 strains of *Bi. glabrata* and 233 strains of *Bi. straminea*.

Davies *et al.* (2001) further found that there was a parasite fitness cost of increased infectivity to the snail host due to a trade-off in infectivity between the different hosts of the schistosome lifecycle (see Fig. 3d). Overall, infectivity to and cercarial production in the snail host were negatively correlated with infectivity to and miracidial production in a mouse definitive host. This unexpected result of a trade-off in fitness between the hosts of the schistosome lifecycle was substantiated in the artificial selection experiment described in Fig. 5. This could therefore explain the presence of polymorphisms in infectivity within such a laboratory *S. mansoni* strain.

Benefits of virulence. Virulence reduces parasite fitness by reducing the duration of infection. Fitness costs of being virulent have been documented in the snail–schistosome system by the reduction in total cercarial production in highly virulent parasite strains (i.e. where snail survival is lower) (Davies *et al.* 2001: Fig. 3). Current theory suggests that virulence may be maintained and promoted where the parasite faces a trade-off between traits which are correlated with fitness, in particular that the fitness costs of high virulence may be offset by the benefits of increased transmission or ability to withstand host defences (Frank, 1996). The evolution of stable virulent parasites thus requires parasite factors that are genetically correlated. A genetic correlation results in linkage between traits that restricts their evolution (Ebert & Herre, 1996) such that, for example, transmission cannot be increased without further increasing virulence. Many evolutionary models propose that virulence is maintained as a side-effect of parasite fecundity, for example if a high concentration of parasites in a host increases the probability of transmission to new hosts or decreases the probability of host recovery, but also increases host death rate (Bull, 1994). In this instance, natural selection is expected to optimise parasite reproductive rates such that the number of new infections is maximized. Davies *et al.* (2001; Fig. 3c) and Davies & Webster (in press; Fig. 4) suggest that virulence in the snail host may be maintained by increased transmission to the definitive host. However, as previously mentioned, variations in reproductive rate in the snail host is not thought to be the mechanism involved.

Multiple infections are a further mechanism that have been proposed to promote the evolution of increased parasite virulence, since intra-host competition can favour high virulence parasite strains of lower potential fitness than less virulent strains wherever they have a local growth or transmission advantage (e.g. Frank, 1996; May & Nowak, 1995; van Baalen & Sabelis, 1995; Nowak & May, 1994; Bonhoeffer & Nowak, 1994; Anita, Levin & May, 1994; Levin & Pimental, 1981). Multiple infections of single snail hosts by more than one schistosome genotype have been detected in both *Bi. glabrata* (Minchella *et al.* 1995) and in *Bu. globulus* (Davies *et al.* 1999; Woolhouse, Chandiwana & Bradley 1990). Preliminary investigations in our laboratory have also suggested the potential for competitive interactions to occur between schistosome strains within snail hosts (Davies, 2000). Demonstration of competition requires evidence that *per capita* parasite success is reduced by the presence of the other parasite line. Groups of snails were therefore exposed to one or other, or a combination of two of the parasite strains developed in Fig. 3. A genetic marker was used to identify resulting progeny in order to estimate the success of each parasite line. The success of parasite M1 sub-strain was shown to be reduced in the presence of the faster growing parasite genotype, M2. This was true irrespective of the relative proportion of M1 sub-strain in the original dose. The success of M2 sub-strain, however, was not affected by the presence of the other line. This study therefore demonstrated evidence of competition between parasite genotypes and asymmetry of competitive interactions (Davies, 2000).

Benefits of resistance. Apart from the obvious benefits of host resistance, in terms of reduced infection and parasite-induced mortality etc. rates, further benefits of resistance may be necessary to maintain resistance genes in the absence of infection. In accordance, potential preferential mate choice towards uninfected (Rupp, 1996), and even resistant-

selected snails, irrespective of their infection status (Webster, 2001 and unpublished data), has been identified in this system, which could account for the maintenance of resistance polymorphisms even amongst unselected laboratory snail stocks.

Therefore, these data serve to demonstrate the potential for costs and benefits of resistance, virulence and infectivity, each of which may affect both the coevolutionary process itself, and the related question regarding the maintenance of polymorphism.

The geographic mosaic theory of coevolution

Subdivision of populations, the spatial pattern of selection, migration and gene flow in hosts and parasites can all influence the coevolutionary process (Dybdahl & Lively, 1996). Such patterns may determine whether or not contact between a host and parasite species will lead to the establishment of a long-term interaction and also the geographic scale over which any such interactions will occur (Burdon & Thrall, 1999). In a number of population genetic surveys (reviewed by Jarne & Theron, this supplement), significant spatial sub-structuring of both snail and schistosome populations has been reported. However, such spatial sub-structuring may vary both within and between seasonal collections, as has recently been shown for *Bi. pfeifferi* in the Zimbabwean highveld (Hoffman *et al.* 1998; Webster *et al.* 2001; Davies *et al.* 2001). Abiotic factors, in particular rainfall and river type, have also been inferred to influence population stability (Webster *et al.* 2001; Kruger *et al.* 2001). Thus differential selection in host–parasite interactions may be expected in different populations of the *Bi. pfeifferi* metapopulation, as would be predicted by the geographic mosaic model of host–parasite coevolution (Thompson, 1994).

Thus in summary, the results of the studies reviewed above suggest that snail–schistosome interactions can provide much needed empirical support for several of the major coevolutionary models and theories, in particular that of the Red Queen hypothesis, Cost–Benefit trade-offs, and the Geographic Mosaic Theory.

CONCLUSIONS AND AREAS FOR FURTHER STUDY

The results of the numerous studies reviewed here have both added a great deal to our understanding of the system and served to highlight the valuable role snail–schistosome interactions may play in exploring general issues of the coevolution of resistance, infectivity and virulence. This may be particularly pertinent as this is an indirectly transmitted macroparasite system, so often overlooked amongst both theoretical and empirical studies.

However, as is always the case in science, the

results produced so far raise as many questions and areas for subsequent research as they answer. For example, in terms of snail–schistosome coevolution and compatibility, cross-species interactions have not yet been considered. In many geographic areas, particular snail species are infected by more than one schistosome species, such as the infection of *Bu. globosus* with human *S. haematobium* and bovine *S. mattheei* in Zimbabwe (Manning *et al.* 1995). Yet, there is currently no information as to whether selection for improved resistance against one species of parasite is associated with improved or reduced defence against other species. Likewise, much remains to be investigated into other potential fitness costs, such as the susceptibility to predation of resistant snails or the fitness costs of being exposed but not infected. Long-term controlled laboratory experiments and field surveys aimed to document coevolution should be developed. Measurements of levels of resistance, infectivity and virulence in natural populations and the existence of polymorphism under differing epidemiological situations is also obviously an important next step in understanding snail–schistosome coevolution. DNA probes developed from molecular markers linked to resistance, infectivity and virulence genes would facilitate such studies. Knight *et al.* (1999) were recently successful in identifying resistance markers, although the population genetic variability and strain-specificity of compatibility documented here (Jarne & Theron, this supplement, Webster *et al.* 2001; Davies *et al.* 1999; Hoffman *et al.* 1998) may suggest that the applicability of such probes to the field would be severely limited. The effort to clarify the role of genetic factors involved in snail–schistosome compatibility is challenging in many respects. Both participants are complex metazoans with numerous chromosomes and relatively large genome size. Furthermore, factors controlling compatibility vary within and between populations and with host age (Bayne & Loker, 1985). One of the greatest challenges will therefore be to integrate genetics, evolutionary biology and population dynamics into models of resistance, infectivity and virulence that can generate hypotheses that can be tested in the field and laboratory on snail–schistosome and other host–parasite systems.

ACKNOWLEDGEMENTS

We are grateful to C. Lively, P. Harvey and D. Ebert for comments on the text, and to the Royal Society (JPW) and the Medical Research Council (CMD) for funding.

REFERENCES

ABDULLAH, A. (1997). The host–parasite relationship of *Schistosoma mansoni* and *Biomphalaria glabrata*: an approach to biological control of schistosomiasis. PhD Thesis. University of Bangor.

ANDERSON, R. M. & MAY, R. M. (1982). Coevolution of hosts and parasites. *Parasitology* **85**, 411–426.

ANITA, R., LEVIN, B. R. & MAY, R. M. (1994). Within host population dynamics and the evolution and maintenance of microparasite virulence. *American Naturalist* **144**, 457–472.

ANTONOVICS, J. & THRALL, P. H. (1994). The cost of resistance and the maintenance of genetic polymorphism in host-pathogen systems. *Proceedings of the Royal Society London, Series B* **257**, 105–110.

BARBOSA, F. S. (1975). Survival and cercaria production of Brazilian *Biomphalaria glabrata* and *B. straminea* infected with *Schistosoma mansoni*. *Journal of Parasitology* **61**, 151–152.

BAYNE, C. J. & LOKER, E. S. (1987). Survival within the snail host. In *The Biology of Schistosomes: From Genes to Latrines* (ed. Rollinson, D. & Simpson, A. J. G.), pp. 321–346. London, Academic Press.

BEHNKE, J. M., LOWE, A., MENGE, D., IRAQI, F. & WAKELIN, D. (2000). Mapping genes for resistance to gastrointestinal nematodes. *Acta Parasitologica* **45**, 1–13.

BLAIR, L., WEBSTER, J. P., DAVIES, C. M. & DOENHOFF, M. J. (in press). An experimental evaluation of the genetic control theory for schistosomiasis. *Transactions of the Royal Society of Tropical Medicine and Hygiene* (in press).

BONHOEFFER, S. & NOWAK, M. A. (1994). Mutation and the evolution of virulence. *Proceedings of the Royal Society London, Series B* **258**, 133–140.

BULL, J. J. (1994). Perspective: Virulence. *Evolution* **48**, 1423–1437.

BURDON, J. J. & THRALL, P. H. (1999). Spatial and temporal patterns in coevolving plant and pathogen associations. *American Naturalist* **153**, S16–S33.

CLAYTON, D. H., LEE, P. L. M., TOMPKINS, D. M. & BRODIE, D. (1999). Reciprocal natural selection on host–parasite phenotypes. *The American Naturalist* **154**, 261–270.

COHEN, L. M. & EVELAND, L. K. (1988). *Schistosoma mansoni*: characterisation of clones maintained by the microsurgical transplantation of sporocysts. *Journal of Parasitology* **74**, 963–969.

COMBES, C. (1997). Fitness of parasites: pathology and selection. *International Journal for Parasitology* **27**, 1–10.

COOPER, L. A., RICHARDS, C. S., LEWIS, F. A. & MINCHELLA, D. J. (1994). *Schistosoma mansoni*: Relationship between low fecundity and reduced susceptibility to parasite infections in the snail *Biomphalaria glabrata*. *Experimental Parasitology* **79**, 21–28.

COUSIN, C., OFORKI, K., ACHOLONU, S., MILLER, A., RICHARDS, C., LEWIS, F. L. & KNIGHT, M. (1995). *Schistosoma mansoni* – Changes in the albumin gland of the *Biomphalaria glabrata* snails selected for non-susceptibility to the parasite. *Journal of Parasitology* **81**, 905–911.

COUSTAU, C., CHEVILLON, C. & FFRENCH-CONSTANT, R. (2000). Resistance to xenobiotics and parasites: can we count the cost? *Trends in Ecology and Evolution* **15**, 378–383.

DAVIES, C. M. (2000). Snail–schistosome interactions and the evolution of virulence. D.Phil. thesis. Dept. of Zoology. University of Oxford.

DAVIES, C. M. & WEBSTER, J. P. (in press). A genetic trade-off of virulence and transmission in a snail–schistosome system. *Transactions of the Royal Society of Tropical Medicine and Hygiene* (in press).

DAVIES, C. M., WEBSTER, J. P., KRUGER, O., MUNATSI, A., NDAMBA, J. & WOOLHOUSE, M. E. J. (1999). Host-parasite population genetics: a cross-sectional comparison of *Bulinus globosus* and *Schistosoma haematobium*. *Parasitology* 119, 295–302.

DAVIES, C. M., WEBSTER, J. P., MUNATSI, A., KRUGER, O., NDAMBA, J., NOBLE, L. R. & WOOLHOUSE, M. E. J. (2001). Schistosome host–parasite population genetics in the Zimbabwean highveld. In *Workshop on Medical and Veterinary Malacology in Africa* (ed. Madsen, H., Appleton, C. C. & Chimbari M.), pp. 65–81. Charlottenlund, Denmark, DBL publications.

DAVIES, C. M., WEBSTER, J. P. & WOOLHOUSE, M. E. J. (2001). Trade-offs in the evolution of virulence of schistosomes – macroparasites with an indirect life-cycle. *Proceedings of the Royal Society London, Series B* 268, 1–7 (in press).

DYBDAHL, M. F. & LIVELY, C. M. (1995). Host–parasite interactions: infection of common clones in natural populations of a freshwater snail (*Potamopyrgus antipodarum*). *Proceedings of the Royal Society London, Series B* 260, 99–103.

DYBDAHL, M. F. & LIVELY, C. M. (1996). The geography of coevolution: comparative population structures for a snail and its trematode parasite. *Evolution* 50, 2264–2275.

DYBDAHL, M. F. & LIVELY, C. M. (1998). Host–parasite coevolution: evidence for rare advantage and time-lagged selection in a natural selection. *Evolution* 52, 1057–1066.

EBERT, D. (1994). Genetic differences in the interactions of a microsporidian parasite and four clones of its cyclically parthenogenetic host. *Parasitology* 108, 11–16.

EBERT, D. & HERRE, E. A. (1996). The evolution of parasitic diseases. *Parasitology Today* 12, 96–101.

FILES, V. S. & CRAM, E. B. (1949). A study of the comparative susceptibility of snail vectors of *Schistosoma mansoni*. *Journal of Parasitology* 35, 555–560.

FRANK, S. A. (1993). Specificity versus detectable polymorphism in host–parasite genetics. *Proceedings of the Royal Society London, Series B* 254, 191–197.

FRANK, S. A. (1994). Co-evolutionary genetics of hosts and parasites with quantitative inheritance. *Evolutionary Ecology* 8, 74–94.

FRANK, S. A. (1996). Statistical properties of polymorphism in host–parasite genetics. *Evolutionary Ecology* 10, 201–317.

FRITZ, R. S. & SIMMS, E. L. (ed). (1992). *Plant Resistance to Herbivores and Pathogens*. Chicago, University of Chicago Press.

FRYER, S. E., OSWALD, R. C., PROBERT, A. J. & RUNHAM, N. W. (1990). The effect of *Schistosoma haematobium* infection on the growth and fecundity of three sympatric species of bulinid snails. *Journal of Parasitology* 76, 557–563.

GANDON, S. & MICHALAKIS, Y. (2000). Evolution of virulence against qualitative or quantitative host resistance. *Proceedings of the Royal Society London, Series B* 267, 985–990.

HAMILTON, W. D. (1980). Sex versus non-sex versus parasite. *Oikos* 35, 282–290.

HAMILTON, W. D., AXELROD, R. & TANESE, R. (1990). Sexual reproduction as an adaptation to resist parasites (a review). *Proceedings of the National Academy of Sciences, USA* 87, 3566–3573.

HOFFMAN, J., WEBSTER, J. P., NDAMBA, J. & WOOLHOUSE, M. E. J. (1998). Extensive genetic variation revealed within *Biomphalaria pfeifferi* from one river system in the Zimbabwean highveld. *Annals of Tropical Medicine and Parasitology* 92, 693–698.

KNIGHT, M., MILLER, A. N., PATTERSON, C. N., ROWE, C. G., MICHAELS, G., CAR, D., RICHARDS, C. S. & LEWIS, F. A. (1999). The identification of markers segregating with resistance to *Schistosoma mansoni* infection in the snail *Biomphalaria glabrata*. *Proceedings of the National Academy of Sciences, USA* 96, 1510–1515.

KRAAIJEVELD, A. R. & GODFRAY, H. C. J. (1999). Geographic patterns in the evolution of resistance and virulence in *Drosophila* and its parasitoids. *American Naturalist* 153, S61–S74 (Suppl.).

KRAAIJEVELD, A. R., VAN ALPHEN, J. J. M. & GODFRAY, H. C. J. (1998). The coevolution of host resistance and parasitoid virulence. *Parasitology* 116, S29–S45.

KRUGER, O., PACKER, M. J., ROBINSON, T. P., NDAMBA, J. & WEBSTER, J. P. (2001). Population genetics of schistosome intermediate hosts incorporating satellite remotely-sensed data. In *Workshop on Medical and Veterinary Malacology in Africa* (ed. Madsen, H., Appleton, C. C. & Chimbari, M.), pp. 169–181. Charlottenlund, Denmark, DBL publications.

LANGAND, J., JOURDANE, J., COUSTAU, C., DELAY, B. & MORAND, S. (1998). Cost of resistance, expressed as a delayed maturity, detected in the host–parasite system *Biomphalaria glabrata/Echinostoma caproni*. *Heredity* 80, 320–325.

LEVIN, S. & PIMENTAL, D. (1981). Selection of intermediate rates of increase in host–parasite systems. *American Naturalist* 117, 308–315.

LIVELY, C. M. (1989). Adaptation by a parasitic trematode to local populations of its snail host. *Evolution* 43, 1663–1671.

LIVELY, C. M. (1999). Migration, virulence and the geographic mosaic of adaptation by parasites. *American Naturalist* 153, S34–S47.

LIVELY, C. M. & APANIUS, V. (1995). Genetic diversity in host–parasite interactions. In *Ecology of Infectious Disease in Natural Populations* (ed. Dobson, A. & Grenfell, B.), pp. 421–449. Cambridge, Cambridge University Press.

LIVELY, C. M. & DYBDAHL, M. F. (2000). Parasite adaptation to locally common host genotypes. *Nature* 405, 679–681.

LO, C. T. & LEE, K. M. (1995). *Schistosoma japonicum*, zoophilic strain, in *Oncomelania hupensis chiui* and *O. h. formosana*: miracidial penetration and comparative histology. *Journal of Parasitology* 81, 708–713.

MANNING, S. D., WOOLHOUSE, M. E. J. & NDAMBA, J. (1995). Geographic compatibility of the freshwater snail *Bulinus globosus* and schistosomes from the Zimbabwe highveld. *International Journal for Parasitology* **25**, 37–42.

MAY, R. M. & NOWAK, M. A. (1995). Co-infection and the evolution of virulence. *Proceedings of the Royal Society London, Series B* **261**, 209–215.

MCMANUS, D. P. & HOPE, M. (1993). Molecular variation in the human schistosomes. *Acta Tropica* **53**, 255–276.

MESSENGER, S. L., MOLINEUX, I. J. & BULL, J. J. (1999). Virulence evolution in a virus obeys a trade-off. *Proceedings of the Royal Society London, Series B* **266**, 397–404.

MICHELSON, E. H. & DUBOIS, L. (1978). Susceptibility of Bahrain populations of *Biomphalaria glabrata* to an allopatric strain of *Schistosoma mansoni*. *American Journal of Tropical Medicine and Hygiene* **27**, 782–786.

MINCHELLA, D. J., LEATHERS, B. K., BROWN, K. M. & MCHANIR, J. N. (1985). Host and parasite counteradaptations: an example from a freshwater snail. *American Naturalist* **126**, 843–854.

MINCHELLA, D. J. & LOVERDE, P. T. (1981). A cost of increased early reproductive effort in the snail *Biomphalaria glabrata*. *American Naturalist* **118**, 876–881.

MINCHELLA, D. J., SOLLENBERGER, K. M. & PEREIRA DE SOUZA, C. (1995). Distribution of schistosome genetic diversity within molluscan intermediate hosts. *Parasitology* **111**, 217–220.

MORAND, S., MANNING, S. D. & WOOLHOUSE, M. E. J. (1996). Parasite–host coevolution and geographic patterns of parasite infectivity and host susceptibility. *Proceedings of the Royal Society London, Series B* **263**, 119–128.

NELSON, G. S. & SAOUD, M. F. A. (1968). A comparison of the pathogenicity of two geographical strains of *Schistosoma mansoni* in rhesus monkeys. *Journal of Helminthology* **17**, 339–362.

NOWAK, M. & MAY, R. M. (1994). Super-infection and the evolution of parasite virulence. *Proceedings of the Royal Society London, Series B* **255**, 81–89.

PAN, C. (1965). Studies on the host–parasite relationship between *Schistosoma mansoni* and the snail *Australorbis glabratus*. *American Journal of Tropical Medicine and Hygiene* **14**, 931–976.

PARAENSE, W. L. & CORREA, L. R. (1963). Variation in susceptibility of populations of *Australorbis glabratus* to a strain of *Schistosoma mansoni*. *Revista do Instituto Medicina Tropica Sao Paulo* **5**, 15–22.

PARSONS, P. A. (1980). Isofemale strains and evolutionary strategies in natural populations. In *Evolutionary Biology* (ed. Hecht, M., Steere, W. and Wallace, B.), pp. 201–235. New York, Plentum Publishing Corporation.

POULIN, R., HECKER, K. & THOMAS, F. (1998). Hosts manipulated by one parasite incur additional costs from infection by another parasite. *Journal of Parasitology* **84**, 1050–1052.

READ, A. F. (1994). The evolution of virulence. *Trends in Microbiology* **2**, 73–76.

RICHARDS, C. S. (1975 a). Genetic factors in susceptibility of *Biomphalaria glabrata* for different strains of *Schistosoma mansoni*. *Parasitology* **70**, 231–241.

RICHARDS, C. S. (1975 b). Genetic factors of biological conditions and susceptibility to infection in *Biomphalaria glabrata*. *Annals of the New York Academy of Sciences* **266**, 394–396.

RICHARDS, C. S. (1975 c). Genetic studies on variation in infectivity of *Schistosoma mansoni*. *Journal of Parasitology* **61**, 233–236.

RICHARDS, C. S. & SHADE, P. C. (1987). The genetic variation of compatibility in *Biomphalaria glabrata* and *Schistosoma mansoni* infection. *Journal of Parasitology* **73**, 1146–1151.

ROLLINSON, D. & SOUTHGATE, V. R. (1987). The genus *Schistosoma*: a taxonomic appraisal. In *The Biology of Schistosomes: From Genes to Latrines* (ed. Rollinson, D. & Simpson, A. J. G.), pp. 1–41. London, Academic Press.

RUPP, J. C. (1996). Parasite-altered behaviour: impact of infection and starvation on mating in *Biomphalaria glabrata*. *Parasitology* **113**, 357–365.

SORCI, G., MOLLER, A. P. & BOULINIER, T. (1997). Genetics of host–parasite interactions. *Trends in Ecology and Evolution* **12**, 196–200.

STURROCK, B. M. (1966). The influence of infection with *Schistosoma mansoni* on the growth rate and reproduction of *Biomphalaria pfeifferi*. *Annals of Tropical Medicine and Parasitology* **60**, 187–197.

STURROCK, R. F. (1973). Field studies on the transmission of *Schistosoma mansoni* and on the bionomics of its intermediate host, *Biomphalaria glabrata* on St. Lucia, West Indies. *International Journal for Parasitology* **3**, 175–194.

STURROCK, R. F. & STURROCK, B. M. (1970). Observations of some factors affecting the growth rate and fecundity of *Biomphalaria glabrata*. *Annals of Tropical Medicine and Parasitology* **64**, 357–363.

THIONGO, F. W., MADSEN, H., OUMA, J. H., ANDREASSEN, J. & CHRISTENSEN, T. O. (1997). Host–parasite relationships in infections with two Kenyan isolates of *Schistosoma mansoni* in NMRI mice. *Journal of Parasitology* **83**, 330–332.

THOMPSON, J. N. (1994). *The Co-evolutionary Process*. Chicago, University of Chicago Press.

THOMPSON, J. N. (1999). Specific hypotheses on the geographic mosaic of coevolution. *American Naturalist* **153**, S1–S14.

TOFT, C. A. & KARTER, J. J. (1990). Parasite–host coevolution. *Trends in Ecology and Evolution* **5**, 326–329.

VAN BAALEN, M. & SABELIS, M. W. (1995). The dynamics of multiple infection and the evolution of virulence. *American Naturalist* **140**, 881–910.

VAN DER KNAAP, W. P. W. & LOKER, E. S. (1990). Immune mechanisms in trematode–snail interactions. *Parasitology Today* **6**, 175–182.

VERA, C., JOURDANE, J., SELIN, B. & COMBES, C. (1990). Genetic variability in the compatibility between *Schistosoma haematobium* and its potential vectors in Niger. Epidemiological implications. *Tropical Medicine and Parasitology* **41**, 143–148.

WARD, R. D., LEWIS, F. A., YOSHINO, T. P. & DUNN, T. S. (1988). *Schistosoma mansoni*: relationship between cercarial production levels and snail host susceptibility. *Experimental Parasitology* **66**, 78–85.

WARREN, K. S. (1967). A comparison of Puerto Rican, Brazilian, Egyptian and Tanzanian strains of *Schistosoma mansoni* in mice: penetration of cercariae, maturation of schistosomes and production of liver disease. *Transactions of the Royal Society of Tropical Medicine and Hygiene* **61**, 795–802.

WEBSTER, J. P. (2001). Compatibility and sex in a snail–schistosome system. *Parasitology* **122**, 423–432.

WEBSTER, J. P., DAVIES, C. M., HOFFMAN, J. I., NDAMBA, J., NOBLE, L. R. & WOOLHOUSE, M. E. J. (2001). Population genetics of the schistosome intermediate host *Biomphalaria pfeifferi* in the Zimbabwean highveld: implications for co-evolutionary theory. *Annals of Tropical Medicine and Parasitology* **95**, 203–214.

WEBSTER, J. P. & WOOLHOUSE, M. E. J. (1998). Selection and strain specificity of compatibility between snail intermediate hosts and their parasitic schistosomes. *Evolution* **52**, 1627–1634.

WEBSTER, J. P. & WOOLHOUSE, M. E. J. (1999). Cost of Resistance: relationship between reduced fertility and increased resistance in a snail–schistosome

host–parasite system. *Proceedings of the Royal Society London, Series B* **266**, 391–396.

WILLIAMS, J. G. K., KUBELIK, A. R., LIVAK, K. J., RAFALSKI, J. A. & TINGEY, S. V. (1990). DNA polymorphisms amplified by arbitrary primers are useful as genetic markers. *Nucleic Acids Research* **18**, 6531–6535.

WOOLHOUSE, M. E. J. (1989). The effect of schistosome infection on the mortality rates of *Bulinus globosus* and *Biomphalaria pfeifferi*. *Annals of Tropical Medicine and Parasitology* **83**, 137–141.

WOOLHOUSE, M. E. J., CHANDIWANA, S. K. & BRADLEY, M. (1990). On the distribution of schistosome infections among host snails. *International Journal for Parasitology* **20**, 325–327.

WOOLHOUSE, M. E. J. & WEBSTER, J. P. (2000). In Search of the Red Queen. *Parasitology Today* **16**, 506–508.

WRIGHT, C. A. (1974). Snail susceptibility or trematode infectivity? *Journal of Natural History* **8**, 545–548.

A perspective on the ecology of trematode communities in snails

G. W. ESCH[1]*, L. A. CURTIS[2] and M. A. BARGER[1]

[1] Department of Biology, Wake Forest University, P.O. Box 7325, Winston-Salem, NC 27109 USA
[2] Cape Henlopen Laboratory, College of Marine Studies, University of Delaware, Lewes, DE 19958 USA

SUMMARY

This paper presents a perspective on the ecology of trematodes in snail hosts based on recent evidence. Because flukes use snails almost obligatorily as first intermediate hosts, we highlight the role of gastropods as keystone species for trematodes and their communities. After reviewing recent developments in the transmission of trematodes to and from snails, we discuss trematode communities within individual snails (infracommunities) and in snail populations (component communities). Results garnered using various protocols at the infracommunity level are reviewed. The few data available, all from marine systems, indicate that low colonization rates characterize infracommunities, suggesting that trematode infracommunities tend to be isolationist in character rather than interactive. The variety of trematode species present in a component community seems to be determined by spatial overlap of definitive hosts. Relative abundance of species in a component community shows little dependence on negative interspecific interactions at the level of the infracommunity. Temporal aspects of trematode communities are related to the life history of the host snail. The component communities of long-lived snails (mostly marine) integrate many infection episodes whereas shorter-lived snails (mostly freshwater) acquire new component communities each time host cohorts turnover.

Key words: Infracommunity, component community, competition, heterogeneity, trematode, gastropod.

INTRODUCTION

Snails and trematodes as keystone taxa

The concept of a keystone species was introduced by Paine (1966). The idea is a relatively simple one. It was used to describe a species whose influence on community structure is so significant that, should the species be eliminated, the structure of the new community would be altered to such an extent it would not resemble that of the original community. In a coevolutionary context, there are no parasites which are so inextricably linked to a single group of hosts as digenetic trematodes are to snails and other molluscs. With the exception of a few marine sanguinicolids that use annelids, a mollusc is the required first intermediate host for all flukes. In this sense then, molluscs can be considered as keystone species for digenetic trematodes. The keystone concept also may be applied to many trematodes in their roles as indicator species. In the life cycles of many digeneans, an obligate, but usually passive, involvement in various predator–prey relationships at different trophic levels is a clear testimonial for many of the important food-web interactions that occur in both aquatic and terrestrial ecosystems. Accordingly, the present paper will attempt to employ, where possible, the keystone species concept as leverage in dealing with different aspects of the ecology of trematode-snail interactions.

We will begin this paper with a treatment of the basic model for trematode life cycles. Following this brief introduction there will be three major focuses. The first will deal with transmission of eggs or miracidia to the snail host and then from the snail to another host. The latter step involves release of cercariae from the snail, which frequently may include precise patterns in the temporal emergence of cercariae. This segment in the parasite's life cycle also may embrace changes in the behaviour of the snail induced by the trematode's intramolluscan stages to ensure spatial and temporal overlap of the parasite and its next host. With many species, cercariae exhibit well-orchestrated, and always fascinating, swimming behaviour, frequently in response to physical or chemical stimuli emanating from the next host. The second major focus will deal with the establishment and maintenance of trematode infracommunities in snail hosts. Most ecologists working in this arena agree that trematode infracommunities can be either interactive or isolationist in character. Some of these investigators have argued that competition, or predation, or both, may be important factors in defining the nature of interactive infracommunities. As will be emphasized, however, the extent to which these structuring forces actually operate is under considerable discussion, and even debate. Finally, the establishment and organization of trematode component communities will be examined. At this level, especially, scale becomes an over-riding determinant. For example, spatial factors may affect success in transmission, and thereby act as a determining element in the establishment of component communities. We also will examine space within the context of landscape

* Corresponding author.

Parasitology (2001), **123**, S57–S75.   © 2001 Cambridge University Press
DOI: 10.1017/S0031182001007697

ecology/epidemiology, a rather old concept, but one that has seen a deserved revival in recent years. Time is another major determinant of component community dynamics and it too can be related to landscape epidemiology in both ecological and evolutionary terms.

Basic life-cycle patterns for snail-trematode systems

Trematode life cycles usually require two, three, or four hosts for successful completion. There are, however, a few bizarre exceptions that do not follow these patterns. For example, Barker & Cribb (1993) reported that *Mesostephanus haliastuiris*, a prohemistomid fluke, is able to produce daughter sporocysts, cercariae, and miracidia, within mother sporocysts, suggesting the possibility of a one-host life cycle. Barger & Esch (2000) reported an opecoelid trematode *Plagioporus sinitsini*, which gives rise to sporocysts in which cercariae, metacercariae, and egg-producing adults may all occur simultaneously. They also provided convincing field evidence indicating an operational one-host life cycle for this parasite. Despite these uncommon deviations, most authorities (for reviews, see Brooks & McClennan, 1993; Poulin, 1998) agree that the three-host cycle is ancestral for digenetic trematodes, with one-, two-, and four-host cycles all being derived. Whatever the number of hosts required in the completion for any of the 25 000 described species of digeneans, a mollusc must serve as the first host in the life cycle, hence the keystone character of molluscs for these parasites (a few marine sanguinicolids that use annelids are exceptions to this rule).

Assuming the trematodes reach the final site of infection in an appropriate definitive host and that successful copulation has occurred, egg production will ensue. Eggs of many digeneans hatch and ciliated, free-swimming miracidia will emerge. Evidence suggests that the swimming behaviour of miracidia initially is random; at some point, however, it will be influenced by chemical agents released from the snail. After seeking out and penetrating the host, the parasite migrates internally to a final site of infection. In some species, the eggs to not hatch but must be consumed by the snail directly.

Within the snail host, the larva transforms into a sporocyst. The primary function of this larval form is to serve as a brood chamber for the next stage, or stages, in the parasite's life cycle. Normally, sporocysts remain fixed at specific internal sites, i.e. the hepatopancreas, gonads, mantle, etc., and produce either daughter sporocysts or rediae. Rediae are morphologically distinct from sporocysts in possessing a mouth and a primitive gut. The presence of a mouth permits the redia to consume host tissue directly.

All intramolluscan embryogenesis is asexual, via a process known as polyembryony. It should be noted that whatever sequential pattern of internal development is followed, it is species specific. It should also be emphasized that all of the asexual development within the molluscan intermediate host represents, in simplest terms, an amplification of the parasite's gene pool. Whereas an adult, a miracidium, and a cercaria of the same species are highly distinct morphologically, they are nonetheless 'vehicles of the same genetic information' (Poulin, 1998).

Some species begin production of cercariae within sporocysts, others within redia. It is from this point in intramolluscan development that one can begin to see the enormous diversity in life-cycle patterns and variation in cercariae emergence and behaviour. It is this diversity and variation that then leads to ecological division among digenetic trematodes. Generally, a cercaria that emerges from the molluscan host possesses a tail and is, therefore, capable of swimming. There are many species-specific variations in the pattern of cercariae production and release, and also numerous exceptions to the swimming 'rule'. Most of the exceptions are related to the absence of a tail and, in turn, these are manifested in many ecological tendencies of trematodes.

Transmission from the snail to the definitive host usually requires a second intermediate host and, afterwards, predator/prey interaction. However, in some species, the second intermediate host is by-passed and the parasite either encysts in the open as a metacercaria or penetrates a definitive host directly. If the parasites encyst in the open, they will be accidentally ingested by the definitive host, usually a grazing mammal of some sort. Direct penetration of a definitive host by cercariae may require special behavioural coordination between the snail, or cercaria, and the definitive host. Trematode life cycles are completed when the parasites, by whatever means, gain access to their appropriate definitive hosts, and maturation and sexual reproduction can take place.

TRANSMISSION

To the snail

Behaviour. The transmission process from the definitive host to a snail invariably involves either a miracidium or an egg. When an egg is released from its definitive host, it may or may not be fully embryonated. If embryonation is required, it will require appropriate environmental conditions (e.g. temperature, light, etc.). In some species (e.g. the schistosomes) hatching of an embryonated egg is prevented until external conditions are favourable for successful transmission. Eggs of *S. mansoni*, for example, are completely inhibited by NaCl concentrations of 0·6%, but hatch readily when NaCl concentration drops below 0·1 %. This osmotic effect prevents premature hatching in the blood stream of the host.

Wright (1959) described host finding by miracidia as a 3-step process. After emerging from the egg, light or gravity, or both, stimulate the miracidium to move into a habitat most likely to be occupied by the snail intermediate host; random swimming then follows. Once positioned properly, miracidia respond to specific chemical stimuli, usually short-chain fatty acids, amino acids, or simple sugars, released by the snail host. Since the miracidium does not feed while it is free swimming, its life span is brief, generally lasting 24–36 hours. Assuming that a miracidium has responded to the appropriate stimulus, it must then locate the snail host, to which it must attach and penetrate (a comprehensive review of these processes is given by Sukhdeo & Mettrick, 1987).

Despite the simple elegance of these host-finding behaviours, the constraints involved in transmission via free-living miracidia are extensive. Species specificity by digeneans for their snail hosts and size preferences within a snail population further restrict successful transmission. Considering all of the potential risks, one might conclude that success in this phase of transmission would not be very high. In natural settings, there is a remarkable consistency in prevalences, usually ranging from 5–10%, with rare peaks as high 60% for a few species (Fernandez & Esch, 1991*a*; Snyder & Esch, 1993; Esch & Fernandez, 1994). Despite what may seem to be relatively low prevalences, the high capacities of intramolluscan reproduction by digeneans compensate nicely.

Time and space. With free-swimming larval stages, there is an obvious extension of the parasite's spatial range. As has been noted, however, eggs of many digeneans do not hatch and must instead be ingested by the snail. In these cases, there is usually extended survivorship of the parasite's eggs, providing a sort of tactical compensation in overall transmission as compared with species using free-living larval stages. Time thereby provides another dimension, in addition to space, to the transmission process (Poulin, 1998).

Whereas temporal and spatial dissemination of miracidia and eggs within a microhabitat are important in the transmission of trematodes to their molluscan hosts, the wider dispersal of digenean eggs is significant in terms of extending the parasites' geographic ranges. This is especially true for digeneans in avian hosts or marine fishes and mammals that move long distances annually. Since most of these migrations are highly seasonal, synchronization in the timing of host and parasite reproductive cycles becomes a cardinal element in the successful transmission of many digeneans.

From the snail

In an excellent and very thorough review Haas (1994) relates that 'physiological analyses of cercarial behaviour have demonstrated great diversity in behaviour related to host-finding and host-recognition, even in species infecting the same host genera, and this may reflect diverse adaptations for a high success of transmission'. In another excellent review of cercaria behaviour, Combes *et al.* (1994) divided adaptations for increasing transmission probability into two categories, one of which describes the activity of parasites in terms of 'host-time' and the other in terms of 'host-space'. Here, we add another that we feel is integral to overall life-cycle success, the actual movement of cercariae to the host once 'host space' is located by the cercaria.

Host time. For many digenean species, the temporal emergence of cercariae from a snail is not random. Rather, it is timed very precisely, almost always being dependent on photoperiod and usually synchronized with the chronobiological behaviour of the next host in the parasite's life cycle (Combes *et al.* 1994). Several classic examples include species of *Schistosoma* that infect a range of mammalian definitive hosts in Africa. All but one of these schistosomes have a single, distinctive shedding peak during what Combes *et al.* (1994) refers to as a nycthemere, or 24 h photocycle. *Schistosoma margrebowiei*, yet another African species, has an ultradian rhythm, with two emergence peaks, one at dawn and again at dusk (Raymond & Probert, 1991). Apparently, the adaptive significance of these shedding patterns is to enhance transmission of the parasite by concentrating the short-lived cercariae within a relatively brief period of time when opportunities for contact with their appropriate definitive hosts are greatest.

Host space. Most mobile cercariae favourably position themselves for transmission by active swimming. This mobility can be attributed to the presence of a tail that is highly variable morphologically. The cercaria tail possesses limited glycogen stores that account, in part, for the brief transmission opportunity at this stage in the parasite's life cycle. Swimming is generally intermittent, with brief, but very intense, periods of 'tail whipping' that may either push, or pull, the cercaria, depending on the species. These flashes of activity are then followed by longer periods of drifting or sinking in the water column. A few species with reduced tail size appear to creep, in 'inch worm' fashion, on the substratum where they are most apt to encounter their benthic-dwelling crustacean hosts.

The capacity to locate within an appropriate 'host space' assumes the ability of cercariae to perceive and respond to certain types of physical provocations. For most cercariae, these include light, gravity and mechanical stimuli. In the case of light and gravity, both positive and negative responses are known to occur. Whatever the nature of the stimulus, the adaptive response value rests in placing the cercariae in a position where contact with the next host in the cycle is most likely to occur. If the host is a benthic-dwelling microcrustacean, for example, then this would require a negative phototaxis or positive geotaxis, or both, on the part of the cercaria. The reverse would be called for in the case of hosts living 'up' in the water column.

Once the parasite is in the vicinity of the next host, it must respond to specific signals emanating from that host or, according to Combes et al. (1994), it must signal the host and cause it to respond in some favourable manner. Stimuli from the host are either of a physical or a chemical nature. In the case of physical stimuli, for example, cercariae are known to respond to water currents such as those created by movement of a fish's fin or the opercular covering the gills. Chemical stimuli are apparently rare, although cercariae of *Echinostoma revolutum* are known to respond (Fried & King, 1989) to a diasylate of *Biomphalaria glabrata*. Cercariae of *Cotylurus flabelliformis* not only respond chemically to their snail hosts, they are able to distinguish between those snails which are infected and those which are not, suggesting an ability to deter high infection intensity which might kill hosts (Fried & King, 1989).

One of the more remarkable behavioural traits to enhance parasite transmission is shown by cercariae of certain azygiid trematodes, most of which are species of *Proterometra* (see LaBeau & Peters, 1995). These furcocercous cercaria are unique in terms of their size, some with tails ranging between four and five mm in length. Also unusual is their swimming behaviour that is strikingly similar to that exhibited by larval mosquitoes. The tail furcae flap up and down, pulling the large cercariae upwards in the water column; when the flapping stops, the large cercariae settle downwards in the water column. The movement up and down is highly suggestive of a mosquito 'wiggler' and continues until the cercaria exhausts the supply of glycogen stored in the tail or the parasite is consumed by an unsuspecting, piscine definitive host. The enormous size of the cercaria apparently precludes the production of large numbers of free-swimming larvae, but the attractive behaviour compensates enough to ensure a high degree of success in transmission (LaBeau & Peters, 1995). Another adaptive quality contributing to successful transmission of these azygiids is progenesis. The body of the parasite, which is embedded in the tail of the cercaria, is usually in full egg production when it is released from the snail and almost always by the time the mime of the cercariae has attracted an unsuspecting piscine definitive host.

THE INFRACOMMUNITY

Introduction

In the context of the infracommunity, it is an individual of the host species that is the keystone. A number of methods have been used to study snail-trematode infracommunities. These include dissections of hosts, examination of site displacement, experimental infections, measurement of cercarial output and tracking of changes in species composition over time. Here we discuss these protocols using Wright's (1973) book as a starting point and covering important work since about the 1960s. More recent reviews of within-snail processes include those of Lie (1973), Combes (1982), Sousa (1992) and Esch & Fernandez (1994).

Wright (1973) wrote of a 'pecking order' among species coinfecting the same snail and captured the essence of much subsequent thought with his Plate XII 3, which shows an echinostome redia consuming a schistosome cercaria. He was citing work later reviewed by Lim & Heyneman (1972). This review has influenced workers on infracommunities of trematodes in snails since and must be mentioned first. Lie, Basch & Umathevy (1965) experimentally infected the freshwater pulmonate *Lymnaea rubiginosa* with various combinations of 2 echinostomes, a xiphidiocercaria, a schistosome and a strigeid. Echinostome rediae caused the demise of sporocyst infections of other trematodes, even those already established. Subsequently, Lim & Heyneman (1972) developed methods of rearing an albino strain of *Biomphalaria glabrata* and its parasites and studied mainly the interactions of *Paryphostomum segregatum* and *Schistosoma mansoni*. Certain redial species tended to be dominant, consuming stages of subordinate species. The main dominance characteristics of rediae included large size, possession of a mouth and a muscular pharynx. Thus, for example, *P. segregatum* could consume the smaller rediae of *Echinostoma lindoense* as well as schistosome sporocysts. They referred to this as direct antagonism. They also identified indirect antagonism, which occurs when parasite A retards the course of infection by B or leads to degeneration of B's stages. The mechanism for this sort of antagonism can still only be speculated upon. The notion of a dominance hierarchy for trematodes in snails, which assumes importance in later studies, had its origin largely in this work.

Kill and dissect

Collecting snails to obtain data on trematode infections is a time-honoured and productive exercise for documenting the variety and frequency of multiple infections. A foundation study in this vein is that of Cort, McMullen & Brackett (1937) in

Table 1. *Field infection profiles for some snail/trematode systems. Included are host species, total hosts examined, total trematode species found, number of infected hosts, and the number of double and triple infections observed. The number of species combinations is indicated in parentheses*

Host snail	Number Examined	Species	Infected*	Double	Triple	Reference
Stagnicola emarginata†	7259	17	4559	511 (25)	18 (9)	Cort *et al.* (1937)
Cerithidea californica‡	12995	17	8680	667 (38)	23 (9)	Martin (1955)
Cerithidea californica	2910	10	448	13 (5)	0	Yoshino (1975)
Cerithidea californica	25854	15	5025	127 (33)	1	Sousa (1993)
Ilyanassa obsoleta‡	5717	8	1467	52 (?)	0	Gambino (1959)
Ilyanassa obsoleta	5025	8	340	14 (?)	0	Vernberg *et al.* (1969)
Ilyanassa obsoleta	14978	6	614	0	0	McDaniel & Coggins (1972)
Ilyanassa obsoleta	11774	9	6010	1305 (16)	143 (7)	Curtis (1997)
Velacumantus autralis‡	1146	2	321	39 (1)	0	Ewers (1960)
Velacumantus autralis	3842	4	1644	40 (?)	1	Walker (1979)
Velacumantus autralis	8883	3	2221	165 (3)	1	Appleton (1983)
Lymnaea stagnalis†	1659	6	801	216 (6)	15 (3)	Bourns (1963)
Buccinum undatum‡	1375	4	203	8 (2)	0	Koie (1969)
Hydrobia stagnorum‡	16323	13	6128+	428 (17)	7 (3)	Vaes (1979)
Planaxis sulcatus†	4542	6	2053	192 (6)	1	Rohde (1981)
Littorina littorea‡	2691	6	1152	88 (7)	0	Lauckner (1980)
Helisoma anceps†	4899	8	1485	69 (2?)	0	Fernandez & Esch (1991 *a*, *b*)
Physa gyrina†	1181	6	406	87 (7)	3 (2)	Snyder & Esch (1993)

* with 1 or more species; consult reference; † freshwater; ‡ marine.

several Michigan lakes (Table 1). It was the first to find frequent multiple infections, including one quadruple. Curtis (1997) found 2 quadruple infections (1 species combination) in the estuarine *Ilyanassa obsoleta* (formerly *Nassarius obsoletus*). These are the only quadruples ever reported; infections of all species were patent and four is probably the limit for infracommunity species richness. Cort *et al.* (1937) were first to analyse frequency data to predict the number of multiple infections expected in a sample, assuming species combine at random (prevalence A × prevalence B × sample size = expected number of AB double infections in the sample). Observed and expected numbers of double infections were then compared. Most occurred about as expected, but some less often than expected (especially *Diplostomum flexicaudum* and *Plagiorchis muris*), suggesting barriers to coinfection.

Cort *et al.* (1937) did not use statistical analyses to compare observed and expected numbers of multiple infections. However, they noted that spatial (or temporal) heterogeneity in prevalence could affect expected numbers of multiples if samples from two or more sites (or times) were combined into one analysis. For example, if parasite A occurred only at site 1 and parasite B only at site 2, a combined-sample analysis would predict double infections. Nevertheless, A and B should not be expected to coinfect any snails.

Table 1 and Kuris & Lafferty (1994) identify a number of other important studies. For example, Martin (1955) dissected *Cerithidea californica*

(⩾ 20 mm length) samples each month for a year from a small basin (∼ 20 × 40 m) in Upper Newport Bay, California, a gathering point for shorebirds. Single and multiple infections were common. He commented that infection longevity might be conducive to multiple infections or that species may occupy different sites within the snail, making coexistence possible. Ewers (1960) tallied occurrences of *Stictodora* spp. and *Austrobilharzia terrigalensis* in an estuarine snail in Australia. Using a random infection model, three times more double infections were observed than expected. He proposed that one species might predispose the snail to infection by the other and that snails eating faecal material from birds with multiple infections might also be involved. Bourns (1963) reported that multiple infections in *Lymnaea stagnalis appressa* in Ontario, Canada, often did not occur at random. Indeed, four combinations occurred significantly more often than expected. He suggested that, once infected by one species, predisposition to/against infection by other species, or inadequate sampling, might have produced his results. Koie (1969) studied a prosobranch along the Danish coast; double infections were few and she concluded they occurred at random. Vernberg, Vernberg & Beckerdite (1969) examined *Ilyanassa obsoleta* from Beaufort, North Carolina, first by observing cercarial emergence, then by dissection of infected snails. One sporocyst species, *Zoogonus lasius* (now *Z. rubellus*) was noted as being involved in many doubles. Another pair, *Himasthla quissetensis* and *Lepocreadium setiferoides* (both redial species, the former an echinostome), were never observed together. They proposed that alterations in thermal acclimation patterns by one species prevented colonization by the other. Lim & Heyneman (1972) rejected this hypothesis in favour of direct antagonism. Yoshino (1975) found moderate prevalence and few double infections in *C. californica*, with the most abundant trematode being involved most frequently. He posited that the paucity of doubly-infected snails was due to direct antagonism. Multiple infections were rare in a *Hydrobia stagnorum* population, but Vaes (1979) saw little redial predation to explain it. Rohde (1981) concluded that species combined at random in an Australian coral reef snail. Lauckner (1980) presented original data on European *Littorina littorea*, observing that most doubles included *Himasthla elongata* and *Renicola roscovita*. Notably, there were no *Cryptocotyle lingua–H. elongata* doubles, even though both were frequent as single infections. Other species pairs also co-occurred rarely, others as expected.

Fernandez & Esch (1991a) studied infra-communities in the pulmonate *Helisoma anceps* in Charlie's pond (North Carolina). Most of the few double infections were *Halipegus occidualis* (redia) with *Haematoloechus longiplexus* (sporocyst). Laboratory observations established the dominance of *H. occidualis*. Nevertheless, it was concluded that spatial and temporal factors made such encounters rare in the host population and competition was seen as unimportant in structuring infracommunities. Snyder & Esch (1993) studied *Physa gyrina* in the same pond. Prevalence was similar, but many multiple infections were observed and dominance was not involved. In fact, congeners of the species that did not co-occur in *H. anceps* coexisted in *P. gyrina*. Based on these results, it can be stated that predictions regarding dominance based on taxonomic relatedness are not always reliable. Infra-community differences in the two pulmonates were attributed to snail behaviour and factors external to the snails. Within-snail antagonisms were also considered of minor importance by Curtis (1997), who examined *Ilyanassa obsoleta* from nine sites in Delaware estuaries from 1981 to 1993. Prevalence ranged from 8·7 to 100%. Multiple infections were observed frequently at high prevalence sites. The sporocyst species, *Zoogonus rubellus*, *Gynaecotyle adunca* and *Austrobilharzia variglandis*, combined most frequently, both with each other and with redial species. Curtis (1995) has determined that *I. obsoleta* is long lived and, as a consequence, multiple infections are able to accumulate in the component community.

Statistical provision for spatial and temporal heterogeneity was made in a series of studies done on an estuarine snail (Kuris, 1990; Sousa, 1990, 1993; Lafferty, Sammond & Kuris 1994). A concern of these authors was whether heterogeneity or com-petition within snails structure trematode communities. Kuris (1990) is influential in this regard. He proposed a method by which the importance of heterogeneity in generating expected frequencies of double infections could be dismissed or embraced. For example, suppose monthly samples have been collected from a given site. Using the random species combination approach, samples are analysed separately and monthly estimates of expected double infections are summed. Then, in another calculation, the total sample (all months combined) is analysed for prevalences and expected doubles. If the sum of monthly expected doubles is greater than expected doubles based on the total sample, then Kuris asserts the total sample analysis underestimates the expected number of multiple infections. Such an inequality would arise when prevalence of both parasites is simultaneously high in some months and low in others. If the sum of monthly expected values is smaller than the expected value based on the total sample, then the total sample analysis overestimates expected multiples. This could arise if prevalences of both parasites are independently variable among months. In either case, analysis of the combined sample would give a distorted expectation of double infections and heterogeneity is, therefore, deemed important. If expected doubles based on the sum of monthly samples equals the expected number based on the total sample, then Kuris says that heterogeneity 'may not be important'. The most clear-cut case would be when prevalences are identical in each month. There are, however, sets of variable prevalences among samples that encompass considerable heterogeneity yet yield nearly equal values. Accordingly, this outcome is equivocal. Mathematically, either expected value may be compared with the observed. However, biologically, it might still be better to recognize heterogeneity among samples. Kuris (1990) offered that, if heterogeneity can not explain the absence of doubles, then competitive exclusion in infracommunities must be the cause.

Kuris (1990) may well have a valid argument in a perfectly sampled world. However, any such analysis is based on the accuracy of prevalence estimates and these could easily be in error. For example, Curtis & Hurd (1983) and Curtis (1997), working with the estuarine *I. obsoleta*, found it difficult to collect representative samples. It was noted, for example, that a slight change in tidal elevation where samples are collected can change prevalences of certain species and, therefore, frequency of multiple infections. Thus, two samples collected in what is thought to be the same 'place' may actually be from different 'places'. Further, if the samples are collected at different times, they produce what is thought to be temporal heterogeneity when it is actually spatial. Reversat & Silan (1991) commented cogently, '... the distribution of this type of helminth is rarely uniform in host populations which are not uniformly distributed either ...'. Sampling strategy must be wisely designed and we rarely possess sufficient knowledge of host/parasite biology to do this automatically.

Kuris (1990) used *Cerithidea californica* and erected a competitive dominance hierarchy for 15 trematodes that infect this species of snail. The hierarchy was based partly on sources of direct evidence. Snails were examined in the laboratory (for release of cercariae) and colour coded according to infecting species, then released, recollected at some later time and re-examined. If one species replaced another, or if a new species colonized an existing infection, this was taken to indicate dominance. Also, dominance was indicated if, on dissection, rediae of 1 species were observed consuming stages of another. Lines of indirect evidence for dominance included possession of rediae (especially if large), taxonomic relatedness, use of large rather than small snails, the gonad as the site of infection and site displacement. Basically, species with large rediae (*Parorchus acanthus*, *Himasthla rhigedana*) were high in the hierarchy and sporocyst species lower. Oddly, schistosome sporocysts, *Austrobilharzia* sp., could coexist with both dominant redial species, but in turn dominated only the poorest competitors.

With the establishment of the dominance hierarchy, Kuris (1990) estimated the impact of competition. Essentially, if A is dominant over B, snails infected by A were not available to be infected by B (corrected for any AB double infections); losses in B due to the presence of A in the component community could be estimated. Estimates of the percent of subordinate infections lost to competition ranged from 6 to 44%. He concluded that competitive interactions were an important structuring force for trematode communities.

Lafferty *et al.* (1994) developed this conclusion further for *C. californica* and Kuris & Lafferty (1994) extended it to other snail-trematode infracommunities. Whereas these papers addressed features of the component community, infracommunity competition was used as the basis to explain them. Lafferty *et al.* (1994) collected snails with a shell height of 25–30 mm (to reduce size heterogeneity) from 5 sites about 50 m apart over 20 days. The life span of *C. californica* is at least 7 years (Sousa, 1993) and these snails would have been several years old. Snails within sites apparently mingled at random and no movement between sites was assumed. Their model also assumed that many more double infections are present initially than are observed in samples of mature snails. They analysed for spatial heterogeneity effects (Kuris 1990) based on their model and concluded that combined samples would overestimate expected double infections, and that subordinate infections were lost to competitions based on their dominance hierarchy. They concluded

that competition within the infracommunity was the major structuring force for the component community. The model underlying this conclusion holds that first a flood of miracidia infects young snails, yielding many multiple infections; competition then occurs and many of these multiples are resolved into single infections. However, the rate at which trematodes actually colonize *C. californica* was not examined and this factor was not considered in their model. A high rate of parasite recruitment among young snails would be required to support the model. Unfortunately, little is known about the rate at which gastropods accumulate trematodes. The only studies directly addressing this issue are those of Sousa (1993), Curtis (1996), and Curtis & Tanner (1999), and results indicate that the rate of parasite recruitment is low.

Kuris & Lafferty (1994) performed a meta-analysis on 62 studies of larval trematodes in snails to see if competition in infracommunities structured these component communities. They erected dominance hierarchies for the sets of trematode species included based on published studies when available, then used taxonomic relationships and other indirect lines of evidence (Kuris, 1990) when not. The analysis included 296180 snails, of which 62942 were infected with one or more species of trematodes. Overall, they estimated that competition eliminated 13% of the infections and concluded it to be a potent structuring force.

Sousa (1990) looked at the component community of trematodes in *C. californica* at two sites in Bolinas Lagoon, California from 1981 to 1988. Recognizing the competitive hierarchy of Kuris (1990), he held that infracommunity interactions were partly responsible for the paucity of multiple infections. He concluded, however, that competition could not determine component community structure because new patches of juvenile snails are always available as a resource. Consequently, trematodes coexist in the component community even if they do not in infracommunities. Sousa (1993) paid more direct attention to the rarity of multiple infections and infracommunity dynamics. Spatial and temporal heterogeneity were deemed important and analyses were based on unaggregated samples. Because double infections were uncommon, Monte Carlo methods were used to estimate the likelihood of doubles being so rare. The analysis showed that multiples were rarer than expected. Pair-wise analyses revealed that individual species did not occur randomly in snails. Redial species tended not to co-occur; they also tended not to co-occur with sporocyst species, although a few did, notably when the schistosome *Austrobilharzia* sp. was involved. Moreover, sporocyst species tended to occur together, as predicted. Observations of *Himasthla*, *Parorchis* and *Echinoparyphium* rediae consuming co-occurring species during dissections were also reported. Consistent with his earlier treatment, Sousa concluded that, whereas larval trematodes may interact negatively, coexistence in the host population is unaffected because there are always new hosts to infect.

Site displacement

Most infections in snails are by a single species that is adapted to occupy preferred sites, often in the gonad-digestive region. In multiple infections, when one species evicts another, distribution of the subordinate species must become progressively restricted until eliminated. One species may also displace another from its preferred site, but the two establish an equilibrium in the snail. These two situations would be difficult to distinguish based on a single observation.

DeCoursey & Vernberg (1974) examined *I. obsoleta* from North and South Carolina by dissection and histology. They found *Zoogonus rubellus* sporocysts often existed side-by-side with *Lepocreadium setiferoides* rediae, but *L. setiferoides* was occasionally displaced. *Austrobilharzia variglandis* appeared to occupy an abnormal site when co-occurring with *Cardiocephalus brandesii*, even though both were present throughout the snail. When the echinostome *Himasthla quissetensis* was present with *A. variglandis*, a few *H. quissetensis* rediae were displaced to sites nearer the head where some were degenerated to encapsulated. If these are consistent interspecific responses, they do not conform to expected dominance relationships since species with rediae, especially echinostomes, should dominate. Curtis & Hubbard (1993) viewed trematode spatial distribution in *I. obsoleta* in Delaware. They examined snail tissues microscopically and found that no species consistently displaced a co-occurring species. In most snails, '... all species present occurred throughout'. Although Vernberg *et al.* (1969) and Curtis (1985) reported that *H. quissetensis* and *L. setiferoides* never co-occurred, they were observed together in the study by Curtis & Hubbard (1993). When together, both were producing cercariae, but the latter species seemed the more vigorous infection.

Yoshino (1975) noted that certain trematode species in *C. californica* occupied the mantle and could coexist with species using the visceral spiral. Further, *Euhaplorchis californiensis* displaced a strigeid from the gonad to the digestive gland. He suggested that the locus of single infections may change with locality of the snail due to variability in the genetic strain of snail or parasite. Cheng, Sullivan & Harris (1973) also suggested this to be the case for *I. obsoleta*. Postulating local genetic variability is problematic when definitive host mobility and, with *I. obsoleta* planktonic larvae, are considered (Curtis & Hubbard, 1993; Gandon & Van Zandt, 1998).

Walker (1979) and Appleton (1983) found that *Austrobilharzia terrigalensis* only infects *Velacumantus australis* when the snail is already infected by another trematode and that it subsequently inhibits, but does not evict, the other species. They speculated that this trematode requires the host's tissues to be altered or the host's defences neutralized by the first infection before it can successfully colonize.

When *Echinoparyphium elegans* and *Schistosoma bovis* co-occur in the freshwater snail *Bulinus truncatus*, the latter's distribution appears to be extended (Mouahid & Mone, 1990). In *Biomphalaria pfeifferi*, Jourdane & Monkassa (1986) observed displacement of *Schistosoma mansoni* primary sporocysts to a deeper position in the snail foot when a dominant species, *Echinostoma caproni*, was also present. Sporocysts that were displaced survived whereas those remaining tended to degenerate. Movement to deeper tissues was interpreted as part of the competitive interaction between the two trematodes and coevolution was suggested.

Experimental infections (and redial habits)

Lim & Heyneman (1972) reviewed the basic work in this area and general findings are discussed in the introduction to this section. Here, we consider the following question. Does consumption by rediae represent an aggression toward individuals of another species for purposes of nutrition, i.e. is it predation *per se*, or is it simply opportunistic browsing? Sousa (1992) entertained the browsing hypothesis and rejected it because (1) rediae seldom consume their own kind, (2) they tend to attach actively to larvae of subordinate species, and (3) mobile rediae tend to aggregate where subordinates are concentrated. Alternatively, it is suggested that rediae are adapted to 'chew' their way through host tissues, and not to consume other trematodes. To this end, they have a mouth, often large size and a sucking pharynx. Moreover, rediae must be adapted to avoid consuming their own kind because it would be maladaptive to do otherwise. Finally, feeding on loosely attached germinal sacs of larval trematodes, especially where concentrated, may be easier than tearing off pieces of host tissue. Redial infections that grow faster in a coinfection (Lim & Heyneman, 1972) may be the result of easily-obtained nutrition. A key issue in deciding which hypothesis, e.g. directed competition through consumption of the other's stages or opportunistic browsing, is closer to the truth is whether the 2 trematode species in question encounter each other frequently enough in nature to respond to each other adaptively. Curtis & Hubbard (1993) noted that, even if multiple infections are quite common collectively, particular pairs seldom coexist. It is thus conceivable that consumption of stages of other species by rediae is a by-product of the manner in which rediae are adapted to interact with the host rather than a direct response to another species of trematode.

Interspecific interaction with cercariae output

A snail presumably represents a limited resource and one infecting species may affect another in terms of how many cercariae are produced. If frequent enough, this could have ecological and evolutionary consequences. The conundrum is that this effect would best be measured in the field, which is problematic given the usual rarity of multiple infections.

DeCoursey & Vernberg (1974) studied cercariae production by 10 *I. obsoleta* in the laboratory every 3 hours for 27 hours, enumerating cercariae produced by *Zoogonus rubellus* alone, *Lepocreadium setiferoides* alone, and in combination. Single infections produced a mean of ~ 3500 cercariae, whereas in double infections a reduction was suggested; *Z. rubellus* released ~ 900 and *L. setiferoides* ~ 1500. Curtis & Hubbard (1993) also measured cercarial production in *I. obsoleta* in a variety of single ($n = 162$), double ($n = 134$), and triple ($n = 65$) infections involving *Himasthla quissetensis*, *L. setiferoides*, *Z. rubellus* and *Gynaecotyle adunca*. Variability in cercarial production was high. A downward trend in output from multiple infections was often seen, but it was not significant. When in combination, these larval trematodes thus do not appear to consistently affect one another.

Species composition changes

Snails can be tested for infection by cercarial release and marked, then released in the field, recaptured later and re-examined. Workers with freshwater snails generally seem confident that release of cercariae will accurately reflect patent infections, although prepatent infections will certainly be missed (Goater *et al.* 1989). Results of Curtis & Hubbard (1990) suggest that, in marine snails at least, caution must be exercised in interpreting release of cercariae. When individuals of *I. obsoleta* were tested in field containers, 59.1% of determinations proved to be erroneous on later dissection. Errors were most frequent among multiply infected hosts.

Fernandez & Esch (1991*a*) reported that *Helisoma anceps* in Charlie's Pond (North Carolina) had few double infections among their marked snails, but were able to observe parasite recruitment as well as loss of active infections (see also Goater *et al.* 1989), sometimes being replaced by other species. Sapp & Esch (1994) used *H. anceps* and *Physa gyrina* caged in the field to test whether interspecific trematode interactions dictated infracommunity structure. Into one set of cages were placed five uninfected and marked *H. anceps* and *P. gyrina*. In another set of enclosures were placed *H. anceps* experimentally infected with *Halipegus occidualis*. It was not possible to experimentally infect *P. gyrina*. Cages were positioned at various depths and distances from shore, as these factors had been shown to affect parasite recruitment. Placing enclosures at least 5 cm from the bottom of the pond prevented colonization by egg-transmitted species, e.g. *H. occidualis* and *H. eccentricus*, the 2 most common parasites in the pond. This manipulation did not affect infracommunity structure as there was no significant difference between prevalence of infections with miracidia-transmitted species in caged and uncaged snails. In effect, there was no competitive release and colonization was unaffected by the absence of common species.

Curtis (1996) released 1400 individually marked and uninfected *I. obsoleta* on Cape Henlopen (Delaware) to study colonization by trematodes. These sentinels were released into an area with high prevalence ($\sim 80\% +$) of infected snails. Of 185 sentinels recovered, 2.7% had single-species infections. After correcting for infections that may have been undetected during initial screening, he concluded that the colonization rate was 1.6% yr^{-1}. This result, coupled with the long life spans of hosts and their infections (Curtis, 1995), led him to conclude that time, not competition, was the most important factor in development of trematode infra- and component communities in this snail.

Sousa (1993) tracked 1170 previously infected *C. californica* for up to 4 years. Of the snails released, 1.5% recruited an additional species and 6.3% switched infections. *Himasthla rhigedana* and *Parorchis acanthus*, dominant trematodes in the Kuris (1990) hierarchy, were responsible for $> 90\%$ of these changes. It is a matter of judgement, but this number of infection changes in snails exposed to new infections for such a considerable time seems modest. Curtis & Tanner (1999) examined colonization of *I. obsoleta* in Delaware. They released 300 marked natives (most of which were infected) onto Cape Henlopen, and 249 uninfected snails and 231 natives (virtually all infected) at Savages Ditch in nearby Rehoboth Bay. Only 16 initially uninfected snails were recovered over three summers and two had become infected; their estimate for colonization rate at Savages Ditch was 6.3% yr^{-1}. This was higher than on Cape Henlopen (Curtis, 1996), but was based on many fewer snails and still quite low. Their final conclusions about changes in infection status among natives involved a number of considerations, but ultimately they asserted that six of 123 (4.9%) recovered natives (both sites) had changed infection status over periods ranging up to 3 years. Based on all evidence to date, species composition changes appear infrequent.

Interactive versus isolationist infracommunity debate

Holmes & Price (1986) developed the concepts of isolationist and interactive parasite communities. They stated that '... competition must be demonstrated as a process and any inference about past evolutionary change through competition must be approached with extreme caution'. We would add that attributing pervasive *ecological* importance to interactions requires equal caution. Demonstrating intra- or interspecific competition requires identifying a limiting resource, or competitive release in the absence of interaction, or both. The theatre of interaction is the individual host and frequent colonization is a primary requirement for interactive communities. New species must be recruited often and must then compete for limited resources. In contrast, low colonization rate is a primary feature of isolationist communities.

The debate about whether parasite infracommunities are interactive or isolationist has rarely been centered on larval trematodes in snails. Certainly in the view of Kuris (1990), Lafferty *et al.* (1994) and Kuris & Lafferty (1994), competition is important and infracommunities are interactive. Sousa (1994) also considered the issue and pointed out that since infections reproduce asexually, infrapopulations are almost always dense. He went on to state that snails offer a limited number of target organs to infect and that there is much evidence that species interact antagonistically when they co-occur. He concluded that 'infracommunities of larval trematodes have the potential to be strongly interactive' and that colonizations of snail hosts from definitive hosts 'are spatially and temporally variable, as are the rates of interspecific interaction among intramolluscan larval stages'. Infracommunities can thus be interactive or isolationist at different times and places, which is probably close to the reality of the matter.

Sousa (1993), Curtis (1996) and Curtis & Tanner (1999) are the only studies to address trematode colonization rate in snails explicitly and all suggest rates are low. The latter two investigations were done in habitats where trematode prevalence was very high and multiple infections frequent. If competition between species is to be frequent, then it should occur under these conditions. However, Curtis & Tanner (1999) observed that, even if species composition changes occurred at a posited rate of 10% yr^{-1}, it would mean that, on average, an infracommunity might see a new species interaction once in 10 years. *Ilyanassa obsoleta* is long lived and snails with the same infection 3 years running (now 5; Curtis, unpublished observation) have been documented. These infracommunities likely persist long enough for a change to occur, but an interactive situation is clearly not indicated. In freshwater systems, however, many snails are short lived and

trematode communities are reset as snail generations turnover annually, or even more frequently (Esch & Fernandez, 1994). Perhaps miracidia colonize snails at higher rates in freshwater. If so, given the widely observed tissue-consumptive behaviour of rediae, freshwater infracommunities should be more interactive. However, data on colonization rates in the field seem not to exist. Further, Fernandez & Esch (1991a), Snyder & Esch (1993) and Sapp & Esch (1994) all held interspecific interactions to be unimportant. In a marine system, Curtis & Hubbard (1993) and Curtis (1997) worked with abundant multiple infections and noted that particular species co-occur infrequently, with time being the important factor in the accumulation of infections. Are trematode infracommunities in snails interactive or isolationist? We think, as does Sousa (1994), they potentially can be interactive. However, to be so requires high colonization rates which, to date, have not been demonstrated. It is probably best to operate under the null hypothesis that they are isolationist until there is substantive field evidence to the contrary.

THE COMPONENT COMMUNITY

The spatial landscape

Spatial effects on trematode component community structure in snails can be examined through a landscape ecology approach. Landscape ecology is concerned with four general properties (Turner, 1989): (1) spatial heterogeneity and how it changes over time; (2) biotic and abiotic exchanges across heterogeneous areas; (3) how heterogeneity affects biotic and abiotic processes; and (4) managing or preserving spatial heterogeneity. Landscape ecology is not new to parasitology, dating back to Pavlovsky's (1966) work (landscape epidemiology) on the distribution of parasitic disease across large areas. However, the techniques and concepts widely used today in landscape ecology have been developed most profusely in the last couple of decades (Sokal & Oden, 1978; Burrough, 1981; Gardner *et al.* 1987; Milne, 1988; Tilman, 1994; Gustafson, 1998).

In a very general sense, a landscape is any heterogeneous or patchy area (With, 1994; Burke, 1997). Though the traditional focus of landscape ecology has been on relatively large areas, encompassing many ecosystems, the approaches and methods are also applicable at much smaller spatial scales. This is useful for parasite ecology because parasites are nested within a hierarchical framework, i.e. the infra-, component, and compound communities. At the largest scale, the species pool of trematodes from which local communities are drawn is called the regional pool. Within a region, there are a number of (more or less) distinct ecosystems; this is the local scale and roughly equivalent to the

Table 2. *Distribution of trematode species in* Ilyanassa obsoleta *from nine collecting sites in and around Delaware Bay (from Curtis, 1997)** × = present; − = absent

Parasite	Site								
	CH	CN	CC	IR	BB	GM	CS	SD	TI
Lepocreadium setiferoides	×	×	×	×	×	×	×	×	×
Zoogonus rubellus	×	×	×	×	×	×	×	×	×
Himasthla quissetensis	×	×	×	×	×	×	−	×	−
Stephanostomum tenue	×	×	×	×	−	×	×	−	−
Stephanostomum dentatum	×	×	×	×	×	−	−	−	−
Gynaecotyla adunca	×	×	×	×	−	−	−	×	−
Austrobilharzia variglandis	×	×	×	×	−	−	−	−	×
Pleurognius malaclemys	−	×	×	−	−	−	×	−	−
Unknown	×	×	−	−	−	−	−	−	−
Diplostomum nassa	×	−	−	−	−	−	−	−	−

* Index of nestedness = $10.54°$ ($P = 0.0003$) based on 'temperature' method of Atmar & Patterson (1993).

compound community. Within each local ecosystem, snail populations, as well as the infective stages of trematodes, are likely to be non-randomly distributed among habitats and habitat types, with some degree of exchange among habitats. At a smaller scale (microgeography of Esch & Fernandez [1994]), snails represent patches of habitat (Sousa, 1993) and infective stages represent patches of propagules. At the smallest scale considered here are processes that might occur within an infected snail, e.g. competition.

Regional scale. The species richness of any trematode component community cannot exceed the upper limit imposed by the regional species pool. Deviations from local areas harbouring all regional species are thought to occur by two processes. First, there is always an element of chance involved in determining the species present in any local community; recruitment is determined by dispersal history and local extinction. Second, a series of biotic and abiotic characteristics in local systems may set a limit to the species richness and composition, either through deterministic interspecific interactions, or through the pattern of spatial covariance of critical resources for each individual community member, or some combination of both.

As suggested earlier, interspecific interactions among trematodes in snails may affect community structure (Kuris, 1990; Fernandez & Esch, 1991*a*; Kuris & Lafferty, 1994; Lafferty *et al.* 1994), but do these processes provide power of prediction at the regional scale? Studies by Lafferty *et al.* (1994) and Curtis (1997) address this question because their work is the result of dedicated effort within single regional species pools. Curtis (1997) surveyed the trematode parasites of *I. obsoleta* occurring at nine sites in and around the Delaware Bay and found strikingly different communities at each of the sites. The data in Table 2 suggest that the pattern of trematode occurrence on this regional scale is nested

among local areas and, although other factors could be involved, that definitive hosts might occur among these sites in a nested pattern. In contrast, the Lafferty *et al.* (1994) study on *C. californica* at five sites on the California coast does not reveal a nested pattern of trematode species composition (Table 3), even though species richness and composition are still rather variable among sites. The absolute scales of investigation in these two studies are rather different, encompassing many km in the Delaware investigation and less than a quarter of a km in California. Nonetheless, at both scales, variety in species composition is the striking pattern.

Studies that incorporate direct observations on the occurrence, much less the abundance, of definitive host taxa are absent on a regional scale. Definitive hosts for trematodes range from the rather locally-restricted fishes to broad-ranging and mobile birds and mammals. Although a number of null hypotheses could be tendered for the study of regional determinants of species richness and composition, the most logical and simple one is that definitive host distributions determine the entire pattern.

Local scale. Rarely can an entire ecosystem, say a pond or estuary, reasonably be considered as a homogeneous environment. Critical resources are neither uniformly nor randomly distributed within local systems. Again, detailed studies regarding the distribution of all hosts in trematode life cycles are largely absent at his scale. However, a number of independent, if not directed, attempts have been made to account for the manner in which spatial heterogeneity affects trematode component community structure at this scale.

The investigation by Bartoli & Holmes (1997) illustrates how spatial heterogeneity can affect transmission, and thus abundance, of trematodes in a snail population. Shown in Fig. 1 are redrawn distributions of the spatial coincidence of the hosts in

Table 3. *Distribution of trematode species in* Cerithidea californica *from five collecting sites on the California coast (from Lafferty* et al. *1994)** × = present; − = absent.

Parasite	Site 2	3	4	1	5
Euhaplorchis californiensis	×	×	×	×	×
Himasthla rhigedana	×	×	×	×	×
Parorchis acanthus	×	×	×	×	−
Echinoparyphium sp.	×	×	×	×	−
Acanthoparyphium spinulosum	×	×	×	−	×
Cyathocotylid	×	×	−	×	×
Small xiphidiocercaria	×	×	×	−	−
Renicola buchanani	×	−	−	−	×
Large xiphidiocercaria	−	−	×	−	−

* Index of nestedness = 10·29° ($P = 0\cdot107$) based on 'temperature' method of Atmar & Patterson (1993).

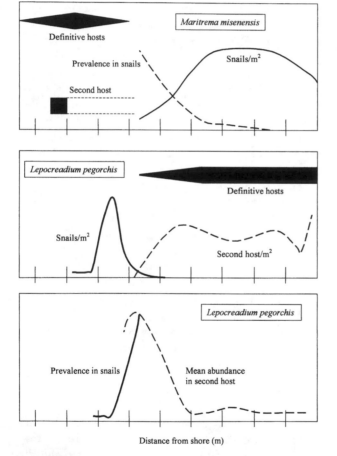

Fig. 1. Distribution of host populations for 2 trematodes, *Maritrema misenensis* (top panel) and *Lepocreadium pegorchis* (middle and bottom panels), along a transect in the Mediterranean Sea off France. Redrawn from Bartoli & Holmes (1997).

the life cycles of two trematodes occurring in two different snail species along a 150 m transect in the Mediterranean waters off France. For both *Maritrema misenensis* and *Lepocreadium pegorchis*, the definitive hosts and the second intermediate

hosts co-vary in a strong, positive manner with respect to spatial distributions. The snails are, however, somewhat negatively associated with the other two hosts in the life cycle. In neither case are these trematodes most prevalent in snails where snails are most dense, rather the focus of infection in snails appears to be determined by areas of spatial overlap of the host populations.

The lesson to be garnered from these observations is that spatial constraints to transmission definitely exist within ecosystems, indicating a way in which structure might be purveyed to trematode component communities. As Bartoli & Holmes (1997) point out, small-scale investigations provide an appropriate window through which constraint on parasite transmission can be viewed. Incorporating such a landscape approach to understanding trematode component communities is a most exciting prospect for future investigations involving these systems.

Many studies document spatial heterogeneity in the richness and species composition of trematode communities in snails at a local level (Appleton, 1983; Curtis & Hurd, 1983; Williams & Esch, 1991; Sapp & Esch, 1994). Snails themselves respond to various environmental stimuli (Harman, 1972; Okland, 1983), producing aggregated distributions among habitats or habitat types, or both. Furthermore, the diversity of snail species both in freshwater and marine environments ensures that a number of different habitat preferences will be represented in an ecosystem. For instance, Laman, Boss & Blankespoor (1984) studied the depth-distributions of seven species of snails in Douglas Lake, Michigan and found striking differences among the snails present, with most species distributed unimodally but differing in the depths at which they were most dense. Such systems in which detailed knowledge of snail distributions are known provide excellent opportunities to determine whether these patterns influence the spatial structure of the trematode component communities.

Curtis & Hurd (1983) demonstrated a sharp boundary in the parasite community structure along the Cape Henlopen sandflat occupied by *I. obsoleta*. In one zone, snails harboured six species of trematodes but in an adjacent zone, only two trematodes were present. Parasitism increased with shell height for snails from both zones; however, in the zone with greater species richness, parasitism was significantly higher in medium-sized snails than in the zone with low species richness. These data suggest that definitive host distributions among these areas were responsible for the differences in trematode species richness among snail sub-populations. Sapp & Esch (1994) gathered point-in-space data for each 1231 *Physa gyrina* and 1532 *Helisoma anceps* collected over a year in Charlie's Pond, North Carolina. Overall prevalence of parasitism in *H. anceps* was negatively associated with water depth where the snails were collected (Williams & Esch, 1991). The same result was observed for overall parasitism in *P. gyrina*, as well as a significant effect of snail size and distance from shore. Individual species of trematodes in this system differed in their relationships with habitat characteristics where snails occurred. The limited mobility of these snails, together with their short life spans (< 1 year for *H. anceps*; about 3 months for *P. gyrina*), makes these point-measures of habitat use particularly important. In more vagile and longer-lived snails, point-measures are less likely to be as informative.

Appleton (1983) examined the prevalence of three trematodes in *Velacumantus australis* at 14 sites in the Swan estuary, Australia. Trematode species richness varied from zero to three among these sites. In general, prevalence of these trematodes was negatively associated with water turbulence and velocity, which Appleton (1983) attributed to the terrace profile of the sites, in addition to the positive effect of aggregating avain definitive hosts in areas sheltered from incoming tides. Bowers (1969), Cannon (1979), James (1968), and Threlfall & Goudie (1977) all found that trematode infections were concentrated in a limited area of the host snail's range, defined by vertical zonation and likely caused by the aggregation of definitive hosts. Appleton's (1983) work also revealed that prevalence of infection in *V. australis* for each trematode was related unimodally to snail density, such that lower-medium densities of snails were associated with the highest levels of infection, whereas extremely dense groups of snails and extremely sparse groups of snails were less heavily infected. This observation is reminiscent of Bartoli & Holme's (1997) work described earlier. Jokela & Lively (1995) found *Microphallus* sp. infections in *Potamopyrgus antipodarum* to be highest in shallower habitats, whereas *Telogaster opisthorchis* was more prevalent in deeper habitats. Prevalence in

a number of rarer trematodes was apparently not related to water depth, and was independent of *Microphallus* sp. infection in *P. antipodarum*. Jokela & Lively (1995) attributed these differences to definitive host behaviour as well.

The emerging theme from these studies of spatial variation in trematode prevalence and component community structure within single ecosystems is one of spatial constraint on transmission from definitive hosts. The complexity of spatial patterns among parasites that can require three or more host organisms during their life cycle could be overwhelming (see first section of this paper). Nevertheless, we think detailed studies at the local level, incorporating data on definitive hosts, snails, other intermediate hosts and habitat characteristics are particularly useful because they provide instant predictions that can be tested by moving to another compound community. Different ecosystems in a region will be characterized by different physical, chemical and biological properties and, thus, provide the requisite variation in putative structuring forces discovered at the local level to test the generality of the model at more regional levels. This protocol, i.e. intensive study in one ecosystem followed by broadening to a regional level, is an approach that could be utilized to determine how space and spatial structure affects trematode component communities, but it has not yet been employed.

Landscape ecology among trematodes in snails, as well as many other parasites, has a second, equally important facet. All organisms modify their environment to some extent. Trematodes, however, not only modify their immediate habitat (snails), they also can affect the positioning of that habitat in an ecosystem. The known cases in which a trematode infection results in altered snail behaviour seem to centre upon, in a functional sense, the bridging of spatial constraints imposed by non-overlapping host distributions. Thus, when considering the effects of spatial heterogeneity on component community structure, it is not always enough to know where all the hosts are, but also if their spatial texture is influenced by the presence of a trematode.

Curtis (1987, 1990) found that *I. obsoleta* infected with *Gynaecotyla adunca*, migrated up onto beaches and sandbars during periods preceding low tides, especially night-time low tides. Semi-terrestrial crustacean beachhoppers, which are the second intermediate host, are most active at night and occur on the beach head. Uninfected *I. obsoleta*, or snails infected with other species of trematodes, do not exhibit this behaviour. Curtis (1987, 1990) interpreted this parasite-induced altered snail behaviour as an adaptation to increase the probability of contact between beachhoppers and cercariae from the snail. During a series of observations, many *G. adunca*-infected snails were seen to make multiple migrations up the beach head, suggesting that this modified

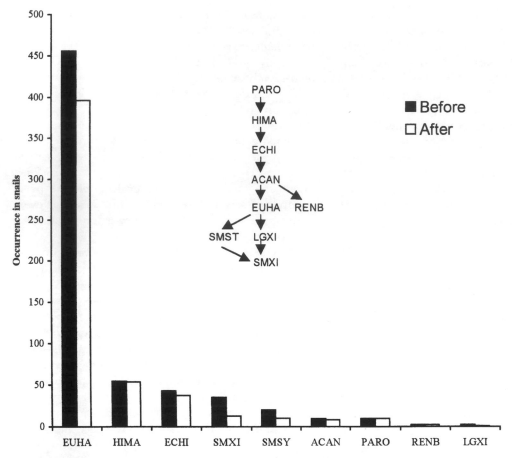

Fig. 2. Abundance of trematodes (number of snails infected) in *Cerithidea californica* as observed (after) and as estimated prior to the effects of competitive exclusion (before). Dominance hierarchy of trematodes shown in inset. Data taken from Lafferty *et al.* (1994).

behaviour is sufficiently complex to allow tracking of tidal cycles. Furthermore, shedding of cercariae is restricted to night-time at low tide. *Gynaecotyla adunca* occurs frequently in multiple infections (Curtis, 1987, 1997), but how these form and why is unknown. Sindermann (1960) found that infection of *I. obsoleta* with a variety of trematodes tended to reduce inshore to offshore migration during the fall, when uninfected snails typically move to low tide zones to overwinter. Movement of snails from deep water to shallow water in spring is coincident with avian definitive host behaviour of *Austrobilharzia variglandis*, providing spatial overlap of the two hosts in the life cycle. Thus, with the *I. obsoleta* system there are examples of different species of trematodes differentially affecting behaviour of the same snail, although each of these responses results in essentially the same microhabitat use.

Infracommunity scale. Because an entire section of this review has already been dedicated to infra-communities, we will not revisit any of the details here. It is clear, however, that the topic of inter-specific interactions among trematodes has received a lion's share of recent focus among researchers in

this field. There is no doubt that these interactions occur and that deterministic outcomes are the result when two trematodes infect the same snail. But, there is no clear consensus on the importance of these interactions at larger scales.

Studies of competition demonstrate that habitat patchiness is sufficient to result in coexistence of competitors that exclude each other at the local (patch) level (Tilman, 1994). Generally, some trade-off is necessary, whereby inferior competitors require dispersal abilities greater than superior competitors for coexistence to be permanent over long time-scales. Although it is conceivable that a superior competitor could have sufficient dispersal ability as to saturate all patches and thus exclude all inferior competitors, no such system has been found in parasitic or free-living systems. Thus, contrary to the original 'paradox of diversity', a focus on spatial patchiness and heterogeneity has resulted in a new paradigm, i.e. diversity is the null hypothesis because patchiness and heterogeneity are ubiquitous. The original non-spatial models of competition (Lotka–Volterra) are not incorrect, they simply apply only at the infra-patch level. Only if the superior competitor is ubiquitous among habitat patches and (at least) dynamically permanent in all patches, do the non-spatial models apply at the "meta"-community level (component community for parasites). Thus, the expectation is that infracommunity-level competition will not be a controlling factor on component communities because of the patchiness of the habitat in which competition occurs. This prediction is supported for all trematode communities in snails that have been studied to date. There is no evidence that interspecific trematode interactions result in exclusion of a trematode species from an ecosystem. This is no doubt due to the overwhelming effect of constant egg input into the systems by wide-ranging definitive hosts and that snails are an abundant and renewable resource.

Competition in individual snails can, however, modify the prevalence of inferior competitors in a snail population. Lafferty *et al.* (1994) estimated that 16% of observed single infections in *C. californica* were actually the result of a resolved coinfection, with species-specific effects ranging from 1·7% to 63·4%, and the most inferior competitors suffering the greatest percentage losses in patch occupancy. However, the overall pattern of abundance relationships among trematode species (community structure) was not seriously altered by competition (Fig. 2). If competition 'scales-up' as a determinant of community structure, it should affect processes (and therefore patterns) at larger scales. It is rare when a trematode component community consists of a superior competitor that also is the most abundant (prevalent) in the assemblage, although there is considerable variation. We know of no case where relative abundances of trematodes in a snail component community are ordered representations of a dominance hierarchy such as that proposed by Kuris (1990) for trematodes in *C. californica*.

Sousa (1993), for instance, found that at Kent Island, California, the two most prevalent trematodes in *C. californica*, *Parorchis acanthus* and *Himasthla rhigedana*, are also the most dominant in the hierarchy. At Pine Gulch Creek, these two species were less prevalent than only the third most competitively dominant trematode *Echinoparyphium* sp. On the other hand, in the study by Lafferty *et al.* (1994) in a different area, *Euhaplorchis californiensis*, fifth in the dominance hierarchy of seven levels, was far and away the most prevalent trematode in *C. californica*, averaging over five sites as 41 times more prevalent that *P. acanthus* (1st in the dominance hierarchy), over 7 times more prevalent that *H. rhigedana* (2nd), and over 10 times more prevalent than *Echinoparyphium* sp. (3rd). *Helisoma anceps* in Charlie's Pond (North Carolina) were most often infected by *Halipegus occidualis*, which is positioned in the middle of the dominance hierarchy of trematodes in this system (Fernandez & Esch, 1991*a*, *b*). The two most competitively dominant trematodes, both echinostomatids, were orders-of-

magnitude less abundant than *H. occidualis*. Similarly, Sapp & Esch (1994) reported that *H. occidualis* occurred at a peak prevalence of 59·5% in *H. anceps*, whereas the remaining species, including the dominant *Echinostoma* sp., were much less prevalent (this echinostome peaked at approximately 6% in prevalence).

It does not appear as though knowledge of a dominance hierarchy produces much predictability among local areas. Prevalent trematodes in one ecosystem may be relatively rare in other ecosystems, apparently independent of competitive ability at the infracommunity level. There may be a trade-off in most local systems between position in the competitive hierarchy and prevalence, thereby ensuring coexistence of trematodes. Whether this trade-off has evolved or is simply happenstance is a matter requiring study. Differences in trematode abundance relationships among systems likely are reflections of the abundance of definitive hosts, intermediate hosts, or both, rather than reflections of competitive ability. Whereas all large-scale patterns and processes necessarily affect smaller-scale patterns and processes, the reverse is not always true. Interspecific interactions can have controlling effects on overall community structure, e.g. keystone predator (Paine, 1966). The interspecific interactions among trematodes in snails, with the results to date, is not one of the systems in which this occurs.

The temporal landscape

The temporal landscape is not fundamentally different from the spatial landscape except in being a single dimension rather than three. The questions essentially are the same, e.g. how is this dimension used by each species in the community and what effect does this pattern of utilization have on patterns and processes at higher and lower levels of organization?

The longevity of snails in a population, coupled with the longevity of trematode infrapopulations, should determine the degree to which a component community is susceptible to seasonal and year-to-year vagaries. The component communities of short-lived hosts (multiple cohorts per season) should be influenced more by processes changing over ecological time-scales than in long-lived hosts (multiple seasons per cohort). This draws an obvious dichotomy between marine and freshwater systems. Marine prosobranchs live for several years to several decades, whereas the majority of freshwater pulmonates live for a few years at most. Freshwater prosobranchs fall somewhere in the middle, having life spans that can approach a decade. In temperate freshwater systems, seasonal variation in the input of trematode eggs to a system is canonized; less is known about the dynamics of egg production and release among adult trematodes in marine vertebrates.

Short-lived hosts might be expected to harbour distinct component communities throughout a single season if the trematodes use definitive hosts with temporally localized breeding and/or activity seasons, e.g. a spring community followed by an autumn community. Twin infection peaks in trematode parasitism have been reported for a number of snails (Probert, 1966; references in Esch & Fernandez, 1994). However, these are largely due to the life history of the snail, i.e. mortality and larval recruitment, although the ephemeral nature of allogenic (those requiring a host outside, and usually ephemeral to, the ecosystem) parasites also is a factor. Snyder & Esch (1993) studied the seasonal dynamics of the component community of *P. gyrina* in Charlie's Pond. Over the summer, *P. gyrina* continually reproduces, such that cohorts are so broadly overlapping that no change in trematode component community composition was observed. Furthermore, most of the definitive hosts in this system are fairly persistent throughout the entire summer; only *Echinostoma* sp., an allogenic trematode, was temporally restricted during the summer. As pointed out previously, Curtis (1996) estimated that the probability of *I. obsoleta*, which lives for decades, recruiting a trematode on the Cape Henlopen sandflat was 1·6% over the summer months. Given that overall prevalence among native snails at Cape Henlopen is around 80%, this value suggests slow accumulation of infections over the 30 or more year life span of this snail and effectively eliminates seasonality as a factor in the temporal landscape.

The probability of local extinction should be one of the most striking factors determining the structure of a component community over time. In general, autogenic trematodes (those completing their life cycles within a single ecosystem) should have lower probabilities of local extinction than allogenic trematodes. Thus, trematodes of migratory birds and wide-ranging mammals may be expected to appear and disappear from local component communities, depending on the movement of their hosts from season to season. Over the course of many years and numerous studies in Charlie's Pond, the trematodes of migratory waterfowl have experienced the greatest local turnover rates (Schotthoefer, 1998). These local extinctions have been associated with the trematode species that otherwise would be the most competitively dominant in the component community of *H. anceps*. In this way, temporal factors are likely to affect the prevalence of interspecific antagonism in infracommunities as well as overall species richness and composition. During the same time period, an autogenic trematode of *P. gyrina* in Charlie's Pond, *Halipegus eccentricus*, also became locally extinct sometime around 1996, and has remained absent from Charlie's Pond since (Schotthoefer, 1998). Autogenic species, which should be more persistent locally, also are less likely to recolonize an ecosystem once local extinction has occurred. Thus, within a regional species pool, and subject to the constraints imposed by local conditions, trematode component communities may be expected to reach a dynamic equilibrium of species richness and composition determined by the values of dispersal and extinction for autogenic and allogenic species (Esch & Fernandez, 1994).

The component communities of long-lived snails should be more resistant to local extinction if trematode infections are rather permanent at the infracommunity level. Even ephemeral visitation by definitive hosts should result in long-term persistence of a trematode in a system where the snail lives for decades and infection with that trematode can persist as long as the snail is alive. Utilizing trematodes in snails as indicator species in long-lived and short-lived snails has variable connotations. A well-done, short-term survey of the trematodes in a long-lived snail should reveal long-term relationships among all the hosts in each trematode life cycle. For instance, Curtis & Tanner (1999) posited that trematode infrapopulations in *I. obsoleta* persist unchanged, on average, for about 10 years. Thus, a survey today of this snail should reflect events that occurred years ago and long-term study of the same snail population may not reveal seasonal differences. In contrast, short-lived hosts can be used as indicators of season-to-season differences in ecosystem-level interactions, e.g. visitation by migratory waterfowl.

The lack of long-term census and quantification of dispersal/extinction probabilities hinder any detailed account of whether the predictions outlined above are valid. The mosaic of hosts that trematodes require for their life cycles makes these parasites valuable indicators of ecosystem-level properties. At the same time, however, this same complexity makes quantification of trematode dynamics very difficult. For instance, the ability to colonize new ecosystems often is described as 'greater' for allogenic trematodes than for autogenic trematodes. Whereas undoubtedly true, this categorization is of little use when trying to place autogenic and allogenic trematodes on a probability continuum. Whether lumping all autogenic parasites and all allogenic parasites into single groups provides the necessary predictive power over longer ecological time scales will require dedicated study over several decades.

REFERENCES

APPLETON, C. C. (1983). Studies on *Austrobilharzia terrigalensis* (Trematoda: Schistosomatidae) in the Swan Estuary, Western Australia: frequency of infection in the intermediate host population. *International Journal for Parasitology* **13**, 51–60.

ATMAR, W. & PATTERSON, B. D. (1993). The measure of order and disorder in the distribution of species in fragmented habitat. *Oecologia* **96**, 373–382.

BARKER, S. C. & CRIBB, T. H. (1993). Sporocysts of *Mesostephanus haliasturis* (Digenea) produce miracidia. *International Journal for Parasitology* **23**, 137–139.

BARGER, M. A. & ESCH, G. W. (2000). *Plagioporus sinitsini* (Digenea: Opecoelidae): A one-host life cycle. *Journal of Parasitology* **86**, 150–152.

BARTOLI, P. & HOLMES, J. C. (1997). A transmission study of two sympatric digeneans: Spatial constraints and solutions. *Journal of the Helminthological Society of Washington* **64**, 169–175.

BOURNS, T. K. R. (1963). Larval trematodes parasitizing *Lymnaea stagnalis appressa* Say in Ontario with emphasis on multiple infections. *Canadian Journal of Zoology* **41**, 937–941.

BOWERS, E. A. (1969). *Cercaria bucephalopsis haimeana* (Lacaze-Duthiers, 1854) (Digenea: Bucephalidae) in the cockle, *Cardium edule* L. in South Wales. *Journal of Natural History* **3**, 409–422.

BROOKS, D. R. & McCLENNAN, D. A. (1993). *Parascript: Parasites and the Language of Evolution*. Smithsonian Institution Press, Washington, D.C.

BURKE, V. J. (1997). The value of descriptive studies in landscape ecology. *US-IALE (U.S. Chapter of the International Association for Landscape Ecology) Newsletter* **13**, 6–8.

BURROUGH, P. A. (1981). Fractal dimensions of landscapes and other environmental data. *Nature* **294**, 240–242.

CANNON, L. R. G. (1979). Ecological observations on *Cerithium moniliferum* Kiener (Gastropoda: Cerithiidae) and its trematode parasites at Heron Island, Great Barrier Reef. *Australian Journal of Marine and Freshwater Research* **30**, 365–374.

CHENG, T. C., SULLIVAN, J. T. & HARRIS, K. R. (1973). Parasitic castration of the marine prosobranch gastropod *Nassarius obsoletus* by sporocysts of *Zoogonus rubellus* (Trematoda): histopathology. *Journal of Invertebrate Pathology* **21**, 183–190.

COMBES, C. (1982). Trematodes: antagonism between species and sterilizing effects on snails in biological control. *Parasitology* **84**, 151–175.

COMBES, C., FOURNIER, A., MONE, H. & THERON, A. (1994). Behaviour in trematode cercariae that enhance parasite transmission: Patterns and Processes. *Parasitology* **109**, S3–S13.

CORT, W. W., McMULLEN, D. B. & BRACKETT, S. (1937). Ecological studies on the cercariae in *Stagnicola emarginata angulata* (Sowerby) in the Douglas Lake region, Michigan. *Journal of Parasitology* **23**, 504–532.

CURTIS, L. A. (1985). The influence of sex and trematode parasites on carrion response of the estuarine snail *Ilyanassa obsoleta*. *Biological Bulletin* **169**, 377–390.

CURTIS, L. A. (1987). Vertical distribution of an estuarine snail altered by a parasite. *Science* **235**, 1509–1511.

CURTIS, L. A. (1990). Parasitism and the movements of intertidal gastropod individuals. *Biological Bulletin* **179**, 105–112.

CURTIS, L. A. (1995). Growth, trematode parasitism, and longevity of a long-lived marine gastropod (*Ilyanassa*

obsoleta). *Journal of the Marine Biological Association United Kingdom* **75**, 913–925.

CURTIS, L. A. (1996). The probability of a marine gastropod being infected by a trematode. *Journal of Parasitology* **82**, 830–833.

CURTIS, L. A. (1997). *Ilyanassa obsoleta* (Gastropoda) as a host for trematodes in Delaware estuaries. *Journal of Parasitology* **83**, 793–803.

CURTIS, L. A. & HUBBARD, K. M. (1990). Trematode infections in a gastropod host misrepresented by observing shed cercariae. *Journal of Experimental Marine Biology and Ecology* **143**, 131–137.

CURTIS, L. A. & HUBBARD, K. M. (1993). Species relationships in a marine gastropod-trematode ecological system. *Biological Bulletin* **184**, 25–35.

CURTIS, L. A. & HURD, L. E. (1983). Age, sex, and parasites: spatial heterogeneity in a sandflat population of *Ilyanassa obsoleta*. *Ecology* **64**, 819–828.

CURTIS, L. A. & TANNER, N. L. (1999). Trematode accumulation by the estuarine gastropod *Ilyanassa obsoleta*. *Journal of Parasitology* **85**, 419–425.

DECOURSEY, P. J. & VERNBERG, W. B. (1974). Double infections of larval trematodes: competitive interactions. In *Symbiosis in the Sea* (ed. Vernberg, W. B.), pp. 93–109. Columbia, S.C., University of South Carolina Press.

ESCH, G. W. & FERNANDEZ, J. (1994). Snail-trematode interactions and parasite community dynamics in aquatic systems: a review. *American Midland Naturalist* **131**, 209–237.

EWERS, W. H. (1960). Multiple infections of trematodes in a snail. *Nature* **186**, 990.

FERNANDEZ, J. & ESCH, G. W. (1991*a*). Guild structure of larval trematodes in the snail *Helisoma anceps*: patterns and processes at the individual host level. *Journal of Parasitology* **77**, 528–539.

FERNANDEZ, J. C. & ESCH, G. W. (1991*b*). The component community structure of larval trematodes in the pulmonate snail *Helisoma anceps*. *Journal of Parasitology* **77**, 540–550.

FRIED, B. & KING, B. (1989). Attraction of *Echinostoma revolutum* cercariae to *Biomphalaria glabrata* dialysate. *Journal of Parasitology* **75**, 55–57.

GANDON, S. & VAN ZANDT, P. A. (1998). Local adaptation and host-parasite interactions. *Trends in Ecology and Evolution* **13**, 214–216.

GAMBINO, J. J. (1959). The seasonal incidence of infection of the snail *Nassarius obsoletus* (Say) with larval trematodes. *Journal of Parasitology* **45**, 440–456.

GARDNER, R. H., MILNE, B. T., TURNER, M. G. & O'NEILL, R. V. (1987). Neutral models for the analysis of broad landscape pattern. *Landscape Ecology* **1**, 19–28.

GOATER, T., SHOSTAK, A. W., WILLIAMS, J. & ESCH, G. W. (1989). A mark–recapture study of trematode parasitism in overwintered *Helisoma anceps* (Pulmonata) with special reference to *Halipegus occidualis* (Hemiuridae). *Journal of Parasitology* **75**, 553–560.

GUSTAFSON, E. J. (1998). Quantifying landscape spatial pattern: what is the state of the art? *Ecosystems* **1**, 143–156.

HARMAN, W. N. (1972). Benthic substrates: their effect on freshwater Mollusca. *Ecology* **53**, 271–277.

HAAS, W. (1994). Physiological analyses of host-finding

behaviour in trematode cercariae: Adaptations for transmission success. *Parasitology* **109**, S15–S29.

HOLMES, J. C. & PRICE, P. W. (1986). Communities of parasites. In *Community Ecology: Patterns and Processes* (ed. Kikkawa, J. & Anderson, D. J.), pp. 187–213. Blackwell Scientific Productions, Oxford.

JAMES, B. L. (1968). The occurrence of *Parvatrema homoeotecnum* James, 1964 (Trematoda: Microphallidae) in a population of *Littorina saxatilis tenebrosa* (Mont.). *Journal of Natural History* **2**, 21–37.

JOKELA, J. & LIVELY, C. M. (1995). Spatial variation in infection by digenetic trematodes in a population of freshwater snails (*Potamopyrgus antipodarum*). *Oecologia* **103**, 509–517.

JOURDANE, J. & MOUNKASSA, J. B. (1986). Topographic shifting of primary sporocysts of *Schistosoma mansoni* in *Biomphalaria pfeifferi* as a result of coinfection with *Echinostoma caproni*. *Journal of Invertebrate Pathology* **48**, 269–274.

KOIE, M. (1969). On the endoparasites of *Buccinum undatum* L. with special reference to the trematodes. *Ophelia* **6**, 251–279.

KURIS, A. (1990). Guild structure of larval trematodes in molluscan hosts: Prevalence, dominance and significance of competition. In *Parasite Communities: Patterns and Processes* (ed. Esch, G. W., Bush, A. & Aho, J.), pp. 69–100. Chapman and Hall, London.

KURIS, A. M. & LAFFERTY, K. D. (1994). Community structure: larval trematodes in snail hosts. *Annual Review of Ecology and Systematics* **25**, 189–217.

LABEAU, M. R. & PETERS, L. E. (1995). *Proterometra autrani* n. sp. (Digenea: Azygiidae) from Michigan's upper penninsula and a key to the species of *Proterometra*. *Journal of Parasitology* **81**, 442–445.

LAFFERTY, K. D., SAMMOND, D. T. & KURIS, A. M. (1994). Analysis of larval trematode communities. *Ecology* **75**, 2275–2285.

LAMAN, T. G., BOSS, N. C. & BLANKESPOOR, H. D. (1984). Depth distribution of seven species of gastropods in Douglas Lake, Michigan. *The Nautilus* **98**, 20–25.

LAUCKNER, G. (1980). Diseases of Mollusca: Gastropoda. In *Diseases of Marine Animals Vol. I* (ed. Kinne, O.), pp. 311–424. John Wiley & Sons, New York.

LIE, K. J. (1973). Larval trematode antagonism: principles and possible application as a control method. *Experimental Parasitology* **33**, 343–349.

LIE, K. J., BASCH, P. F. & UMATHEVY, T. (1965). Antagonism between two species of larval trematodes in the same snail. *Nature* **206**, 422–423.

LIM, H. K. & HEYNEMAN, D. (1972). Intramolluscan inter-trematode antagonism: a review of factors influencing the host-parasite system and its possible role in biological control. *Advances in Parasitology* **10**, 191–268.

MARTIN, W. E. (1955). Seasonal infections of the snail, *Cerithidea californica* Haldeman, with larval trematodes. In *Essays in the Natural Sciences in Honor of Captain Allan Hancock, Allan Hancock Foundation*, pp. 203–210. Los Angeles, University of Southern California Press.

MCDANIEL, J. S. & COGGINS, J. R. (1972). Seasonal larval trematode infection dynamics in *Nassarius obsoletus*

(Say). *Journal of the Elishia Mitchell Scientific Society* **88**, 55–57.

MILNE, B. T. (1988). Measuring the fractal geometry of landscapes. *Applied Mathematics and Computation* **27**, 67–79.

MOUAHID, S. & MONE, H. (1990). Interference of *Echinoparyphium elegans* with the host-parasite system *Bulinus truncatus–Schistosoma bovis* in natural conditions. *Annals of Tropical Medicine and Parasitology* **84**, 341–348.

OKLAND, J. (1983). The factors regulating the distribution of freshwater snails (Gastropoda) in Norway. *Malacologia* **24**, 277–288.

PAINE, R. T. (1966). Food web complexity and species diversity. *American Naturalist* **100**, 65–75.

PAVLOVSKY, E. N. (1966). *Natural Nidality of Transmissable Diseases*. Urbana, University of Illinois Press.

POULIN, R. (1998). *Evolutionary Ecology of Parasites*. London, Chapman and Hall.

PROBERT, A. J. (1966). Studies on the incidence of larval trematodes infecting the freshwater molluscs of Llandorse Lake, South Wales. *Journal of Helminthology* **40**, 115–130.

RAYMOND, K. & PROBERT, A. J. (1991). The daily cercarial emission rhythm of *Schistosoma margrebowiei* with particular reference to dark period stimuli. *Journal of Helminthology* **65**, 159–168.

REVERSAT, J. & SILAN, P. (1991). Comparative population biology of digenes and their 1st intermediate host mollusc – the case of 3 *Helicometra* (Trematoda, Opecoelidae) endoparasites of marine prosobranchs (Gastropoda). *Annals de Parasitologie Humaine et Compare* **66**, 219–225.

ROHDE, K. (1981). Population dynamics of two snail species, *Planaxis sulcatus* and *Cerithium moniliferum* and their trematode species at Heron Island, Great Barrier Reef. *Oecologia* **49**, 344–352.

SAPP, K. K. & ESCH, G. W. (1994). The effects of spatial and temporal heterogeneity as structuring forces for parasite communities in *Helisoma anceps* and *Physa gyrina*. *American Midland Naturalist* **132**, 91–103.

SCHOTTHOEFER, A. M. (1998). Spatial variation in trematode infections and fluctuations in component community composition over the long-term in the snails *Physa gyrina* and *Helisoma anceps*. M.S. Thesis, Department of Biology, Wake Forest University, Winston-Salem, North Carolina.

SINDERMANN, C. J. (1960). Ecological studies of marine dermatitis-producing schistosome larvae in northern New England. *Ecology* **41**, 678–684.

SNYDER, S. D. & ESCH, G. W. (1993). Trematode community structure in the pulmonate snail *Physa gyrina*. *Journal of Parasitology* **79**, 205–215.

SOKAL, R. R. & ODEN, N. L. (1978). Spatial autocorrelation in biology 2: some biological implications and four examples of evolutionary and ecological interest. *Biological Journal of the Linnaean Society* **10**, 229–249.

SOUSA, W. P. (1990). Spatial scale and the processes structuring a guild of larval trematode parasites. In *Parasite Communities: Patterns and Processes* (ed. Esch, G. W., Bush, A. & Aho, J.), pp. 41–67. London, Chapman and Hall.

SOUSA, W. P. (1992). Interspecific interactions among larval trematode parasites of freshwater and marine snails. *American Zoologist* **32**, 583–592.

SOUSA, W. P. (1993). Interspecific antagonism and species coexistence in a diverse guild of larval trematode parasites. *Ecological Monographs* **63**, 103–128.

SOUSA, W. P. (1994). Patterns and processes in communities of helminth parasites. *Trends in Ecology and Evolution* **9**, 52–57.

SUKHDEO, M. R. & METTRICK, D. F. (1987). Parasite behavior: understanding platyhelminth responses. *Advances in Parasitology* **26**, 73–144.

THRELFALL, W. & GOUDIE, R. I. (1977). Larval trematodes in the rough periwinkle, *Littorina saxatilis* (Olivi), from Newfoundland. *Proceedings of the Helminthological Society of Washington* **44**, 229–232.

TILMAN, D. (1994). Competition and biodiversity in spatially structured habitats. *Ecology* **75**, 2–16.

TURNER, M. G. (1989). Landscape ecology: the effect of pattern on process. *Annual Review of Ecology and Systematics* **20**, 171–197.

VAES, M. (1979). Multiple infection of *Hydrobia stagnorum* with larval trematodes. *Annals de Parasitologie* **54**, 303–312.

VERNBERG, W. B., VERNBERG, F. J. & BECKERDITE, F. W. (1969). Larval trematodes: double infections in common mud-flat snail. *Science* **164**, 1287–1288.

WALKER, J. C. (1979). *Austrobilharzia terrigalensis*: a schistosome dominant in interspecific interactions in the molluscan host. *International Journal for Parasitology* **9**, 137–140.

WILLIAMS, J. A. & ESCH, G. W. (1991). Infra- and component community dynamics in the pulmonate snail *Helisoma anceps*, with special emphasis on the hemiurid trematode, *Halipegus occidualis*. *Journal of Parasitology* **77**, 246–253.

WITH, K. A. (1994). Using fractal analysis to assess how species perceive landscape structure. *Landscape Ecology* **9**, 25–36.

WRIGHT, C. A. (1959). Host location by trematode miracidia. *Annals of Tropical Medicine and Parasitology* **53**, 288–292.

WRIGHT, C. A. (1973). *Flukes and Snails*. New York, Hafner Press.

YOSHINO, T. P. (1975). A seasonal and histologic study of larval Digenea infecting *Cerithidea californica* from Goleta Slough, Santa Barbara County, California. *Velliger* **18**, 156–161.

Seasonality in the transmission of schistosomiasis and in populations of its snail intermediate hosts in and around a sugar irrigation scheme at Richard Toll, Senegal

R. F. STURROCK[1], O.-T. DIAW[2], I. TALLA[3], M. NIANG[3], J.-P. PIAU[3] and A. CAPRON[4]

[1] Department of Infectious and Tropical Diseases, London School of Hygiene & Tropical Medicine, Keppel Street, London WC1E 7HT, UK
[2] Service de Parasitologie, (L.N.E.R.V.), Institut Sénégalais de Réserches Agricoles, BP 2057, Dakar, Sénégal
[3] Programme ESPOIR, District Médicale de Richard Toll, BP 394, Richard Toll, Sénégal
[4] Institut Pasteur de Lille, 1 Rue du Professor Calmette, 59000 Lille CEDEX, France

SUMMARY

Irrigation for intensive sugar cultivation started in the early 1980s at Richard Toll, some 100 km from the mouth of the Senegal River. Infections with *Schistosoma mansoni* were first seen in late 1988. This study records quantitative snail surveys for over 3 years from 1992 at sites representing different habitats in and around the irrigation scheme. Populations of both *Biomphalaria pfeifferi* (the intermediate host of *S. mansoni*) and *Bulinus* spp. (mainly *B. truncatus*, the local host of *S. bovis*) peaked in late 'spring' or early 'summer', depending on the habitat, and then remained low until the following 'spring'. *B. pfeifferi* favoured smaller, man-made habitats with most transmission between May and August each year. The less abundant *Bulinus* spp. favoured larger natural and man-made habitats with most *S. bovis* transmission between April and July. *S. mansoni* infections were more, but *S. bovis* infections were less abundant than other trematodes in their respective snail hosts. Ecological changes in the early 1980s due to sugar irrigation pre-dated similar, more widespread changes in the late 1980s when the completion of dams across the Senegal River prevented seasonal rain fed floods and sea water intrusion. *S. mansoni* has since spread rapidly around Richard Toll. The incompatibility of the local *S. haematobium* strains with the dominant bulinid snails has so far prevented an epidemic of urinary schistosomiasis at Richard Toll, but the invasion of similar downstream habitats by susceptible *B. globosus* is worrying. The principal control measure, chemotherapy, given in the 'winter' would minimise the rate of reinfection. It could be reinforced by judicious mollusciciding within the sugar irrigation scheme but not elsewhere.

Key words: Schistosomiasis, *S. mansoni*, *S. haematobium*, *S. bovis*, *Biomphalaria pfeifferi*, *Bulinus* spp., population biology, transmission, temperature, seasonality.

INTRODUCTION

Human surveys along the middle and lower reaches of the Senegal River Basin reported isolated *Schistosoma haematobium* foci but no *S. mansoni* in the 1970s and early 1980s (Jobin, Negron-Aponte & Michelson, 1976; Monjour *et al.* 1981; Chaine & Malek, 1983; Vercruysse, Southgate & Rollinson, 1985; Malek & Chaine, 1989). *Bulinus* spp. transmitted *S. haematobium* focally throughout the region and animal schistosomes – predominantly *S. bovis* and, possibly, *S. curassoni* (Rollinson *et al.* 1990; Vercruysse *et al.* 1994). Only two small, isolated colonies of *Biomphalaria pfeifferi* were detected (Chaine & Malek, 1983). *S. mansoni* was first reported at Richard Toll in 1988 (Talla *et al.* 1990). As diagnostic facilities improved, the number of cases increased rapidly to epidemic proportions (Talla, Kongs & Verlé, 1992). The epidemic had been preceded by significant agricultural changes in the region and the construction of the Manantali Dam, upstream in Mali, and the Diama Dam near the mouth of the Senegal River.

Traditional agriculture along the middle and lower reaches of the river had relied on seasonal floods (Diop *et al.* 1994), similar to basin irrigation in Egypt (Abdel-Waheb, 1982). Silt-laden flood water from seasonal rains upstream reached the area between July and September each year, giving high water levels with fast flows. Water levels dropped as the flood passed and, when the flow eventually stopped, sea water intruded more than 100 km from the river mouth causing the saline soil conditions and water unsuitable for irrigating many crops.

Large-scale irrigation was introduced for commercial sugar cultivation at Richard Toll in the early 1980s. Fresh flood water was diverted through the newly constructed Canal Taouey to the natural Lac de Guiers, south of Richard Toll, and released later to irrigate some 10000 hectares of sugar cane. Completion of the lower dam prevented sea water intrusion after 1985, making water available throughout the year to irrigate salt-sensitive crops outside the sugar scheme. Water in the upper dam finally reached spillway level in 1991, severely altering the annual flood cycle, rendering ineffective traditional irrigation and creating a need for modern irrigation schemes, especially for rice (Diop *et al.* 1994).

Parasitology (2001), **123**, S77–S89.   © 2001 Cambridge University Press
DOI: 10.1017/S0031182001008125

These agricultural and hydrological changes were accompanied by major agro-industrial developments at Richard Toll. Massive immigration increased the population at least five-fold to ∼ 50000 by 1990 with no parallel infrastructure improvement, particularly of sanitation, water supplies and housing (Handschumacher *et al.* 1992, 1994). An influx of refugees from Mauritania in 1989/1990 aggravated the situation. The net result of all these factors was a substantial change in the ecology of the lower Senegal River Valley favouring *S. mansoni* transmission around Richard Toll. Control has so far relied entirely on praziquantel treatment variously for symptomatic cases, for seasonal sugar workers leaving the region, and for systematic treatment in some districts of the town and its environs (Talla *et al.* 1992; Gryseels *et al.* 1994; Stclma *et al.* 1995; Picquet *et al.* 1996, 1998; Ernould, Ba & Sellin, 1999 *a*).

Diaw *et al.* (1990, 1991) reported large *B. pfeifferi* populations with high *S. mansoni* infection rates in a cross-sectional survey of different habitats around Richard Toll. Belot, Geerts & Diouf (1993) reported much lower infection rates from an incomplete, one-year longitudinal survey of irrigation canal sites outside the town boundaries in 1989/90 and concluded that transmission was continuous with little evidence of seasonality. Deme (1993), continuing surveys in similar sites in 1991/92, found rather more seasonal variation.

This paper reports the findings of fortnightly snail surveys of a variety of habitats from May 1992 to July 1995 at sites selected mostly for extensive human water contact. The object was to detect any seasonality of *S. mansoni* transmission by *B. pfeifferi*, with simultaneous observations on *Bulinus* spp., mainly *B. truncatus* (Syn. *B. truncatus rohlfsi* – Brown, 1994).

MATERIALS AND METHODS

Study area

Richard Toll (16·38° N; 15·78° W; alt. < 10 m) lies on the southern bank of the Senegal River 100 km east of its mouth on the Atlantic Ocean (Talla *et al.* 1990; Handschumacher *et al.* 1994). Briefly, the town is T-shaped with 7 main districts linked by two minor districts to the village of Ndombo to the south. The Senegal River forms the northern boundary across the T. The new, man-made Canal Taouey cuts south through the natural, meandering Marigot Taouey down the stem of the T, from the river to the Lac de Guiers.

Open canals and drains from the sugar irrigation scheme run through or beside most districts and provide the only reliable source of domestic water for many people. Water is siphoned into some house-holds, but many people continue to collect it by hand and to have direct contact with sites on the margins of natural and man-made water bodies in and around the town for washing clothes, personal washing, bathing, swimming, recreation, fishing, and washing cars and domestic animals. Many houses have some rudimentary sanitation (Handschumacher *et al.* 1994) but there is abundant evidence of promiscuous defaecation at almost every water contact site. Women of all ages openly use these sites despite a strong Mohammedan influence.

Climate

Senegal lies in the Sahel Zone but the cold Canary Current cools its coast. Winds off the Sahara cause extensive dust and sand storms between April and June. There is a limited rainy season between July and September when thunderstorms give localised but unreliable rainfall. Relative humidity is low for the most of the year. Mean air temperatures at Richard Toll are several degrees higher than at the coast, ranging from < 20 °C in January to > 40 °C in September, with a diurnal range of 5–20 °C depending on the cloud or dust cover.

Sugar irrigation

Water was lifted by pumps at the northern end of the Canal Taouey into two open, elevated main canals to two irrigation areas: one area to the east of the town, the other to the west and extending south to the northern shore of the Lac de Guiers. The sugar refinery draws water from the western main canal, the outfall of which is used for rice irrigation beyond the sugar scheme boundary. Water is fed by gravity from the main canals to a network of secondary and tertiary canals whence it is siphoned through plastic pipes into furrows in the individual fields. Excess water is led by porous, sub-surface pipes to deep, open main drains to be pumped either into the river or the Lac de Guiers, or back into the main canals if its salinity is low enough. Concrete junction boxes control water distribution throughout the irrigation network. The earth-lined canals and drains have to be cleared regularly (by hand, mechanically or with herbicides) to prevent aquatic and intrusive marginal vegetation impeding water flows.

The larger canals flow continuously except in the rare periods of heavy rainfall but minor canals are filled only as needed in each 12–14 day irrigation cycle, and surface water remains on the fields for only one or two days. (Excess water persists longer to leach out unwanted salts during reclamation of saline land.) Fields are cultivated and replanted on a three year cycle. Different sugar cane cultivars, ripening at different rates, are grown in blocks in each field to provide a continuous supply of cane for the refinery except in the rainy season when it closes for maintenance. Consequently, secondary canals are rarely dry for more than 10–12 days and residual pools in the junction boxes provide refuges for snails between irrigation cycles.

Selection of snail sampling sites and sampling procedures

Snail sites were chosen mostly for easy access within the town on the basis of evidence of extensive human water contact and the presence of snails, including infected snails, during pilot surveys. Some remote sites, less subject to human interference, were also included to represent the conditions within the sugar fields. Six sites were included at Ndombo to give transmission information for a separate study (Gryseels *et al.* 1994). Sampling began in May 1992 at 24 sites (2 river, 3 Marigot, 5 Canal Taouey, 6 main canal, 7 secondary canal and 1 drain). Eight more were added in 1994 (2 Canal Taouey, 5 Marigot and 1 secondary canal).

At each site, a marked length of the bank about 25 m long was searched for 10 min by two technicians using snail scoops (Olivier & Schneiderman, 1956; Ouma *et al.* 1989). *B. pfeifferi* and *Bulinus* spp. were placed in labelled plastic containers. Prevailing site conditions were noted and sub-surface water temperature measured at the end of each search. (N.B. At Ndombo the searches were extended to 20 min to increase snail recoveries).

Sites were sampled fortnightly between 07·00 and 10·00 h: 1–8 on day 1, 9–16 on day 2, 17–24 on day 3 and 25–32 on day 4. Snails were taken to a field laboratory, washed in clean water and exposed individually in glass tubes in about 10 ml of filtered water. A 30–40 min outdoor exposure to indirect sunlight preceded exposure to a low wattage electric lamp in an air-conditioned laboratory. Cercariae were detected by eye and their identity confirmed using a low power dissecting microscope. Snails were kept overnight for a final examination and then destroyed.

Cercariae were classified as either schistosome (furcocercous) cercariae or cercariae of other trematodes. Several isolates of furcocercous cercariae from *B. pfeifferi* were passaged through mice to confirm the identification as *S. mansoni* (Sène, 1994; Fallon *et al.* 1995). This process failed with furcocercous cercariae from *Bulinus* spp. which were assumed to have been predominantly *S. bovis* as most were from *B. truncatus*: the few (< 1%) from the rarely collected *B. forskalii* and *B. globosus* may have been *S. bovis* or *S. haematobium* respectively (Vercruysse *et al.* 1985; Diaw *et al.* 1990; Rollinson *et al.* 1990). Amphistome cercariae and xiphidocercariae were also shed by both *B. pfeifferi* and *Bulinus* spp. but it was not possible to identify them further.

Data handling and analysis

After each collection, data from individual site field forms were entered onto computer spreadsheets for analysis using standard statistical packages. Snail counts were transformed to $\log_{10}(x+1)$ and percentages to angles for the full data from 85 collections at 24 sites. Because the data were unbalanced with respect to the number of samples per habitat, only one-way analysis of variance with available statistics packages was possible, giving an excessive number of degrees of freedom (df) for the error term. To explore interactions between the two main factors (habitat and collection time) by two-way analysis of variance, the full data were simplified into a balanced summary table (Sturrock *et al.* 1994; Fulford *et al.* 1996). The original data sets were divided into two subsets comprising alternate collections (time samples) with the collection numbers coded by the month in which they were made. A summary table was then derived comprising 144 replicated observations (2×12 months $\times 6$ habitats). Because both analyses gave essentially the same results (with one exception – see below), for main factor effects, only analyses of the summary data are reported in the text, and habitat $\times$ time interactions only where they were significant at $P < 0.05$.

Mean monthly schistosome-to-total trematode infection ratios were calculated from the summary data (Sturrock *et al.* 1994). The generally piecemeal chemotherapy campaigns during this study were unlikely to have affected these ratios at Richard Toll so they will form a base-line for any changes due to future, systematic campaigns.

Monthly transmission potential indices (MTPI) were also derived from the summary tables (Klumpp, 1982). The mean number of infected snails collected per site in a month was expressed as a proportion of annual mean totals to give a general picture of transmission within a calendar year. For analysis, the logit transformation was used to normalise the data, adding or subtracting respectively 0·000001 from extreme values of 0 and 1 to bring them within the permitted limits of > 0 to < 1. (N.B. The sum of 12 untransformed monthly proportions equals 1 for each habitat but the sum of the transformed monthly proportions does not. Habitats with continuous, low infection rates score less than those with intermittent, but high, infection rates.)

RESULTS

Table 1 summarises the full data from the 85 fortnightly collections as means by habitat. Also shown are the results of one-way analyses of variance on the full data for differences between habitats. Fig. 1 illustrates the fortnightly collection means for temperature and for both *B. pfeifferi* and *Bulinus*

Table 1. *Geometric means (GM) by habitat of water temperatures and recoveries of B. pfeifferi and Bulinus spp.: total and infected with schistosomes or other trematodes*

Habitat	Senegal River	Canal Taouey	Primary canal	Secondary canal	Drain	Marigot Taouey	Overall
No. of sites	2	5	6	7	1	2	24
No. of collections	170	425	510	595	85	255	2040
(a) Water temperature							
Mean	22·30	23·36	23·11	22·51	23·23	23·56	22·99 ns
(SD)	(6·0)	(6·1)	(5·9)	(7·0)	(6·2)	(6·1)	(6·3)
(b) *B.pfefferi* infected with *S. mansoni*							
Log(x+1)	0·061	0·203	0·377	0·220	0·412	0·230	0·252***
(SD)	(0·19)	(0·37)	(0·55)	(0·41)	(0·53)	(0·23)	(0·44)
GM	0·15	0·60	1·38	0·66	1·59	0·70	0·79
(c) *B. pfeifferi* with other trematode infections							
Log(x+1)	0·028	0·247	0·269	0·209	0·288	0·085	0·174***
(SD)	(0·13)	(0·26)	(0·48)	(0·38)	(0·41)	(0·21)	(0·36)
GM	0·07	0·26	0·86	0·62	0·94	0·22	0·49
(d) Total *B. pfeifferi* collected							
Log (x+1)	0·148	0·332	0·591	0·574	0·831	0·366	0·490***
(SD)	(0·31)	(0·46)	(0·69)	(0·71)	(0·63)	(0·47)	(0·62)
GM	0·41	1·15	2·90	2·75	5·78	1·33	2·09
(e) *Bulinus* infected with *S. bovis*							
Log(x+1)	0·039	0·062	0·058	0·026	0·057	0·024	0·044***
(SD)	(0·17)	(0·19)	(0·19)	(0·14)	(0·18)	(0·10)	(0·16)
GM	0·10	0·14	0·14	0·06	0·14	0·06	0·11
(f) *Bulinus* with other trematode infections							
Log(x+1)	0·075	0·075	0·074	0·049	0·096	0·027	0·062**
(SD)	(0·19)	(0·20)	(0·20)	(0·19)	(0·22)	(0·10)	(1·19)
GM	0·19	0·14	0·19	0·12	0·25	0·06	0·16
(g) Total *Bulinus* collected							
Log(x+1)	0·558	0·550	0·327	0·255	0·538	0·330	0·383***
(SD)	(0·46)	(0·44)	(0·45)	(0·42)	(0·52)	(0·37)	(0·43)
GM	2·45	2·55	1·13	0·80	2·45	1·14	1·42

Log(x+1) = mean of raw counts (x) transformed to $\log_{10}(x+1)$.
(SD) = Standard deviation.
ns, **, ***: P for one-way analysis of variance of habitat means > 0·05, 0·01 and < 0·001 respectively.

spp. and their respective trematode infections. Years 1, 2 and 3 start in May 1992, 1993 and 1994, respectively. For descriptive purposes, it is convenient to use 'seasons' normally associated with northern latitudes when discussing these results, even though Richard Toll lies within the tropics.

Water temperature

Mean water temperatures (Table 1(a)) for the full data from the first 76 surveys of the initial 24 sites (no measurements were possible for the last 9 collections) did not differ by habitat from analysis of the full data ($F_{5,1776df} = 1·7$, $P > 0·05$). In contrast, using the summary data suggested that the river and the secondary canals were significantly cooler than the rest ($F_{5,72df} = 6·2$; $P < 0·001$). However, this habitat difference may be an artifact due to bias in the measurement of water temperatures: river sites were always sampled early in the morning, before they were warmed by the sun, and many secondary canals were dry in the heat of the summer.

Water temperature (Fig. 1A) showed a clear seasonal variation peaking in the summer/autumn 'flood' season (August to November), when the river

was full of warm water from inland, and were minimal in mid-winter and early spring (January to April). There was a significant difference by month ($F_{11,72} = 247·6$; $P < 0·001$).

Basic B. pfeifferi results

Total snails. Overall geometric mean (GM) *B. pfeifferi* recoveries varied considerably by habitat (Table 1(d)): primary canals, secondary canals and the drain yielded substantially more snails than the Canal Taouey and Marigot, and the river yielded even less ($F_{5,72} = 84·0$; $P < 0·001$). GM recoveries also showed clear seasonal variations (Fig. 1C). Peak numbers varied between years with summer populations (May to August) always greater than the rest ($F_{11,72} = 28·8$; $P < 0·001$) after rapid population growth in spring and early summer, followed by a decline when high mid-summer temperatures set in. Minimal populations persisted through the winter before the cycle was repeated. A significant month × habitat interaction (data not shown) was due mainly to river populations peaking rather earlier than the rest and disappearing through much of the winter ($F_{55,72} = 2·8$; $P < 0·001$).

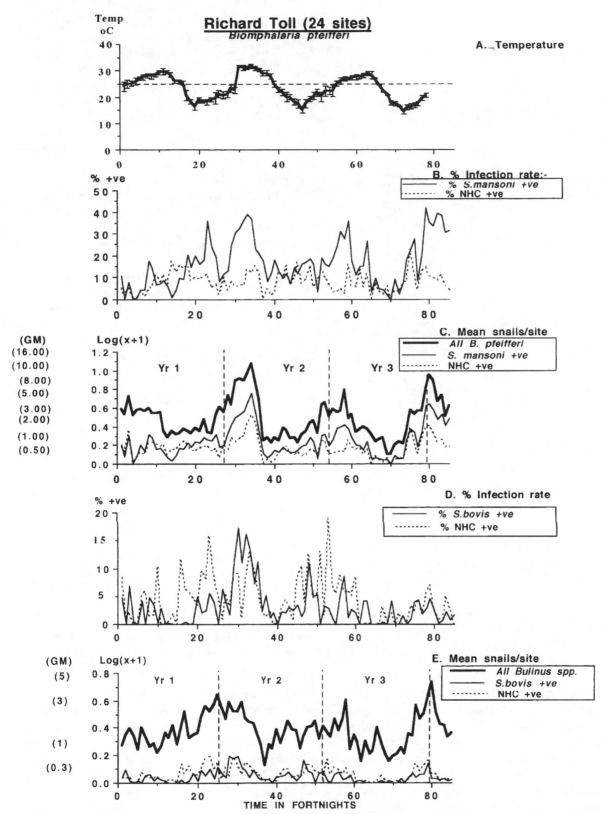

Fig. 1. Fortnightly means for 85 collections starting in May 1992 from all 24 original sites. A. water temperature in °C with standard deviations; B. arithmetic mean infection rates of *B. pfeifferi* with *S. mansoni* and other (non human) trematodes (NHC); C. geometric mean recoveries of *B. pfeifferi*: total and infected with *S. mansoni* or other trematodes; D. arithmetic mean infection rates of *Bulinus* spp. with *S. bovis* and NHC; and E. geometric mean recoveries of *Bulinus* spp.: total and infected with *S. bovis* or other trematodes.

S. mansoni-*infected* B. pfeifferi. *S. mansoni* infection rates mirrored total *B. pfeifferi* numbers (Fig. 1B) with peak rates in mid-summer dropping to a trough in mid-winter. The differences between months were significant ($F_{11,72} = 4.42$, $P < 0.01$). The same applied to habitats (data not shown) with

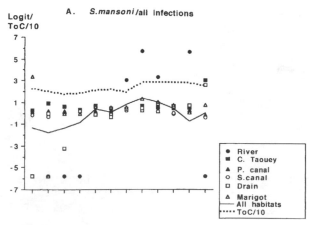

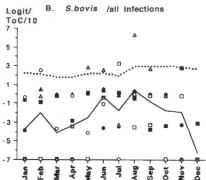

Fig. 2. Schistosome/(schistosome + other trematode) ratios from summary data plotted as logits by month and habitat. A. *S. mansoni* in *B. pfeifferi* and B. *S. bovis* in *Bulinus* spp. The solid line is the mean monthly ratio over all six habitats; the broken line is the mean, monthly water temperature in °C × 10⁻¹.

the highest rates in the primary canals, Canal Taouey and secondary canals ($F_{5,72} = 11\cdot95$, $P < 0\cdot001$). A significant interaction between these factors on *S. mansoni* infection rates was due to peak rates in the river occurring earlier than in other sites ($F_{55,72} = 2\cdot23$, $P < 0\cdot01$). Infection rates are sensitive to small sample size, even when transformed to angles, and the number of infected snails collected can be a more reliable indicator of transmission intensity.

There were significant differences in *S. mansoni*-infected snail numbers by habitat (Table 1(b)): primary canals and the drain yielded many more than the Canal Taouey, Marigot and secondary canal, with even less from the river ($F_{5,72} = 55\cdot4$, $P < 0\cdot001$). Variations over time were also significant ($F_{11,72} = 320\cdot8$, $P < 0\cdot001$). Infected snails were far more abundant in summer than in winter, with intermediate values in spring and autumn. The month × habitat interaction was significant ($F_{55,72} = 2\cdot9$, $P < 0\cdot001$). Again, this was due to the earlier summer peak in the river from which snails all but disappeared in winter. *S. mansoni* infections were relatively scarce in the secondary canals, even in July and August, despite large snail populations in 1993 and 1994. *B. pfeifferi* occurred irregularly in the drain but had high infection rates if present in mid-summer.

B. pfeifferi *infected with other trematodes*. *B. pfeifferi* non-human cercarial (NHC) infection rates were like those of *S. mansoni*, (Fig. 1B), but were generally lower and the pattern during the year was a little different ($F_{11,72} = 4\cdot32$, $P < 0\cdot01$). Besides a late spring/early summer peak there was another in the autumn. There was a significant habitat effect (data not shown), with the highest rates in the primary canals, Canal Taouey and secondary canals ($F_{5,72} = 26\cdot07$, $P < 0\cdot001$).

Considering the actual numbers of NHC-infected *B. pfeifferi*, they were usually less common than *S. mansoni*-infections (Table 1(c)), especially in the summer months (Fig. 1C). They varied significantly by month with more infected snails in early summer and autumn than in spring and winter ($F_{11,72} = 10\cdot1$, $P < 0\cdot001$). Their abundance also varied by habitat with significantly more from primary canal, secondary canals and the drain than from the other sites ($F_{5,72} = 66\cdot7$, $P < 0\cdot001$).

Basic Bulinus *spp. results*

Total snails. The data for *Bulinus* spp. are summarised in Table 1(g) and Fig. 1E. Cyclical fluctuations over time resembled *B. pfeifferi*, except that both the 'spring' build up and the mid-summer decline generally began rather earlier, especially in year 2, the hottest year. The normally lower abundance of *Bulinus* spp. compared with *B. pfeifferi* is clear from Figs 1C and 1E, even at peak populations. Variation in snail numbers by month was significant ($F_{11,72} = 6\cdot17$, $P < 0\cdot001$). So, too, was variation in abundance by habitat (Table 1(g)) with snails significantly more abundant in the river and Canal Taouey than in secondary canals and the Marigot ($F_{5,72} = 14\cdot56$; $P < 0\cdot001$). The month × habitat interaction (data not shown) was significant ($F_{55,72} = 3\cdot23$; $P < 0\cdot001$). Spring population peaks were not discernible in the primary canals and Marigot, and autumn/winter peaks did not occur in the drain or secondary canals.

S. bovis *infections of* Bulinus *spp*. *S. bovis* infection rates varied during the year (Fig. 1D), peaking in mid-summer but dropping to a trough in mid-winter, and by habitat (Table 1(e)). Infection rates were higher in summer than in winter months ($F_{11,72} = 4\cdot32$, $P < 0\cdot01$). Rates were highest in the primary canals, Canal Taouey and secondary canals ($F_{5,72} = 8\cdot06$, $P < 0\cdot001$).

Unlike *S. mansoni*, *S. bovis* infected snails were less abundant than other trematode infections in most habitats (Table 1(c) and 1(f)). The habitat pattern differed slightly compared with the infection rates, with higher recoveries from the Canal Taouey, primary canal and drain than from the Marigot and secondary canals ($F_{5,72} = 2\cdot6$, $P < 0\cdot05$). The numbers varied significantly by month with peak recoveries in July ($F_{11,55} = 3\cdot95$; $P < 0\cdot001$).

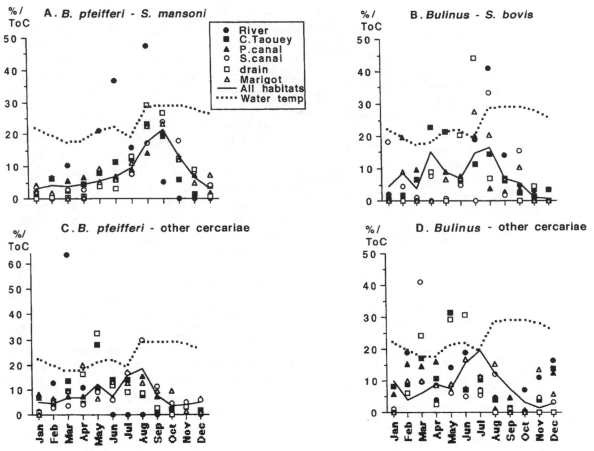

Fig. 3. Monthly transmission potential indices (MTPI) from summary data plotted as percentages by month and habitat. A. *S. mansoni* in *B. pfeifferi*; B. *S. bovis* in *Bulinus* spp. C. other trematodes (NHC) in *B. pfeifferi* and D. other trematodes in *Bulinus* spp. The solid line is the mean MTPI over all six habitats; the broken line is the mean, monthly water temperature in °C.

Other trematode infections of Bulinus *spp.* Other trematode infection rates in *Bulinus* spp. varied during the year (Fig. 1D), with rather more pronounced peaks in late spring/early summer than for NHC infected *B. pfeifferi*, but with a similar mid-winter trough ($F_{11,72} = 2.54$, $P < 0.01$). There were significant differences by habitat (data not shown), with the highest rates from the primary canals, Canal Taouey and secondary canals ($F_{5,72} = 7.04$, $P < 0.001$).

GM recoveries of snails infected with other trematodes (Table 1 (f)) showed significant effects by habitat: infections were least abundant in the Marigot ($F_{5,72} = 3.44$, $P < 0.05$). They were significantly higher (Fig. 1E) in mid-summer than in the winter, with intermediate values in the spring and autumn ($F_{11,72} = 4.21$, $P < 0.001$). The summer peaks occurred rather later than those of *S. bovis*.

Relative infection rates

S. mansoni *and other trematode infections.* S. *mansoni* infections accounted for over 60% of all *B. pfeifferi* trematode infections but the ratio *S. mansoni*/all infections varied during the year and by habitat (Fig. 2A, where equal proportions give a

value of 0 on the logit scale). The difference between months was significant. *S. mansoni* predominated in the summer but was relatively scarcer in the winter ($F_{11,72} = 2.7$, $P < 0.01$). The river rates were higher overall than those from other habitats ($F_{5,72} = 3.9$, $P < 0.01$) and the significant month × habitat interaction ($F_{55,72} = 1.9$, $P < 0.05$) was due to the much greater difference between summer and winter recoveries from the river compared with other habitats.

S. bovis *and other trematode infections.* S. *bovis* accounted for less than 40% of all trematode infections detected in *Bulinus* spp. (*ct. S. mansoni* in *B. pfeifferi*). Overall *S. bovis*/all trematode ratios (Fig. 2B) showed an annual pattern similar to *S. mansoni* (Fig. 2A), peaking in early summer when they briefly exceeded zero (in the logit scale), and dropping in winter. There were significant differences by month ($F_{11,72} = 2.8$, $P < 0.01$) and by habitat ($F_{5,72} = 3.6$, $P < 0.01$). Ratios in the Canal Taouey and primary canals tended to be higher overall than those of the river and drain, with exceptionally high summer peaks in the Canal Taouey, the drain and the Marigot.

Monthly transmission potential indices (MTPI)

For B. pfeifferi. The mean MTPI for *S. mansoni* (Fig. 3A) showed a summer peak and nearly 60% of the transmission occurred between May and September. The difference between months was significant ($F_{11,72} = 8.51$, $P < 0.001$). So, too, was the pattern for different habitats in the angular transformation ($F_{5,72} = 22.62$; $P < 0.001$): the values were higher for the river than for other habitats. The month $\times$ habitat interaction was significant ($F_{55,72} = 2.04$; $P < 0.01$). In the river and the drain, the proportion infected was high when snails were present but snails were often absent for long periods.

The MTPI for NHC in *B. pfeifferi* (Fig. 3C) showed a minor peak in May as well as the main summer peak in August. Over half the annual transmission occurred between May and August. The July and August indices were greater than those from October to February ($F_{11,72} = 3.53$; $P < 0.001$). The river MTPI were significantly less than from other habitats ($F_{5,72} = 15.40$; $P < 0.001$) in most months.

For Bulinus *spp.* The MTPI of *S. bovis* (Fig. 3B) had a minor peak in April as well as the main June/July summer transmission peak. These peaks contributed nearly two thirds of the annual transmission. The differences were significant by month: the April, July and August values were higher than those of other months ($F_{11,72} = 3.88$; $P < 0.001$). The habitat effect was also significant. The river and drain indices were less than those from the Canal Taouey and the primary canal ($F_{5,72} = 8.12$; $P < 0.001$). There was a significant month $\times$ habitat interaction: spring peaks occurred in primary canals and Canal Taouey, and autumn peaks in the river and secondary canals ($F_{55,72} = 2.14$; $P < 0.01$).

For other trematodes in *Bulinus* spp., the MTPI peaked in July (Fig. 3D) with over 60% of the transmission in spring and early summer (June to August) when values were higher than those from September to April ($F_{11,72} = 6.91$; $P < 0.001$). The drain was significantly less important than other habitats ($F_{5,72} = 7.81$; $P < 0.001$).

DISCUSSION

In this study, snails collected were not returned after examination for cercial infections. Such repeated depletion may eventually reduce the field snail populations in small, closed habitats. However, this was not a problem at Richard Toll where the habitats were large and open.

One-way analysis of variance on the full data agreed (with one exception) with two-way analyses of the summary data on the main factor (month and habitat) effects, giving confidence in the significance of interactions detected by the latter, where month was a surrogate for water temperatures because of the annual temperature cycle.

Belot *et al.* (1993) concluded that *S. mansoni* transmission occurred throughout the year at Richard Toll, but this does not preclude seasonality in its intensity which could be caused just by variations in *B. pfeifferi* abundance. This study shows such annual cycles in snail numbers and the intensity of both *S. mansoni* and *S. bovis* transmission.

The seasonal cycle in the numbers of *B. pfeifferi* and *Bulinus* spp. at Richard Toll resembled that in Egyptian irrigation schemes (Dazo, Hairston & Dawood, 1966): all snail populations grew rapidly in spring/early summer, dropped sharply in the heat of the summer and remained relatively low until the following spring. However, this pattern was modified by habitat, a phenomenon also noted more recently by Shaw *et al.* (1999). *B. pfeifferi* was more abundant in smaller, man-made habitats (primary and secondary canals) but peaked later in the larger, natural (river and Marigot) and man-made (Canal Taouey) habitats where wave action and/or substantial flows restricted snails to sheltered microhabitats among vegetation. *Bulinus* spp. preferred larger, natural habitats where their populations peaked earlier than *B. pfeifferi* with, sometimes, a secondary, autumn peak.

The effects of temperature on snail survival and reproduction are well known (Appleton, 1977; Dazo *et al.* 1966; Shiff, 1964; Sturrock, 1966; Sturrock & Sturrock, 1972) as, too, are its effects on the rate of development of intramolluscan stages of schistosomes (Foster, 1964; Pflüger, 1981). Water temperature (month), habitat type and their interaction also affects the behaviour of the definitive hosts of the schistosomes, influencing the type, degree and frequency of contact with and contamination of transmission sites by humans as well as animals transmitting non-human schistosomes and other snail-borne trematodes.

Fish, amphibians, reptiles, birds and rodents were seen at most water bodies in and around Richard Toll. Cattle, goats and horses had unrestricted access to the river, Marigot and Canal Taouey where *S. bovis* infections were most common, but were discouraged (not always successfully) from canals and drains to prevent damage to their earthen banks. Almost every accessible water body was also utilised by people throughout the year, although casual observation suggested more prolonged contact in the hotter, summer months, especially among children.

For most of the year, *S. mansoni* was more common than other trematodes in *B. pfeifferi*, with a single transmission peak between May and August depending on the habitat. Nearly 80% of all *S. mansoni*-infected snails were found in this period; mostly from the Canal Taouey and the primary canals with only a few from the river. In contrast, *S.*

% annual transmission by month, and possible extrinsic factors

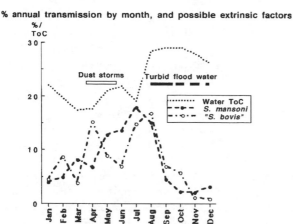

Fig. 4. Monthly transmission potential indices (MTPI) plotted as percentages for *S. mansoni* in *B. pfeifferi* (bold broken line) and *S. bovis* in *Bulinus* spp. (broken line), together with other seasonal factors that may affect transmission. The dotted line is mean monthly water temperature in °C; the solid bars indicate the periods when flood water is passing and the open bar shows the period when dust storms occur.

bovis, less common than other trematodes in *Bulinus* spp., except in the Marigot and secondary canals, was most abundant in the primary canals, the Canal Taouey and river. Its seasonality resembled that of *S. mansoni* but peak summer transmission occurred slightly earlier, with minor autumn peaks in some habitats.

The local strain of *S. mansoni* is very compatible with local *B. pfeifferi* and has a normal midday peak of cercarial shedding (Tchuem Tchuente *et al.* 1999). It is also compatible with local rodents (Sène *et al.* 1996). However, naturally-infected wild rodents were found only from September to November by Sène (1994) and Sène *et al.* (1996). Presumably these infections were acquired two or three months earlier during the late spring/summer transmission peak. The lack of natural infections in rodents in spring and early summer suggests little winter transmission, even though infected snails were sometimes found in small numbers during the cooler winter months. As infections were detected by forced shedding in the laboratory, cercariae may not have been shed in the cooler field sites. Low temperatures would also have slowed the development of infections and hence the number of cercariae produced. Elsewhere in the region, both Ernould *et al.* (1999*a*) and Shaw *et al.* (1999) reported seasonality in the transmission of *S. mansoni* but De Clercq *et al.* (1999) observed it throughout the year round the Lac de Guiers. Cercariometry (Théron, 1986) or animal exposures (Prentice & Ouma, 1984) might help to resolve these contradictions.

The relative abundance of schistosomes and other trematodes is interesting. In addition to schistosome cercariae, *B. pfeifferi* and *Bulinus* spp. each shed at least two other trematodes. It was not possible to

identify them but numerous potential definitive hosts (see above) had access to most sites. *Bulinus* spp. were historically widespread along the Senegal River and would already have acquired a trematode fauna, including schistosomes (which remain in the minority) before the recent hydrological changes. In contrast, the widespread abundance of *B. pfeifferi* has developed only recently. *S. mansoni*, introduced by man, is the commonest trematode: perhaps the situation will change as other trematodes such as amphistomes and paramphistomes start to utilise this newly available host.

The dynamics of snail populations and parasite transmission are affected by many factors other than water temperature. A long-term shift in water conditions from acid to alkaline since completion of the dams favours both processes (Southgate, 1997). Some other seasonal factors are indicated in Fig. 4. Seasonal dust and sand storms from the Sahara in the spring months choke aquatic vegetation, smothering its growth and the development of snail eggs. The same effect could be due to silt-laden flood water, as in Sudanese irrigation schemes (Babiker *et al.* 1985). Although the Manantali Dam has reduced the natural seasonal flood (Diop *et al.* 1994), warm and silt-laden water from downstream rainfall run-off still reaches Richard Toll in mid-summer co-inciding with the declining snail populations from mid-summer onwards. Atypical irrigation routines when the sugar factory is closed in July and August may affect snail populations adversely. So, too, will dredging and weed clearance to maintain water flows in canals and drains. Periodic release of waste products from the sugar factory into nearby secondary canals temporarily eliminated snail populations, but only locally in sites remote from human contact.

Long-term ecological changes consequent on the water resource developments are still in progress in the Senegal River Valley: some may be permanent, others transient. At present, the changes have favoured the spread and growth of *B. pfeifferi* populations so that *S. mansoni* is the main problem at Richard Toll. During these studies, there was no evidence of increased haematuria in the local residents, nor were significant numbers of *S. haematobium* eggs found, suggesting that urinary schistosomiasis is not a threat at Richard Toll. *B. senegalensis*, known to transmit the local strain of *S. haematobium* (Picquet *et al.* 1996), occurred sporadically but does not seem to thrive in the newly created habitats. Nor does the local strain of *B. globosus*. At Richard Toll, the commonest bulinid snail, *B. truncatus*, seems better adapted to the new habitats, though less so than *B. pfeifferi*. It poses no immediate threat (Southgate, 1997), unless a compatible strain of the parasite is imported from elsewhere in West Africa (Vera, 1992). *B. truncatus* is a greater threat to domestic animals as putative *S.*

bovis infections were quite common. The greatest danger is that B. globosus could eventually invade canals at Richard Toll, allowing the introduction of a local, compatible strain of S. haematobium from existing nearby foci in the Senegal River Delta (Verlé et al. 1994; Ernould et al. 1999b).

Extrapolation of the findings at Richard Toll to other areas of the Senegal River Basin requires some caution. Before 1980, there were essentially only two types of habitat available for freshwater snails in the lower Senegal River basin: the Senegal River and its tributaries or temporary, rain-fed pools on elevated laterite soils away from the river. Both are subject to the seasonal effects already discussed. The pools represent more extreme conditions and snails colonising them require considerable powers of aestivation to survive for many months when they dry out. B. pfeifferi is only moderately tolerant of droughts and was originally restricted to a few, stable sites around the Lac de Guiers in numbers too small to support transmission or to allow its introduction by infected immigrants. Bulinus spp., especially B. senegalensis, were better adapted to these conditions and were widely enough distributed to support low level transmission of schistosomes and other trematodes. Extensive development of rice irrigation in the region has been followed by local upsurges in S. haematobium transmission (Verlé et al. 1994; Shaw et al. 1999; Picquet et al. 1998).

Chaine & Malek (1983) predicted that any human schistosome problem that the proposed dams might create would involve S. haematobium. This prediction seemed reasonable in the light of experience in Ghana after the construction of the Volta Dam (Paperna, 1969; Klumpp, 1982) but overlooked the new types of habitats that would be created along the Senegal River. The development of sugar irrigation at Richard Toll provided a foretaste of the effects of the dams on the Senegal River.

At Richard Toll, natural habitats around the Lac de Guiers were stabilised and protected from sea water intrusion, allowing B. pfeifferi populations to proliferate and invade the irrigation scheme as water was pumped via the Canal Taouey into the canal system. B. pfeifferi populations exploded in the relatively stable conditions within the canal system. A parallel invasion of infected people detonated the subsequent S. mansoni epidemic because conditions were ideal for transmission. There was no S. haematobium epidemic as in Ghana because B. truncatus, the only bulinid snail that thrived in the newly created habitats, is incompatible with the S. haematobium strain at Richard Toll (Picquet et al. 1996).

What happens next outside the sugar irrigation scheme depends on the types of habitat created there by construction of the dams. From this study, the spread of S. mansoni seemed inevitable. The Senegal River and its tributaries had been stabilised into a freshwater lake stretching back from the Diama Dam to Richard Toll and beyond. Numerous canals have been constructed to serve new, extensive rice irrigation schemes. Whether or not rice fields form important transmission sites (Mather, 1984), they provide suitable habitats for B. senegalensis (Vercruysse et al. 1985) with a risk of spreading S. haematobium transmission, and their canals could provide suitable habitats for B. pfeifferi favouring the spread of S. mansoni. This is no longer theoretical speculation. B. pfeifferi colonies are now established in Senegal in the river and canals well downstream from Richard Toll where S. mansoni is now established (Talla et al. 1992; Ernould & Ba, 1994; Gryseels et al. 1994; Picquet et al. 1996; Ernould et al. 1999a). In Mauritania, both B. pfeifferi and S. mansoni are established opposite Richard Toll up to 50 km from the Senegal River (Diop et al. 1994). Ominously, S. haematobium transmission through B. globosus has now been found in canals serving a rice irrigation schemes between Richard Toll and the coast (Verlé et al. 1994) and in other habitats (Ernould et al. 1999b). Perhaps Chaine & Malek's original prediction may still come true.

So far, attempts to contain the spread of schistosomiasis in and around Richard Toll have relied predominantly on treatment with praziquantel (Southgate, 1997). The findings of this study suggest that population-based chemotherapy for S. mansoni should avoid the transmission season between May and September because of the high risk of rapid reinfection. Snail numbers and infection rates are minimal between November and February when human infections acquired in the previous transmission season will be easy to diagnose and susceptible to drugs. This view is supported for S. haematobium on the Lac de Guiers, where transmission is light and highly seasonal, and De Clercq et al. (1999) reported effective treatments in both January 1996 and March 1997. The result was less satisfactory for S. mansoni in the face of much heavier and more prolonged transmission. Since human S. mansoni infections were much more intense than those of S. haematobium, infections in April 1996 probably represent worms surviving treatment: two or more treatments would be needed to dislodge them before the main transmission season later in the year (Picquet et al. 1998). In areas where both human schistosomes occur, ecological differences in the (different) habitats of their respective snail hosts may affect the timing and intensity of transmission seasons. This may be an alternative explanation to inter-specific competition between adult worms in the definitive host for the differential effects of praziquantel treatment on the two species (Ernould et al. 1999a).

It is unlikely that chemotherapy alone will ever eradicate transmission. Safe, well-maintained water supplies and improved sanitation, reinforced by education, would reduce transmission of *S. mansoni* and many other diseases besides, but their provision needs considerable long-term investment and commitment. Progress towards a vaccine is being made but it is likely to be many years before one will be available for widespread use. In the meantime, chemotherapy could be supplemented with well timed mollusciciding of canals on sugar and rice irrigation schemes to interrupt transmission at an acceptable cost. The use of molluscicides elsewhere is precluded by its prohibitive expense and unacceptable fish kills.

ACKNOWLEDGEMENTS

This work was performed as part of the ESPOIR programme for research and control of schistosomiasis in the Senegal River Basin which received financial support from the Commission of the European Communities Research Programme 'Life Sciences and Technologies for Developing Countries and the Region Nord Pas de Calais. We wish to acknowledge the diligent work of our field teams, in particular M. Johnson and O. Sow, and the Compagnie Sucrière Sénégalais for their cooperation in the field work.

REFERENCES

ABDEL-WAHAB, M. F. (1982). *Schistosomiasis in Egypt*. pp. 62–65. Florida, Boca Raton, CRC Press.

APPLETON, C. C. (1977). The influence of temperature on the life-cycle and distribution of *Biomphalaria pfeifferi* (Krauss, 1948) in south-eastern Africa. *International Journal for Parasitology* **7**, 335–345.

BABIKER, A., FENWICK, A., DAFALLA, A. A. & AMIN, M. A. (1985). Focality and seasonality of *Schistosoma mansoni* transmission in the Gezira Irrigated Area, Sudan. *Journal of Tropical Medicine and Hygiene* **88**, 65–73.

BELOT, J., GEERTS, S. & DIOUF, M. (1993). Observations on the population dynamics of snail hosts for schistosomiasis in the Delta of the Senegal River. *Journal of Molluscan Studies* **59**, 7–13.

BROWN, D. S. (1994). *Freshwater Snails of Africa and their Medical Importance*. 2nd Edition. p. 228, London. Taylor & Francis.

CHAINE, J. P. & MALEK, E. A. (1983). Urinary schistosomiasis in the Sahelian region of the Senegal River Basin. *Tropical and Geographical Medicine* **35**, 249–256.

DAZO, B. C., HAIRSTON, N. G. & DAWOOD, I. K. (1966). The ecology of *Bulinus truncatus* and *Biomphalaria alexandrina* and its implication for the control of bilharzia in the Egypt-49 control area. *Bulletin of the World Health Organization* **35**, 339–356.

DE CLERCQ, D., VERCRUYSSE, J., PICQUET, M., SHAW, D. J., DIOP, M., LY, A. & GRYSEELS, B. (1999). The epidemiology of a recent focus of mixed *Schistosoma haematobium* and *S. mansoni* infections around the Lac de Guiers in the Senegal River Basin. *Tropical Medicine and International Health* **4**, 544–550.

DEME, I. (1993). Epidémiologie des bilharzioses humaines et animales dans le Delta du bassin du Fleuve Sénégal: étude malacologique. MSc thesis, Prince Leopold Institute, Antwerp, Belgium.

DIAW, O. T., VASSILIDES, G., SEYE, M. & SARR, Y. (1990). Proliférations de mollusques et incidence sur les trématodoses dans la Region du delta et du lac de Guiers après la construction du barrage de Diama sur le fleuve Sénégal. *Revue de l'élevage et de Médecine Vétérinaire de Pays Tropicaux* **43**, 499–502.

DIAW, O. T., VASSILIDES, G., SEYE, M. & SARR, Y. (1991). Epidémiologie de la bilharziose intestinale à *Schistosoma mansoni* à Richard Toll (Delta du fleuve Sénégal): étude malacologique. *Bulletin de la Société de Pathologie Exotique* **84**, 174–183.

DIOP, M., JOBIN, W. R., ADREIN, N. G., ARFAA, F., AUBEL, J., BERTLOIMINOR, S., KLUMPP, R. & ROSENMAAL, J. (1994). *Senegal River Basin Health Master Plan Study*. 163 pp., Washington, WASH Field Report 453, USAID.

ERNOULD, J.-C. & BA, K. (1994). Cartographies du bas delta du fleuve Sénégal: resultats préliminaire (avril 1994). L'eau et la santé dans les contextes du dévelopment. ORSTOM-St Louis. ORSTOM/ES/DK/72.94.

ERNOULD, J. C., BA, K. & SELLIN, B. (1999*a*). The increase of intestinal schistosomiasis after praziquantel treatment in a *Schistosoma haematobium* and *S. mansoni* mixed focus. *Acta Tropica* **73**, 143–152.

ERNOULD, J. C., BA, K. & SELLIN, B. (1999*b*). The impact of the local water-development programme on the abundance of the intermediate hosts of schistosomiasis in three villages of the Senegal River Delta. *Annals of Tropical Medicine and Parasitology* **93**, 135–145.

FALLON, P. G., STURROCK, R. F., CAPRON, A., NIANG, M. & DOENHOFF, M. J. (1995). Diminished susceptibility to praziquantel in a Senegal isolate of *Schistosoma mansoni*. *American Journal of Tropical Medicine and Hygiene* **53**, 61–62.

FOSTER, R. (1964). The effect of temperature on the development of *Schistosoma mansoni* Sambon 1907 in the intermediate host. *Journal of Tropical Medicine and Hygiene* **67**, 289–292.

FULFORD, A. J. C., BUTTERWORTH, A. E., DUNNE, D. W., STURROCK, R. F. & OUMA, J. H. (1996). Some mathematical issues in assessing the evidence for acquired immunity to schistosomiasis. In *Models for Infectious Diseases. Their Structure and Relation to Data* (ed. Isham, V. & Medley, G.), pp. 139–159. Cambridge, Cambridge University Press.

GRYSEELS, B., STELMA, F. F., TALLA, I., VAN DAM, G. J., POLLMAN, K., SOW, S., DIAW, M., STURROCK, R. F., DOEHRING-SCHWERDTFEGER, E., DECAM, C., NIANG, M. & DEELDER, A. M. (1994). Epidemiology, immunology and chemotherapy of *Schistosoma mansoni* infections in a recently exposed community in Senegal. *Tropical and Geographic Medicine* **46**, 209–219.

HANDSCHUMACHER, P., DORSAINVILLE, R., DIAW, O.-T., HEBRARD, G., NIANG, M. & HERVE, J.-P. (1992). Constraints climatiques et amenagements hydroliques à propos de l'epidémie de bilharziose intestinale de Richard-Toll (Sénégal) ou la modification des risques sanitaire en milieu sahelian. In *Risques Pathalogiques, Rythmes et Paroxysmes Climatiques* (ed. Besancenot, J. P.), pp. 287–295. Paris, John Libbey Eurotext.

HANDSCHUMACHER, P., TALLA, I., HERVÉ, J.-P., DUPLANTIER, J. M., SENE, M., DIAW, O.-T. & HEBRARD, G. (1994). *Petit Atlas Informatise de Richard Toll*.

Environment Urbain et Bilharziose Intestinal. Dakar, ORSTOM.

JOBIN, W. R., NEGRON-APONTE, H. & MICHELSON, E. H. (1976). Schistosomiasis in the Gorgol Valley of Mauritania. *American Journal of Tropical Medicine and Hygiene* **25**, 587–594.

KLUMPP, R. K. (1982). A study on the transmission of *Schistosoma haematobium* in Volta Lake, Ghana. PhD thesis. London University, pp. 307–315.

MALEK, E. A. & CHAINE, J. P. (1989). Effects of the developments in the Senegal River Basin on the prevalence and spread of schistosomiasis. In *Demography and Vector-borne Diseases* (ed. Service, M. W.), pp. 181–192, Boca Raton, Florida, CRC Press.

MATHER, T. H. (1984). Environmental management for vector control in rice fields. *Food and Agriculture Organization Drainage paper 41.* Rome.

MONJOUR, L., NIEL, G., MOGAHED, A., SIDATT, M. M. & GENTILINI, M. (1981). Geographic distribution of bilharziasis in the Senegal River Valley (sero-epidemiologic evaluation – 1973). *Annales de la Société Bélgiques de Médecine Tropicale* **61**, 453–460.

OLIVIER, L. & SCHNEIDERMANN, M. (1956). A method for estimating the density of aquatic snail populations. *Experimental Parasitology* **5**, 109–117.

OUMA, J. H., STURROCK, R. F., KLUMPP, R. K. & KARIUKI, H. C. (1989). A comparative evaluation of snail sampling and cercariometry to detect *Schistosoma mansoni* transmission in a large-scale, longitudinal field study in Machakos, Kenya. *Parasitology* **94**, 349–355.

PAPERNA, I. (1969). Study of an outbreak of schistosomiasis in the newly formed Volta Lake in Ghana. *Zeitschrift für Tropenmedizin und Parasitologie* **21**, 339–353.

PFLÜGER, W. (1981). Experimental epidemiology of schistosomiasis. II. Prepatency of *Schistosoma mansoni* in *Biomphalaria glabrata* at diurnally fluctuating temperatures. *Zeitschrift für Parasitenkunde* **66**, 221–229.

PICQUET, M., ERNOULD, J. C., VERCRUYSSE, J., SOUTHGATE, V. R., MBAYE, A., SAMBOU, B., NIANG, M. & ROLLINSON, D. (1996). The epidemiology of human schistosomiasis in the Senegal River Basin. *Transactions of the Royal Society of Tropical Medicine and Hygiene* **90**, 340–346.

PICQUET, M., VERCRUYSSE, J., SHAW, D. J., DIOP, M. & LY, A. (1998). Efficacy of praziquantel against *Schistosoma mansoni* in northern Senegal. *Transactions of the Royal Society of Tropical Medicine and Hygiene* **92**, 90–93.

PRENTICE, M. A. & OUMA, J. H. (1984). Field comparison of mouse immersion and cercariometry for assessing the transmission potential of water containing cercariae of *Schistosoma mansoni*. *Annals of Tropical Medicine and Parasitology* **78**, 169–174.

ROLLINSON, D., SOUTHGATE, V. R., VERCRUYSSE, J. & MOORE, P. J. (1990). Observations on natural and experimental interactions between *Schistosoma bovis* and *S. curassoni* from West Africa. *Acta Tropica* **47**, 101–114.

SÈNE, M. (1994). Étude de la schistosomiasis intestinale chez les rongeurs suavages à Richard Toll (Sénégal): suivi de l'infestation naturelle et transmission

expérimentale. Thèse de 3° cycle, Université Cheikh Anta Diop, Dakar. 87 pp.

SÈNE, M., DUPLANTIER, J. M., MARCHAND, B. & HERVÉ, J. P. (1996). Susceptibility of rodents to infection with *Schistosoma mansoni* in Richard Toll (Senegal). *Parasite* **3**, 321–326.

SHAW, D. J., VERCRUYSSE, J., PICQUET, M., SAMBOU, B. & LY, A. (1999). The effect of different treatment regimens on the epidemiology of seasonally transmitted *Schistosoma haematobium* infections in four villages in the Senegal River Basin, Senegal. *Transactions of the Royal Society of Tropical Medicine and Hygiene* **93**, 142–150.

SHIFF, C. J. (1964). Studies on *Bulinus (Physopsis) globosus* in Rhodesia, I. The influence of temperature on the intrinsic rate of natural increase. *Annals of Tropical Medicine and Parasitology* **58**, 94–105.

SOUTHGATE, V. R. (1997). Schistosomiasis in the Senegal River Basin: before and after the construction of the dams at Diama, Senegal, and Manantali, Mali, and future prospects. *Journal of Helminthology* **71**, 125–132.

STELMA, F. F., TALLA, I., SOW, S., KONGS, A., NIANG, M., POLLMAN, K., DEELDER, A. M. & GRYSEELS, B. (1995). Efficacy and side effects of praziquantel in an epidemic focus of *Schistosoma mansoni*. *American Journal of Tropical Medicine and Hygiene* **53**, 167–170.

STURROCK, R. F. (1996). The influence of temperature on the biology of *Biomphalaria pfeifferi* (Krauss), an intermediate host of *Schistosoma mansoni*. *Annals of Tropical Medicine and Parasitology* **60**, 100–105.

STURROCK, R. F., KLUMPP, R. K., OUMA, J. O., BUTTERWORTH, A. E., FULFORD, A. E., KARIUKI, H. C., THIONGO, F. W. & KOECH, D. (1994). Observations on the effects of different chemotherapy strategies on the transmission of *Schistosoma mansoni* in Machakos District, Kenya, measured by long-term snail sampling and cercariometry. *Parasitology* **109**, 443–453.

STURROCK, R. F. & STURROCK, B. M. (1972). The influence of temperature on the biology of *Biomphalaria glabrata* (Say), intermediate host of *Schistosoma mansoni* on St Lucia, West Indies. *Annals of Tropical Medicine and Parasitology* **66**, 385–398.

TALLA, I., KONGS, A., VERLÉ, P., BELOT, J., SARR, S. & COLL, A. M. (1990). Outbreak of intestinal schistosomiasis in the Senegal River Basin. *Annales de la Société Belgiques de Médecine Tropicale* **70**, 173–180.

TALLA, I., KONGS, A. & VERLÉ, P. (1992). Preliminary study of the prevalence of human schistosomiasis in Richard Toll (Senegal River Basin). *Transactions of the Royal Society of Tropical Medicine and Hygiene* **86**, 182.

TCHUEM TCHUENTE, L. A., SOUTHGATE, V. R., THÉRON, A., JOURDANE, J., LY, A. & GRYSEELS, B. (1999). The compatability of *Schistosoma mansoni* and *Biomphalaria pfeifferi* in northern Senegal. *Parasitology* **118**, 595–603.

THÉRON, A. (1986). Cercariometry and the epidemiology of schistosomiasis. *Parasitology Today* **2**, 61–63.

VERA, C. (1992). Contribution à l'étude de la variabilité génétique des schistosomes et leurs hôtes intermediaries: polymorphisme de la compatabilité entre diverse populations de *Schistosoma haematobium*,

S. bovis et *S. curassoni* et les bulins hôtes potentiels en Afrique de l'Ouest. PhD thesis, Université Montpellier. ORSTOM, Paris. 303 pp.

VERCRUYSSE, J., SOUTHGATE, V. R. & ROLLINSON, D. (1985). The epidemiology of human and animal schistosomiasis in the Senegal River Basin. *Acta Tropica* **42**, 249–259.

VERCRUYSSE, J., SOUTHGATE, V. R., ROLLINSON, D., DE CLERCQ, D., SACKO, M., DE BONT, J. & MUNGOMBA, L. M.

(1994). Studies on the transmission and schistosome interactions in Senegal, Mali and Zambia. *Tropical and Geographic Medicine* **46**, 220–226.

VERLÉ, P., STELMA, F., DÈSREUMAUX, P., DIENG, A., DIAW, O., NIANG, M., SOW, S., TALLA, I., STURROCK, R. F., GRYSEELS, B. & CAPRON, A. (1994). Preliminary study of urinary schistosomiasis in the delta of the Senegal River Basin, Senegal. *Transactions of the Royal Society of Tropical Medicine and Hygiene* **88**, 401–405.

Contributions to and review of dicrocoeliosis, with special reference to the intermediate hosts of *Dicrocoelium dendriticum*

M. Y. MANGA-GONZÁLEZ*, C. GONZÁLEZ-LANZA, E. CABANAS *and* R. CAMPO

Consejo Superior de Investigaciones Científicas (CSIC), Estación Agrícola Experimental, Apdo. 788, 24080 León, Spain

SUMMARY

An epidemiological study on dicrocoeliosis caused by *Dicrocoelium dendriticum* was carried out on sheep, molluscs and ants in the mountains of León province (NW Spain) between 1987–1991. The results concerning the intermediate hosts and a review of some aspects of dicrocoeliosis are summarized. Mollusc collection for the helminthological study was random throughout the study area at fortnightly intervals. Twenty-nine Gastropoda species were identified. *D. dendriticum* infection was only detected in 2·98% of the 2084 *Helicella itala* examined and in 1·06% of 852 *H. corderoi*. The highest infection prevalence was detected in *H. itala* in September and in *H. corderoi* in February. Daughter sporocysts with well-developed cercariae predominated in spring and autumn. Infection prevalence increased with mollusc age and size. Ants were collected from anthills or plants to which they were attached. The behaviour of ants in tetania was followed. Twenty-one Formicidae species were identified, but only the following harboured *D. dendriticum*: *Formica cunicularia* (1158 examined specimens, 0·69% infection prevalence, 2–56 metacercariae per ant); *F. sanguinea* (234, 1·28%, 2–63); *F. nigricans* (1770, 4·97%, 1–186); *F. rufibarbis* (288, 6·59%, 2–107). In a flat area close to León town, 95·39% of the 2085 *F. rufibarbis* specimens collected in tetania contained metacercariae (1–240) in the abdomen. These were used for parasite characterization by isoelectric focusing and to infect lambs and hamsters. Only one brainworm per ant was found.

Key words: *Dicrocoelium dendriticum*, mollusc and ant intermediate hosts.

INTRODUCTION

Dicrocoelioses are parasite infections caused by the species of the genus *Dicrocoelium* Dujardin, 1845 (Trematoda, Digenea), although mainly by *Dicrocoelium dendriticum* (Rudolphi, 1819) Looss, 1899. This parasite, which cycles in land molluscs and ants, is located in the bile ducts and gall bladders of numerous species of domestic and wild mammals, mainly ruminants, which act as definitive hosts in several countries in Europe, Asia, America and North Africa (Malek, 1980). *D. dendriticum* is a very common species in ruminants in the Iberian peninsula (Cordero, Castañón & Reguera, 1994). Moreover, this parasite can also occasionally infect humans (Mohamed & Mummery, 1990). Infection of the definitive hosts occurs by ingesting the ants which harbour infective metacercariae.

The genus *Dicrocoelium* is included in the subfamily Dicrocoeliinae Looss, 1899, of the Dicrocoeliidae Family Odhner, 1911. According to La Rue (1957), this family belongs to the superfamily Plagiorchioidea, suborder Plagiorchiata, order Plagiorchiida and superorder Epitheliocystida. The most important species of this genus which infect ruminants are: *Dicrocoelium dendriticum*; *D. hospes* Looss, 1907, *D. chinensis* Tang & Tang, 1978 and *D. suppereri* Hinaiday, 1983 (syn. *D. orientalis*

Sudarikov et Ryjikov, 1951). *D. dendriticum* is found in America, Asia, North Africa and Europe (Malek, 1980) including practically all of the Iberian Peninsula (Cordero *et al.* 1994). The second, the third and the fourth species have been found in Africa (Lucius, 1981), Asia (Tang *et al.* 1983) and the old Soviet Union and Austria (Hinaidy, 1983), respectively.

This paper mainly concerns dicrocoeliosis produced by *D. dendriticum* as it is the most widespread amongst the ruminants of several countries. For a long time this parasite was confused with an immature form of *Fasciola hepatica*, as both trematodes are frequently found together in the liver of ruminants. Its description thus came late. The synonymy of this parasite is complex, due to the different generic and specific denominations received (Mapes, 1951; Schuster, 1987).

Dicrocoeliosis causes irritation of the mucosa of the large bile ducts in definitive hosts. This explains the proliferation and increase in the secretion of the glandular cells as well as cholangitis and cholangiectasis of the septal and hepatic bile ducts, granulomatous type portal hepatitis associated with portal, septal and perisinusoidal fibrosis and vacuolar degeneration of the hepatocytes (Dhar & Singh, 1963; Wolff, Hauser & Wild, 1984; Sanchez-Campos *et al.* 1996; Ferreras *et al.* 1997). This disease is generally chronic and the immune response does not protect from reinfection. Experimental studies carried out in lambs infected with 1000 and

* Corresponding author: Tel: +34 987 317156. Fax: +34 987 317161. E-mail: y.manga@eae.csic.es

Parasitology (2001), **123**, S91–S114.   © 2001 Cambridge University Press
DOI: 10.1017/S0031182001008204

3000 *D. dendriticum* metacercariae showed that the first egg elimination took place between days 49 and 79 post-infection (Campo, Manga-González & González-Lanza, 2000), while the first detection of IgG antibodies by ELISA technique was observed on day 30 p.i. Maximum antibody levels were obtained 60 days p.i. and remained high until the experiment ended 180 day p.i. (González-Lanza *et al.* 2000). Moreover the studies of the liver and hepatic lymph nodes (from the lambs slaughtered 2 months p.i.) immunolabelled by avidin-biotin complex system showed that the parasite induced a humoral and cell-mediated local immune response that contributed to the inflammation observed but did not seem effective for the destruction of the parasite (Ferreras *et al.* 2000).

The economic and health significance of dicrocoeliosis is partly due to the direct losses occasioned by the confiscation of altered livers (Del Rio, 1967; Lukin, 1980; Karanfilovski, 1983) and also the indirect ones caused by the digestive disorders derived from the hepatobiliary alterations caused by these parasites, such as decreased animal weight (Boray, 1985), growth delay (Hohorst & Lämmler, 1962), reduced milk production (Cavani *et al.* 1982), amongst others. Moreover, the additional costs incurred by the application of anthelminthic treatments, to which the animals must be subjected, have to be considered.

The life cycle of *Dicrocoelium dendriticum* is extremely complex because land molluscs and ants are required as first and second intermediate hosts, respectively. Until Krull & Mapes (1952, 1953) managed to complete the life cycle for the first time, numerous studies were carried out over more than a century to try elucidate it (reviewed by Mapes, 1951 and Del Río, 1967). The adults of the genus *D. dendriticum* live in the liver and bile ducts of the definitive hosts where they lay their embryonated eggs which pass through the intestine to be eliminated in the faeces. Egg hatching and miracidium liberation only occur in the intestine of numerous species of land molluscs that act as first intermediate hosts. The miracidium penetrates the intestinal wall of the mollusc and settles in the hepatopancreas, where it becomes a mother sporocyst, which takes the shape of the spaces between hepatopancreatic lobules because it has no wall itself. This larval stage produces sacciform daughter sporocysts with their own wall in which cercariae are formed when they are well developed. These abandon the sporocysts when they are mature and migrate to the respiratory chamber of the mollusc where they are covered in slime. The slimeballs are eliminated through the pneumostoma by the respiratory movements of the snail. When these slimeballs are ingested by different species of ants, which act as second intermediate hosts, the cercariae cross the craw of the ants, lose their tail and one of them

(sometimes 2 or 3), called the "brainworm", settles in the suboesophageal ganglion of the ant and the rest become metacercariae in the abdomen. When the temperature falls, the brainworm alters the behaviour of the ant by causing tetania of its mandibular muscles. Due to this the ant remains temporarily attached to grass and this promotes ingestion by the definitive host. The mature abdominal metacercariae excyst in the intestine, the young flukes migrate to the liver through the opening of the common bile duct (sometimes the portal circulation) and become adult worms in the bile duct. When these are mature, they lay eggs which exit in the faeces of the host and this allows the life cycle to begin again. A more detailed description of the adult worm and of the different stages of the parasite can be found in Manga-González (1999) and in Manga-González & Quiroz-Romero (1999).

The role played by molluscs in the epidemiology of dicrocoeliosis is very important as *D. dendriticum* egg hatching and miracidium liberation only occur in the intestine of the molluscs that act as intermediate hosts. Moreover, the parasite multiplies enormously by asexual reproduction inside them (numerous cercariae can be formed from one ingested egg). This increases the possibilities of parasite transmission. Since Piana (1882) first encountered the long-tailed cercaria in *Helix carthusiana* (= *Monacha* (*M.*) *cartusiana*), which was later described by Von Linstow (1887) as *Cercaria vitrina* (from *Zebrina detrita*) and associated with *Dicrocoelium dendriticum*, many studies have been carried out to discover the mollusc species which act as first intermediate hosts for this parasite. More than 100 mollusc species (Gastropoda, Pulmonata, Stylommatophora) have been found receptive to *D. dendriticum* under natural and laboratory conditions (some of them are mentioned in the Discussion below). It can be deduced from this that *D. dendriticum* shows markedly little specificity as regards its first intermediate host. In addition the parasite can develop in various mollusc species in the same area: Manga-González (1987, 1992) found 11 species of Helicidae infected with the parasite in the province of León (Spain). The life history of the mollusc intermediate hosts is of great epidemiological interest, as regards both the ingestion periods of *D. dendriticum* eggs, dependent on the molluscs' activity, and the survival of the parasite in them. Species, age and nutritional state of the molluscs, infective dose, ambient temperature and relative humidity, amongst other aspects, all influence the development of larval stages of this digenean in the first intermediate hosts. Gómez *et al.* (1996) observed a higher infection percentage and faster development of *D. dendriticum* in *Cernuella* (*Xeromagna*) *cespitum arigonis* than in *Cernuella* (*Cernuella*) *virgata*, although both species had been tested at the same time with identical doses of eggs

and kept in the same conditions (20 °C and 40%
relative humidity).

The importance of ants in the epidemiology of
dicrocoeliosis is mainly due to their abundance, wide
distribution and the fact that the alteration in their
behaviour, caused by the presence of the parasite in
the brain, make their ingestion by definitive hosts
easy when the infected ones are in tetania on plants.
Since Krull & Mapes (1952) showed that an ant
(*Formica fusca*) acted as second intermediate host of
D. dendriticum, various authors have endeavoured to
discover the species of ants which act as secondary
intermediate hosts for *D. dendriticum* in nature and
under experimental conditions, the prevalence and
intensity of the infection, the behaviour of infected
ants and the risk period for infection of the definitive
hosts, amongst other aspects. At least 21 Formicidae
species mainly from *Formica* genus have been
described as receptive to this parasite in different
countries. Some of these species are mentioned in
the Discussion section.

Tegelström, Nilsson & Wyoni (1983) used isoelectric focusing to study the proteins of the ant
species *Formica rufa*, *F. polyctena* and *F. pratensis*, in
head and thorax homogenate. These authors
detected the presence of about 68 bands of general
proteins and 18 of non-specific esterases among the
three ant species. However, according to our information, no studies using isoelectric focusing on
thin-layer polyacrylamide gel techniques have been
carried out to characterize the larval stages of *D.
dendriticum* and their detection in ants, although
they have been done for the larval stages found in the
molluscs (Campo *et al.* 1992, and other unpublished
information) and for the adult parasite (Campo *et al.*
1998).

The application of efficacious dicrocoeliosis control measures, which have not been satisfactory so far
(Eckert & Hertzberg, 1994), requires good diagnosis
and knowledge of the epidemiology of the disease.
Nevertheless, integrated studies of the *D.
dendriticum* transmission process are
scarce – possibly due to the long length of its life
cycle and its great complexity. So, as a basis for the
design of strategic effective control programmes, we
decided to throw light on the epidemiological model
in an area in the mountains of León province (Spain)
where we had previously detected the maximum
infection prevalence in randomly sampled sheep
(73·7%) and the highest values of the maximum
(5340) and mean (398·8±5) numbers of eggs per
gram (Manga-González, González-Lanza & Del-
Pozo, 1991). It is necessary to take into account that
the epidemiology of this parasitosis is influenced by
local conditions such as: existence of definitive hosts
(domestic or wild) receptive to the parasite, farming
model, animal handling, presence, biology and
ethology of the molluscs and ants which act as first
and second intermediate hosts, meteorological

factors, soil type and vegetation. Simultaneous
studies on definitive and intermediate *D. dendriticum*
hosts were carried out in order to discover the
shedding kinetic of eggs per marked sheep, the
mollusc and ant species acting as intermediate hosts
and various aspects of their life history, the dynamics
of their infection and degree of development of the
larval stages they harbour, the transmission period to
the definitive hosts via detection in the field of ants
with infective metacercariae, and the influence of the
biotic and abiotic factors on the dynamic of the
mollusc and ant populations on the degree of
development of the larval stages and on the infection
rate of the definitive and intermediate hosts. This
paper reports on the data relative to the molluscs,
ants and their infection by *D. dendriticum*.

In addition, results relating to the prevalence and
intensity of *D. dendriticum* metacercariae in *Formica
rufibarbis* Fabricius, 1794 (Formicidae) collected in
tetania from around the town of León (Spain) are
included. This was in order to obtain metacercariae
to carry out the morphological and isoenzymatic
study and experimentally infect lambs and hamsters
in order to study different aspects of experimental
dicrocoeliosis.

MATERIALS AND METHODS

Molluscs

The study was carried out over two consecutive
years from June, 1987 to May, 1989 in the valley of
Redipollos (U.T.M. co-ordinates: 30TUN1663;
altitude 1100 to 1400 m), situated in the upper basin
of the Porma in the province of León (NW Spain), in
an area measuring 560 hectares used as pasture by a
communal flock of 120 sheep. The climate is
continental within the Mediterranean-Atlantic transition. Before beginning sampling, a survey was done
in the area which allowed us to distinguish the
following habitats and their extension (%): pasture
on limestone (46%), limestone wall (19%), pasture
on acid soil (1%), pasture-*Genista hispanica* complex
(22%) and *Arctostaphylus uva-ursi* areas (13%)
where the botanical composition was studied. Three
zones were established in the valley according to the
movement of the sheep flock throughout the year:
Zone A, known as "Peña El Castillo", situated to the
west of Redipollos, with slopes facing the 4 compass
points and used by the flock between December and
April; Zone B, a slope to the south close to the
village, facing north-west and used as pasture
throughout the year; Zone C, a slope to the east quite
far away from the village, facing north and used by
the flock between April and December. The area
covered by the different habitats in the whole study
area and in the three separate zones was estimated
using a planimeter.

Collections to discover the existing mollusc

species, their distribution and various aspects of their biology and ecology were carried out on three plot types, in different zones, habitats and orientations of the Redipollos valley. The number and the size of the plots were: ten 4 × 4 m, forty-two 1 × 1 m, one hundred and one 0·5 × 0·5 m. Biotic and abiotic variables were checked in each plot at monthly intervals. These results have not yet been published.

Snail collection for the helminthological study was carried out at random over the area at fortnightly intervals. A distance of 10 m from the plots established for the mollusc population dynamic study was respected. The monthly values of environmental temperature and precipitation came from the two nearest stations and were supplied by the staff of the Duero basin Meteorological Service (Valladolid, Spain). The average environmental temperatures and precipitation in the different months were: 1·5 °C, 190·9 mm in January; 1·7 °C, 111·8 mm in February; 3·4 °C, 81 mm in March; 3·3 °C, 203·4 mm in April; 9·2 °C, 120·7 mm in May; 11 °C, 89·8 mm in June; 14·3 °C, 94·2 mm in July; 13·9 °C, 38·4 mm in August; 11 °C, 67·6 mm in September; 8·1 °C, 262·5 mm in October; 4·7 °C, 117 mm in November; and 1·9 °C, 108 mm in December. The live molluscs were taken to the laboratory and the different species were identified, the specimens were measured and their degree of maturity noted. Material fixation and later examination of the molluscs were done as specified by Manga-González (1983). The determination of the mollusc species was carried out primarily based on conchological and anatomical traits, mainly of the different genital apparatus structures. The publications of Riedel (1972), Kerney, Cameron & Jungbluth (1983), Manga-González (1983), Castillejo (1998), Wiktor (2000), amongst others, were consulted. The helminthological study was done according to Manga-González (1987). The molluscs were anaesthetised and dissected in order to detect *D. dendriticum* daughter sporocysts in the hepatopancreas using a stereomicroscope and to study their degree of development using an optical microscope. Infection intensity was semi-quantitatively estimated on the basis of the lesser or greater extent to which the hepatopancreas was invaded by sporocysts. The length and width of all the sporocysts up to 100 extracted complete from each mollusc were measured under the microscope and, if the number was higher, 10% of the rest were also measured. The developmental stage of these sporocysts was studied for each of the infected molluscs *in vivo*. Determination of parasite species was carried out by means of morphoanatomical, chaetotaxic and isoelectric focusing studies and also by comparing the larval stages with those obtained by experimental infection. In order to discover the possible influence of the collection month of the molluscs studied, their age (established according to

the development degree of the genitalia and several shell characteristics), zone and habitat from which they came, on the infection prevalence by *D. dendriticum*, the data were analysed using the χ^2 test and the 2 × 2 contingency tables.

Ants

Three types of sampling were carried out to collect ants in the mountain area of Redipollos, which has already been described for the mollusc study. The first consisted of introducing a 2·5 cm diameter by 9 cm long polyurethane tube, in duplicate at ground level, into 40 of the 101 plots measuring 0·5 × 0·5 m established to study the mollusc populations. This type of sampling was only carried out on two occasions (October and November, 1988) due to the unsatisfactory results obtained. These tubes, filled with a mixture of formol at 4 parts per thousand and glycerine, acted as traps for the insects which fell into them and were removed after 8 days. The second type of sampling was carried out at the same time as the collection of molluscs for the helminthological study. A large number of ants was collected directly from 53 anthills found in the different habitats, 23 in zone A, 16 in zone B and 14 in zone C. The third type of sampling was carried out by individually collecting ants which were in tetania in the area surrounding the anthills. In addition, for several days various *F. pratensis* anthills were controlled in zone C to detect ants in tetania and record their behaviour. The number of ants in tetania, the type of plants used to support them, the position of the ants, the geographical orientation, the hours spent in tetania and the times when they started activity, as well as the temperature and relative humidity at ground level at those times were all recorded in this study.

For the studies on experimental dicrocoeliosis in lambs and hamsters and for the enzymatic characterization of *D. dendriticum* metacercariae, between 1992 and 1998 ants of the *F. rufibarbis* species in tetania were collected from 9 sites in the flat area near the town of León (Spain). They were situated in the lower basin of the rivers: Porma (Mellanzos, San Cipriano del Condado and Valdelafuente), Torio (Navatejera) and Bernesga (Grulleros and Villa de Soto), as well as the banks of the river Esla (Villafalé, Palanquinos and Valencia de Don Juan) near where it joins these tributaries. These sites, at altitudes of between 763 and 903 m, were chosen because *D. dendriticum* infection of molluscs and sheep had previously been detected there. Ants were collected from May to October, mainly in the early hours of the morning and sometimes in the afternoons. The collection time, ambient temperature and relative humidity at ground level, general weather conditions and type of plant were also noted down at the time of collection.

Table 1. *Prevalence of infection by* D. dendriticum *of the Gastropoda species studied*

Mollusc species	Number examined	Percentage infected
Subclass PULMONATA		
Order STYLOMMATOPHORA		
Agriolimacidae		
Deroceras (*Agriolimax*) *agreste* (Linnaeus)	9	0
Deroceras (*Agriolimax*) *reticulatum* (Müller)	24	0
Arionidae		
Arion (*Arion*) *ater* (Linnaeus)	18	0
Arion (*Arion*) *lusitanicus* (Mabille)	4	0
Arion (*Arion*) *rufus* (Linnaeus)	21	0
Arion (*Microarion*) *intermedius* (Normand)	1	0
Buliminidae		
Jaminia (*Jaminia*) *quadridens* (Müller)	35	0
Chondrinidae		
Chondrina kobelti cliendentata Gittenberger	151	0
Cochlicopidae		
Cochlicopa lubrica (Müller)	45	0
Helicidae*		
Cepaea (*Cepaea*) *nemoralis* (Linnaeus)	71	0
Hygromiidae		
Helicella (*Helicella*) *itala* (Linnaeus)	2084	2·98
Helicella corderoi Gittenberger & Manga	852	1·06
Oestophora sp.	1	0
Oestophorella buvinieri (Michaud)	12	0
Pupillidae		
Lauria (*Lauria*) *cylindracea* (Da Costa)	7	0
Pupilla (*Pupilla*) *muscorum* (Linnaeus)	3	0
Pyramidulidae		
Pyramidula rupestris (Draparnaud)	25	0
Valloniidae		
Vallonia costata (Müller)	3	0
Vallonia excentrica Sterki	3	0
Vertiginidae		
Truncatellina cylindracea (Férussac)	2	0
Vertigo pygmaea (Draparnaud)	1	0
Vitrinidae		
Phenacolimax (*Gallandia*) *annularis* (Studer)	5	0
Vitrina pellucida (Müller)	50	0
Zonitidae		
Nesovitrea hammonis (Ström)	5	0
Vitrea contracta (Westerlünd)	4	0
Aegopinella nitidula (Draparnaud)	2	0
Subclass PROSOBRANCHIA		
Order MESOGASTROPODA		
Cyclophoridae		
Cochlostoma sp.	130	0
Total	3568	1·98

* Some authors, like Vaught (1989), split the old Helicidae family into Helicidae and Hygromiidae.

The ants collected live were transferred to the laboratory in appropriately labelled polyethylene flasks. They were then kept in artificial anthills until their dissection in the following days. Only a small number of *F. rufibarbis* specimens collected in tetania were kept alive until they died a natural death in order to obtain information on survival.

The ants collected in Redipollos from different plots and in different samplings were anaesthetised and measured. Dissection was under the stereo-microscope examining the head, thorax and abdomen separately in Petri dishes containing a saline solution (154 mM NaCl). The *D. dendriticum* brainworm and metacercariae found in each ant were extracted and counted. A sample of them was studied '*in vivo*' under the optical microscope. The rest of the material was kept in 70 % alcohol.

The helminthological study of the 2085 *F. rufibarbis* ants collected in tetania in the flat area was done in the same way. Most metacercariae obtained

Fig. 1. *Helicella corderoi*, 1st intermediate host for
Dicrocoelium dendriticum. Scale: 1 mm.

were used to infect lambs and hamsters. The rest of
the metacercariae and ants were placed in liquid
nitrogen to carry out later isoenzymatic analyses, or
in 70% alcohol to carry out morphological studies.
In 25 encysted metacercariae extracted from the
abdomen the maximum body width and length and
the thickness of the cyst wall were measured using an
ocular micrometer connected to an optical microscope. In addition, the same measurements were
made (except that of the thickness of the cyst wall) of
25 metacercariae excysted in distilled water and then
fixed in 70% alcohol.

In order to characterize *D. dendriticum*
metacercariae, isoelectric focusing in thin-layer polyacrylamide gel technique was used, as described in
Campo *et al.* (1998) for *D. dendriticum* adults. A total
of 1608 metacercariae, distributed among samples
consisting of 15–185 metacercariae each – all from
the abdomen of the same ant – were analysed.
The activity of the following enzymes was studied:
lactate dehydrogenase (LDH, EC 1.1.1. 27),
glucose phosphate isomerase (GPI, EC 5.3.1.9),
phosphoglucomutase (PGM, EC 2.7.5.1) and acid
phosphatase (AcP, EC 3.1.3.2), in 1608, 924, 1458
and 1608 metacercariae, respectively, extracted from
33 *F. rufibarbis* ants from 4 sites in the flat area: 13
from Grulleros, 11 from San Cipriano del Condado,
8 from Villafalé and 1 from Valencia de Don Juan.
All the enzymatic activity bands observed were considered to interpret the results obtained, but they
were differentiated as strong bands (S), weak bands
(W) and very weak bands (V W).

RESULTS

Molluscs

Twenty-six species of Gastropoda were found, most
of them on pasture on limestone, but infection by *D.*

dendriticum was only detected in *Helicella* (*H.*) *itala*
and *H. corderoi* (Table 1, Fig. 1). Both species, which
were the most abundant, also hosted larval stages
of Brachylaimidae spp. (Digenea) and Proto-
strongylidae spp. (Nematoda), although this paper
only includes infection by *D. dendriticum*.

Helicella itala. The abundance of this species was
greater in spring, fell in summer and increased again
in autumn, although it was still lower than in spring.
The young (very young and young) specimens were
more abundant in spring whilst the adults were more
so in autumn. A more variable situation was observed
on considering the data for each zone and habitat in
detail. The highest abundance of *H. itala* was
detected in zone B, followed by zone A and in the
pasture on limestone habitat. A large number of very
young *H. itala* was observed in zone C in April and
May. Specimens of this species were mainly found in
pasture areas in spring and sheltering between soil
and rock in moss and other plants, principally the
Genista hispanica shrub and *Eringium campestre*
thistle, in summer. In autumn a high percentage of
specimens were again found on pastureland. Active
molluscs were recorded in spring, September and
October and molluscs withdrawn into their shells,
but without an epiphragm, in every month of the
year. Molluscs with an epiphragm were observed,
mainly in summer and winter.

D. dendriticum daughter sporocysts were
harboured in the hepatopancreas by 2·98% of a total
2084 specimens of *H. itala* examined (Table 1).
As can be seen in Fig. 2 infection was recorded in
every month except February. The sampling of
February 1989 could not be done because of the bad
meteorological conditions (20 days of snow and 12 of
frost during the month). The highest prevalence was
observed in September. By means of the chi-square
test, statistically significant differences were
observed among the sampling months with regard to
infection prevalence ($\chi^2 = 21\cdot903$; $P \leqslant 0\cdot05$; g.l. =
11).

Infection prevalence increased with their age and
shell diameter (Table 2). Statistically significant
differences in the infection prevalence between age
groups were observed on using χ^2 analysis ($\chi^2 =
31\cdot143$; $P \leqslant 0\cdot005$; g.l. = 2). On applying the 2×2
contingency tables statistically significant differences
were detected between the young and intermediate,
young and adults, intermediate and adults. The
highest infection prevalence was detected in the
young molluscs in September (7·69%) and in the
adults in May (30·43%).

The examined molluscs from different zones and
habitats and their prevalence of infection are
summarized in Table 3. By means of χ^2 analysis
statistically significant differences were observed in
the infection prevalence between the pasture on

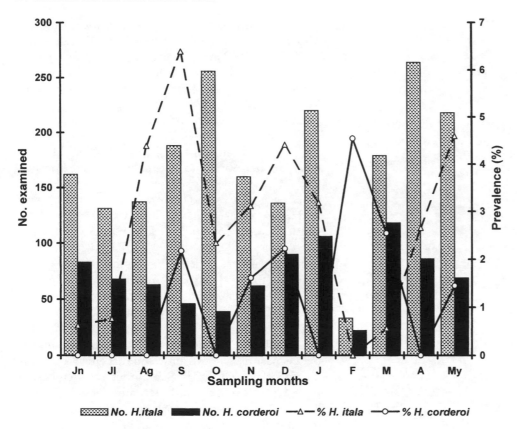

Fig. 2. Infection prevalence of *Helicella itala* and *Helicella corderoi* by *D. dendriticum* throughout the sampling period (June, 1987–May, 1989: the data for each month in the 2 years are shown jointly).

Table 2. Infection prevalence of *H. itala* and *H. corderoi* by *D. dendriticum* according to their age and shell diameter (size)

Age/size(mm)	H. itala		H. corderoi	
	Exam.	Infect. (%)	Exam.	Infect. (%)
Young	786	0·63	300	0·33
Intermediate	303	2·97	25	0·00
Adult	995	4·82	527	1·72
Size 2–4	—	—	27	0·00
Size 4–6	—	—	481	0·42
Size 6–8	—	—	344	2·03
Size 3–6	29	0·00	—	—
Size 7–10	486	0·41	—	—
Size 11–14	1152	2·60	—	—
Size 15–18	417	7·19	—	—

Exam. = number examined; Infect. (%) = % infected.

Table 3. Infection prevalence of *H. itala* by *D. dendriticum* according to the zones and habitats

Origin/habitat	Molluscs	
	Examined	Infected (%)
Zone A	1172	3·25
Zone B	822	2·19
Zone C	90	6·77
Pasture on limestone	1548	2·71
Pasture – *Genista hispanica*	536	3·73
Zone A/Pasture on limestone	1172	3·25
Zone B/Pasture on limestone	306	0·33
Zone C/Pasture on limestone	71	4·17
Zone B/Pasture – *Genista hispanica*	516	3·29
Zone C/Pasture – *Genista hispanica*	21	15

limestone and pasture-*Genista hispanica* complex habitats belonging to each zone ($\chi^2 = 18\cdot292$; $P \leqslant 0\cdot005$; g.l. = 4). On applying the 2×2 contingency tables statistically significant differences were detected between the pasture A–pasture B; pasture A–*Genista* C; pasture B–*Genista* B; pasture B–pasture C; *Genista* B–*Genista* C habitats.

The very reduced and concentrated infection at one or various points of the hepatopancreas corresponded to the presence of daughter sporocysts containing barely evolved germinal masses (Fig. 3), whilst sporocysts with well-developed cercariae were mainly observed in the infection spread throughout the hepatopancreas, even invading other organs (Figs. 4,5,6). However, some molluscs were found with the parasite at different development stages at the same time. The time when the different de-

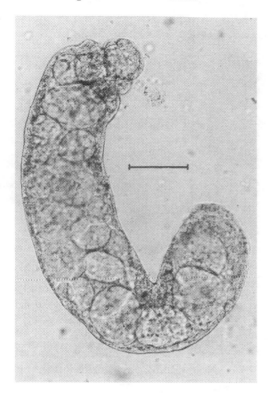

Fig. 3. *Dicrocoelium dendriticum* daughter sporocyst containing germinal masses. Scale bar: 100 μm.

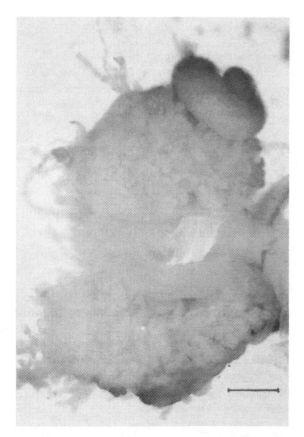

Fig. 4. *Helicella itala* hepatopancreas with many well-developed daughter sporocysts (white sacciform elongated structure) spread throughout the whole hepatopancreas. Scale bar: 1 mm.

velopment degrees of the *D. dendriticum* sporocysts were found in the whole study area and the age of the molluscs which harboured them are summarized in Fig. 7. The highest percentage of molluscs containing well-developed cercariae was detected in May and October and that of the one harbouring germinal mass in September.

According to the measurements corresponding to the 1000 daughter sporocysts studied at different stages of development, the length varied between 560 μm (in a sporocyst with undifferentiated germinal mass) and 4160 μm (in a sporocyst with developed cercariae) ($\bar{x} = 1986 \pm 0.0084$ SE) and the width between 150 and 501 μm ($\bar{x} = 246 \pm 0.008$).

Helicella corderoi. Specimens of this species were collected through the year but the abundance was, in general, greater at the end of autumn and in winter. Adult specimens were clearly predominant in September and November and young ones in May and October. *H. corderoi* was always more abundant in the pasture on the limestone habitat, and predominantly in Zone C in September and October, in Zone B in November and December and in Zone A during the rest of the year. Active specimens were found in October, March and April. Molluscs withdrawn into their shells but without an epiphragm were observed in all the months of the year and those with an epiphragm predominated in summer. *H. corderoi* showed a preference for pasture, except in summer, when it chose the soil–rock transition area; in October it left this location to be on limestone and bare soil, seeking out pasture again when the bad weather started.

D. dendriticum daughter sporocysts were found in the hepatopancreas of 1·06 % of the 852 *H. corderoi* specimens examined (Table 1). On considering the whole sampling period, infection was only detected in six months and the highest prevalence was observed in February although a few specimens were examined (Fig. 2).

Prevalence increased with age and shell diameter (Table 2). Only one young *H. corderoi* was seen infected in February. The greatest infection prevalence was detected in the adults in March. Although specimens from the three study zones and the different habitats were examined, infected molluscs were only found from zone A and from pasture on limestone habitats.

The only young *H. corderoi* found infected contained daughter sporocysts with germinal masses, concentrated at one point of the hepatopancreas. Most of the infected adults contained daughter sporocysts with cercariae migrating to different parts of the hepatopancreas. Nevertheless, in one of them collected in May sporocysts with well-developed cercariae invaded all the hepatopancreas and the genital apparatus. Taking into account the measurements carried out on 500 daughter *D. dendriticum*

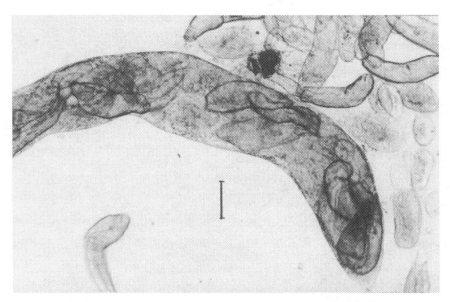

Fig. 5. *Dicrocoelium dendriticum* daughter sporocyst containing well-developed cercariae and some free cercariae. Scale bar: 150 μm.

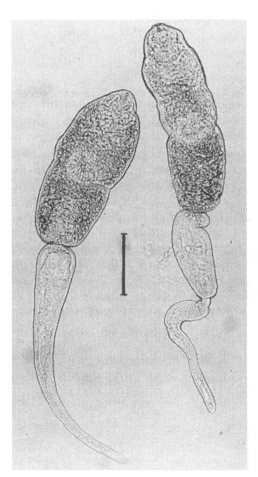

Fig. 6. *Dicrocoelium dendriticum* cercariae. Scale bar: 100 μm.

sporocysts, at different stages of development, the length varied between 200 and 1 540 μm ($\bar{x} = 833 \pm$ 0·246 SE), and the width between 20 and 170 μm ($\bar{x} = 86 \pm 0·026$).

Ants

In the mountain area. The results obtained with tubes placed in the earth were unsatisfactory because both the number of ant species captured and the specimens collected from each was very small. We used 160 traps and ants were only found in 29 of the 80 placed in October, 1988, belonging to the species *Formica sanguinea*, *F. cunicularia* and *F. pratensis*. The maximum number of ants per trap was 12 ($\bar{x} = 2·59$).

In the second sampling type the ants were taken directly from 53 anthills: 23 in zone A, 16 in zone B and 14 in zone C. In this way 24 ant species, all belonging to the family Formicidae were collected (Table 4). These species are grouped by subfamilies according to Collingwood (1978). The helminthological study was carried out on 5993 of the 6555 specimens gathered. As a result we found *D. dendriticum* metacercariae in ants collected between April and November from 14 anthills: 2 *Formica cunicularia* and 3 *F. pratensis* in zone A; 2 *F. rufibarbis* in zone B; 3 *F. sanguinea* and 4 *F. pratensis* in zone C.

The third sampling type allowed us to detect infected ants in tetania in the surroundings of 29 nests: 2 *F. pratensis* and 9 *F. rufibarbis* (Fig. 8) in May 1990, in zone A; 5 *F. rufibarbis* in May 1990, 4, 1 and 8 *F. pratensis* in May 1990, August 1990 and June 1991, respectively, in zone C. The infected ants were found attached to various plants, such as *Anthyllis vulneraria*, *Poterium sanguisorba*, *Lotus corniculatus*, *Plantago lanceolata*, *Genista hispanica* and various gramineae species, like *Bromus erectus* and *Festuca rubra*, situated in the area surrounding the anthills. The number of workers in tetania observed in each anthill was variable and oscillated

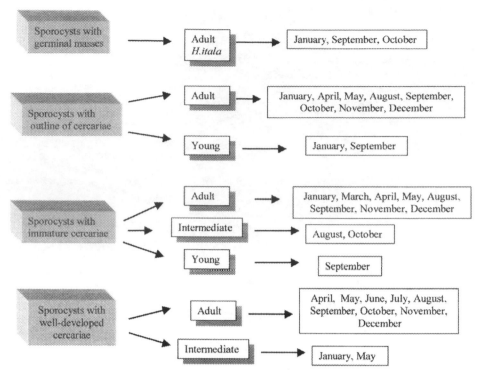

Fig. 7. Schematic diagram of the time in which the different development degrees of *Dicrocoelium dendriticum* daughter sporocysts were observed in *Helicella itala* collected in the whole study area.

between 1 and 9 (average 5). The most frequently occurring position for these workers was head down and on plants in the south-eastern part of the anthill. The behaviour of infected *F. pratensis* ants in relation to time and ambient temperature at ground level around some of the anthills examined in zone C is shown in Fig. 9. Fifty-five of these ants were helminthologically examined.

The *F. cunicularia*, *F. pratensis*, *F. rufibarbis* and *F. sanguinea* infection prevalence and intensity by *D. dendriticum* metacercariae are included in Tables 4 and 5, respectively. In addition, on dissecting various specimens, only one brainworm of the parasite per ant was observed in the suboesophageal ganglion, which was not encysted (Fig. 10). We observed the scars produced by the cercariae on crossing the craw in their flight towards the head, but we did not do a recount.

In the flat area. In the flat area sites close to the town of León active *Formica rufibarbis* ants were observed between March and November and in tetania from May to October. The ambient temperature (at ground level) at the time of collection, carried out between May and September, oscillated between 16 and 28 °C and the relative humidity from 64 to 77 %. The ants in tetania were found attached to plants of the species *Lotus corniculatus*, *Plantago lanceolata*, *Taraxacum dens-leonis*, *Trifolium pratense* and *Medicago sativa*, although mainly in the last one. We managed to collect several dozen ants in tetania from one single *M. sativa* plant, usually attached to the flowers, up to 8 or 10 specimens per flower. It was

not necessary to locate the nests of *F. rufibarbis*, which are barely visible, to find the ants infected with *D. dendriticum* attached to the plants. Some of the ants collected in tetania survived in the artificial nests in the laboratory for 6·5 months (from the end of August, 1992 to the middle of March, 1993).

A total of 2085 specimens of *F. rufibarbis* were examined (length 4·2–5·7 mm; $\bar{x} = 4·95 \pm 0·10$) and metacercariae were found in the abdomen (Figs. 11, 12) of 95·39 % of these specimens (Table 4). The origin of the ants and the infection percentages were: 1584 (95·26 % infected) from Villafalé, 369 (96·20 %) from Grulleros, 28 (96·42 %) from Valencia de Don Juan, 27 (100 %) from San Cipriano del Condado, 27 (100 %) from Navatejera, 21 (90·47 %) from Mellanzos, 13 (92·30 %) from Villa de Soto, 13 (84·61 %) from Valdelafuente, 3 (66·66 %) from Palanquinos. The number of metacercariae per ant abdomen (Table 5), similar every month, varied between $48·31 \pm 2·19$ in June and $59·77 \pm 1·50$ in August. The size of these encysted metacercariae is shown in Table 6. The length of metacercariae excysted in distilled water was 412·80–624 μm ($\bar{x} = 510·33 \pm 9·36$) and the width from 115·20 to 201·60 μm ($\bar{x} = 161·66 \pm 4·40$).

Metacercaria isoenzymatic determination. To interpret the results obtained by using isoelectric focusing, of *D. dendriticum* metacercariae from *F. rufibarbis*, strong (S), weak (W) and very weak (V W) enzymatic activity bands were considered.

The LDH activity in the 1608 *D. dendriticum* metacercariae analysed was observed in the pH range

Table 4. Prevalence of infection by *D. dendriticum* metacercariae in the abdomen of the Formicidae species studied from the mountains, except *F. rufibarbis* **

Ant species	Number examined	Percentage infected
Dolichoderinae		
Tapinoma ambiguum Emery	287	0
Tapinoma erraticum (Latreille)	26	0
Formicinae		
Formica cunicularia Latreille	1158	0·69
Formica fusca (Linnaeus)	68	0
Formica polyctena (Förster)	250	0
Formica pratensis Retzius	2030	4·33
*Formica pratensis**	55	100
Formica pressilabris (Nylander)	390	0
Formica rufibarbis Fabricius	288	6·59
Formica rufibarbis **	2085	95·39
Formica sanguinea Latreille	234	1·28
Lasius alienus (Förster)	150	0
Lasius flavus (Frabicius)	60	0
Lasius niger (Linnaeus)	65	0
Lasius umbratus (Nylander)	31	0
Polyergus rufescens Latreille	5	0
Proformica depilis Bondroit	17	0
Myrmicinae		
Aphaenogaster gibbosa (Latreille)	64	0
Leptothorax pardoi Tinaut	95	0
Leptothorax unifasciatus Latreille	152	0
Myrmica sabuleti Meinert	183	0
Myrmica shencki Emery	4	0
Tetramorium caespitum (Linnaeus)	229	0
Tetramorium hispanicum Emery	5	0
Tetramorium impurum (André)	45	0
Tetramorium semilaeve (André)	157	0
Total	8133	20·39

* Collected in tetania.
** Collected in the flat area in tetania.

Fig. 8. Infected *Formica rufibarbis* specimens, 2nd intermediate host for *Dicrocoelium dendriticum*, in tetania attached to a plant in the field. The real size of the ants is about 5 mm.

6·40–7·13. Four enzymatic types were found designated as: LDH-1 (pH, band intensity: 6·40, W; 6·60, W; 6·72, W; 6·86, W; 7·8, W), LDH-2 (6·60, W; 6·72, W; 6·86, W; 7·08, V W; 7·13, W), LDH-3 (6·40, S; 7·08, V W) and LDH-4 (6·40, V W; 6·72, W; 6·86, V W; 7. 08, S).

On studying the GPI activity 4 enzymatic types were found in 924 *D. dendriticum* metacercariae: GPI-1 (6·27, V W; 6·43, V W; 6·54, S; 6·69, W; 6·80, V W; 6·83, V W), GPI-2 (6·27, V W; 6·43, V W; 6·54, S; 6·72, S; 6·80 S), GPI-3 (6·54, S; 6·72, V W; 6·80, V W) and GPI-4 (6·54, W; 6·72, W; 6·80, W). The activity bands were in a pH range between 6·27 and 6·83.

The PGM activity on analysing 1458 *D. dendriticum* metacercariae was detected in a pH range 6·20 – 6·60. Two enzymatic models were observed: PGM-1 (6·20, V W; 6·60, S) and PGM-2 (6·20, S; 6·40, S)

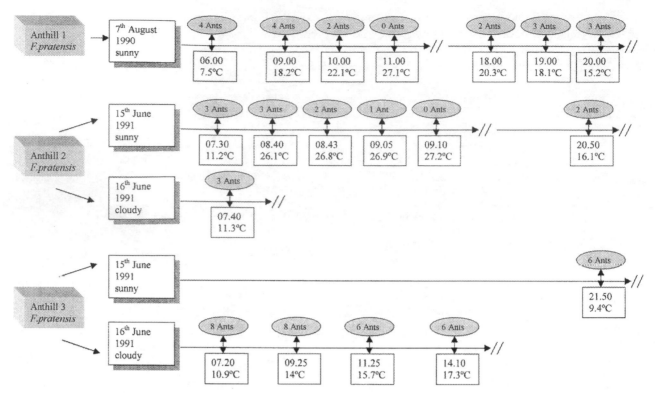

Fig. 9. Schematic diagram of the number of ants in tetania in relation to the time of day and ambient temperature at ground level.

On examining 1608 *D. dendriticum* metacercariae AcP activity was detected between pH 5·86–5·92. Although the bands observed are diffuse, 4 enzymatic models could be identified: AcP-1 (5·86, V W; 5·91, S; 5·92, S), AcP-2 (5·92, S), AcP-3 (5·91, V W; 5·92, V W) and AcP-4 (5·91, S).

In the enzymatic systems studied different models were generally observed in metacercariae extracted from the same ant, from ants originating from different locations and, sometimes, from those collected in the same place.

DISCUSSION

Molluscs

According to our results (Table 1), *H. itala* and *H. corderoi* are the species responsible for *D. dendriticum* transmission in the study area. Nevertheless, the importance of *H. itala* in dicrocoeliosis epidemiology is greater for being more widely distributed and more abundant, not only in the study area but also in the whole of León province, northern Spain and west and central Europe, while *H. corderoi* is an endemic species in the mountain area of León province. Moreover the size of *H. itala* specimens is bigger so they could harbour more parasites. The description of both species and different data on their distribution, biology and ecology are recorded in Manga-González (1983).

From our observations it seems that *H. itala* and *H. corderoi* abundance is conditioned by the existence of appropriate temperature and relative hu-

midity values, as occurs in spring and autumn, although temperature is a more limiting factor than humidity during the rest of the year. The temperature-humidity binomial has a great influence on mollusc abundance for most authors (Staikou, Lazaridou-Dimitriadou & Kattoulas, 1989, amongst others).

Our results concerning the highest abundance of *H. itala* adults in autumn and young ones in spring coincides with the observations of Manga-González (1983, 1987) and Schuster (1993) in *H. itala* and *H. obvia*, respectively. Although it is difficult to establish a *H. corderoi* development pattern with the results we obtained, the predominance of adult specimens in November coincides with the observations of Manga-González (1983) for this species in the upper Torio basin (León province, Spain). Ortiz de Zárate (1950) stated that *Helicella* species are annual and generally do not develop completely until autumn.

Locomotor activity in land pulmonates depends on various external environmental factors, such as light intensity, relative humidity, temperature, soil moisture and time of day. Moreover, endogenous circadian rhythmicity in locomotor activity has been demonstrated (Riddle, 1983). From our results it seems that *H. itala* and *H. corderoi* activity is influenced by temperature and relative humidity at the same time, and the formation of the epiphragm in *H. corderoi* is mainly conditioned by high temperatures and low precipitation, whilst low temperatures also have an influence in the case of *H.*

Table 5. *Comparison of the infection intensity by* D. dendriticum *metacercariae reported here in the ants and those mentioned by other authors*

Authors	Ant species	Infection range	Intensity x̄ ± s. e.
Own observations	*Formica cunicularia*	2–56	26·0 ± 15·89
	F. pratensis	1–186	97·5 ± 20·80
	F. rufibarbis*	2–107	28·3 ± 7·90
	F. rufibarbis**	1–240	56·4 ± 0·93
	F. sanguinea	2–63	42·0 ± 20·00
Krull & Mapes (1953)	*F. fusca*	6–128	—
Klesov & Popova (1958)	*F. pratensis*	—	251
Hohorst (1962)	*F. cunicularia*	—	> 300
Del Río (1967	*F. pratensis*	1–67	22
	F. rufibarbis	,,	,,
	F. sanguinea	,,	,,
Denev *et al.* (1970)	*F. fusca*	1–580	—
	F. pratensis	,,	—
Jonlija *et al.* (1972)	*F. lugubris*	1–212	—
	F. pratensis	,,	—
	F. polyctena	,,	—
Badie *et al.* (1973)	*F. cunicularia*	2–300	—
	F. nigricans***	,,	—
Srivastava (1975)	*F. pratensis*	176–443	—
Kalkan (1976)	*F. rufibarbis*	1–117	—
Paraschivescu (1976)	*F. cunicularia*	11–173	—
	F. pratensis	,,	—
Paraschivescu *et al.* (1976)	*F. pratensis*	5–165	—
Angelovski & Iliev (1978)	*Cataglyphus bicolor*▲	5–155	—
	F. rufibarbis	,,	—
	F. cunicularia	,,	—
Badie (1979)	*F. cunicularia*	—	> 200
	F. nigricans***	—	
Paraschivescu (1981*a*)	*F. cunicularia*	21–31	—
	F. sanguinea	2–87	—
	F. pratensis	11–173	—
	F. fusca	6–329	—
Spindler *et al.* (1986)	*F. polyctena*	1–53	—
Schuster (1991)	*F. rufibarbis*	2–127	38
	F. pratensis	2 213	76

* Collected in the mountains; ** collected in the flat area; *** = *F. pratensis* according to Seifert (1992); ▲ = *Cataglyphis*; ,, = Figures given for all species without specification.

itala. Bonavita (1961) stated that the formation of an epiphragm is due to a combination of different environmental temperature and relative humidity values. The presence of this temporary structure, observed by us mainly in winter and summer in *H. itala* and in summer in *H. corderoi*, is important for the epidemiology of dicrocoeliosis as it means that the molluscs do not eat the parasite eggs. Nevertheless, active molluscs or those withdrawn into their shells but without an epiphragm, were detected in every month of the year.

Specimens of *H. itala* were mainly found in pasture areas in spring and sheltering between soil and rock in moss and other plants, principally the *Genista hispanica* shrub and *Eryngium campestre* thistle, in summer. The preference of *H. itala* for *Eryngium campestre* in summer was also mentioned by Engel (1957). In autumn a high percentage of specimens was again found on pastureland. Wolda, Zweep & Schuitema (1971) stated this could indicate the beginning of the search for a place to hibernate,

as grazing land provides favourable sheltering areas in winter; Schuster (1993) also recorded this migration. Riddle (1983) mentioned that the slugs and some land snails occupy fairly protected habitats in which daily and seasonal extremes in air temperature are largely avoided.

When we compared the percentage of infection by *D. dendriticum* obtained by us in *H. itala* (2·98 %) and in *H. corderoi* (1·06 %) with those given by other authors in different species of molluscs and countries (Table 7), great variability could be observed. If we focus only on *H. itala* our percentage was lower than those obtained by Del Rio (1967), Tarry (1969) and Manga-González (1987, 1992) and higher than those given by Rozman, Jonlija & Mustapic (1971) and Rozman, Gradjanin & Cankovic (1974). The infection prevalence of *H. corderoi* was lower than that obtained for *H. itala* in this study and considerably below that obtained by Manga-González (1992) studying *H. corderoi* from various mountain areas in León province (Spain). Nevertheless, we must point

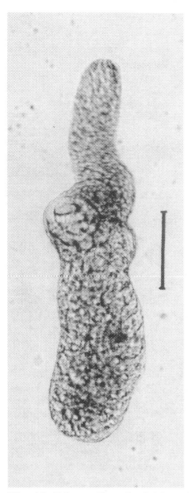

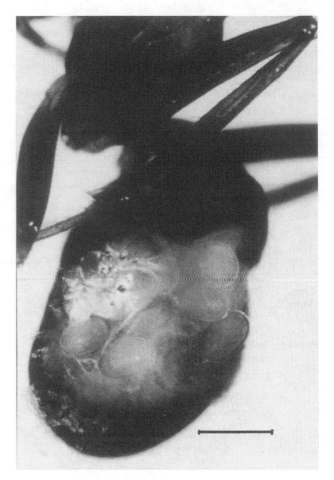

Fig. 10. *Dicrocoelium dendriticum* brainworm extracted from the *Formica pratensis* suboesophageal ganglion. Scale bar: 100 μm.

Fig. 11. *Dicrocoelium dendriticum* metacercariae specimens inside the *Formica rufibarbis* open abdomen. Scale bar: 400 μm.

out that the infection prevalence in both species of molluscs could be higher because according to experimental laboratory studies (González-Lanza *et al.*, 1997) the infection is not visible under the stereomicroscope until at least 50 days post-infection. Moreover, in experimental studies carried out under controlled field conditions (unpublished research) the minimum period for detecting the parasite under the stereomicroscope varied between 2 and 9 months p.i. depending on the infection month.

According to our results (Fig. 2) the highest prevalence for *H. itala* was obtained in September, May and December, while it was in February and March for *H. corderoi*. This species seem to be very well adapted to the low temperatures. Nevertheless, it is very risky to establish a *D. dendriticum* transmission model through this species due to the low number of infected specimens found and also due to the lack of publications that could help to interpret our data. There is no unanimity amongst the authors about the dynamics of the mollusc infection, probably due to the different species studied and also the different environmental conditions. So Dementev & Karabaev (1968) recorded the highest infection prevalence in May

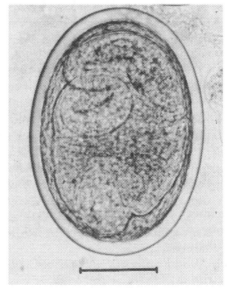

Fig. 12. *Dicrocoelium dendriticum* metacercaria extracted from the *Formica sanguinea* abdomen. Scale bar: 100 μm.

and the lowest in September in Kazakhstan; Kalkan (1970) obtained the highest prevalence in summer and autumn in Turkey; Kuz'movich & Kostinik (1975) gave an almost constant infection rate

Table 6. Size of *D. dendriticum* encysted metacercariae obtained from the ant abdomen reported here and by other authors

| Authors | Ant species | *D. dendriticum* encysted metacercaria size (μm) | | |
		Length	Width	Cyst wall thickness
Own observations	*Formica rufibarbis*	336–461 (380)	240–369 (283)	5–12 (9)
Krull & Mapes (1952)	*F. fusca*	325–465	—	10–40
Groschaft (1961)	*F. fusca glebaria**	288–440	192–328	—
	F. pratensis	”	”	—
	F. sanguinea	”	”	—
Fromunda *et al.* (1965)	*F. glebaria**	185–306	146–204	48–50
	*F. nigricans***	148–250	92–111	”
Del Rio (1967)	*F. pratensis*	280–320 (298)	150–210 (181)	15–22 (19)
	F. rufibarbis	371–424 (404)	260–310 (288)	19–37 (25)
	F. sanguinea	380–430 (408)	267–310 (288)	19–38 (24)
Odening (1969)	*F. rufibarbis fuscorufibarbis*	243–400 (365)	200–270 (250)	16–23 (20)
Fromunda *et al.* (1973)	*F. cunicularia*	172–579	91–321	5·4
Kalkan (1976)	*F. rufibarbis*	368–448	288–336	80–96
Paraschivescu *et al.* (1976)	*F. pratensis*	174–572	90–320	5·4
Angelovski & Iliev (1978)	*Cataglyphus bicolor*▲	240–440	280–330	20
	F. fusca	”	”	”
	F. rufibarbis	”	”	”
Paraschivescu (1981*a*)	*F. pratensis*	481–579	260–311	—

* = *Formica cunicularia*; ** = *Formica pratensis* according to Seifert (1992); ▲ = *Cataghyphis*; ” = Figures given for all species without specification.

throughout the year in Ukraine; Manga-González (1987) observed the highest prevalence in October, December and May in Spain and Schuster & Neumann (1988) found the highest infection prevalence in spring and the lowest in summer with it increasing again in autumn in Germany.

Infection prevalence increased with the age and size of the molluscs (Table 3), a tendency that had already been observed in field (Manga-González, 1987 and Schuster, 1993) and experimental conditions (Alunda & Rojo, 1983). The highest *H. itala* infection prevalence was detected in the young molluscs in September and in the adults in May. These results partially coincide with those obtained by Manga-González (1987), who recorded the highest infection rate for young of *H. itala* in October and that for adults in October and May.

On taking our results into account, it seems that the infection rate (Fig. 2) and the degree of development (Fig. 7) of the sporocysts which *H. itala* hosts did not follow any clear pattern throughout the year or in the different zones and orientations. This could be due to the fact that viable egg ingestion by the molluscs is related to the time in which contamination of the pastures takes place. So, Zone A was grazed by animals between December and April, thus the pasture was contaminated by parasite eggs then. Contamination favours infection of the molluscs as soon as they become active in spring, with a consequent higher infection rate in September, since several months must pass before the infection becomes visible under the stereo-

microscope, as previously mentioned. In contrast, livestock graze in zone C between April and December, so pasture contamination and mollusc infection must occur during that period, which explains the infection detected in the adult specimens in the following spring. In zone B livestock graze practically throughout the year, so contamination is continuous, since animals eliminate the eggs uninterruptedly (Manga-González *et al.* 1991; González-Lanza, Manga-González & Del-Pozo, 1993), although the period of greatest elimination is detected in autumn and, above all, in winter. This meant that infected molluscs were found in almost every month of the year. The exact date when the molluscs were infected is difficult to estimate, but we believe that all the molluscs found infected from the beginning of the year until the end of spring were infected the previous year, although at different times according to the degree of development observed in the parasites. The molluscs that harboured scarcely developed sporocysts from the end of summer until December must have been infected in spring or at the beginning of summer, while those containing sporocysts with well-developed cercariae in the same period could be infected at the beginning of spring or even in the previous year.

The fact that sometimes different developmental stages of the parasites were found in a mollusc at the same time – as had already been observed by Del Rio (1967) and Manga-González (1987) in naturally infected specimens and by González-Lanza *et al.* (1997) in experimentally infected ones – seems to

Table 7. Infection prevalence of some Stylommatophora molluscs by *D. dendriticum* obtained by various authors

Mollusc species	Prevalence (%)	Country	Authors
Bradybaenidae			
Bradybaena alma-atini	7·4–0·9	Kazakhstan	Dementev & Karabaev (1968)
B. lantzi	6·4	Kazakhstan	Dementev (1968)
B. plectotropis var. *phaeozona*	3·2	Kazakhstan	Dementev (1968)
B. rubens	0·6	Kazakhstan	Dementev (1968)
B. semenovi	1·9	Kazakhstan	Dementev (1968)
Eulota maacki (1)	28	Russia	Ovcharenko (1964)
Clausiliidae			
Clausilia bidentata	1·5	Germany	Schmidt (1967)
Laciniaria varnensis	21·4	Bulgaria	Denev *et al.* (1970)
Cochlicopidae			
Cionella lubrica (2)	28	USA	Mapes (1951)
C. lubrica	5–20	Spain	Del Río (1967)
C. lubrica	10	France	Badie (1979)
Buliminidae			
Chondrula microtraga	23·3	Bulgaria	Denev *et al.* (1970)
C. tridens	16·5	Bulgaria	Denev *et al.* (1970)
Jaminia tridens (3)	5·6	Bosnia-Herzegovina	Jonlija *et al.* (1973)
J. potaniniana	0·7	Kazakhstan	Dementev (1968)
Zebrina detrita	20·5	Bulgaria	Denev *et al.* (1970)
Z. detrita	5·6	Bosnia-Herzegovina	Jonlija *et al.* (1973)
Z. hohenackeri (4)	62	Caucasus	Bocharova (1983)
Helicidae*			
Cepaea nemoralis	5·6	Bosnia-Herzegovina	Jonlija *et al.* (1973)
C. nemoralis	0·72	Spain	Manga-González (1992)
C. vindobonensis	5·6	Bosnia-Herzegovina	Jonlija *et al.* (1973)
Hygromiidae			
Cernuella (*C.*) *virgata*	1·36	Spain	Manga-González (1992)
C. virgata	1·0	Turkey	Kalkan (1970)
C. (*Microxeromagna*) *vestita*	5·17	Spain	Manga-González (1992)
C. (*Xeromagna*) *cespitum arigonis*	2·77	Spain	Manga-González (1992)
Cochlicella acuta	2·1	France	Badie & Rondelaud (1987)
C. barbara	20	Spain	Manga-González (1992)
C. ventricosa (5)	0·5	France	Badie & Rondelaud (1987)
Helicella (*H.*) *candicans*	4·3	Turkey	Kalkan (1970)
H. corderoi	7·84	Spain	Manga-González (1992)
H. itala	50	Spain	Del Río (1967)
H. itala	6·02	U. K.	Tarry (1969)
H. itala	0·05	Bosnia-Herzegovina	Rozman *et al.* (1974)
H. itala	5·68	Spain	Manga-González (1987,1992)
H. jamuzensis	1·03	Spain	Manga-González (1992)
H. madritensis	1·86	Spain	Manga-González (1992)
H. (*H.*) *obvia*	4·30	Turkey	Kalkan (1970)
H. obvia	4·3	Bulgaria	Denev *et al.* (1970)
H. obvia	20·67	Germany	Schuster (1992)
H. ordunensis	2·13	Spain	Manga-González (1992)
Helicopsis krynickii (7)	2·6	Turkey	Kalkan (1970)
H. (*Xeropicta*) *derbentina* (6)	4·0	Turkey	Kalkan (1970)
Monacha carthusiana (8)	1·0	Turkey	Kalkan (1970)
M. (*M.*) *cartusiana*	2·39	Spain	Manga-González (1992)
Theba carthusiana (8)	1·5	Bulgaria	Denev *et al.* (1970)
Zonitidae			
Aegopis acies	1·47	Bosnia-Herzegovina	Rozman *et al.* (1974)

(1) = *Bradybaena maacki*; (2) = *Cochlicopa lubrica*; (3) = *Chondrula tridens*; (4) = *Napaeopsis hohenackeri*; (5) = *Cochlicella barbara*; (6) = *Xeropicta derbentina*; (7) = *Xeropicta krynickii*; (8) = *Monacha* (*M.*) *cartusiana*. * See the explanation given in Table 1.

indicate that either the mollusc was infected at a different time or that larval development of eggs ingested at the same time is diachronic.

Molluscs containing sporocysts with well-developed cercariae were found throughout nearly the whole year (Fig. 7), although the highest percentages were detected in May and October. The highest values of these percentages were observed in April and October by Tverdokhlebov (1970), in winter and spring by Manga-González (1987) and in spring

Table 8. Infection prevalence of some Formicidae ants by *D. dendriticum* metacercariae obtained by various authors

Ant species	Prevalence (%)	Country	Authors
Formicinae			
Cataglyphis aenescens	2·4–1·82	Russia	Urazbaev (1979)
Cataglyphus bicolor (1)	0·37	Macedonia	Angelovski & Iliev (1978)
Formica cinerea	3·21	Central Caucasus	Popov & Kalitina (1962)
F. fusca	31	USA	Krull & Mapes (1953)
F. fusca	0·8–18	Russia	Vershinin (1957)
F. fusca	0·71–11·11	Bulgaria	Veselinov (1962)
F. fusca	3–9	Bulgaria	Denev *et al.* (1970)
F. fusca	0·24	Macedonia	Angelovski & Iliev (1978)
F. fusca	95·71*	Germany	Schuster & Neumann (1988)
F. fusca glebaria (2)	0·20	Czech Republic	Groschaft (1961)
F. glebaria (2)	0·04	Rumania	Fromunda *et al.* (1965)
F. lugubris	0·1	Bosnia-Herzegovina	Jonlija *et al.* (1973)
	95·8*		
F. lugubris	0·14	Bosnia-Herzegovina	Rozman *et al.* (1974)
	71·9*		
F. mesasiatica	1·3–2·08	Russia	Urazbaev (1979)
F. nigricans (3)	0·11	Rumania	Fromunda *et al.* (1965)
F. pratensis	28	Kazakhstan	Dementev (1968)
F. pratensis	33·4	Bulgaria	Denev *et al.* (1970)
F. pratensis	83·38*	Bulgaria	Srivastava (1975)
F. pratensis	0·09–0·45	Ukraine	Klesov & Popova (1958, 1959)
F. pratensis	0·38	Czech Republic	Groschaft (1961)
F. pratensis	0·18	Spain	Del Río (1967)
F. pratensis	0·1	Bosnia-Herzegovina	Jonlija *et al.* (1973)
	95·8*		
F. pratensis	1·04	Bosnia-Herzegovina	Rozman *et al.* (1974)
	70·4*		
F. picea	0·57	Central Caucasus	Popov & Kalitina (1962)
F. polyctena	0	Bosnia-Herzegovina	Jonlija *et al.* (1973)
	15·9*		
F. polyctena	0	Bosnia-Herzegovina	Rozman *et al.* (1974)
	16·8*		
F. rufibarbis	0·50	Spain	Del Río (1967)
F. rufibarbis	0·47	Serbia	Vujic (1972)
F. rufibarbis	2–3	Lebanon and Syria	Tohmé & Tohmé (1977)
F. rufibarbis	0·92	Macedonia	Angelovski & Iliev (1978)
F. sanguinea	0·50	Spain	Del Río (1967)
F. subpilosa	13·4	Tajikistan	Dil'man (1978)

(1) = *Cataglyphis*; (2) = *Formica cunicularia*; (3) = *Formica pratensis* according to Seifert (1992). * Collected in tetania.

and autumn by Schuster (1993). This means that slimeballs could be shed mainly during the mentioned periods. Veselinov (1962) observed that *Zebrina detrita* shed slimeballs from May to July in Bulgaria; Hohorst & Lämmler (1962) detected them in Germany from May to October.

The size of *D. dendriticum* daughter sporocysts from *H. corderoi* was smaller than that of the specimens of *H. itala* studied in this paper and in another studied by Manga-González (1987), possibly due to the smaller size of the *H. corderoi* specimens.

Ants

According to our results *F. cunicularia, F. rufibarbis, F. pratensis* and *F. sanguinea* act as second intermediate hosts of *D. dendriticum* in the Porma mountain region and *F. rufibarbis* also in the flat

study area. The ant infection prevalence obtained by us (Table 4) and by other authors (Table 8) varied among the different species but mainly due to the type of sampling. So infection of *F. pratensis* collected directly from the anthills by us was very low (4·05%) in comparison with that obtained in ants gathered in tetania (100%); the same applied to *F. rufibarbis* (6·59% and 95·39%, respectively). This coincides with what was seen in *F. lugubris, F. pratensis* and *F. polyctena* by Jonlija, Milka & Cankovic (1973) and Rozman *et al.* (1974).

The fact that not all the ants collected in tetania hosted metacercariae of *D. dendriticum* in the abdomen could be because not all the brains of the dissected ants were examined. So it cannot be ruled out that they only contained the brainworm (or worms) which alter their behaviour. Non-infection could also be due to some ants not in tetania being collected amongst others in that stage on the same plant. It also has to be considered that infection by the fungus *Entomophthora* produces behaviour in ants similar to that caused by *D. dendriticum*, which makes them climb plants and grip them in the same way (Loos-Frank & Zimmermann, 1976).

The number of *D. dendriticum* metacercariae per ant found by us and by other authors (Table 5) varied among the different species and even within the same one. This variability could be due to: the time of year, as it is higher in summer (Paraschivescu, Hurghisiu & Popescu, 1976); the different affinity of the ant species for slimeballs (Loos-Frank, 1978); the type of vegetation and the ant species (Paraschivescu, 1978); the size of the abdomen of the different formicidae, greater in those that have a large abdomen (Kalkan, 1976); and the different species of ant, the size of each ant – the number of metacercariae in the abdominal cavity of ants is proportional to the amount of space available to them – and, moreover, possibly ecological and behavioural causes (Schuster, 1991).

We only found one *D. dendriticum* brainworm in the ventral part of the suboesophageal ganglion of the ant specimens examined, as did Anokhin (1966) in *F. pratensis*. Hohorst & Graefe (1961) and Hohorst (1962) confirmed the existence of a worm in the suboesophageal ganglion but sometimes in cases of superinfections, up to 2 or 3 metacercariae can be found in the brain of the same ant. Carney (1969) also comments that one, two or three *D. dendriticum* cercariae without tails cross the craw, arrive at the haemocele and then the suboesophageal ganglion where they encyst. Romig, Lucius & Frank (1980) using histological examination of the ant (*F. polyctena* and *F. rufa*) brains observed that most ants infected with *D. dendriticum* have only one brainworm which settles in the suboesophageal ganglion in 90 % of cases. There the larva was mainly situated in the ventral part in close contact with the origin of the mandibular nerves. These authors also found *D. dendriticum* larvae in the optic and antennal lobes and between the antennal lobe and the suboesophageal ganglion. Therefore they do not consider it clear whether the alteration in the behaviour of infected ants occurs because of a mechanical influence on the surrounding nerve tissue by the brainworm, since the larva located in the optic lobules or other parts of the brain induces the same changes in behaviour as the brainworm situated in the suboesophageal ganglion. This made them think it more likely that metabolic products influence the nervous system of the ant. The substances could be produced in the brainworm and the metacercariae from a certain stage of maturity. We could not find a cyst wall in the parasite larvae situated in the suboesophageal ganglion, as happened in most of the cerebral stages found by Romig *et al.* (1980).

The number of *F. pratensis* ant specimens in tetania found by us in the area around each anthill (1–9, x̄ = 5) was slightly lower than that observed by Badie (1979) (x̄ = 7–8) and much lower than those found by Paraschivescu (1981*b*) in *F. pratensis* (2–57) and Spindler, Zahler & Loos-Frank (1986) in *F. polyctena* (70). The most frequently occurring position of *F. pratensis* in tetania, fixed on different plant species around the anthill was head down, coinciding with what was indicated by Badie & Rondelaud (1988) in their studies of *F. nigricans* (= *F. pratensis*) and *F. cunicularia* in France. Spindler *et al.* (1986) found half of the ants (*F. polyctena*) on grasses forming a ring around the nest; most of the others were found up to 3·65 m from the nest, and some up to 14 m. By following the infected ants marked with coloured nail varnish, they observed that the choice of specific plant species as a support depended on relative abundance. Moreover the authors observed that most of the ants (55 %) attached themselves to plants with flowers and mainly preferred the blossoms (74 %). We also observed this, mainly in the numerous specimens of *F. rufibarbis* collected in the flat area. Paraschivescu (1981*a*) found ants on plants situated at a distance of one metre from the top of the nest, on different organs, such as the main nerve of the lower face of the foliar limb, on the corolla and on the edge of the limb. Paraschivescu & Raicev (1980) observed that in *F. pratensis* in tetania there were non-rhythmical movements of legs, gaster and antennae followed by rest periods. They found by marking ants in tetania and the plant parts that some of the ants remain for a longer period of time and return to the same place on the plant, even after 4 days, but others are not constant.

The changes in behaviour of the infected ants are mainly regulated by fluctuations in the ambient temperature, due to which the availability of metacercariae to the grazing animals has a circadian rhythm. The tetania of the infected ants normally occurs when solar intensity and temperature decrease at the end of the afternoon and disappears in the morning when insolation and temperature increase. The alteration in ant behaviour favours ingestion of the parasite by the definitive host. We usually observed the greatest number of ants in tetania in the early hours of the morning and late in the afternoon (Fig. 9), as did most authors. However, on cloudy or warmer days we also detected ants in tetania at the end of the morning and beginning of the afternoon (17–20 °C). These results coincide with what was indicated by Spindler *et al.* (1986) who once observed that 10 % of the 70 ants which were in tetania at 6·30 (at 4 °C) continued in that condition between 14·00 and 15·00 h, when the highest temperature was reached (17·5 °C). In addition these authors observed that, even in the middle of the day when the temperature was 20·5 °C, new ants arrived and attached themselves. They think that most of the ants attach themselves either at nightfall or (less likely) in the early morning. For Badie *et al.* (1973) the number of infected ants (*F. cunicularia* and *F. nigricans*) fixed to grass increased at low temperatures (9 to 10 °C) and decreased at higher temperatures (about 24 to 26 °C). The highest temperature at which we observed ants in tetania, 26·9 °C in the mountains and 28 °C in the flat area, was higher than that stated by Jonlija *et al.* (1972) 21 °C, Paraschivescu (1983) 18–20 °C, Schuster (1991) below 18 °C and Spindler *et al.* (1986) 20 °C, amongst others.

We observed active ants between March and November, infected ones from April to November in those collected from the nest and in tetania between May and October. The ants could be infected throughout their activity period, as sporocysts with well-developed cercariae were observed in January and from April to December. In Germany, Hohorst & Lämmler (1962) observed most of the attached ants (*F. rufibarbis* and *F. cunicularia*) in spring and autumn, collecting the former in the middle of March and the latter at the end of October. In Kazakhstan Dementev & Karabaev (1968) found the highest *D. dendriticum* infection values in spring and autumn and the minimum in summer in *F. pratensis*. In Bosnia-Herzegovina, Jonlija *et al.* (1972) observed the largest number of infected ants in September and October and Rozman *et al.* (1974) observed the greatest number in tetania in April. This decreased during the summer and increased again in September. In France, Badie (1975) observed that the maxima obtained in June and August (1973) and in April and September (1974) were preceded by heavy rains. Badie (1978) comments that it is not possible to consider the hypothesis that two well-defined ant infection periods exist because, although he observed infected ants (*F. cunicularia* and *F. nigricans*) in tetania between March and November, the infection rate varied with the year and month. The highest number of attached ants was detected in April 1974, June 1973, July 1975 and June 1976. In Russia, Urazbaev & Sumakovich (1981) observed the highest number in tetania between May and July, but considered that the definitive hosts can be infected throughout the grazing season (May–September). In the Republic of Georgia, Chitiashvili (1982) saw ants in tetania between April and June. In Germany, Paraschivescu (1983) mainly observed them in spring and Schuster (1991) collected them April to October.

Parasite transmission to the definitive hosts only occurs at the times when the ants are not hibernating. Nevertheless, the survival of the metacercariae in hibernating ants plays an important role in the epidemiology of dicrocoeliosis. Badie (1978) believes that some of the ants detected in March–April were infected at the end of autumn and during the winter of the previous year. Tarry (1969) states that the metacercariae can remain in the abdomen of the ant

for a year or more without affecting its survival and thus the parasite spends the winter and makes the pasture infective again in spring. In our study, specimens of *F. rufibarbis* collected in the field in tetania survived in the laboratory, kept and fed in artificial nests, for up to 6·5 months, whilst Paraschivescu (1981*c*) states that the infected ants starve and die near the anthill and the same occurs when they are transferred to the laboratory and kept at room temperature. According to Dementev (1979) none of the ants collected in tetania survived for 25 days. Spindler *et al.* (1986) stated that the ants do not readily die from starvation and can be kept without food for many days provided that the conditions are moist enough. Wolff (1976) carried out autopsies (50 days after exposure) on a group of ovines, which remained on *D. dendriticum*-contaminated Swiss pastures, and others from groups kept there for only 19 days, at different times during the grazing season. Wolff suggested that infection occurs continually, although the highest rate was recorded in April – May, due to the ingestion of metacercariae which survived the winter, and the lowest in October.

The size of *D. dendriticum* encysted metacercariae from *F. rufibarbis* collected in tetania was similar to that given by other authors for metacercariae extracted from different ant species and the thickness of the cyst wall was, in general, lower (Table 6). Schuster (1991) calculated the volume of metacercarial cysts and noted that it depended mainly on the burden per ant and the size of the ant, but was not influenced by the species or time of collection. Rosicky & Groschaft (1982) stated that the cyst size depended on the number of encysted metacercariae in the intermediate host. The cyst wall consists of 4 layers: an external fibrous one (0·42 μm thick), a hyaline one (26·66 μm), an internal fibrous one (0·46 μm) and an internal lipoprotein membrane, which immediately surrounds the larvae (Tverdokhlebov, 1984).

As regards the size of the excysted metacercariae (length 412·80 – 624 μm, $\bar{x} = 510·33$ μm; width 115·20 – 201·60 μm, $\bar{x} = 161·66$ μm), the lengths were similar to or slightly under those observed by Klesov & Popova (1959), 426 μm, and Krull & Mapes (1953), 540 – 635 μm, and below those of Fromunda, Popescu & Paraschivescu (1973), 758 μm, Paraschivescu *et al.* (1976), 753 μm, and Paraschivescu (1981*a*), 758 μm. The mean width was above that given by Klesov & Popova (1959), 80 μm, and slightly above those stated by Paraschivescu *et al.* (1976), Fromunda *et al.* (1973) and Paraschivescu (1981*a*), 150, 151 and 151 μm, respectively.

Metacercaria isoenzymatic study

On comparing the results obtained in this paper on the LDH, GPI, PGM and AcP activity in the metacercariae extracted from the abdomen of *F. rufibarbis* with those recorded by Campo *et al.* (1998) in *D. dendriticum* adults, it seems that all the activity bands forming the metacercariae patterns correspond to those of some of the patterns found in the adults. However, in general a more reduced number of activity bands and a lower band intensity were detected in the larval stages than in the adults. This coincides with what was observed by Wright, Rollinson & Goll (1979) who, on comparing the GPI patterns for the adult worms of *Schistosoma haematobium*, *Paramphistomum microbothrium* and *Echinostoma revolutum* with those of their respective larval stages, showed that the same bands of activity were present, although in the second and third species there were some additional bands and in all 3 the greater enzyme concentration in the adult extracts gave more intense patterns.

The activity band repeated in all the LDH models in the metacercariae was that situated at pH 7·08. This coincides with what was observed in the adults (Campo *et al.* 1998), although it was generally more intense in the case of the latter. As regards GPI, the common bands in the three models observed in the metacercariae were at pH 6·80 and 6·54, the latter being more intense, for both the metacercariae and the adults. In PGM the band at pH 6·20 was observed in both models obtained in the metacercariae and was also the most repeated and intense in the 5 models of the adults. The AcP bands were observed in the metacercariae in a very narrow pH range between 5·91 and 5·92, except a very weak AcP-1 model situated at 5·86. However, a band at pH 5·70 was always detected in the adults, although it was very weak.

The PGM-2 model is exactly the same in the metacercariae as in the adults and the PGM-1 of the former is differentiated from the PGM-4 of the latter only in the fact that the activity band at pH 6·20 is strong in the adults and very weak in the metacercariae. The GPI-2 model is common to both parasite development stages, except as regards the intensity of the 6·43 band, which is very weak in the metacercariae and strong in the adults.

In the isoenzymatic systems studied different patterns were generally observed in metacercariae extracted from the same ant, from ants from different places and sometimes from those collected in the same place. On the basis of the results obtained, the best enzymatic systems for the characterization of the *D. dendriticum* metacercariae from *F. rufibarbis* specimens were LDH, GPI and PGM.

Approach to dicrocoeliosis epidemiology

According to our field and experimental research the following epidemiological model can be deduced. *D. dendriticum* egg elimination by livestock occurs throughout the year, although the highest elimination takes place at the end of autumn and in winter – when the temperatures are lower. Considering that low temperatures do not affect egg survival, pasture contamination is very high at the end of winter and in spring. This facilitates egg ingestion by the molluscs which start to be active and are very abundant in spring. The molluscs infected at the beginning of this period could shed slimeballs with cercariae at the end of summer and during autumn, whilst those infected later can shed slimeballs the following year beginning in spring, if they survive the harsh winter. Approximately 45 days later the cercariae ingested by ants will have become infected metacercariae for the definitive hosts. This will allow the parasite cycle to be completed when the ants are ingested by ruminants on grazing, during the active period of the ants between March and November. Nevertheless, some infected ants survive hibernation in their nests during the winter and they are responsible for definitive host infection at the beginning of the spring. In the following months and until November, the risk of ruminant infection increases, due to the fact that, together with the above-mentioned ants, others infected by ingestion of the slimeballs shed by the infected molluscs that survive the winter, or by the ones that ingested eggs of the parasite at the end of winter or beginning of spring, will also be available in the pasture.

Therefore, the ingestion of the infective metacercariae (contained in the ants) by the definitive hosts and the number of *D. dendriticum* adult worms in the liver of the animals increase with the activity period of the ants. As a consequence of this, egg elimination reaches the highest values in January–February, that is, about two months after the last ingestion of the infected ants, before hibernation starts.

ACKNOWLEDGEMENTS

We wish to express our deep gratitude to: M. L. Carcedo, C. Espiniella and M. P. Del Pozo (CSIC, León) for their technical assistance; C. Otero, biologist, for her enthusiastic collaboration, mainly on ants; Dr Wiktor (Wroclaw, Poland) and Drs Gómez and Altonaga (Bilbao, Spain) for their help in the identification of some mollusc species; Drs Espadaler (Barcelona, Spain) and Collingwood (Leeds, U.K.) for ant identification; Dr Espadaler for helping us to see ants in tetania and the brainworm in the suboesophageal ganglion. This study was supported by the "Junta de Castilla y León" (Project No. 0701/89), by the British Council-Spanish Ministry of Education and Science (Joint Research Programme 1991–1993) and by the Spanish CICYT (Projects No. AG92-0588, 1FD97-0776 and 1FD97-1313-CO2-02).

REFERENCES

ALUNDA, J. M. & ROJO, F. A. (1983). Effect of infection rate and host age on the intramolluscan development of *Dicrocoelium dendriticum*. *Helminthologia* **20**, 251–258.

ANGELOVSKI, T. & ILIEV, A. (1978). The species of ants and their role in biology of *Dicrocoelium dendriticum* in SR Macedonia. *Proceedings of the Second European Multicolloquium of Parasitology*, 1–6 September, 1975, Trogir, Yugoslavia. Association of Yugoslav Parasitologists, Belgrade, 163–166.

ANOKHIN, I. A. (1966). [Diurnal cycle of the activity and behaviour of ants (*Formica pratensis* Retz.) invaded by metacercaria of *Dicrocoelium lanceatum* during the grazing period). *Zoologicheskie Zhurnal* **45**, 687–692.

BADIE, A. (1975). Cycle annuel d'activité des fourmis parasitées par les métacercaires de *Dicrocoelium lanceolatum* (Rudolphi, 1819). *Annales de Recherches Vétérinaires* **6**, 259–269.

BADIE, A. (1978). La dicroceliose ovine: incidence des facteurs climatiques et contribution à la mise au point d'une méthode de prévision. *Annales de Parasitologie Humaine et Comparée* **53**, 373–385.

BADIE, A. (1979). La petite douve *Dicrocoelium lanceolatum* ou *Dicrocoelium dendriticum*–ses hôtes intermédiaires. *Pâtre* **268**, 51–54. .

BADIE, A. & RONDELAUD, D. (1987). Les mollusques hôtes intermédiares de *Dicrocoelium lanceolatum* Rudolphi. A propos de quinze années d'observations. *Bulletin de la Société Française de Parasitologie* **5**, 105–108.

BADIE, A. & RONDELAUD, D. (1988). Les fourmis parasitées par *Dicrocoelium lanceolatum* Rudolphi in Limousin. Les relations avec le support végétal. *Revue de Médecine Vétérinaire* **139**, 629–633.

BADIE, A., VINCENT, M., MOREL-VAREILLE, C. & RONDELAUD, D. (1973). Cycle de *Dicrocoelium dendriticum* (Rudolphi, 1819) en Limousin. Ethologie des fourmis parasitées par les métacercaires. *Comptes Rendus des Séances de la Société de Biologie* **167**, 725–727.

BOCHAROVA, M. M. (1983). (Foci of dicrocoeliasis in the mountain forest-steppe of the northern slopes of the central Caucasus). In *Fauna i Ékologiya Zhivotnykh Severnykh Sklonov Tsentral'nogo Kavkaza*. (*Mezhvuzovskii Sbornik*). Ordzhonikidze; USSR; Severo-Osetinskii Gosudarstvennyi Institut, 10–13

BONAVITA, D. (1961). Conditions de la production de l'épiphragme chez quelques mollusques Hélicides. *Comptes rendus hebdomadaires des séances de l'Académie des sciences* (*Paris*) **253**, 3101–3102.

BORAY, J. C. (1985). Flukes of domestic animals. In *Parasites, Pests and Predators*. (*World Animal Science, B2*) (ed. Gaafar, S. M., Howard, W. E. & Marsh, R. E.), pp. 179–218. Amsterdam, Elsevier Science Publishers B. V.

CAMPO, R., MANGA-GONZÁLEZ, M. Y. & GONZÁLEZ-LANZA, C. (2000). Relationship between egg output and parasitic burden in lambs experimentally infected with different doses of *Dicrocoelium dendriticum* (Digenea). *Veterinary Parasitology* **87**, 139–149.

CAMPO, R., MANGA-GONZÁLEZ, M. Y., GONZÁLEZ-LANZA, C. & DEL-POZO, P. (1992). Detección isoenzimática de las primeras fases larvarias de *Dicrocoelium dendriticum* (Trematoda) en moluscos hospedadores intermediarios infestados experimentalmente. *Cuadernos de Investigación Biológica* **17**, 44.

CAMPO, R., MANGA-GONZÁLEZ, M. Y., GONZÁLEZ-LANZA, C., ROLLINSON, D. & SANDOVAL, H. (1998). Characterization

of adult *Dicrocoelium dendriticum* by isoelectric focusing. *Journal of Helminthology* **72**, 109–116.

CARNEY, W. P. (1969). Behavioral and morphological changes in carpenter ants harboring dicrocoeliid metacercariae. *The American Midland Naturalist* **82**, 605–611.

CASTILLEJO, J. (1998). *Guía de las Babosas Ibéricas*. Santiago de Compostela (España), Real Academia Galega de Ciencias.

CAVANI, C., LOSI, G., MANFREDINI, M., PAVONCELLI, R. M., PIETROBELLI, M. & RESTANI, R. (1982). Ricerche sull'influenza della dicroceliasi sulle caratteristiche quantitative e qualitative della produzione di latte in pecore. *Obiettivi e Documenti Veterinari* **3**, 59–63.

CHITIASHVILI, B. G. (1982). [Effect of temperature and relative humidity on the activity of infected ants]. *Nauchnye Trudy, Gruzinskii Zootekhnichesko-Veterinarnyi Institut* **127**, 32–34.

COLLINGWOOD, C. A. (1978). A provisional list of Iberian Formicidae with a key to the worker caste (Hym. Aculeata). *EOS* **52**, 65–95.

CORDERO, M., CASTAÑÓN, L. & REGUERA, A. (1994). *Índice-catálogo de Zooparásitos Ibéricos*. Universidad León (España), Secretariado de Publicaciones.

DEL RIO, J. (1967). Epizootiología de la dicroceliosis en la provincia de León. *Anales de la Facultad de Veterinaria de León* **13**, 211–253.

DEMENTEV, I. S. (1968). (Epizootiology of dicrocoeliasis of sheep in south-eastern Kazakhstan). *Materiali Seminara-Soveshchaniya po Borbe s Gel'mintozami sel'.-khoz. Zhivotnikh v Chimkente, Alma-Ata*, 56–58.

DEMENTEV, I. S. (1979). [The role of abiotic ants from the genus *Formica* in the epizootiology of dicrocoeliasis]. *Materialy respublikanskogo seminara po bor'be s parazitarnymi boleznyami sel'skokhozyaistvennykh zhivotnykh, posvyashchennogo 100-letiyu so dnya rozhdeniya Akademika K. I. Skryabina. Alma-Ata, USSR*, 42–47.

DEMENTEV, I. S. & KARABAEV, D. K. (1968). (Infection of terrestrial molluscs and ants by *Dicrocoelium dendriticum* larvae in southeastern Kazakhstan). *Vestik sel'-skokhozyaistvennoi Nauki, Alma-Ata* **12**, 51–55.

DENEV, I., SAVOVA, S., STOIMENOV, K., TAHIROV, B., DONEV, A., KASSABOV, R. & PETKOV, P. (1970). [An investigation into the dicrocoeliosis in north-eastern Bulgaria. II. On the intermediary and additional hosts of *Dicrocoelium lanceatum* Stiles et Hassall, 1896]. *Veterinamomeditsinski Nauki, Sofia* **7**, 23–31.

DHAR, D. N. & SINGH, H. S. (1963). Pathology of liver in dicrocoeliasis. *Indian Journal of Veterinary Science and Animal Husbandry* **33**, 200–210.

DIL'MAN, P. N. (1978). [*Formica subpilosa* Ruzs, a new second intermediate host of *Dicrocoelium lanceatum* Stiles et Hassall, 1895 (*D. lanceolatum*)]. *Trudy Nauchno-Issledovatel'skogo Veterinarnogo Instituta Tadzhikskoi SSR* **8**, 88–89.

ECKERT, J. & HERTZBERG, H. (1994). Parasite control in transhumant situations. *Veterinary Parasitology* **54**, 103–125.

ENGEL, H. (1957). Oekologisch-faunitische Studien im Rhône-delta, unter besonderer Berücksichtigung der Mollusken. *Bonner Zoologische Beiträge* **8**, 5–55.

FERRERAS, M. C., GARCÍA-IGLESIAS, M. J., MANGA-GONZÁLEZ,

M. Y., PÉREZ, C., CAMPO, R., GONZÁLEZ-LANZA, C., ESCUDERO, A. & GARCÍA-MARÍN, J. F. (1997). Lesiones hepáticas en corderos infectados experimentalmente con *Dicrocoelium dendriticum*. *Acta Parasitológica Portuguesa* **4**, 86.

FERRERAS, M. C., MANGA-GONZÁLEZ, M. Y., PÉREZ, C., GARCÍA-IGLESIAS, M. J., CAMPO, R., GONZÁLEZ-LANZA, C., ESCUDERO, A. & GARCÍA-MARÍN, J. F. (2000). Local immune response to experimental ovine dicrocoeliosis. *Acta Parasitologica* **45**, 198.

FROMUNDA, V., PARASCHIVESCU, D. & POPESCU, S. (1965). Cercetari privind gazdele complementare pentru *Dicrocoelium lanceatum* in Republica Socialista Romania. *Lucrarile ICVB Pasteur* **4**, 269–280.

FROMUNDA, V., POPESCU, S. & PARASCHIVESCU, D. (1973). Contributions à la connaissance des formes larvaires et pré-imago du trematode *Dicrocoelium lanceatum* (Stiles et Hassall, 1896). *Archiva Veterinaria* **10**, 85–91.

GÓMEZ, B. J., MANGA-GONZÁLEZ, M. Y., ANGULO, E. & GONZÁLEZ-LANZA, C. (1996). Alteraciones histológicas producidas por *Dicrocoelium dendriticum* (Trematoda) en dos especies de *Cernuella* (Mollusca) sacrificadas a los tres meses post-infestación. *Iberus* **14**, 189–195.

GONZÁLEZ-LANZA, C., MANGA-GONZÁLEZ, M. Y., CAMPO, R. & DEL-POZO, P. (1997). Larval development of *Dicrocoelium dendriticum* in *Cernuella* (*Xeromagna*) *cespitum arigonis* under controlled laboratory conditions. *Journal of Helminthology* **71**, 311–317.

GONZÁLEZ-LANZA, C., MANGA-GONZÁLEZ, M. Y., CAMPO, R., DEL-POZO, P., SANDOVAL, H., OLEAGA, A. & RAMAJO, V. (2000). IgG antibody response to ES or somatic antigens of *Dicrocoelium dendriticum* (Trematoda) in experimentally infected sheep evaluated by ELISA. *Parasitology Research* **86**, 472–479.

GONZÁLEZ-LANZA, C., MANGA-GONZÁLEZ, M. Y. & DEL-POZO, P. (1993). Coprological study of the *Dicrocoelium dendriticum* (Digenea) egg elimination by cattle in highland areas in León Province, Northwest Spain. *Parasitology Research* **79**, 488–491.

GROSCHAFT, J. (1961). Mravenci doplnkoví mezihostitelé motolice kopinaté (*Dicrocoelium dendriticum* Rudolphi 1819)]. *Ceskoslovenská Parasitologie* **8**, 151–165.

HINAIDY, H. K. (1983). *Dicrocoelium suppereri* nomen novum (syn. *D. orientalis* Sudarikov et Ryjikov 1951), ein neuer Trematode für die Parasitenfauna Österreichs. *Zentralblatt für Veterinärmedizin B* **30**, 576–589.

HOHORST, W. (1962). Die Rolle der Ameisen im Entwicklungsgang des Lanzettegels (*Dicrocoelium dendriticum*). *Zeitschrift für Parasitenkunde* **22**, 105.

HOHORST, W. & GRAEFE, G. (1961). Ameisen-obligatorische Zwischenwirte des Lanzettegels (*Dicrocoelium dendriticum*). *Naturwissenschaften, Berlin* **48**, 229–230.

HOHORST, W. & LÄMMLER, G. (1962). Experimentelle Dicrocoeliose-Studien. *Zeitschrift für Tropenmedizin und Parasitologie* **13**, 377–397.

JONLIJA, R., CANKOVIC, M. & ROZMAN, M. (1972). Examination of ants the second intermediary hosts of *Dicrocoelium lanceatum* in Bosnia and Hercegovina, I. *Veterinaria* **21**, 317–322.

JONLIJA, R., MILKA, R. & CANKOVIC, M. (1973). Ispitivanje infestiranosti puzeva i mrava razvojnim oblicima

Dicrocoelium lanceatum u Bosni i Hercegovini II. *Veterinaria, Sarajevo* **22**, 449–470.

KALKAN, A. (1970). *Dicrocoelium dendriticum* (Rudolphi, 1819) Looss, 1899 in Turkey. 1. Field studies of intermediate and final hosts in the South Marmara region, 1968. *British Veterinary Journal* **127**, 67–75.

KALKAN, A. (1976). [*Dicrocoelium dendriticum* in Turkey. 2. Observations on second intermediate hosts (ants) in the South Marmara Region]. *Etlik Veteriner Bakteriyoloji Enstitüsü Dergisi* **4**, 11–37.

KARANFILOVSKI, G. (1983). Economic losses caused by fascioliasis and dicrocoeliasis to livestock production in Pljevlja and surrounding area. *Veterinaria Yugoslavia* **32**, 253–260.

KERNEY, M. P., CAMERON, R. A. D. & JUNGBLUTH, J. H. (1983). *Die Landschnecken Nord-und Mitteleuropas.* Hamburg und Berlin, Verlag Paul Parey.

KLESOV, M. D. & POPOVA, Z. G. (1958). [The biology of *Dicrocoelium dendriticum* (Stiles & Hassall, 1896) – the agent of dicrocoeliasis in ruminants]. *Zoologicheski Zhurnal* **37**, 504–510.

KLESOV, M. D. & POPOVA, Z. G. (1959). [Study of the biology of *Dicrocoelium* and the epizootiology of dicrocoeliasis in ruminants]. *Nauchnie Trudi Ukrainski Nauchno-Issledovatelski Institut Eksperimentalnoi Veterinarii* **25**, 5–18.

KRULL, W. H. & MAPES, C. R. (1952). Studies on the biology of *Dicrocoelium dendriticum* (Rudolphi, 1819) Looss, 1899 (Trematoda: Dicrocoeliidae), including its relation to the intermediate host, *Cionella lubrica* (Müller). VII. The second intermediate host of *Dicrocoelium dendriticum*. *Cornell Veterinarian* **42**, 603–604.

KRULL, W. H. & MAPES, C. R. (1953). Studies on the biology of *Dicrocoelium dendriticum* (Rudolphi, 1819) Looss, 1899 (Trematoda: Dicrocoeliidae), including its relation to the intermediate host, *Cionella lubrica* (Müller). IX. Notes on the cyst, metacercaria, and infection in the ant, *Formica fusca*. *Cornell Veterinarian* **43**, 389–410.

KUZ'MOVICH, L. G. & KOSTINIK, I. M. (1975). (The occurrence of *Dicrocoelium lanceatum* Stiles et Hassall, 1896 in *Helicopsis instabilis*). In *Problemy Parazitologii. Materialy VIII nauchnoi konferentsii parazitologov UkSSR* **1**, 289–291.

LA RUE, G. (1957). Parasitological reviews. The Classification of Digenetic Trematoda: A Review and a New System. *Experimental Parasitology* **6**, 306–349.

LOOS-FRANK, B. (1978). Zum Verhalten von Ameisen der Gattung *Formica* (Hymenoptera: Formicidae) gegenüber Schleimballen des Kleinen Leberegels *Dicrocoelium dendriticum* (Digenea: Dicrocoeliidae) und über Infektionsbedingte Veränderungen ihrer Hämolymphe. *Entomologica Germanica* **4**, 12–23.

LOOS-FRANK, B. & ZIMMERMANN, G. (1976). Über eine dem Dicrocoelien-Befall analoge Verhaltensände-rung bei Ameisen der Gattung *Formica* durch einen Pilz der Gattung *Entomophthora*. *Zeitschrift für Parasitenkunde* **49**, 281–289.

LUCIUS, R. (1981). Untersuchungen zur Biologie, Pathologie und Ökologie von *Dicrocoelium hospes* Looss, 1907 (Trematodes, Dicrocoeliidae). Dissertation zur Erlangung des Grades eines Doktors der Naturwissenschaften vorgelegt der Fakultät II (Biologie) der Universität Hohenheim.

LUKIN, A. K. (1980). (The prevalence and economic loss due to dicrocoeliasis in ruminants). *Trudy Saratovskoi Nauchno-Issledovatel'skoi veterinarnoi Stantsii* **14**, 76–79.

MALEK, E. A. (1980). *Snail-Transmitted Parasitic Diseases. Vol I & II*, Florida, CRC Press, Inc.

MANGA-GONZÁLEZ, M. Y. (1983). *Los Helicidae (Gastropoda, Pulmonata) de la Provincia de León.* Institución "Fray Bernardino de Sahagún", Excma. Diputación Provincial de León, Consejo Superior de Investigaciones Científicas (CECEL), León (España)

MANGA-GONZÁLEZ, M. Y. (1987). Some aspects of the biology and helminthofaune of *Helicella* (*Helicella*) *itala* (Linnaeus, 1758) (Mollusca). Natural infection by Dicrocoeliidae (Trematoda). *Revista Ibérica de Parasitología*, Volumen extraordinario Enero 1987, 131–148.

MANGA-GONZÁLEZ, M. Y. (1992). Some land molluscs species involved in the life cycle of *Dicrocoelium dendriticum* (Trematoda) in the wild in the province of León (NW Spain). In *Abstracts of the 11th International Malacologial Congress, Siena* (ed. Giusti, F. & Manganelli, G.), pp. 248–249. Italy, Unitas Malacological. University of Siena.

MANGA-GONZÁLEZ, M. Y. (1999). Trematodos. In *Parasitología Veterinaria* (ed. Cordero-del-Campillo, M. & Rojo-Vázquez, F. A.), pp. 79–104. Madrid, McGraw-Hill. Interamericana.

MANGA-GONZÁLEZ, M. Y., GONZÁLEZ-LANZA, M. C. & DEL-POZO, P. (1991). Dynamics of the elimination of *Dicrocoelium dendriticum* (Trematoda, Digenea) eggs in the faeces of lambs and ewes in the Porma Basin (León, NW Spain). *Annales de Parasitologie Humaine et Comparée* **66**, 57–61.

MANGA-GONZÁLEZ, M. Y. & QUIROZ-ROMERO, H. (1999). Dicroceliosis. In *Parasitología Veterinaria* (ed. Cordero-del-Campillo, M. & Rojo-Vázquez, F. A.), pp. 272–282. Madrid, McGraw-Hill. Interamericana.

MAPES, C. R. (1951). Studies on the biology of *Dicrocoelium dendriticum* (Rudolphi, 1819) Looss, 1899 (Trematoda: Dicrocoeliidae), including its relation to the intermediate host, *Cionella lubrica* (Müller). I. A study of *Dicrocoelium dendriticum* and *Dicrocoelium* infection. *Cornell Veterinarian* **41**, 382–432.

MOHAMED, A. R. E. & MUMMERY, V. (1990). Human dicrocoeliosis. Report on 208 cases from Saudi Arabia. *Tropical and Geographical Medicine* **42**, 1–7.

ODENING, K. (1969). Der Lanzettegel oder Kleine Leberegel (*Dicrocoelium dendriticum*). *Merkblätter über Angewandte Parasitenkunde und Schädlingsbekämpfung* (*Supplement Angewandte Parasitologie*, **10**) No. 16, 265–281.

ORTIZ DE ZÁRATE, A. (1950). Observaciones anatómicas y posición sistemática de varios helícidos españoles. III. (Especies de los subgéneros *Candidula, Helicella sensu stricto, Xerotricha, Xeromagna* y *Pseudoxerotricha* nov. subg.). *Boletín de la Real Sociedad Española de Historia Natural* **48**, 21–85.

OVCHARENKO, D. A. (1964). (The biology of *Dicrocoelium dendriticum* in deer preserves in the Far East). *Vestnik Leningradskogo Universitata. Seriya Biologii* **19**, 35–39.

PARASCHIVESCU, D. (1976). (Dynamics of tetany in ants in

some pastures infested with *Dicrocoelium* in Romania). *Studii si Comunicari, Muzeul de Stiintele Naturii Bacau* **9**, 65–77.

PARASCHIVESCU, D. (1978). Dynamics of formicids (Hym., Formicidae) in tetany in some meadows with dicrocelioza in Romania. *Travaux du Muséum d'Histoire Naturell 'Grigore Antipa'* **19**, 321–323.

PARASCHIVESCU, D. (1981a). Etho-ökologische Untersuchungen an einigen Formiciden-Arten als Zwischenwirte von *Dicrocoelium dendriticum.* *Waldhygiene* **14**, 65–72.

PARASCHIVESCU, D. (1981b). Untersuchungen zur künstlichen Infektion von Schaflämmern mit dem Kleinen Leberegel *Dicrocoelium dendriticum* Rudolphi über den Zwischenwirt *Formica pratensis* Retz. *Waldhygiene* **14**, 73–78.

PARASCHIVESCU, D. (1981c). Experimentelle Untersuchungen zur Unterbrechung der Infektionskette beim Kleinen Leberegel *Dicrocoelium dendriticum* Rudolphi (Trematoda) durch Schutz der Nesthügel des Zwischenwirtes *Formica pratensis* Retz. (Formicidae). *Waldhygiene* **14**, 79–84.

PARASCHIVESCU, D. (1983). Untersuchungen zur Temperaturabhangigkeit der Tetaniephase von *Formica pratensis* bei einer Infektion mit Stadien des Kleinen Leberegels *Dicrocoelium dendriticum.* *Waldhygiene* **15**, 21–24.

PARASCHIVESCU, D., HURGHISIU, I. & POPESCU, S. (1976). Bioecologic and biochemical research upon Formicidae complementary hosts of the *Dicrocoelium lanceatum* fluke (Stiles and Hassall, 1896). *Archiva Veterinaria* **11/12**, 159–178.

PARASCHIVESCU, D. & RAICEV, C. (1980). Experimental ecological investigations on the tetany of the species *Formica pratensis* complementary host of the trematode *Dicrocoelium dendriticum. Travaux du Muséum d'Histoire Naturelle "Grigore Antipa"* **22**, 299–302.

PIANA, P. (1882). Le cercarie nei molluschi studiata in rapporto colla presenza del *Distoma epatico* e del *Distoma lanceolato* nel fegato del ruminanti domestici. *Clinica Veterinaria Milano* **5**, 306–314.

POPOV, K. K. & KALITINA, A. I. (1962). (The development of *Dicrocoelium dendriticum* on high altitude grazing grounds of central Caucasus). *Zoologicheski Zhurnal* **41**, 1793–1797.

RIDDLE, W. A. (1983). Physiological ecology of land snails and slugs. In *The Mollusca. Ecology* (ed W. D. Russell-Hunter) **6**, 431–461. Orlando (Florida), Academic Press, Inc.

RIEDEL, A. (1972). Zur Kenntnis der Zonitidae (Gastropoda) Spaniens. *Annales Zoologici, Warszawa* **29**, 115–145.

ROMING, T., LUCIUS, R. & FRANK, W. (1980). Cerebral larvae in the second intermediate host of *Dicrocoelium dendriticum* (Rudolphi, 1819) and *Dicrocoelium hospes* Looss, 1907 (Trematodes, Dicrocoeliidae). *Zeitschrift für Parasitenkunde* **63**, 277–286.

ROSICKY, B. & GROSCHAFT, J. (1982). Dicrocoeliosis. In *CRC Handbook Series in Zoonoses. Section C: Parasitic Zoonoses. Volume III* (ed. Hillyer, G. V. & Hopla, C. E.), pp. 33–52. Boca Raton, Florida, USA, CRC Press, Inc.

ROZMAN, S., GRADJANIN, M. & CANKOVIC, M. (1974). The transitory hosts of *Dicrocoelium dendriticum* in the mountain area of Bosnia and Herzegovina. *Proceedings of the Third International Congress of Parasitology, Munich* vol. **1**, 513–514. Vienna, FACTA Publication.

ROZMAN, S., JONLIJA, R. & MUSTAPIC, A. (1971). Ispitivanje suhozemnih puzeva na razvajne oblike *Dicrocoelium lanceolatum*. *Acta Parasitologica Yugoslavica* **2**, 99–103.

SÁNCHEZ-CAMPOS, S., TUÑÓN, M. J., GONZÁLEZ, P., CAMPO, R., FERRERAS, M. C., MANGA-GONZÁLEZ, M. Y. & GONZÁLEZ-GALLEGO, J. (1996). Effects of experimental dicrocoeliosis on oxidative drug metabolism in hamster liver. *Comparative Biochemistry and Physiology* **115C**, 55–60.

SCHMIDT, F. (1967). Zur Kentnis der Trematodenlarven aus Landmollusken. ll. Über eine Cercarie (Trematoda: Digenea: Dicrocoeliidae) aus *Clausilia bidentata* (Ström, 1765) und die zugehörige Metacercarie aus Isopoden. *Zeitschrift für Parasitenkunde* **29**, 85–102.

SCHUSTER, R. (1987). Ein geschichtlicher Überblick zur Namensgebung des Lanzettegels *Dicrocoelium dendriticum*. *Angewandte Parasitologie* **28**, 205–206.

SCHUSTER, R. (1991). Factors influencing the metacercarial intensity in ants and the size of *Dicrocoelium dendriticum* metacercarial cyst. *Journal of Helminthologie* **65**, 275–279.

SCHUSTER, R. (1992). Zur Beeinflussung von *Helicella obvia* durch *Dicrocoelium* – Parthenitae. *Angewandte Parasitologie* **33**, 61–64.

SCHUSTER, R. (1993). Infection patterns in the first intermediate host of *Dicrocoelium dendriticum*. *Veterinary Parasitology* **47**, 235–243.

SCHUSTER, R. & NEUMANN, B. (1988). Zum jahreszeitlichen Auftreten von *Dicrocoelium dendriticum* in Zwischenwirten. *Angewandte Parasitologie* **29**, 31–36.

SEIFERT, B. (1992). *Formica nigricans*, 1909 – An ecomorph of *Formica pratensis* Retzius, 1783 (Hymenoptera, Formicidae). *Entomologica Fennica* **2**, 217–226.

SPINDLER, E. M., ZAHLER, M. & LOOS-FRANK, B. (1986). Behavioural aspects of ants as second intermediate hosts of *Dicrocoelium dendriticum*. *Zeitschrift für Parasitenkunde* **72**, 689–692.

SRIVASTAVA, G. C. (1975). The intensity of infection in naturally infected *Formica pratensis* with the metacercariae of *Dicrocoelium dendriticum* in relation to their size. *Journal of Helminthology* **49**, 57–64.

STAIKOU, A., LAZARIDOU-DIMITRIADOU, M. & KATTOULAS, M. E. (1989). Behavioural patterns of the edible snail *Helix lucorum* L. in the different seasons of the year in the field. *Haliotis* **19**, 129–136.

TANG, C., TANG, Z., TANG, L., CUI, Q., LU, H. & QIAN, Y. (1983). (Studies on the biology and epizootics of *Dicrocoelium chinensis* in the eastern inner Mongol Autonomous region). *Acta Zoologica Sinica* **29**, 340–349.

TARRY, D. W. (1969). *Dicrocoelium dendriticum*: the life cycle in Britain. *Journal of Helminthology* **43**, 403–416.

TEGELSTRÖM, H., NILSSON, G. & WYONI, P. I. (1983). Lack of species differences in isoelectric focused proteins in the *Formica rufa* group (Hymenoptera, Formicidae). *Hereditas* **98**, 161–165.

TOHMÉ, H. & TOHMÉ, G. (1977). Les hôtes intermédiares du cycle évolutif de la petite Douve du foie du mouton au Liban et en Syrie. *Annales de Parasitologie Humaine et Comparée* **52**, 1–5.

TVERDOKHLEBOV, P. T. (1970). (Significance of different types of pastures in the infection of animals by *Dicrocoelium*). *Byulleten' Vsesoyuznogo Instituta Gel'mintologii im. K. I. Skryabina* **4**, 147–149.

TVERDOKHLEBOV, P. T. (1984). [Morphology of the cyst wall and excystment mechanisms of *Dicrocoelium lanceatum* metacercariae]. In *Gel'minty sel's kokhozyaistvennykh i okhotnich'e-promyslovykh zhivotnykh* (ed. Sonin, M. D.), pp. 189–198. Moscow, USSR, Nauka.

URAZBAEV, G. A. (1979). [The first and second intermediate hosts of *Dicrocoelium lanceolatum* in the interior of Tyan' Shan and the seasonal variations in the prevalence of infection]. *Byulleten' Vsesosyuznogo Instituta Gel'mintologii im. K. I. Skryabina* **24**, 60–65.

URAZBAEV, G. A. & SUMAKOVICH, E. E. (1981). (The intermediate and additional hosts of *Dicrocoelium lanceatum* in the inner Tien-Shan region and the seasonal dynamics of their infectivity). *Byulleten' Vsesosyuznogo Instituta Gel'mintologii im. K.l. Skryabina* **24**, 60–65.

VAUGHT, K. C. (1989). *A Classification of the Living Mollusca*, Florida, American Malacologists, Inc.

VERSHININ, I. I. (1957). (Epizootiology of *Dicrocoelium* infections of sheep and its biology in the Kaluga region). *Trudi Moskovskoi Veterinarnoi Akademii* **19**, 3–15.

VESELINOV, G. D. (1962). (Study of the development of *Dicrocoelium dendriticum* in Bulgaria). *Izvestiya na Tsentralnata Khelmintologichna Laboratoriya Sofia* **7**, 127–135.

VON LINSTOW, D. (1887). Helminthologische Untersuchungen, *Cercaria vitrina n.sp. Zoologische Jahrbücher. Abteilung für Systematik Oekologie und Geographie der Tiere (Jena)* **3**, 105–106.

VUJIC, B. (1972). [Biology of *Dicrocoelium lanceolatum* (Rudolphi 1803) in some mountainous and flat areas in Serbia (II)]. *Veterinarski Glasnik* **26**, 827–833.

WIKTOR, A. (2000). Agriolimacidae (Gastropoda: Pulmonata) – A Systematic Monograph. *Annales Zoologici* **49**, 347–590.

WOLDA, H., ZWEEP, A. & SCHUITEMA, K. A. (1971). The role of food in the dynamics of populations on the land snail *Cepaea nemoralis*. *Oecologia* **7**, 361–381.

WOLFF, K. (1976). Zur Epizootiologie der Dicrocoeliose des Schafes. *Berliner und Münchener Tierärztliche Wochenschrift* **89**, 272–276.

WOLFF, K., HAUSER, B. & WILD, P. (1984). Dicrocoeliose des Schafes: Untersuchungen zur Pathogenese und zur Regeneration der Leber nach Therapie. *Berliner und Münchener Tierärztliche Wochenschrift* **97**, 378–387.

WRIGHT, C. A., ROLLINSON, D. & GOLL, P. H. (1979). Parasites in *Bulinus senegalensis* (Mollusca: Planorbidae) and their detection. *Parasitology* **79**, 95–105.

Fasciola hepatica and lymnaeid snails occurring at very high altitude in South America

S. MAS-COMA*, I. R. FUNATSU *and* M. D. BARGUES

Departamento de Parasitología, Facultad de Farmacia, Universidad de Valencia, Av. Vicent Andrés Estellés s/n, 46100 Burjassot-Valencia, Spain

SUMMARY

Fascioliasis due to the digenean species *Fasciola hepatica* has recently proved to be an important public health problem, with human cases reported in countries of the five continents, including severe symptoms and pathology, with singular epidemiological characteristics, and presenting human endemic areas ranging from hypo- to hyperendemic. One of the singular epidemiological characteristics of human fascioliasis is the link of the hyperendemic areas to very high altitude regions, at least in South America. The Northern Bolivian Altiplano, located at very high altitude (3800–4100 m), presents the highest prevalences and intensities of human fascioliasis known. Sequences of the internal transcribed spacers ITS-1 and ITS-2 of the nuclear ribosomal DNA of Altiplanic *Fasciola hepatica* and the intermediate snail host *Lymnaea truncatula* suggest that both were recently introduced from Europe. Studies were undertaken to understand how the liver fluke and its lymnaeid snail host adapted to the extreme environmental conditions of the high altitude and succeeded in giving rise to high infection rates. In experimental infections of Altiplanic lymnaeids carried out with liver fluke isolates from Altiplanic sheep and cattle, the following aspects were studied: miracidium development inside the egg, infectivity of miracidia, prepatent period, shedding period, chronobiology of cercarial emergence, number of cercariae shed by individual snails, survival of molluscs at the beginning of the shedding process, survival of infected snails after the end of the shedding period and longevity of shedding and non-shedding snails. When comparing the development characteristics of European *F. hepatica* and *L. truncatula*, a longer cercarial shedding period and a higher cercarial production were observed, both aspects related to a greater survival capacity of the infected lymnaeid snails from the Altiplano. These differences would appear to favour transmission and may be interpreted as strategies associated with adaptation to high altitude conditions.

Key words: Human fascioliasis, *Fasciola hepatica*, *Lymnaea truncatula*, rDNA ITS-1 and ITS-2, fluke larval development, snail survival, adaptation strategies, high altitude, Bolivia, Andean countries.

INTRODUCTION

Fascioliasis due to the digenean species *Fasciola hepatica* is a well-known veterinary problem worldwide. However, studies carried out in recent years have shown it to be an important public health problem as well (Chen & Mott, 1990; WHO, 1995; Mas-Coma, Bargues & Esteban, 1999). Human cases have been reported in countries of the five continents (Esteban, Bargues & Mas-Coma, 1998), with severe symptoms and pathology being observed (Chen & Mott, 1990; Mas-Coma & Bargues, 1997; Mas-Coma *et al.* 2000). Human endemic areas range from hypo- to hyperendemic (Mas-Coma, Esteban & Bargues, 1999) with characteristic epidemiology (Mas-Coma, 1998).

F. hepatica is a parasite originally associated with Europe, where it is almost exclusively transmitted by the species *Lymnaea truncatula* (see review by Oviedo, Bargues & Mas-Coma, 1996). Although *L. truncatula* may be found even up to 2600 m altitude

in Europe, *F. hepatica* is a parasite typical of lowlands and is never found at high altitudes in Europe. This different altitudinal distribution is mainly related to temperature, as it is known that *F. hepatica* larval development is arrested below 10 °C (Mas-Coma & Bargues, 1997), although a lack of compatibility between the fluke and the populations of lymnaeids from high altitudes has also been suggested (Oviedo, Bargues & Mas-Coma, 1995).

Several lowland regions presenting fascioliasis health problems are known, such as human hypo-endemic zones in France, Corsica and Chile, meso-endemic areas in Portugal and repetitive epidemics in Cuba and Iran (see review by Esteban *et al.* 1998), and recently even hyperendemic areas in the Nile Delta, Egypt (Curtale *et al.* 2000). Interestingly, however, in the Andean countries human fascioliasis hyperendemic areas appear to be linked to very high altitude regions of 3000–4100 m (Esteban *et al.* 1998; Mas-Coma *et al.* 1999). Fascioliasis transmission at lower altitudes, between 1000 and 2500 m, is known in different zones of Europe (Manga-Gonzalez, Gonzalez-Lanza & Otero-Merino, 1991; Pareau *et al.* 1994), Africa (Van Someren, 1946; Bergeon & Laurent, 1970; Scott & Goll, 1977; Loker *et al.* 1993) and Asia (Kendall, 1954; Kendall

* Corresponding author: Prof. Dr Santiago Mas-Coma, Departamento de Parasitologia, Facultad de Farmacia, Universidad de Valencia, Av. Vicent Andres Estelles s/n, 46100 Burjassot-Valencia, Spain. Tel: +34-96-386-42-98. Fax: +34-96-386-47-69. E-mail: S.Mas.Coma@uv.es

Parasitology (2001), **123**, S115–S127.   © 2001 Cambridge University Press
DOI: 10.1017/S0031182001008034

& Parfitt, 1959; Morel & Mahato, 1987), as well as in Central (Perez-Reyes, Jimenez-Nava & Varela-Ramirez, 1985) and South America (Morales & Pino, 1981, 1983).

Among the latter, the Northern Bolivian Altiplano, located at the very high altitudes between 3800–4100 m, presents the highest prevalences and intensities of human fascioliasis known, affecting mainly children. Up to 72 % and 100 % prevalences have been recorded in given localities according to coprological and immunological survey methods, respectively (Hillyer *et al.* 1992; Mas-Coma *et al.* 1995, 1999; Bjorland *et al.* 1995; Esteban *et al.* 1997*a, b,* 1999; O'Neill *et al.* 1998), and more than 5000 eggs per g of faeces have been observed (Esteban *et al.* 1997*a, b,* 1999).

A multidisciplinary project was undertaken to understand the transmission and epidemiology of fascioliasis in the Northern Bolivian Altiplano. Sheep and cattle have been shown to be the main reservoir host species inhabiting the endemic region with high infection prevalences and intensities (Ueno & Morales, 1973; Ueno *et al.* 1975; Mas-Coma *et al.* 1995, 1997; Hillyer *et al.* 1996; Buchon *et al.* 1997; Grock *et al.* 1998). Studies showed that pigs and donkeys may be considered as reservoir hosts of secondary importance and that horses, llamas and alpacas (Mas-Coma *et al.* 1997), as well as rabbits, wild hares, domestic guinea pigs and wild rodents (Fuentes *et al.* 1997) do not significantly participate in the transmission of the disease.

The endemic zone proved to be stable, isolated and apparently unable to extend from its present outline, boundaries being marked by geographic, climatic and soil-water chemical characteristics. Altiplanic lymnaeids were found to inhabit mainly permanent water bodies, which enables parasite transmission during the whole year (Mas-Coma *et al.* 1999). The transmission foci appear patchily distributed and linked to the presence of appropriate water bodies according to an irregular geographic distribution within the endemic area. The likelihood of transmission can be assessed by means of forecast indices based on meteorological characteristics and appropriately modified for high altitude (Fuentes *et al.* 1999) and even more accurately by the normalized difference vegetation index obtained from remote sensing data (Fuentes, Malone & Mas-Coma, 2001).

In the Northern Bolivian Altiplano a confluence of several factors takes place which mitigates the negative effects of the very high altitude, mainly the cold temperatures and the high evapotranspiration rates. Among such positive factors there are (a) the nearness to the equator with increasing temperature, (b) the neighbourhood of Lake Titicaca moderating temperatures and increasing humidity, (c) the existence of numerous fresh water bodies deriving from the thaw of the perpetual snow amounts of the Eastern Andean Chain which combined with the existence of shallow phreatic layers assure the presence of permanent water collections for the survival of lymnaeids, (d) the absence of shadow because of the lack of trees and shrubs permitting a marked daily increase of the temperature of the water bodies, and (e) the scarcity of land leads human inhabitants to a marked dependence on livestock including various potential definitive host species (Mas-Coma *et al.* 1999).

Nevertheless, a detailed study demonstrated that the very high altitude climatic characteristics of this endemic area markedly differ from those of fascioliasis endemic lowland areas, such as those of the original European geographical distribution of *F. hepatica*. In the Northern Altiplano the following points are worth mentioning regarding the life cycle of *F. hepatica* and the transmission of the disease. The temperature has no marked seasonal character; the mean environmental temperature is very low throughout the year, about 10 °C or a little lower, which is not appropriate for the fluke development. There are large variations in temperature within a daily 24 hour period, both free-living and intra-molluscan larval stages of the parasite needing to adapt to such severe changing daily conditions. The rainfall distribution is seasonal, with a long dry season coinciding with the lowest minimum temperatures and a long wet season in which rainfall is concentrated, thus differing significantly from the European endemic areas in which fascioliasis is traditionally biseasonal, appearing in spring and autumn. Evapotranspiration is very high, temporary water bodies being of very short duration, mainly in the arid period, explaining why Altiplanic lymnaeids are almost always restricted to permanent water collections and transmission can occur throughout the year, contrary to what is typical of fascioliasis in Europe, where *L. truncatula* is markedly amphibious and develops mainly in temporary water collections. Lastly, the solar radiation is high not only because of altitude, but also because of the absence of shade as the consequence of lack of trees and shrubs (Fuentes *et al.* 1999).

Three important questions immediately arise: (a) how has the liver fluke adapted to maintain its life cycle at the extreme conditions associated with this very high altitude zone? (b) What strategies have been developed by the parasite to reach very high transmission rates in such an *a priori*-inhospitable environment? (c) Has this adaptation been reflected in both parasite and intermediate snail hosts at genotype and phenotype levels? To analyse these questions different studies have been performed. The aim of this paper is to present the results obtained in studies on: (1) the genetic characterization of both Altiplanic liver fluke and lymnaeid snails, and (2) the larval stage development of the liver fluke from the Northern Bolivian Altiplano.

For the first objective, the sequences of the first

and second internal transcribed spacers (ITS-1 and ITS-2) of the nuclear ribosomal DNA were obtained and analysed. These spacers were selected taking into account their usefulness for species, subspecies and population differentiation in parasites and vectors in general (Mas-Coma, 1999). For the second objective, experimental infections of Altiplanic lymnaeids kept in the laboratory under standardised rearing conditions with liver fluke isolates from Altiplanic sheep and cattle were carried out to establish the transmission characteristics.

MATERIALS AND METHODS

rDNA ITS sequencing

Liver fluke materials. Adult worms found in naturally infected hosts from the Northern Bolivian Altiplano and Spain were used for ITS-1 and ITS-2 sequencing (Table 1). For sequence comparison purposes, the following rDNA ITS-2 were used: (1) *F. hepatica* from: Australia, Hungary, Mexico and New Zealand (Adlard *et al.* 1993); unknown origin (GenBank Accession No. LO7844) (Michot *et al.* 1993); Australia (Hashimoto *et al.* 1997); Uruguay (Acc. No. AB010974) (Itagaki & Tsutsumi, 1998); (2) *F. gigantica* from Indonesia and Malaysia (Adlard *et al.* 1993); Malaysia (Hashimoto *et al.* 1997); Zambia isolate I (Acc. No. AB010975), Zambia isolate II (Acc. No. AB010976) and Indonesia (Acc. No. AB010977) (Itagaki & Tsutsumi, 1998); (3) *Fasciola* sp. from: Japan (Adlard *et al.* 1993); Japan (Hashimoto *et al.* 1997); Japan isolate I (Acc. No. AB010978) and Japan isolate II (Acc. No. AB010979) (Itagaki & Tsutsumi, 1998).

Lymnaeid snail materials. Snail specimens from the Northern Bolivian Altiplano and different geographic localities of Europe and Morocco were used for ITS-1 and ITS-2 sequencing (Table 1). The absence of infection by helminth parasites was always verified prior to the selection of the lymnaeids for molecular techniques. Taking helminth microhabitats into account, the region of the foot was chosen as the only snail part to be used for DNA extraction.

Molecular techniques. *F. hepatica* adults washed extensively in PBS (37 °C) were subsequently fixed in 70% ethanol. Snail feet were also fixed in 70% ethanol and stored at 4 °C for several weeks before DNA extraction according to the phenol-chloroform method (Sambrook, Fritsch & Maniatis, 1989) following the protocol of Bargues & Mas-Coma (1997).

The fragments corresponding to the ITS-1 and ITS-2 of each trematode and lymnaeid were ampli-

fied by the Polymerase Chain Reaction (PCR) using 4–6 μl of genomic DNA for each 50 μl PCR reaction, according to methods outlined previously (Almeyda-Artigas, Bargues & Mas-Coma, 2000; Marcilla *et al.* 2001). The PCR amplification was performed using primers designed in conserved positions of 18S, 5·8S and 28S rRNA genes of several eukaryote Metazoa species. For the ITS-1 and ITS-2 regions, the primers used were as described by Almeyda-Artigas *et al.* (2000). Only one additional primer, LT1 (forward) 5′-TCGTCTGTGTGAGGGTCG was designed for amplification and sequencing purposes.

Amplifications were generated in a GeneAmp PCR system 9600 (Perkin Elmer, Norwalk, CT, USA), by 30 cycles of 30 sec at 94 °C, 30 sec at 50 °C and 1 min at 72 °C, preceded by 30 sec at 94 °C and followed by 7 min at 72 °C. Ten microliters of the reaction mixture were examined by 1% agarose gel electrophoresis, followed by ethidium bromide staining.

Primers and nucleotides were removed from PCR products by purification on Wizard™ PCR Preps DNA Purification System (Promega, Madison, WI, USA) according to the manufacturer's protocol and resuspended in 50 μl of 10 mM TE buffer (pH 7·6). The final DNA concentration was determined by measuring the absorbance at 260 and 280 nm.

Sequencing was performed on both strands by the dideoxy chain-termination method, and with the Taq dye-terminator chemistry kit for ABI PRISM 377 (Perkin Elmer, Foster City, CA), using PCR primers. For sequence alignment the CLUSTAL-W version 1·8 (Thompson. Higgins & Gibson, 1994) was used.

Experimental life cycle studies

Liver fluke materials. Living eggs of *F. hepatica* were obtained from sheep and cattle gall-bladders after bile filtration. Gall-bladders from naturally infected sheep and cattle hosts of the Northern Bolivian Altiplano endemic area were obtained in the slaughterhouses of Batallas and El Alto, respectively. Eggs were washed three times with natural water and transported to the laboratory of Valencia, where they were stored in fresh water and complete darkness at 4 °C until required.

Lymnaeid snail materials. Lymnaeid snails used were collected in the Huacullani zone, where fascioliasis is highly prevalent in humans (Esteban *et al.* 1997*a*, 1999: varying between 31·2 and 38·2% in children, depending on year periods; with intensities of up to 5064 eggs per g of faeces), sheep (Grock *et al.* 1998: 70·0%) and cattle (Buchon *et al.* 1997: 45·2%). Lymnaeids were transported under isothermal conditions to the laboratory of Valencia, where they were adapted to experimentally controlled conditions

Table 1. Populations of liver fluke and lymnaeid populations chosen for sequencing of rDNA ITS-1 and ITS-2. *Lymnaea* sp. morph I (= *L. viatrix sensu* Ueno *et al.* 1975) and morph II (= *L. cubensis sensu* Ueno *et al.* 1975) according to Oviedo, Bargues & Mas-Coma (1995)

Populations	Geographic location	Country	ITS-1		ITS-2	
			Acc. No.	bp length	Acc. No.	bp length
Liver fluke	Batallas, Northern Altiplano	Bolivia	AJ243016	433	AJ272053	364
Fasciola hepatica	Castellon	Spain	AJ243016	433	AJ272053	364
Lymnaea sp. morph I	Tambillo, Northern Altiplano	Bolivia	AJ272052	504	AJ272051	401
Lymnaea sp. morph II	Batallas, Northern Altiplano	Bolivia	AJ272052	504	AJ272051	401
Lymnaea truncatula	Javalambre, Castellon	Spain	AJ243018	504	AJ296271	401
Lymnaea truncatula	Beira	Portugal	AJ243018	504	AJ296271	401
Lymnaea truncatula	Monaccia, Corsica island	France	AJ243018	504	AJ296271	401
Lymnaea truncatula	Albufera, Sueca, Valencia	Spain	AJ243018	504	AJ243017	401
Lymnaea truncatula	Benicasim, Castellon	Spain	AJ243018	504	AJ243017	401
Lymnaea truncatula	Minho	Portugal	AJ243018	504	AJ243017	401
Lymnaea truncatula	Le Taulard, Laussane	Switzerland	AJ243018	504	AJ243017	401
Lymnaea truncatula	Site Oudesa	Morocco	AJ296270	504	AJ243017	401

in 2000 ml fresh water in standard breeding containers at 20 °C, 90 % relative humidity and 12 h/12 h light/darkness photoperiod in precision climatic chambers (Heraeus-Vötsch HPS-1500 and HPS-500). The water was changed weekly and lettuce added *ad libitum*.

Experimental procedures. A continuous temperature of 20 °C was selected and used throughout because of two reasons: (1) lymnaeids from the Bolivian Northern Altiplano proved to adapt to 20 °C better than to higher temperatures; (2) a 20 °C temperature has been largely used by other authors in experiments with the European *F. hepatica/L. truncatula* model, thus allowing comparisons of results.

Liver fluke eggs in fresh water were maintained under complete darkness at 20 °C to start the embryogenic process. Embryogenesis was followed at intervals of 4 days by counting eggs presenting an incipient morula, eggs including an advanced morula, eggs with outlined miracidium, and fully embryonated eggs containing a developed miracidium.

Developed miracidia were forced to hatch by placing fully embryonated eggs under light and used for the experimental infection of snails. Only 4–5 mm-long first generation snails born in the laboratory were used for experiments. A total of 32 and 25 lymnaeids were infected monomiracidially by the Altiplanic sheep and cattle isolates, respectively, by exposing each snail to 1 miracidium for 4 hours in a small Petri dish containing 2 ml of fresh water. Snails were afterwards returned to the same standard conditions in the climatic chamber (2000 ml containers, 20 °C, 90 % r.h., 12 h/12 h light/darkness, dry lettuce *ad libitum*) until day 30 post-infection, in which they were again isolated in Petri dishes to allow daily monitoring of cercarial shedding by

individual snails. Lettuce was provided *ad libitum* to each snail in a Petri dish during both shedding and post-shedding periods until death of the snail.

The chronobiology of the cercarial shedding was followed by counting metacercariae in each Petri dish. To test the viability and infectivity of metacercariae, they were stored in natural water in total darkness until required. The storage temperature selected was 4 °C according to the mean winter temperature obtained in five meteorological stations of the Bolivian Altiplano endemic region in the 1949–1990 year period (Mas-Coma *et al.* 1999). This temperature has also been used for storing metacercariae by other authors, thus enabling comparative analysis with lowland liver fluke isolates. Male Wistar rats (Iffa Credo, Barcelona, Spain) aged 4–5 weeks were used for infection with metacercariae inoculated orally by means of a gastric gavage. Rats were housed in Micro-Isolator boxes (Iffa Credo, Barcelona, Spain) and maintained in a pathogen-free room, electrically heated with a 12 h/12 h light/darkness cycle (conditions in compliance with the European Agreement of Strasbourg, 18 March 1986). Food and water were provided *ad libitum*. Infection prevalence and intensity (number of worms successfully developed in each rat) were determined by dissection (Valero & Mas-Coma, 2000).

RESULTS

rDNA ITS-1 and ITS-2 sequences

New nucleotide sequence data reported in this paper on both fasciolids and lymnaeids are available in the EMBL, GenDank and DDBJ databases under the accession numbers listed in Table 1.

Liver fluke. The complete sequence of the ITS-1 showed no nucleotide difference between the Northern Bolivian Altiplano and Spain. It was 433 bp long,

Table 2. Nucleotide differences found in the comparison of the ITS-2 sequences of *Fasciola* populations [[a] = after Itagaki & Tsutsumi (1998); [b] = after Hashimoto *et al.* (1997); [c] = after Aadlard *et al.* (1993); pb seq. = number of pb sequenced; bp numbers in parentheses indicate partial sequences; * = position not sequenced]

Species	Origin	pb seq.	Positions 1	51	53	78	97	101	115	210	221	234	273	279	287	327	330	337
F. hepatica	Bolivia	364	G	T	G	T	G	T	C	T	T	T	C	C	C	T	T	G
F. hepatica	Spain	364	G	T	G	T	G	T	C	T	T	T	C	C	C	T	T	G
F. hepatica	Uruguay[a]	364	C	T	G	T	G	T	C	T	T	T	C	C	T	T	T	G
F. hepatica	Australia[b]	362	C	T	G	T	G	T	C	T	T	T	C	C	C	T	T	G
F. hepatica	Australia[c]	(292)	*	T	—	T	—	T	C	T	T	T	C	C	C	T	T	G
F. hepatica	Hungary[c]	(264)	*	*	*	T	—	T	C	T	T	T	C	C	C	T	T	G
F. hepatica	Mexico[c]	(292)	*	T	—	T	—	T	C	T	T	T	C	C	T	T	T	G
F. hepatica	New Zealand[c]	(292)	*	T	—	T	—	T	C	T	T	T	C	C	C	T	T	G
F. gigantica	Zambia I[a]	362	C	T	G	T	G	T	C	T	T	A	C	C	C	T	—	G
F. gigantica	Zambia II[a]	362	C	T	G	T	G	T	C	T	T	C	T	T	C	T	—	A
F. gigantica	Indonesia[a]	362	C	T	G	T	G	T	C	C	T	C	T	T	C	T	—	A
F. gigantica	Indonesia[c]	(214)	*	*	*	*	*	*	*	C	T	C	T	T	C	—	T	A
F. gigantica	Malaysia[c]	(252)	*	*	*	*	—	—	C	C	T	C	T	T	C	—	T	A
F. gigantica	Malaysia[b]	362	C	T	G	T	G	T	C	C	T	C	T	T	C	T	—	A
Fasciola sp.	Japan[c]	(225)	*	*	*	*	*	*	—	C	C	C	T	T	C	—	T	A
Fasciola sp.	Japan I[a]	362	C	T	G	T	G	T	C	T	T	T	C	C	C	T	T	G
Fasciola sp.	Japan II[a]	362	C	T	G	T	G	T	C	C	C	C	T	T	C	T	—	A
Fasciola sp.	Japan[b]	362	C	T	G	T	G	T	C	C	T	C	T	T	C	T	—	A

Table 3. Nucleotide contents and differences found in the comparison of the sequences of ITS-1 and ITS-2 of lymnaeid populations from the Northern Bolivian Altiplano and Europe

Lymnaeid populations	ITS-1 (504 bp long) Positions % GC	74	75	132	ITS-2 (401 bp long) Positions % GC	55	149
Lymnaea sp. morph I Bolivia	57·5	G	T	T	58·6	T	T
Lymnaea sp. morph II Bolivia	57·5	G	T	T	58·6	T	T
L. truncatula Javalambre, Beira, Corsica	57·5	A	G	T	59·1	G	C
L. truncatula Sueca, Benicasim, Minho, Le Toulard	57·5	A	G	T	58·8	G	T
L. truncatula Morocco	57·7	A	G	C	58·8	G	T

with a 51·9 % GC content (see Table 1). As it is the first time that the sequence of this spacer of the liver fluke has been obtained, no comparison with populations from other areas can be made.

The ITS-2 sequence was also identical between the Northern Bolivian Altiplano and Spain. Its complete sequence was 364 bp long, with a 48·3 % GC content (see Table 1). The comparison with the ITS-2 sequences of *F. hepatica* from other geographic origins only showed a few nucleotide differences consisting in insertions/deletions and C/T transitions (Table 2). More substantial differences (mostly found at the 3′ end) were noted between our sequence and that reported by Michot *et al.* (1993), although the latter has been already criticised by different authors (i.e. Hashimoto *et al.* 1997). Interestingly, the ITS-2 sequence from Bolivia and Spain differ from all other presently known at least in one position (Table 2).

According to the ITS-2 sequence comparison obtained in the 364-bp-long alignment, the liver fluke from Bolivia and Spain both belong to the species *F. hepatica*. *F. hepatica* and *F. gigantica* differ at only 5 sites (positions 210, 234, 273, 279 and 337 in Table 2). The classification of the fasciolid designated Zambia I by Itagaki & Tsutsumi (1998) as belonging to *F. gigantica* becomes doubtful.

Lymnaeid snails. Length and GC content of both ITS-1 and ITS-2 sequences of all lymnaeid populations studied are noted in Table 3. No nucleotide difference between the two morphs I and II from the Northern Bolivian Altiplano in both spacers was found, but differ in a few nucleotide positions from all other populations studied from Europe and Morocco (Table 3).

At the level of ITS-1, all European populations appear identical and differ in only one nucleotide

Table 4. Results obtained in infection experiments of Altiplanic
Lymnaea truncatula with sheep and cattle isolates of *Fasciola hepatica*
from the Northern Bolivian Altiplano (means in parentheses)

Host isolate	Sheep	Cattle
Day presenting the maximum % of eggs with advanced morula	12/88·9%	8/86·7%
Day in which the first outlined miracidium appeared inside egg	18	18
Day presenting the maximum % of eggs with outlined miracidium	36/37·3%	38/56·2%
Day in which the first developed miracidium appeared inside egg	38	24
Day presenting the maximum % of eggs with developed miracidium	58/16·4%	46/24·9%
Isolate infectivity (% snails infected)	69·2	39·1
Prepatent period (days)	48–58 (49·6)	49–58 (51·0)
Shedding period (days)	47–88 (71·5)	42–85 (74·1)
No. cercariae shed per individual snail	384–562 (451·8)	151–589 (446·2)
Snails surviving until shedding (% snails)	86·6	92·0
Snail survival after end of shedding period (days)	1–132 (44·5)	1–133 (55·4)
Postinfection longevity in shedding snails (days)	95–192 (157·1)	91–268 (165·0)
Longevity in non-infected snails (days post-infection)	98–182 (175·0)	100–209 (145·0)

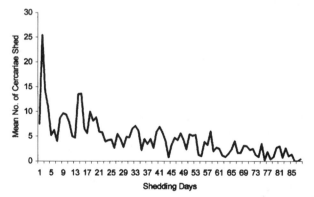

Fig. 1. Chronobiology of cercarial shedding of the sheep isolate of *Fasciola hepatica* by *Lymnaea truncatula*, both from the Bolivian Northern Altiplano. Shedding period analysed from the day of the emergence of the first cercaria by each snail.

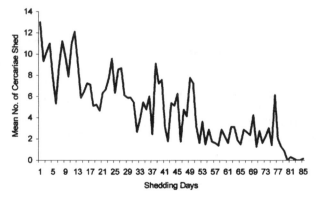

Fig. 2. Chronobiology of cercarial shedding of the cattle isolate of *Fasciola hepatica* by *Lymnaea truncatula*, both from the Bolivian Northern Altiplano. Shedding period analysed from the day of the emergence of the first cercaria by each snail.

position from that of Morocco. At the level of ITS-2, given European populations (Javalambre, Beira and Corsica) differ from the remaining from Europe and that of Morocco in one C/T transition (Table 3).

Life cycle characteristics of sheep and cattle isolates

Results of embryogenesis inside the egg, lymnaeid snail infection, intramolluscan parasite larval development and influences of the latter on snail survival are noted in Table 4. Embryogenesis

followed a similar pattern in both sheep and cattle isolates from the Northern Bolivian Altiplano, although development in eggs of the cattle isolate was somewhat faster. The prepatent period in lymnaeid snails and the cercarial shedding period, as well as the number of cercariae shed per lymnaeid individual appeared similar in both sheep and cattle isolates, although the snail infectivity of the latter was lower than that of the sheep isolate. The host isolate did not appear to influence lymnaeid survival, results obtained before shedding started and after

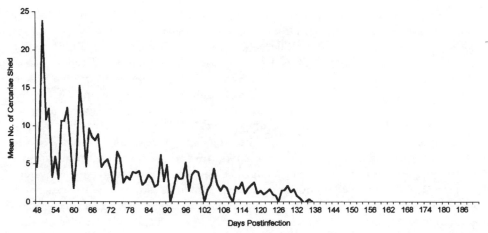

Fig. 3. Chronobiology of cercarial shedding of the sheep isolate of *Fasciola hepatica* by *Lymnaea truncatula*, both from the Bolivian Northern Altiplano. Shedding period analysed from the day of the miracidial infection. Prepatent period not shown. Curve followed up to the death of shedding snails.

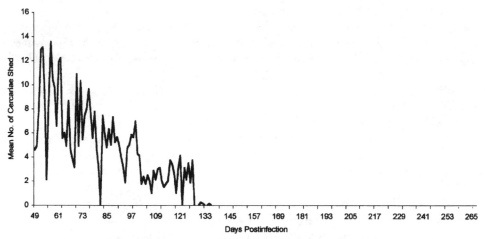

Fig. 4. Chronobiology of cercarial shedding of the cattle isolate of *Fasciola hepatica* by *Lymnaea truncatula*, both from the Bolivian Northern Altiplano. Shedding period analysed from the day of the miracidial infection. Prepatent period not shown. Curve followed up to the death of shedding snails.

shedding ended being similar in both isolates. Interestingly, longevity after the day of miracidial infection was also similar in both infected and non-infected lymnaeids.

The chronobiological patterns of cercarial emergence in the sheep and cattle isolates are shown in Figures 1 and 3, and 2 and 4, respectively. It appears to be similar in both isolates. When the shedding period is analysed from the day of the emergence of the first cercaria by each snail (Figs 1, 2), it appears that the shedding process appears as an irregular succession of waves in which a slowly decrease of the daily number of cercariae is envisaged, the higher acrophases taking place in the first days of the emergence period. When the shedding period is analysed from the day of the miracidial infection (Figs 3, 4), a similar decreasing multiwave pattern is detected. In these curves of Figs 3 and 4, days occurred in which all snails failed to shed any cercaria. When analysing the emergence curves per individual snail (figures not shown), a total of 6 shedding pauses are evident. In the sheep isolate

there is a first pause between days 60 and 73 post-infection, a second between days 90 and 93, a third on days 100–103, a fourth on days 108–112, a fifth on days 117–121, and a last sixth one on days 125–126. In the cattle isolate there is a first pause on days 63–69 post-infection, a second on days 80–84, a third on days 92–96, a fourth on days 100–106, a fifth on days 111–113, and a final sixth one on days 117–119. Consequently, there are about 20–30 days between the two first pauses and afterwards the pauses become separated by an elapse which appears to follow a slowly progressive reduction from 10 to 6 days.

No significant difference between sheep and cattle isolates appeared in metacercarial infectivity assays (statistical comparisons with the Chi-Square test using SPSS 6·1; values considered statistically significant when $P < 0.05$). In the sheep isolate, 11 from a total of 14 rats (78·6 %) became infected when inoculated with 20 metacercariae each, and between 1 and 8 adult worms were obtained per rat (mean 3·6). In the cattle isolate, 6 from a total of 8 rats

(75%) became infected when inoculated with the same metacercarial dose, giving rise to 1–2 adult flukes per rat (mean 1·7).

DISCUSSION

ITS-1 and ITS-2 of the liver fluke

According to molecular clock estimations based on ITS sequences (see review in Bargues *et al.* 2000), the very few nucleotide differences between ITS sequences of liver flukes from different regions of the world suggest a very recent geographical dispersal of the parasite. This agrees with previous assumptions indicating that *F. hepatica* is a parasite transmitted by *L. truncatula* original of Europe (Oviedo *et al.* 1995), where it has even been found in prehistoric human populations living at the end of the Meso-lithic period, 5000–5100 years ago (Aspöck, Auer & Picher, 1999). From this European geographical area it has spread by the exportation of European livestock to the five continents where it has adapted to other lymnaeid species (Mas-Coma & Bargues, 1997).

ITS sequence comparison confirms that the liver fluke of the Northern Bolivian Altiplano is *F. hepatica*. Moreover, the fact that ITS-1 and ITS-2 sequences of *F. hepatica* from the Altiplano are identical to those of *F. hepatica* from Spain and differ from those of other world regions suggests that the Bolivian liver fluke is derived from Iberian liver flukes, probably a recent introduction at the time of Spanish colonization or somewhat later.

Despite the absence of ITS sequence differences, phenotypical studies demonstrated that slight differences can be found in the allometry of body measurements between the *F. hepatica* adult from highland and that from lowland populations of Bolivian and Spanish sheep, respectively (Valero *et al.* 1999). Moreover, the Bolivian sheep and cattle liver fluke populations have a smaller uterus size than that of European populations (Valero, Panova & Mas-Coma, 2000). These uterus differences between the liver fluke population of highlands and lowlands have been tentatively related to high altitude in-fluences, although it cannot be discounted that to some extent this may be attributable to intraspecific variability (Valero *et al.* 2000). As known, high altitude environmental factors exert an influence on vertebrates, and those born and living at high altitude show morphological and physiological charac-teristics different from those inhabiting low altitudes. The development of *F. hepatica*, mainly because of its tissue migration and haematophagous diet (Dawes & Hughes, 1964; Boray, 1969), may be influenced by changes such as hypoxia, alterations in immune response, elevated haematocrit levels, differences in blood oxygen pressure values and blood viscosity and elimination of dissolved gases, especially N_2, from the blood (Valero *et al.* 2000).

It is known that oxygen is still required for egg production in *F. hepatica* (Mansour, 1958; Bjorkman & Thorsell, 1963; McGonigle & Dalton, 1995). Thus, high altitude hypoxic conditions could be the origin of a reduced egg production by the flukes. Moreover, although the uterus in digeneans has traditionally not been considered as a storage organ but mainly an organ adapted to the developmental time of the eggs (in fasciolids, eggs are layed unembryonated, the miracidium beginning its de-velopment once the egg is deposited in freshwater), our recent experimental studies with rats have demonstrated that there is a direct relation between *F. hepatica* uterus size and the number of eggs shed per g of faeces (Valero *et al.*, unpublished). In the Northern Bolivian Altiplano, climatic conditions, freshwater body characteristics and lymnaeid ecol-ogy enable fascioliasis transmission to take place throughout the year (Mas-Coma *et al.* 1999; Fuentes *et al.* 1999), so that egg storage is *a priori* not needed as in the northern hemisphere latitudes where fascioliasis transmission is typically seasonal (Valero *et al.* 2000).

ITS-1 and ITS-2 of the lymnaeid snails

ITS sequence analyses of the lymnaeid morphs I and II prove that there is only one species in the Northern Bolivian Altiplano. Moreover, the very few differences between the ITS sequences of the Bolivian lymnaeids and *L. truncatula* populations from Europe and Morocco indicate that the snail species involved in the transmission of human and animal fascioliasis in this Andean endemic area is *L. truncatula*. At any rate, the nucleotide differences in both ITS-1 and ITS-2 sequences allow us to differentiate the Altiplano population from all other European and African populations studied. As in the case of the liver fluke, ITS results suggest that fascioliasis transmitting snails of the Altiplano were introduced from Europe, most probably imported by Spanish colonizers.

These results do not agree with those obtained by Ueno *et al.* (1975), who mentioned two different lymnaeid forms transmitting fascioliasis in the Northern Bolivian Altiplano, which they classified as *Lymnaea viatrix* and *L. cubensis*. However, our results are in agreement with those obtained in more recent studies on shell morphology and visceral mass anatomy (Oviedo *et al.* 1995; Samadi *et al.* 2000), which suggested that what Ueno *et al.* (1975) classified as two different species are in fact only extreme morphs within a large intraspecific varia-bility (Oviedo *et al.* 1995; Samadi *et al.* 2000), morphs I and II of Oviedo *et al.* (1995) cor-responding to *L. viatrix* and *L. cubensis* of Ueno *et al.* (1975), respectively. Further molecular confirmation was obtained by the absence of nucleotide differences in the complete sequence of the 18S rDNA gene between both extreme morphs and between them and *L. truncatula* from Europe (Bargues & Mas-Coma, 1997; Bargues *et al.* 1997). At any rate, this gene is largely more conserved and is usually not very useful for the differentiation of species which are very similar, such as *L. viatrix*, *L. cubensis* and *L. truncatula* (Bargues *et al.*, unpublished); moreover, its validity for lymnaeid taxonomy and phylogeny has recently been questioned owing to single nucleotide polymorphisms detected in *L. natalensis* and intraspecific *versus* interspecific divergence levels it shows in certain groups (Stothard *et al.* 2000), although further studies on the 18S rDNA gene of lymnaeids are evidently needed. Results obtained in isoenzyme comparison analyses also suggested the link between Altiplanic lymnaeids and *L. truncatula* from western Europe. Additionally, isoenzyme analyses proved that all lymnaeid populations inhabiting the Altiplanic endemic area are monomorphic, a clonicity related to selfing reproduction processes and a single founding population (Jabbour-Zahab *et al.* 1997). Recent studies on genetic diversity and population structure of lymnaeids by means of polymorphic DNA microsatellite analyses also confirmed the presence of a single *L. truncatula* genotype on the Bolivian Altiplano (Meunier *et al.* 2001). Interestingly, moreover, differences detected in all these comparative studies between the Altiplanic lymnaeids and *L. truncatula* from Europe were always non-significant or non-existent at all.

Life cycle characteristics of Altiplanic *Fasciola hepatica isolates*

The embryonation times observed in the sheep and cattle isolates of *F. hepatica* fit in the ranges mentioned by several authors when tested at 20 °C (Kendall, 1965; Diez-Baños & Rojo-Vazquez, 1976; Foreyt & Todd, 1978). As known, the development of the miracidium inside egg is arrested below 9 °C and above 37 °C and has a duration between 9 and 161 days depending upon the temperature, the range 20–25 °C offering the optimum for the hatching of a higher number of miracidia (Roberts, 1950; Rowcliffe & Ollerenshaw, 1960; Valenzuela, 1979; Wilson, Smith & Thomas, 1982).

Infection percentages and prepatent period in monomiracidial infections may also be considered as normal at 20 °C when compared to similar studies carried out with *F. hepatica* isolates and *L. truncatula* specimens from European zones (Roberts, 1950; Pantelouris, 1965; Kendall, 1965, 1970; Odening, 1971; Boray, 1982; Oviedo *et al.* 1996). The prepatent period detected in the Altiplanic material studied also fits in the known range (43·1–58·2 days) for the European *F. hepatica/L. truncatula* in the nature (Rondelaud & Dreyfuss, 1997).

Infectivity of metacercariae from Altiplanic *F. hepatica* does not significantly differ from that of the liver fluke in lowlands of other countries (Valero & Mas-Coma, 2000). However, the results obtained in the shedding period, the number of cercariae shed and the snail survival must be emphasized. The duration of the shedding period in the Bolivian *F. hepatica/L. truncatula* appears to be very long. In European *F. hepatica* experimentally infecting *L. truncatula* snails of the same size as ours under the same constant conditions of 20 °C and 12 h/12 h photoperiod, the patent period lasted only $46 \pm 27·6$ days (Dreyfuss & Rondelaud, 1994). Similarly, results obtained in nature show that the patent period in Europe ranges between 5·0 and 9·3 days in the winter generation and 18·3–40·3 days in the summer generation (Rondelaud & Dreyfuss, 1997).

The cercarial shedding pattern detected in the Bolivian *F. hepatica/L. truncatula* model does not disagree with the patterns observed by other authors. Kendall & McCullough (1951), Bouix-Busson, Rondelaud & Combes (1985), Audousset *et al.* (1989) and Dreyfuss & Rondelaud (1994) reported that cercarial emission was discontinuous with shedding waves separated by periods of rest. The infradian-type rhythm with a periodicity of 7 days and 5–7 waves described by Audousset *et al.* (1989) when working under variable temperatures has been put into doubt by Dreyfuss & Rondelaud (1994) who worked under constant conditions of temperature and photoperiod. These latter authors observed that snails shed their cercariae in 1 to 14 waves, although the majority of snails produced their parasites in 4 and 5 waves. Consequently, the six pauses separating 8 waves is understandable taking into account the longer shedding period in the Bolivian fluke/snail model. The only discrepancy between our results and those of these authors is found at the level of the distribution of cercarial numbers throughout the shedding period, as in the Bolivian model the highest daily numbers of cercariae are clearly shed in the first weeks whereas in the European model acrophases appear delayed.

The number of cercariae shed per individual lymnaeid in the Bolivian material studied is also very high, a phenomenon related to the marked length of the shedding process. Working under the same experimental conditions, a mean of only 238·5 metacercariae per snail was found by Dreyfuss & Rondelaud (1994). Always working with the same European *F. hepatica/L. truncatula* model, Dreyfuss *et al.* (1999) obtained only $114·9 \pm 80·3$ metacercariae per snail infected monomiracidially, and they observed that the duration of the shedding and the number of metacercariae were independent of the number of miracidia used for the infection of each individual lymnaeid, although single-miracidium infections were most effective because of the much higher snail survival rate and despite the mean number of cercariae shed being the same as in multi-miracidial infections.

Lymnaeid snail survival in the experiments carried out with *F. hepatica* and *L. truncatula* from the Northern Bolivian Altiplano are worth mentioning. Differences in survival of different geographical strains of *L. truncatula* to *F. hepatica* infection have already been described (Gold, 1980). The longevity of the experimentally infected Bolivian molluscs after the moment of the miracidial infection is usually longer than the same survival period observed in the European species *L. truncatula*, in which a postinfection longevity of 70 days post-infection is usually observed (Kendall, 1953; Hodasi, 1972 *a, b*; Wilson & Denison, 1980; Rondelaud & Barthe, 1986), with a maximum of 16 weeks described once (Hodasi, 1972 *a, b*), and even longer than that known in other American lymnaeids such as 119 days postinfection for *Lymnaea viatrix* (Venturini, 1978) and 113·4 days postinfection for *Lymnaea bulimoides* (Jay & Dronen, 1985). The capacity of Altiplanic lymnaeids to survive up to more than 4 months after the end of the shedding period is surprising, as in Europe, snails die either during the shedding period, immediately after shedding ends or shortly after it. Dreyfuss & Rondelaud (1994) found that of 102 snails shedding on the first day, the number drastically reduced to only 56 on the second day and subsequently decreased on day 76 to four snails.

CONCLUSIONS

The results on rDNA ITS-1 and ITS-2 sequences suggest that the human and animal fascioliasis high endemic area of the Northern Bolivian Altiplano is the consequence of a recent introduction of both *F. hepatica* and *L. truncatula* from Europe. In this double importation, parasite and lymnaeid, have followed a process of adaptation from the European lowlands to the Bolivian highlands. When analysing the different aspects of the life cycle of *F. hepatica* from the Northern Bolivian Altiplano by comparison with that of *F. hepatica* in *L. truncatula* of Europe, no phase is observed in which the parasite development appears to be negatively modified for the transmission. Several aspects can be shown to be similar to those in Europe, such as duration of miracidium development inside the egg, the lymnaeid infection capacity of miracidia, the duration of the prepatent period and the infectivity of metacercariae.

However, results obtained show that certain aspects favour transmission, such as the longer cercarial shedding period and the higher cercarial production, both related to the greater survival capacity of infected lymnaeid snails. The absence of survival differences between parasitised and non-parasitised molluscs suggests a better parasite-host adaptation in the Bolivian Northern Altiplano. This phenomenon may be related to the clonal charac-

teristics of the *L. truncatula* populations of the Northern Bolivian Altiplano. The initial founding snail individual or few individuals imported from Europe that have given rise by selfing to the numerous monomorphic populations that today inhabit the Altiplano endemic area were most probably snails that were very susceptible for *F. hepatica* infection. These snails would have genetically transmitted their high susceptibility to their descendants by almost absolute predomination of autofecundation, suggesting a large and homogeneous susceptibility of the different Altiplanic *L. truncatula* populations to the liver fluke.

ACKNOWLEDGEMENTS

This work was supported by funding from the STD Programme of the Commission of the European Communities (DG XII: Science, Research and Development) (Contract No. TS3-CT94–0294), Brussels, EU, the Programme of Scientific Cooperation with Latin America, Instituto de Cooperación Iberoamericana, Agencia Española de Cooperación Internacional (I.C.I.-A.E.C.I.), Madrid, Spain, and DGICYT Proyects No. UE96–0001 and PM97–0099 of the Spanish Ministry of Education and Culture, Madrid, Spain. R. Angles, W. Strauss and P. Buchon (La Paz, Bolivia) are acknowledged for technical assistance in the Bolivian slaughterhouses. Dr M. A. Valero is acknowledged for the metacercarial infectivity assays. The authors also acknowledge the facilities provided by the Instituto Nacional de Laboratorios de Salud (INLASA) of La Paz, Bolivia. This work was carried out while the second autor (I.R.F.) was the recipient of a predoctoral fellowship from the MUTIS Scholarship Program of the Instituto de Cooperación Iberoamericana of the Agencia Española de Cooperación Internacional (ICI-AECI), Madrid.

REFERENCES

ADLARD, R. D., BARKER, S. C., BLAIR, D. & CRIBB, T. H. (1993). Comparison of the second internal transcribed spacer (ribosomal DNA) from populations and species of Fasciolidae (Digenea). *International Journal for Parasitology* **23**, 422–425.

ALMEYDA-ARTIGAS, R. J., BARGUES, M. D. & MAS-COMA, S. (2000). ITS-2 rDNA sequencing of *Gnathostoma* species (Nematoda) and elucidation of the species causing human gnathostomiasis in the Americas. *Journal of Parasitology* **86**, 537–544.

ASPÖCK, H., AUER, H. & PICHER, O. (1999). Parasites and parasitic diseases in prehistoric human populations in Central Europe. *Helminthologia* **36**, 139–145.

AUDOUSSET, J. C., RONDELAUD, D., DREYFUSS, G. & VAREILLE-MOREL, C. (1989). Les emissions cercariennes de *Fasciola hepatica* L. chez le mollusque *Lymnaea truncatula* Müller. A propos de quelques observations chronobiologiques. *Bulletin de la Société Française de Parasitologie* **7**, 217–224.

BARGUES, M. D., MANGOLD, A. J., MUÑOZ-ANTOLI, C., POINTIER, J. P. & MAS-COMA, S. (1997). SSU rDNA characterization of lymnaeid snails transmitting human fascioliasis in South and Central America. *Journal of Parasitology* **83**, 1086–1092.

BARGUES, M. D., MARCILLA, A., RAMSEY, J., DUJARDIN, J. P., SCHOFIELD, C. J. & MAS-COMA, S. (2000). Nuclear rDNA-based molecular clock of the evolution of Triatominae (Hemiptera: Reduviidae), vectors of Chagas disease. *Memorias do Instituto Oswaldo Cruz* **95**, 567–573.

BARGUES, M. D & MAS-COMA, S. (1997). Phylogenetic analysis of lymnaeid snails based on 18S rDNA sequences. *Molecular Biology and Evolution* **14**, 569–577.

BERGEON, P. & LAURENT, M. (1970). Différences entre la morphologie testiculaire de *Fasciola hepatica* et *Fasciola gigantica*. *Revue d'Elevage et de Médecine Vétérinaire des Pays Tropicaux* **23**, 223–227.

BJORLAND, J., BRYAN, R. T., STRAUSS, W., HILLYER, G. V. & MCAULEY, J. B. (1995). An outbreak of acute fascioliasis among Aymara Indians in the Bolivian Altiplano. *Clinical Infectious Diseases* **21**, 1228–1233.

BJORKMAN, N. & THORSELL, W. (1963). On the fine morphology of the egg-shell globules in the vitelline glands of the liver fluke (*F. hepatica*). *Experimental Cell Research* **32**, 153–156.

BORAY, J. C. (1969). Experimental fascioliasis in Australia. *Advances in Parasitology* **8**, 95–210.

BORAY, J. C. (1982). Fascioliasis. In *Handbook Series in Zoonoses. Section C. Parasitic Zoonoses. Volume III* (ed. Hillyer, G. V. & Hopla, C. E.), pp. 71–88. Boca Raton, Florida, CRC Press.

BOUIX-BUSSON, D., RONDELAUD, D. & COMBES, C. (1985). L'infestation de jeunes *Lymnaea glabra* Müller par *Fasciola hepatica* L. Les caractéristiques des émissions cercariennes. *Annales de Parasitologie Humaine et Comparée* **60**, 11–21.

BUCHON, P., CUENCA, H., QUITON, A., CAMACHO, A. M. & MAS-COMA, S. (1997). Fascioliasis in cattle in the human high endemic region of the Bolivian Northern Altiplano. *Research and Reviews in Parasitology* **57**, 71–83.

CHEN, M. G. & MOTT, K. E. (1990). Progress in assessment of morbidity due to *Fasciola hepatica* infection: a review of recent literature. *Tropical Diseases Bulletin* **87**, R1–R38.

CURTALE, F., HAMMOUD, E. S., EL WAKEEL, A., MAS-COMA, S. & SAVIOLI, L. (2000). Human fascioliasis, an emerging public health problem in the Nile Delta, Egypt. *Research and Reviews in Parasitology* **60**, 3–4.

DAWES, B. & HUGHES, D. L. (1964). Fascioliasis: the invasive stages of *Fasciola hepatica* in mammalian hosts. *Advances in Parasitology* **2**, 97–168.

DIEZ-BAÑOS, M. A. & ROJO-VAZQUEZ, F. A. (1976). Influencia de la temperatura en el desarrollo de los huevos de *Fasciola hepatica* L. *Anales de la Facultad de Veterinaria de León* **22**, 65–75.

DREYFUSS, G. & RONDELAUD, D. (1994). *Fasciola hepatica*: a study of the shedding of cercariae from *Lymnaea truncatula* raised under constant conditions of temperature and photoperiod. *Parasite* **4**, 401–404.

DREYFUSS, G., VIGNOLES, P., RONDELAUD, D. & VAREILLE-MOREL, C. (1999). *Fasciola hepatica*: characteristics of infection in *Lymnaea truncatula* in relation to the number of miracidia at exposure. *Experimental Parasitology* **92**, 19–23.

ESTEBAN, J. G., BARGUES, M. D. & MAS-COMA, S. (1998). Geographical distribution, diagnosis and treatment of human fascioliasis: a review. *Research and Reviews in Parasitology* **58**, 13–42.

ESTEBAN, J. G., FLORES, A., AGUIRRE, C., STRAUSS, W., ANGLES, R. & MAS-COMA, S. (1997a). Presence of very high prevalence and intensity of infection with *Fasciola hepatica* among Aymara children from the Northern Bolivian Altiplano. *Acta Tropica* **66**, 1–14.

ESTEBAN, J. G., FLORES, A., ANGLES, R. & MAS-COMA, S. (1999). High endemicity of human fascioliasis between Lake Titicaca and La Paz valley, Bolivia. *Transactions of the Royal Society of Tropical Medicine and Hygiene* **93**, 151–156.

ESTEBAN, J. G., FLORES, A., ANGLES, R., STRAUSS, W., AGUIRRE, C. & MAS-COMA, S. (1997b). A population-based coprological study of human fascioliasis in a hyperendemic area of the Bolivian Altiplano. *Tropical Medicine and International Health* **2**, 695–699.

FOREYT, W. J. & TODD, A. C. (1978). Experimental infection of lymnaeid snails in Wisconsin with miracidia of *Fascioloides magna* and *Fasciola hepatica*. *Journal of Parasitology* **64**, 1132–1134.

FUENTES, M. V., COELLO, J. R., BARGUES, M. D., VALERO, M. A., ESTEBAN, J. G., FONS, R. & MAS-COMA, S. (1997). Small mammals (Lagomorpha and Rodentia) and fascioliasis transmission in the Northern Bolivian Altiplano endemic zone. *Research and Reviews in Parasitology* **57**, 115–121.

FUENTES, M. V., MALONE, J. B. & MAS-COMA, S. (2001). Validation of a mapping and predicting model for human fascioliasis transmission in the Northern Bolivian Altiplano using remote sensing data. *Acta Tropica*, in press.

FUENTES, M. V., VALERO, M. A., BARGUES, M. D., ESTEBAN, J. G., ANGLES, R. & MAS-COMA, S. (1999). Analysis of climatic data and forecast indices for human fascioliasis at very high altitude. *Annals of Tropical Medicine and Parasitology* **93**, 835–850.

GOLD, D. (1980). Growth and survival of the snail *Lymnaea truncatula*: effects of soil type, culture medium and *Fasciola hepatica* infection. *Israel Journal of Zoology* **29**, 163–170.

GROCK, R., MORALES, G., VACA, J. L. & MAS-COMA, S. (1998). Fascioliasis in sheep in the human high endemic region of the Northern Bolivian Altiplano. *Research and Reviews in Parasitology* **58**, 95–101.

HASHIMOTO, K., WATANOBE, T., LIU, C. X., INIT, I., BLAIR, D., OHNISHI, S. & AGATSUMA, T. (1997). Mitochondrial DNA and nuclear DNA indicate that the Japanese *Fasciola* species is *F. gigantica*. *Parasitology Research* **83**, 220–225.

HILLYER, G. V., SOLER DE GALANES, M., BUCHON, P. & BJORLAND, J. (1996). Herd evaluation by enzyme-linked immunosorbent assay for the determination of *Fasciola hepatica* infection in sheep and cattle from the Altiplano of Bolivia. *Veterinary Parasitology* **61**, 211–220.

HILLYER, G. V., SOLER DE GALANES, M., RODRIGUEZ-PEREZ, J., BJORLAND, J., SILVA DE LAGRAVA, M., RAMIREZ GUZMAN, S. & BRYAN, R. T. (1992). Use of the Falcon Assay Screening Test – Enzyme-Linked Immunosorbent Assay (FAST–ELISA) and the Enzyme-Linked Immunoelectrotransfer Blot (EITB) to determine the prevalence of human Fascioliasis in the Bolivian Altiplano. *American Journal of Tropical Medicine and Hygiene* **46**, 603–609.

HODASI, J. K. M. (1972*a*). The effects of *Fasciola hepatica* on *Lymnaea truncatula*. *Parasitology* **65**, 359–369.

HODASI, J. K. M. (1972*b*). The output of cercariae of *Fasciola hepatica* by *Lymnaea truncatula* and the distribution of metacercariae on grass. *Parasitology* **65**, 431–456.

ITAGAKI, T. & TSUTSUMI, K. (1998). Triploid form of *Fasciola* in Japan: genetic relationships between *Fasciola hepatica* and *Fasciola gigantica* determined by ITS-2 sequence of the nuclear rDNA. *International Journal for Parasitology* **28**, 777–781.

JABBOUR-ZAHAB, R., POINTIER, J. P., JOURDANE, J., JARNE, P., OVIEDO, J. A., BARGUES, M. D., MAS-COMA, S., ANGLES, R., PERERA, G., BALZAN, C., KHALLAAYOUNE, K. & RENAUD, F. (1997). Phylogeography and genetic divergence of some lymnaeid snails, intermediate hosts of human and animal fascioliasis, with special reference to lymnaeids from the Bolivian Altiplano. *Acta Tropica* **64**, 191–203.

JAY, J. M. & DRONEN, N. O. (1985). The effects of parasitism and long term cutaneous respiration upon the survival and egg production of the lymnaeid snail *Stagnicola bulimoides techella*. *Proceedings of the Helminthological Society of Washington* **52**, 204–205.

KENDALL, S. B. (1953). The life-history of *Limnaea truncatula* under laboratory conditions. *Journal of Helminthology* **27**, 17–28.

KENDALL, S. B. (1954). Fascioliasis in Pakistan. *Annals of Tropical Medicine and Parasitology* **48**, 307–313.

KENDALL, S. B. (1965). Relationships between the species of *Fasciola* and their molluscan hosts. *Advances in Parasitology* **3**, 59–98.

KENDALL, S. B. (1970). Relationships between the species of *Fasciola* and their molluscan hosts. *Advances in Parasitology* **9**, 251–258.

KENDALL, S. B. & MCCULLOUGH, F. S. (1951). The emergence of the cercariae of *Fasciola hepatica* from the snail *Limnaea truncatula*. *Journal of Helminthology* **25**, 77–92.

KENDALL, S. B. & PARFITT, J. W. (1959). Studies on susceptibility of some species of *Lymnaea* to infection with *Fasciola gigantica* and *F. hepatica*. *Annals of Tropical Medicine and Parasitology* **53**, 220–227.

LOKER, E. S., HOFKIN, B. V., MKOJI, G. M., MUNGAI, B., KIHARA, J. & KOECH, D. K. (1993). Distributions of freshwater snails in southern Kenya with implications for the biological control of Schistosomiasis and other snail-mediated parasites. *Journal of Medical and Applied Malacology* **5**, 1–20.

MANGA-GONZALEZ, Y., GONZALEZ-LANZA, C. & OTERO-MERINO, C. B. (1991). Natural infection of *Lymnaea truncatula* by the liver fluke *Fasciola hepatica* in the Porma Basin, León, NW Spain. *Journal of Helminthology* **65**, 15–27.

MANSOUR, T. E. (1958). The effect of serotonin on phenol oxidase from the liver fluke, *Fasciola hepatica*, and from other sources. *Biochimica et Biophysica Acta* **30**, 492–500.

MARCILLA, A., BARGUES, M. D., RAMSEY, J. M., MAGALLON-GASTELUM, E., SALAZAR-SCHETTINO, P. M., ABAD-FRANCH, F., DUJARDIN, J. P., SCHOFIELD, C. J. & MAS-COMA, S. (2001). The ITS-2 of the nuclear rDNA as a

molecular marker for populations, species and phylogenetic relationships in Triatominae (Hemiptera: Reduviidae), vectors of Chagas disease. *Molecular Phylogenetics and Evolution* **18**, 136–142.

MAS-COMA, S. (1998). Human fascioliasis in Europe and Latin America. In *Infectious Diseases and Public Health. A Research and Clinical Update* (ed. Angelico, M. & Rocchi, G.), pp. 297–313. Philadelphia, L'Aquila, Balaban Publishers.

MAS-COMA, S. (1999). Los espaciadores transcritos internos (ITSs) del ADN ribosomal como marcadores en sistemática, ecología, evolución y filogenia de parásitos y vectores. In *XIV Congreso Latinoamericano de Parasitología* (FLAP, Acapulco, México, 11–16 Octubre 1999), Abstracts, 5–6.

MAS-COMA, S., ANGLES, R., ESTEBAN, J. G., BARGUES, M. D., BUCHON, P., FRANKEN, M. & STRAUSS, W. (1999). The human fascioliasis high endemic region of the Northern Bolivian Altiplano. *Tropical Medicine and International Health* **4**, 454–467.

MAS-COMA, S., ANGLES, R., STRAUSS, W., ESTEBAN, J. G., OVIEDO, J. A. & BUCHON, P. (1995). Human fasciolasis in Bolivia: a general analysis and a critical review of existing data. *Research and Reviews in Parasitology* **55**, 73–93.

MAS-COMA, S. & BARGUES, M. D. (1997). Human liver flukes: a review. *Research and Reviews in Parasitology* **57**, 145–218.

MAS-COMA, S., BARGUES, M. D. & ESTEBAN, J. G. (1999). Human Fasciolosis. In *Fasciolosis* (ed. Dalton, J. P.), pp. 411–434. CAB Wallingford, Oxon, UK, International Publishing.

MAS-COMA, S., BARGUES, M. D., MARTY, A. M. & NEAFIE, R. C. (2000). Hepatic Trematodiases. In *Pathology of Infectious Diseases, Vol. 1 Helminthiases* (ed. Meyers, W. M., Neafie, R. C., Marty, A. M. & Wear, D. J.), pp. 69–92. Washington D.C., Armed Forces Institute of Pathology and American Registry of Pathology.

MAS-COMA, S., ESTEBAN, J. G. & BARGUES, M. D. (1999). Epidemiology of human fascioliasis: a review and proposed new classification. *Bulletin of the World Health Organization* **77**, 340–346.

MAS-COMA, S., RODRIGUEZ, A., BARGUES, M. D., VALERO, M. A., COELLO, J. R. & ANGLES, R. (1997). Secondary reservoir role of domestic animals other than sheep and cattle in fascioliasis transmission in the Northern Bolivian Altiplano. *Research and Reviews in Parasitology* **57**, 39–46.

MCGONIGLE, S. & DALTON, J. P. (1995). Isolation of *Fasciola hepatica* haemoglobin. *Parasitology* **111**, 209–215.

MEUNIER, C., TIRARD, C., HURTREZ-BOUSSES, S., DURAND, P., BARGUES, M. D., MAS-COMA, S., POINTIER, J. P., JOURDANE, J. & RENAUD, F. (2001). Lack of molluscan host diversity and the transmission of an emerging parasitic disease in Bolivia. *Molecular Ecology*, in press.

MICHOT, B., DESPRES, L., BONHOMME, F. & BACHELLERIE, J. P. (1993). Conserved secondary structures in the ITS-2 of trematode pre-rRNA. *Federation of European Biochemical Societies (FEBS)* **316**, 247–252.

MORALES, G. & PINO, L. A. (1981). *Lymnaea cubensis* Pfeiffer, 1839, hospedador intermediário de *F. hepatica* en la zona alta de los Andes trujillanos,

Venezuela. *Boletín de la Dirección de Malariología y Sanidad Ambiental* **21**, 39.

MORALES, G. & PINO, L. A. (1983). Infection de *Lymnaea cubensis* par *Fasciola hepatica* dans une région d'altitude, au Venezuela. *Annales de Parasitologie Humaine et Comparée* **58**, 27–30.

MOREL, A. M. & MAHATO, S. N. (1987). Epidemiology of fascioliasis in the Koshi hills of Nepal. *Tropical Animal Health and Production* **19**, 33–38.

ODENING, K. (1971). *Der Grosse Leberegel und Seine Verwandten.* Luther Stadt, A. Ziemsen Verlag Wittenberg.

O'NEILL, S. M., PARKINSON, M., STRAUSS, W., ANGLES, R. & DALTON, J. P. (1998). Immunodiagnosis of *Fasciola hepatica* (fascioliasis) in a human population in the Bolivian Altiplano using purified cathepsin L cysteine proteinase. *American Journal of Tropical Medicine and Hygiene* **58**, 417–423.

OVIEDO, J. A., BARGUES, M. D. & MAS-COMA, S. (1995). Lymnaeid snails in the human fascioliasis high endemic zone of the Northern Bolivian Altiplano. *Research and Reviews in Parasitology* **55**, 35–43.

OVIEDO, J. A., BARGUES, M. D. & MAS-COMA, S. (1996). The intermediate snail host of *Fasciola hepatica* on the Mediterranean island of Corsica. *Research and Reviews in Parasitology* **56**, 217–220.

PANTELOURIS, E. M. (1965). *The Common Liver Fluke, Fasciola hepatica L.* (Ed.) Oxford, Pergamon Press.

PAREAU, M., RONDELAUD, D., DREYFUSS, G., BOUTEILLE, B. & DARDE, M. L. (1994). *Fasciola hepatica* Linné: observations épidémiologiques sur les cas de distomatose humaine dans le Limousin (France) par rapport à l'altitude des cressonières naturelles. *Bulletin de la Société Française de Parasitologie* **12**, 167–174.

PEREZ-REYES, R., JIMENEZ-NAVA, J. J. & VARELA-RAMIREZ, A. (1985). Fascioliasis en el Estado de Chihuahua, México. I Susceptibilidad de *Fossaria modicella* (Say, 1825), huésped intermediario local. *Revista Latino-Americana de Microbiología* **27**, 367–372.

ROBERTS, E. W. (1950). Studies on the life-cycle of *Fasciola hepatica* (Linnaeus) and of its snail host, *Limnaea* (*Galba*) *truncatula* (Müller), in the field and under controlled conditions in the laboratory. *Annals of Tropical Medicine and Parasitology* **44**, 187–206.

RONDELAUD, D. & BARTHE, D. (1986). Les générations rédiennes de *Fasciola hepatica* L. Premières observations chez des Limnées tronquées en fin de cycle parasitaire. *Bulletin de la Société Française de Parasitologie* **4**, 29–38.

RONDELAUD, D. & DREYFUSS, G. (1997). Variability of *Fasciola* infections in *Lymnaea truncatula* as a function of snail generation and snail activity. *Journal of Helminthology* **71**, 161–166.

ROWCLIFFE, S. A. & OLLERENSHAW, C. B. (1960). Observations on the bionomics of the egg of *Fasciola hepatica*. *Annals of Tropical Medicine and Parasitology* **54**, 172–181.

SAMADI, S., ROUMEGOUX, A., BARGUES, M. D., MAS-COMA, S., YONG, M. & POINTIER, J. P. (2000). Morphological studies of lymnaeid snails from the human fascioliasis endemic zone of Bolivia. *Journal of Molluscan Studies* **66**, 31–44.

SAMBROOK, J., FRITSCH, E. F. & MANIATIS, T. (1989). *Molecular Cloning. A Laboratory Manual.* 2nd Ed.

Vols. I, II & III. Cold Spring Harbor, New York, USA, Cold Spring Harbor Laboratory.

SCOTT, J. M. & GOLL, P. H. (1977). The epidemiology and anthelminthic control of ovine fasciliasis in the Ethiopian Central Highlands. *British Veterinary Journal* **133**, 273–280.

STOTHARD, J. R., BREMOND, P., ANDRIAMARO, L., LOXTON, N. J., SELLIN, B., SELLIN, E. & ROLLINSON, D. (2000). Molecular characterization of the freshwater snail *Lymnaea natalensis* (Gastropoda: Lymnaeidae) on Madagascar with an observation of an unusual polymorphism in ribosomal small subunit genes. *Journal of Zoology* **252**, 303–315.

THOMPSON, J. D., HIGGINS, D. G. & GIBSON, T. J. (1994). CLUSTAL W: improving the sensitivity and progressive multiple sequence alignment through sequence weighting, positions-specific gap penalties and weight matrix choice. *Nucleic Acids Research* **22**, 4673–4680.

UENO, H., ARANDIA, R., MORALES, G. & MEDINA, G. (1975). Fascioliasis of livestock and snail host for *Fasciola* in the Altiplano region of Bolivia. *National Institute of Animal Health Quarterly* **15**, 61–67.

UENO, H. & MORALES, G. (1973). Fasciolicidal activity of diamphenetide and niclofolan against *Fasciola hepatica* in sheep in the Altiplano Region of Bolivia. *National Institute of Animal Health Quarterly* **13**, 75–79.

VALENZUELA, G. (1979). Estudio epidemiológico sobre el desarrollo de huevos de *Fasciola hepatica* en el medio ambiente de Valdivia, Chile. *Boletín Chileno de Parasitología* **34**, 31–35.

VALERO, M. A., MARCOS, M. D., COMES, A. M., SENDRA, M. & MAS-COMA, S. (1999). Comparison of adult liver flukes from highland and lowland populations of Bolivian and Spanish sheep. *Journal of Helminthology* **73**, 341–345.

VALERO, M. A. & MAS-COMA, S. (2000). Comparative infectivity of *Fasciola hepatica* metacercariae from isolates of the main and secondary reservoir animal host species in the Bolivian Altiplano high human endemic region. *Folia Parasitologica* **47**, 17–22.

VALERO, M. A., PANOVA, M. & MAS-COMA, S. (2001). Development differences in the uterus of *Fasciola hepatica* between livestock liver fluke populations from Bolivian highland and European lowlands. *Parasitology Research*, in press.

VAN SOMEREN, V. D. (1946). The habitats and tolerance ranges of *Lymnaea* (*Radix*) *caillaudi*, the intermediate snail host of the liver fluke in East Africa. *Journal of Animal Ecology* **15**, 170–197.

VENTURINI, L. M. (1978). Revisión del ciclo biológico de *Fasciola hepatica*. *Analecta Veterinaria* **10**, 13–19.

WILSON, R. A. & DENISON, J. (1980). The parasitic castration and gigantism of *Lymnaea truncatula* infected with the larval stages of *Fasciola hepatica*. *Zeitschrift für Parasitenkunde* **61**, 109–119.

WILSON, R. A., SMITH, G. & THOMAS, M. R. (1982). Fascioliasis. In *The Population Dynamics of Infectious Diseases: Theory and Applications* (ed. Anderson, R. M.), pp. 263–361. London-New York, Chapman and Hall.

WORLD HEALTH ORGANIZATION (1995). *Control of Foodborne Trematode Infections.* WHO Technical Report Series No. 849. WHO, Geneva.

Multiple strategies of schistosomes to meet their requirements in the intermediate snail host

M. de JONG-BRINK*, M. BERGAMIN-SASSEN *and* M. SOLIS SOTO†

Research Institute Neurosciences, Vrije Universiteit, Faculty of Biology, De Boelelaan 1087, 1081 HV Amsterdam, The Netherlands

SUMMARY

The results of the studies on our model combination *Trichobilharzia ocellata–Lymnaea stagnalis*, presented in this review, lead to the conclusion that schistosomes use multiple strategies to reach their goals, i.e. to propagate and to continue their life cycle. They have to escape from being attacked by the internal defence system (IDS) of the snail host and to profoundly affect the host's energy flow, of which reproduction and growth are the main determinants, for their own benefit. These physiological changes they establish mainly by interfering with the two regulatory systems in the snail host, the IDS and the neuroendocrine system (NES). Moreover, these two regulatory systems clearly interact with each other. Parasitic E/S products affect the host's IDS both in a direct and an indirect way. The neuropeptides or neuropeptide-like substances that are secreted by parasite glands into the host directly suppress haemocyte activity in the snail. The indirect effects include effects of (1) peptides from connective tissue cells and (2) neuropeptides from NES and/or IDS. Parasitic E/S products also induce the effects on energy flow in the host. These E/S products act either directly on a target, as shown for the inhibiting effect of the parasite on the development of the male copulation organ, or on the NES regulating reproductive activity, e.g. on gene expression. Indirect effects of E/S products on the NES (hormone-receptor interaction, electrical activity) are mediated by a factor from connective tissue cells, presumably belonging to the IDS. The physiological changes in the snail host are obviously of vital importance for the parasites, since they make use of different strategies to bring them about.

Key words: schistosomes, snail host, internal defence, neuroendocrine regulation, parasite E/S products, energy flows, host reproduction and growth.

INTRODUCTION

The first problem compatible schistosomes have to solve when they have entered their intermediate snail host is to adapt to the host's 'milieu interieur' and to avoid being attacked by the immune system. Secondly, they need to obtain enough energy for their own maintenance, growth and multiplication and to acquire space to accommodate the increasing number of developing parasites. Finally, it is important for them not to severely damage or kill the host. To overcome these problems and to meet their requirements the parasite not only adapts to the situation in the host but also modulates its defence and affects its physiology.

These effects on the physiology, especially those on reproduction and growth, of the snail host vary greatly. They depend for example on the parasite–host combination studied, the size/age of the host at the time of exposure to miracidia and the stage of infection at the time of observation. Under natural conditions, at the population level, fluctuating environmental factors make the picture even more complicated. This implies that an experimental set-up is required to investigate how schistosomes exert the effects on the physiology of the snail host. As an experimental parasite–snail combination we have chosen *Trichobilharzia ocellata–Lymnaea stagnalis* because it combines high parasite productivity and clear physiological effects in the host.

The physiological effects of *T. ocellata* on its snail host are most prominent in two stages of infection, the early stage, when miracidia transform into mother (primary) sporocysts, and the later stage when cercariae are differentiating within daughter (secondary) sporocysts and emerge from the host, the shedding stage. In the early stage of infection *T. ocellata* not only has immunomodulatory effects but also inhibits the development of the reproductive tract of the snail host. Inhibition of development of the reproductive tract starts immediately after infection and is already obvious one week post-infection in snails exposed at a size of 8–10 mm (Sluiters, 1981). If snails are infected at a size of 2–3 mm (1 week after hatching), the development of the reproductive tract is almost completely inhibited (De Jong-Brink, 1992). After a period, in which neither suppression nor activation of the internal defence could be detected, the internal defence appeared to become affected again at the stage when cercariae emerge from daughter sporocysts causing

* Corresponding author: Tel: 31-20-4447105. Fax: 31-20-4446968. E-mail: mdejong@bio.vu.nl
† Present address: Departamento de Histologia, Facultad de Medicina, UANL Apartado Postal 1563, Monterrey, N.L., Mexico.

Parasitology (2001), **123**, S129–S141. © 2001 Cambridge University Press
DOI: 10.1017/S0031182001008149

mechanical and lytic host tissue damage (Amen *et al.* 1992). From the time cercariae are differentiating in daughter sporocysts the parasites start to affect the host's reproductive activity and growth, being the main energy-consuming processes. Although *T. ocellata* stimulates growth of the host, this does not cost much energy as the parasite mainly causes an increase of the wet weight, not of the dry weight (Joosse & Van Elk, 1986). In this way the parasite not only provides itself with enough energy but also with space for its offspring. Similarly, the effects on reproductive activity also depend on the developmental stage of the host at the time of exposure. If subadult snails (shell height 19–22 mm), which are not yet laying egg masses but have already a well-developed reproductive tract, become infected (to attain this they have to be exposed to many more miracidia than juvenile snails) they start producing egg masses before non-infected controls (Schallig *et al.* 1991). As soon as differentiating cercariae are present within the sporocysts, oviposition is completely inhibited in the snail host (Schallig *et al.* 1991). It is unknown how *T. ocellata* causes acceleration of reproduction in *Lymnaea* infected as subadults. This holds also for other digeneans inducing this phenomenon (see e.g. Minchella & LoVerde, 1981; Thornhill, Jones & Kusel, 1986; Lafferty, 1993 *a, b*). Acceleration of maturation might be a mechanism that guarantees reproduction and hence future generations before the snails become (completely) sterilized (see also Stearns, 1989).

In this review we will focus on the question of how *T. ocellata* induces these physiological changes in their intermediate snail host. It has been demonstrated that *T. ocellata*, like other schistosomes and endoparasites (see e.g. Hurd, 1990; Beckage, 1997; Beckage & Gelman, 2001), affects immunological and physiological responses in the host (Van der Knaap & Loker, 1990; De Jong-Brink, 1992, 1995; De Jong-Brink *et al.* 1997, 1999*b*). The parasite excretes/secretes factors (E/S products) by which it actively interferes with the two regulatory systems in the host, the immune system (IS), in invertebrates, preferably called internal defence system (IDS), and the neuroendocrine system (NES). Yoshino, Boyle & Humphries (this supplement) deal with the molecular interaction at the schistosome–snail host interface, i.e. locally, whereas this review focuses on effects exerted more systemically. The molecules released into the host can affect the host's regulatory systems directly and/or indirectly. Indirect effects on the NES may, for example, be exerted by factors derived from the IDS and vice versa. This means that changes in activity of one of the regulatory systems may affect the other. Molecular biological techniques are very helpful to detect parasite-induced changes in the regulatory systems concerned. In addition, functional studies are needed to answer the question why the parasite affects the expression of certain host genes, *viz* to reveal that changes in gene expression are specific and clearly related to the affected physiological processes.

An important conclusion based upon the data obtained with our model combination and presented in this review is that parasites make use of multiple strategies to reach their goals. This indicates that the changes they bring about in the physiology of their intermediate host are of vital importance. They are a '*conditio sine qua non*'.

HOW SCHISTOSOMES ENSURE THEY ARE NOT ATTACKED BY THE INTERNAL DEFENCE

Snails have an open blood circulatory system with only one circulating blood cell type, the haemocyte (Amen *et al.* 1992). These phagocytic haemocytes, which are able to distinguish foreign, non-self material, can move freely from the circulatory system into the connective tissue and vice versa. They are the primary effector cells of the IDS that are able to encapsulate, kill and eliminate invaders such as non-compatible parasites. Killing is effected by haemocyte-mediated cytotoxicity comprising both non-oxidative and oxidative killing mechanisms (Bayne, Buckley & De Wan, 1980; see Núñez, Adema & De Jong-Brink, 1994).

Special physico-chemical and dynamic properties of parasite and haemocyte surfaces play a role in establishing immunological compatibility. In addition, parasites have the capacity to mimic or acquire host antigens and to interfere with the host's internal defence activities, with haemocytes being the effector cells. Cercariae leaving the daughter sporocyst make use of 'masking'. Components of the host's haemolymph were found at their surface as soon as they leave the sporocysts (see De Jong-Brink, 1995). Schistosomes can modulate a variety of haemocyte functions such as motility (Lodes & Yoshino, 1990), protein synthesis (Lodes, Connors & Yoshino, 1991), phagocytosis (Connors & Yoshino, 1990) and bacterial clearance (Núñez *et al.* 1994). Modulation of haemocyte activity in the initial stages of infection is of vital importance: in this period the fate of a parasite is determined. It is possible to study the immunomodulatory role of E/S products released by the parasites in the early stages of infection. After they have been released into a simple medium (Schallig *et al.* 1990) during *in vitro* transformation of miracidia, they can be purified and the fractions obtained can be tested for their effects on haemocyte activity.

Direct effects on haemocyte activity

Phagocytosis of zymosan particles by haemocytes from *L. stagnalis* infected with *T. ocellata* (*in vivo* studies) was activated at the early stage of infection

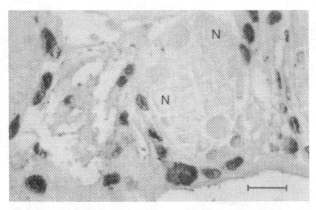

Fig. 1. Light-micrograph showing anti-molluscan defence molecule (MDM) positive granular cells in the connective tissue around the central nervous system (CNS) of *Lymnaea stagnalis*. N, neurons in one of the ganglia. (Bar = 150 µm.)

(1·5–6 h post exposure; p.e.) but in subsequent stages (12–72 h p.e.) this activity of haemocytes became suppressed (Amen *et al.* 1992). In a later period (72–96 h p.e.) the activity was back to normal again. This was confirmed by studying the bacterial clearance activity of haemocytes from these early stages of infection. In these experiments not only haemocytes but also plasma from parasitized snails was used and tested on haemocytes from non-infected snails. Distinction was made between cell- and plasma-associated effects. Following exposure to the parasite, initially both cellular and humoral components of the internal defence appeared to be activated and subsequently suppressed (Núñez *et al.* 1994).

During *in vitro* transformation of miracidia into mother sporocysts, first a fraction (ca. 2 kDa) is released that stimulates haemocyte activity. Subsequently, the E/S products were dominated by a fraction (ca. 40 kDa) which suppresses haemocyte activity (Núñez *et al.* 1997). In a later stage (comparable to the *in vivo* situation 72–96 h p.e.) both fractions were released in very low but equal amounts. This reveals that no effect on haemocyte activity was found anymore in this period. The described *in vivo* effects of parasites on haemocyte activity in their snail host can also be explained by the sequence of E/S product released by *T. ocellata* in its susceptible snail host, *L. stagnalis*. The E/S products have first a stimulating and then a predominantly suppressive effect on haemocyte activity. *S. mansoni*, on the other hand, also caused an initial activation of *Lymnaea* haemocytes but was not able to suppress this activity in a later period (12–72 h) after infection. Similarly, the 2 kDa fraction of the E/S products of *T. ocellata* also activated haemocytes from the incompatible snail species *Planorbarius corneus*. This early activation of haemocyte activity most likely reflects the non-specific process of ciliated plate elimination. The suppressive effect of the 40 kDa fraction, on the other hand, was only found in compatible combinations (Núñez & De Jong-Brink, 1997). So, the suppressing E/S fraction determines, at least in part, whether a trematode–snail association is compatible.

In vitro studies also revealed an indirect effect of schistosomes on haemocyte activity, mediated by factors derived from the snail's central nervous system (CNS) (Amen & De Jong-Brink, 1992). At that time it was not yet clear whether these brain-derived factors were neuronal factors (neuro-peptides, biogenic amines) and/or comprised other unknown factors from connective tissue cells around the CNS or from (micro) glial cells within the CNS (Sonetti *et al.* 1994).

Indirect effects on haemocyte activity

In this subsection we will show that factors derived from connective tissue cells as well as from neurons in the CNS are mediators in the parasite-induced effects on haemocyte activity.

Connective tissue. In the connective tissue of *Lymnaea* several cell types have been described and for some of them it has been suggested that they function in the production of blood/haemolymph proteins (pore cells and granular cells), in the elimination of small-sized molecules/particles (pore cells) or in phagocytosis (tissue-dwelling haemocytes). Apart from that of haemoglobin/haemocyanin, the function of the blood proteins is unknown (Sminia, 1972; Van der Knaap & Loker, 1990).

Differential screening of cDNA libraries of CNS from *T. ocellata* infected and non-infected snails revealed an interesting transcript: a cDNA encoding a protein called Molluscan Defence Molecule (MDM; Hoek *et al.* 1996). It appeared to be down-regulated in parasitized snails. The molecular structure of MDM has a similar overall organization as hemolin, an insect immunoprotein belonging to the Ig super-family (Sun *et al.* 1990). This MDM molecule is synthesized in cells of the connective tissue, the granular cells (Fig. 1). It enhances phagocytic activity of haemocytes: roughly 35 % of haemocytes were phagocytosing zymosan particles in the absence of MDM and 46 % in the presence of MDM (Lageweg & De Jong-Brink, unpublished). This might explain why it is important for the parasite to induce down-regulation of the MDM encoding gene 2, 6 and 8 weeks after infection (Fig. 2). Up-regulation of the gene was only found at 5 h p.e. If up-regulation of the gene results in a higher titer of MDM at this time of infection, MDM possibly also plays a role in activating haemocytes to eliminate the ciliated plates.

Another cDNA encoding for a protein that appeared to be involved in the internal defence in the snail host is derived from the same connective tissue

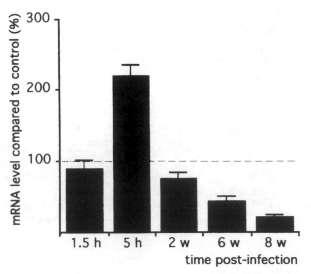

Fig. 2. MDM mRNA levels in the CNS, excised from *Lymnaea stagnalis* at several time points post-infection with *Trichobilharzia ocellata* as percentage of the levels in non-infected controls (= 100%) at the same time points. After a significant increase at 5 h post-infection the MDM mRNA levels show a decrease at all time points studied.

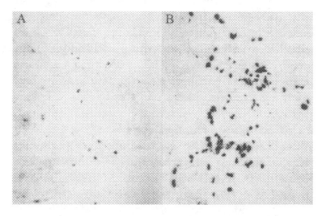

Fig. 3. Low-power micrograph showing the granularin gene (*in situ* hybridization; black dots are granular cells) in the connective tissue sheath around the CNS of a non-parasitized (A) and a parasitized (B) snail. Note the difference: not only more cells express the gene but also the intensity per cell has increased in parasitized snails.

cell type as MDM, the granular cells. This proteinaceous product is called granularin. Granularin is a 62 amino acid protein with sequence homology to various extracellular matrix proteins, e.g. thrombospondin and collagen. In particular, cysteine spacing in granularin is strikingly conserved with domains in these proteins (Smit *et al*. unpublished). It also shows similarity with the von Willebrand factor produced by endothelial cells in vertebrates. This glycoprotein facilitates coagulation of platelets during clot formation (Gartner & Hiatt, 1997). Also granularin appeared to have an effect on phagocytic activity of haemocytes. In contrast to MDM,

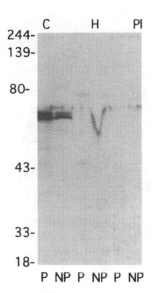

Fig. 4. Western-blots of a-MDM positive material in CNS (C), haemocytes (H) and plasma (Pl) from parasitized (P) and non-parasitized (NP) snails.

granularin appeared to inhibit haemocyte phagocytic activity (29% of haemocytes appeared to phagocytose zymosan particles in the absence of granularin and 19% in the presence of granularin). This explains why the encoding gene is up-regulated (Fig. 3) in parasitized snails from 1·5 h p.e. onwards. Apparently, the parasites simultaneously induce up- and down-regulation of two different genes within one cell type. As parasites affect granularin gene expression within 90 min it can be supposed that (an) E/S product(s) induce(s) these effects on gene expression directly. If that is the case it will be possible to identify the E/S factor(s) concerned.

The observation that both MDM and granularin are involved in the internal defence supports the idea that these cells in the connective tissue belong to the IDS of the snail host. As the cDNAs of both precursor molecules encode for a signal peptide these products are released. However, it is not very likely that MDM circulates in the haemolymph as the anti-MDM positive double band present on a Western blot of CNS extracts was lacking in haemolymph (plasma) proteins (Fig. 4). Therefore, we assume that the secreted MDM remains in the vicinity of the granular cells and affects haemocyte activity only very locally. In this respect MDM differs from the fibrinogen-related proteins (FREPs) circulating in haemolymph of *B. glabrata* infected with *Echinostoma paraensei*. These FREPs, lectins presumably functioning in binding of non-self, also contain regions with sequence similarity to Ig superfamily members (Adema *et al*. 1997).

Although we do not know whether granularin circulates in the haemolymph its molecular structure suggests that it, like MDM, acts locally. Experiments to study the functional aspects of MDM and granularin in more detail are in progress.

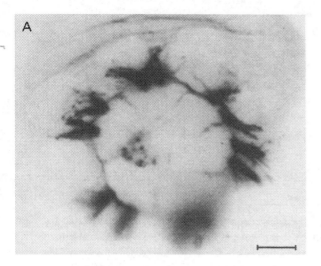

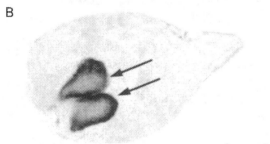

Fig. 5. Whole-mount preparations of miracidia of
Trichobilharzia ocellata (A) and *Schistosoma mansoni* (B)
immunostained with anti-Melanocyte Stimulating
Hormone (α-MSH). The two big cephalic glands of *S.
mansoni* miracidia (B) are positive. (Bar = 12 μm). In *T.
ocellata* (A) 6 neurons in the central part of the nervous
mass and their ciliated nerve endings at the periphery of
the miracidia are anti-α-MSH positive. The cephalic
glands (not visible in picture A) did not stain with this
antiserum. (Bar = 6·5 μm).

NES factors. Indications for the involvement of
neuroendocrine factors in regulating the internal
defence in invertebrates are accumulating. Haemocytes or immunocytes appear to have receptors
for opioid peptides. Opioids stimulate chemotaxis
between haemocytes and the release of mammalian-
like cytokines (interleukin-1 (Il-1), Il-6 and tumor
necrosis factor α (TNFα)) by haemocytes.

It has been shown that opioid peptides are derived
from haemocytes and from the related glial cell type
(microglia) in the CNS (Stefano *et al.* 1996; Stefano,
Salzet & Fricchione, 1998). The question remains
whether the NES itself is also a source of opioid
peptides which might play a role in regulating
haemocyte activity. As far as the CNS of *Lymnaea* is
concerned, no pro-opiomelanocortin (POMC) gene
could be demonstrated in neurons in the CNS,
although they were clearly immunopositive with
antibodies to ACTH, one of the POMC-derived
neuropeptides. This can be explained by cross
reactivity between a certain amino acid sequence of
ACTH and a corresponding amino acid sequence of

the prohormone of these neurons (Boer *et al.* 1979;
Bogerd *et al.* 1991).

A few observations on the presence of a POMC
gene and the encoded opioid peptides in schistosomes are very interesting in this respect. Duvaux-
Miret *et al.* (1990, 1992) demonstrated the presence
of β-endorphin and a POMC-related gene in *Schistosoma mansoni*. This was the first demonstration of a
POMC-related gene transcribed in an invertebrate.
Our studies (unpublished) showed that the two
cephalic or accessory gland cells of miracidia of
S. mansoni were immunopositive with an antiserum
to α-melanocyte-stimulating hormone (α-MSH),
whereas these glandular cells of *T. ocellata* did not
stain with this antiserum (Fig. 5). This suggests
that α-MSH is released from these glands into the
snail host and might affect *B. glabrata* haemocyte
activity and not those from *L. stagnalis*. This supposition has been confirmed in two experiments
in which the effect of synthetic α-MSH was tested
on phagocytotic activity of haemocytes from the two
species, *L. stagnalis* and *B. glabrata*. The peptide
had no effect on phagocytic activity of *L. stagnalis*
haemocytes, whereas it inhibited phagocytosis of
zymosan particles by *B. glabrata* haemocytes (Fig.
6 A and B). Also in vertebrates anti-inflammatory
actions have been ascribed to α-MSH (Lipton &
Catania, 1997).

Besides this opioid peptide, other neuropeptides
supposedly play a role in the internal defence.
Therefore, it is very interesting to investigate
whether the products encoded by genes which are
either up- or down-regulated in the brain of
parasitized snails (*L. stagnalis*) might have an effect
on the internal defence, *viz.* on haemocyte activity.
We have investigated some of the neuropeptides of
which the encoding gene expression appeared to be
obviously affected by parasitization: up-regulation
of the genes encoding FMRFamide (FMRFa),
LFRFamide (LFRFa), and Pedal Peptide (PP)
throughout infection and of the *Lymnaea* Neuropeptide Y (LyNPY) encoding gene coinciding with
differentiation of cercariae in daughter sporocysts
(Hoek *et al.* 1997). The escape glands in cercariae of
both *T. ocellata* and *S. mansoni* also gave a positive
immunostaining with an antiserum to FMRFa (Fig.
7). As these glandular cells are supposed to secrete
this material while migrating through the snail host,
it seemed very interesting to test FMRFa on
phagocytotic activity of both *Lymnaea* and *Biomphalaria* haemocytes. The data (Fig. 6A and B)
demonstrated that FMRFa inhibits haemocyte activity of both species. This indicates that the
neuropeptide FMRFa plays an important immunosuppressive role throughout infection.

The other neuropeptides tested, PP and LyNPY,
did not significantly affect *Lymnaea* haemocyte
activity. For LyNPY a modulatory role in the
internal defence would have been in line with the

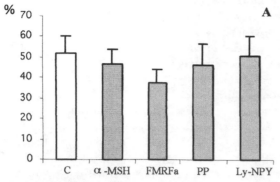 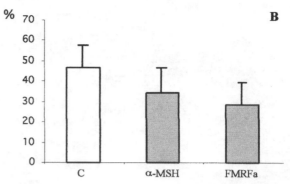

Fig. 6. The percentages (average ± s.d.) of haemocytes (from haemolymph of 3 pools, 10 snails per pool; 1500 haemocytes per pool for *Lymnaea* (A); ca. 300–700 haemocytes for *Biomphalaria* (B)) phagocytosing zymosan particles after having been incubated for 30 min in only Hepes Buffered Saline (HBS; control) or in HBS in the presence of one of the following neuropeptides (10^{-7} M): α-MSH, FMRFamide (FMRFa), Pedal Peptide (PP) or *Lymnaea* Neuropeptide Y (LNPY). The phagocytic activity of haemocytes of *L. stagnalis* (A) was only significantly suppressed by FMRFa, those of *B. glabrata* (B) by both α-MSH and FMRFa.

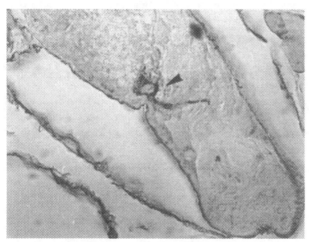

Fig. 7. Whole-mount preparation of a cercaria of *T. ocellata* immunostained with anti-FMRFa. The escape glands (arrow head) are immunopositive for anti-FMRFa. The same holds for the glands in cercariae of *S. mansoni*.

effect NPY has on the IS in mammals: it enhances, for example, proliferation of lymphocytes in the *lamina propria* of the human colon (Elitsur *et al.* 1994).

In summary, it has been demonstrated that *T. ocellata* makes use of multiple strategies to escape from being attacked by the IDS of its snail host. The parasitic E/S products have a direct effect on haemocyte activity and, simultaneously, also an indirect effect. The parasites or their E/S products cause up- or down-regulation of genes in connective tissue cells encoding peptides affecting the internal defence. In addition they induce changes in expression of neuropeptide-encoding genes in neurons/neuroendocrine cells in the host's CNS. Some of these neuropeptides also appear to function in the internal defence. In line with this the products of parasite glands (the cephalic glands of miracidia and the escape glands of cercariae), which are secreted into the snail host, contain neuropeptides

(neuropeptide-like material) also suppressing the internal defence in the snail host.

HOW SCHISTOSOMES OBTAIN THE ENERGY AND SPACE THEY NEED

It seems a very advantageous strategy for schistosomes to inhibit the development of the reproductive tract as soon as they have entered their juvenile snail host. In the next part we will discuss how the effects on the development of the reproductive tract are brought about.

Inhibition of the development of the reproductive tract

Development of the reproductive tract in snails is supposed to be under neuroendocrine control. In the hermaphroditic snail *Lymnaea* the development of the female part of the tract is regulated by the endocrine dorsal bodies (DBs; Geraerts & Joosse, 1975). It is, however, still not clear how the development of the male part is regulated. The development of the male copulation organ in another snail species, the prosobranch *Crepidula forniculata*, is directed by both stimulating and inhibiting factors from the pedal ganglia (Le Gall, 1981). As these male factors have not been identified it is not clear whether they are neuronal in origin. Nevertheless we expected that parasite interference with the NES should underly these inhibiting effects on the development of the reproductive tract in snails which had been infected at a very young stage (one week after hatching; shell height of 2–3 mm). However, *in vitro* experiments have shown that parasitic E/S products inhibit directly – *viz.* not mediated by snail factors produced outside the copulation organ – mitotic divisions in the 'Anlage' of the male copulation organ. This was demonstrated for special, rather large cells, which are localized in a limited area at the tip of the penis of the male copulation organ.

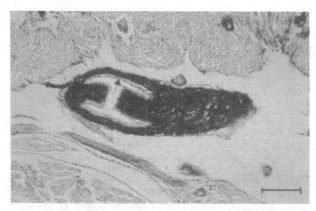

Fig. 8. Histological section of the male copulation organ (praeputium) of an immature *L. stagnalis* immunostained with an antiserum to *Lymnaea* epidermal growth factor (L-EGF). The connective tissue between the muscle cells is positive. Note the unstained epithelial layer lining the lumen (arrow). (Bar = 200 μm).

These cells especially showed immunostaining with a monoclonal antibody to proliferating cell nuclear antigen (PCNA), an S-phase-specific cell-cycle marker. The cells appeared to be smooth muscle cells and as they have the capacity to divide we considered them to be 'myoblasts' (De Jong-Brink *et al.* 1999*b*). As these myoblasts can also be found in the copulation organ of adult snails, they probably enable, as a kind of growth cone, continuous growth of the penis in proportion to the body growth. It was concluded that the effect of parasite E/S products on these myoblasts was exerted directly and not mediated by factors derived from the CNS including its connective tissue sheath. No difference was observed between the number of PCNA-positive myoblasts in an 'Anlage' cultured in a medium with E/S products, (1) lacking any other factors, (2) containing excised CNS or (3) in a medium pre-incubated with CNS of snails, the so called 'conditioned' medium (Ridgeway *et al.* 1991). This makes it rather unlikely that the NES is involved in maintaining these mitotic divisions (growth) and in establishing the inhibiting effects of the parasitic E/S products on mitotic divisions of myoblasts in the penis.

These data, however, do not exclude the possibility that the parasite E/S products also may be acting indirectly. They may, for example, inhibit the stimulating effect of endogenous growth factors on mitotic divisions of the myoblasts in the copulatory organ. In that case, *Lymnaea* epidermal growth factor (L-EGF) might be a good candidate as immunocytochemical studies have revealed the presence of L-EGF in the copulation organ of immature snails (Fig. 8; De Jong-Brink & Van Rijn, unpublished results). The encoding gene is identical to the one expressed both in the albumen gland and in the CNS of the snails (Hermann *et al.* 2000). Experiments to investigate the involvement of en-

dogenous L-EGF in mediating the inhibiting effect of parasite E/S products on the development of the male copulation organ are in progress.

It is rather difficult to investigate whether the development of the other parts of the reproductive tract are affected in a way similar to that of the penial complex because mitotic divisions are much more difficult to detect. As far as the female tract is concerned, observations by Sluiters, Roubos & Joosse (1984) have shown that parasitic infection causes a clear activation of DB activity in *Lymnaea*. This indicates that the interference of the parasite with the development of the female reproductive tract is also exerted at the level of the target organs. Even if in all cases growth factors play a key role, the involvement of neuroendocrine factors in regulation of growth factor activity has to be considered. However, even in a penial complex of a mature snail, in which L-EGF cannot be detected any more, the encoding gene can be activated by extirpation of the organ and keeping it overnight in snail Ringer. This was concluded from the observation that L-EGF could be demonstrated immunocytochemically in the extirpated organ.

A secondary effect of underdevelopment of a reproductive target organ. The question arises whether inhibition of the development of reproductive target organs is reflected at the level of the development of central neurons innervating or regulating these targets.

In vertebrates it is known that the mechanism to adapt neuronal potential or capacity to the size of their target is to vary the number of innervating neurons. Initially the target is innervated by a superfluous number of neurons and subsequently the number is reduced by apoptosis (Patterson, 1992). Survival of neurons (motorneurons, sensory neurons) depends on target-derived neurotrophic factors which are retrogradely transported to the neuronal cell bodies. So, the size of the target is reflected by the amount of neurotrophic factors and hence by the number of surviving neurons.

In snails and members of other groups (amphibians, fish) which grow continuously during their life, the main mechanism by which neuronal capacity increases post-hatching is that the neurons gradually obtain a higher degree of polyploidy by the phenomenon of endomitoses. This is reflected by a stepwise increase in cell and nuclear size (Boer *et al.* 1977). So, we hypothesized that the consequence of underdeveloped reproductive target organs in parasitized snails is reflected by an inhibited development, degree of polyploidy, of the central neurons innervating or regulating these targets.

This supposition was investigated for (1) motor-neurons in the lobus anterior of the right cerebral ganglion innervating the male copulation organ and hence controlling male copulation behaviour and (2)

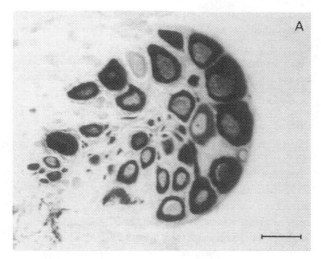

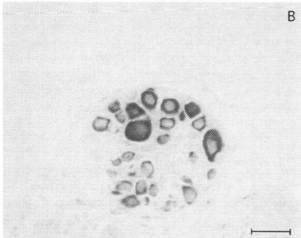

Fig. 9. Histological sections of the anterior lobe of the right cerebral ganglion from the CNS of a nonparasitised (A) and of parasitized snail (B) 9 weeks after parasitisation, immunostained with anti-APGWamide (APGWa). Note that the lobe of the parasitised animal is much smaller than that of the non-parasitized one. The number and size of individual neurons immunopositive with a-APGWa have also decreased considerably due to parasitisation. (Bar = 50 μm).

the caudo-dorsal cells (CDCs), neuroendocrine cells in the cerebral ganglia producing neuropeptides which regulate ovulation, egg laying and accompanying behaviour. The data obtained (De Lange et al. 2001) have demonstrated that already early in infection especially these motorneurons and the neuroendocrine CDCs were smaller and that fewer cells were found to express neuropeptide genes as compared to those in non-parasitized controls (Fig. 9). As far as the penis-innervating motorneurons are concerned, the balance between the size of the target and the development of the motorneurons depends on the nervous connection with the target. As demonstrated in vertebrates the effect on the motorneurons might be exerted by a growth factor from the copulation organ retrogradely transported to the neurones. Because the copulation organ of

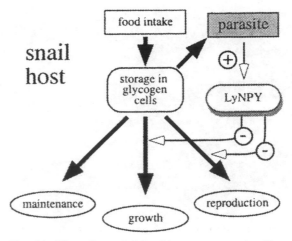

Fig. 10. The effect of the schistosome parasite T. ocellata on the energy flow in the Lymnaea snail host. By stimulating expression of the LyNPY encoding gene in the host the parasite stops energy flows from the storage sites to reproduction and growth. In this way the saved energy becomes available for the parasite.

snails up to 15 mm showed positive immunostaining with an antiserum to L-EGF (see Fig. 8) we suppose that L-EGF might be a good candidate. The effect on the neuroendocrine CDCs is probably mediated by unknown, target-derived humoral factor(s) since they have no physical connection with their target.

So, apart from being interesting to study as such, parasitization has been an excellent tool to show that underdevelopment of reproductive target organs has also consequences for the development, viz. the capacity, of the neurons innervating or regulating these organs.

Interference with neuroendocrine regulation of reproduction and growth

Reproduction and growth are the main determinants in the Dynamic Energy Budget (DEB) model proposed by Kooijman (2000; see Fig. 10). By inhibiting reproductive activity and enhancing abnormal growth of its snail host T. ocellata converts energy in its own direction in the time when they need much energy and space. However, the enhanced growth is not very costly as it can be primarily ascribed to an increase of the wet weight, whereas the dry weight density decreases (Joosse & Van Elk, 1986). Redirection of the main energy flow towards the parasite is reflected by a high production rate of cercariae.

T. ocellata affects reproduction and growth in this stage of infection by interfering with the NES regulating these processes in the host. For that purpose they make use of different strategies. In the first place they induce synthesis and release of a factor, schistosomin, from cells in the connective tissue, the teloglial cells, into the haemolymph. This peptide of 79 amino acids appears to interfere with the neuroendocrine regulation of both reproduction and growth. It inhibits not only the action of gonadotropic hormones upon peripheral reproductive target organs but also the electrophysiological activity of the central neuroendocrine cells, the caudodorsal cells (CDCs), which regulate ovulation, egg laying and the accompanying behaviour (Hordijk *et al.* 1992; De Jong-Brink *et al.* 1997). Although it is unknown whether schistosomin also interferes with the effects of growth and metabolism regulating hormones on their peripheral target cells/organs it is clear that it acts in a direct way on the neuroendocrine light green cells (LGCs). These cells are involved in regulation of growth. They are the source of molluscan insulin-like peptides (MIPs; Smit *et al.* 1988). The excitability of the LGCs increases in the presence of schistosomin.

Unfortunately, we do not know the function of schistosomin in non-parasitized snails. Although it is produced in connective tissue cells, which may belong to the IDS, it does not directly affect phagocytotic activity of haemocytes (Buijs and De Jong-Brink, unpublished data). Other, possibly indirect effects of schistosomin on the internal defence have not yet been studied. However, it clearly plays an important role in parasitic interference with neuroendocrine regulation of reproduction and growth. This function of schistosomin reminds us of that of cytokines in vertebrates.

Differential screening of the cDNA libraries of CNS from both parasitized and non-parasitized snails (Hoek *et al.* 1997) revealed changes in gene expression of two interesting, neuropeptide-encoding genes. Firstly, the caudodorsal cell hormone (CDCH) encoding gene, which appeared to be down-regulated coinciding with the onset of a continuous and high production and release of cercariae (6 weeks p.e. onwards). It is not clear whether the upregulation of the CDCH gene as found in a very early stage of infection is related to the observed phenomenon that egg laying is initially enhanced in snails which had been infected as subadults. Secondly, up-regulation of the *Lymnaea* Neuropeptide Y homologue (LyNPY) encoding gene which started around 6 weeks p.e. As NPY is known to play a central role in regulation of energy budgeting with food intake, reproduction and growth as the main determinants in vertebrates (Frankish *et al.* 1995; White & Martin, 1997) we have investigated whether a similar role can be ascribed to LyNPY in the snail host.

In studies on the function of LyNPY in parasitized snails we assumed that up-regulation of the LyNPY encoding gene coincided with elevation of the titer of this neuropeptide in the snail host. This situation was mimicked by that synthetic LyNPY was administered to non-parasitized snails by implanting a slow-release pellet containing LyNPY (long-term effect) or by a single bolus injection (short-term

effect). In both cases administration of LyNPY caused a profound inhibition of egg mass production and suppression of growth in a dose-dependent manner. In contrast to the role of NPY in regulating food consumption in mammals, LyNPY did not affect the amount of food consumed in *Lymnaea*. However, when the LyNPY titer had returned to normal and reproduction and growth were resumed, a significant increase in food consumption was observed. This short hyperphagic period apparently served to replenish the energy stores, in this case glycogen stored in vesicular connective tissue cells, which had become depleted, to restart the energy requiring activities (De Jong-Brink *et al.* 1999*a*).

The observation that LyNPY suppresses growth does not seem to be in accordance with the enhanced growth observed in parasitized snails. This indicates that the effect of parasites on growth of their snail host is a more complicated phenomenon. Supposedly more/other factors play a role in establishing the peculiar effects parasites have, causing an increase of the wet weight and not of the dry weight of their snail host. The neuroendocrine light yellow cells (LYCs; Hoek *et al.* 1992; Smit, Hoek & Geraerts, 1993; Boer & Montagne-Wajer, 1994; Boer *et al.* 1994) might be involved as well, as these cells are supposed to play a role in regulating cardiovascular functions, blood pressure, ion and water balance, egg laying and feeding.

Data obtained for *Biomphalaria* infected with *S. mansoni* showed that oviposition was re-established in these snails by administering 5HT (Manger, Christensen & Yoshino, 1996). The fact that this does not happen when parasitized *Lymnaea* are injected with 5HT can be explained by assuming that the *Biomphalaria* snails had been infected after differentiation of the reproductive tract had passed a certain level. *Biomphalaria* presumably also has neuroendocrine cells within its CNS producing neuropeptides which regulate oviposition. A similar group of neuroendocrine cells appeared to be immunopositive with an antiserum to the ovulation hormone produced in the CDCs of *Lymnaea* (Roubos & Van de Ven, 1987). Therefore we assume that 5HT acts as a neurotransmitter on these neuroendocrine cells in *Biomphalaria* resulting in resumption/increase of egg laying.

SUMMARY AND CONCLUSIONS

The data presented clearly show that schistosome parasites do not apply only one strategy to continue their life cycle in their intermediate snail host. They make use of multiple strategies to interfere with the two regulatory systems in their host resulting in the physiological host changes they require. This implies that it is of vital interest for them to induce these changes in the snail host, which on the one hand prevent them from being attacked by the IDS and on the other hand profoundly affect the host's energy flow for their own benefit.

To circumvent being attacked by the IDS, transforming miracidia of *T. ocellata* release E/S products which have both direct and indirect effects on the effector cells of the internal defence, the haemocytes of *L. stagnalis*. The E/S products not only have direct effects on haemocyte activity but they also affect expression of genes encoding for two factors produced and released by the granular connective tissue cells, MDM and granularin. These factors, which resemble certain molecules occurring in vertebrates, function in the internal defence of the snail host. Parasitic infections have revealed the involvement of factors not only from the IDS but also from the NES in regulating the internal defence. It is very interesting in this respect that exocrine glands of schistosome parasites also secrete molecules similar or identical to host neuropeptides into the snail host, which contribute to the suppression of the IDS.

To exert their effects on reproduction and growth, the parasites apply different strategies. (1) Immediately after infection they inhibit differentiation and development of the reproductive tract. For the development of the male copulation organ it has been shown that E/S products from transforming miracidia have a direct, not mediated by the NES, effect on development of this target organ. (2) As soon as developing cercariae are present in daughter sporocysts they induce the release of schistosomin, a factor from teloglial cells in the connective tissue of the host. This factor affects neuroendocrine regulation of both reproduction and growth. (3) In the same stage of infection the parasite causes changes in gene expression: the LyNPY encoding gene is simultaneously up-regulated in this energy requiring stage. Experimental elevation of the LyNPY titer in non-parasitized snails explains why the LyNPY gene is up-regulated: it appears to stop reproduction and growth. This means that the energy flow in the host is redirected towards the parasite. Also the ovulation hormone, CDCH, encoding gene was affected. It was down-regulated from 6 weeks post exposure onwards. This might, however, also have resulted from schistosomin causing an inhibitory effect on the CDCs. The possibility that schistosomin is also responsible for up-regulation of the LyNPY encoding gene in LyNPY positive neurons has still to be studied.

It is quite surprising that, to our knowledge, the molecular structure of none of these E/S factors interfering with the IDS and/or the NES of the snail host has been identified up till now. As far as the E/S products released in the early stage of infection are concerned, this seems possible as the products can be obtained from *in vitro*-transformed miracidia. It is, on the other hand, very difficult to study the effects

of E/S products released from *T. ocellata* daughter sporocysts, with differentiating cercariae, because it is as yet impossible to isolate these daughter sporocysts from *Lymnaea* tissue (see also Amen & Meuleman, 1992). However, recent data showing that co-culture of miracidia and *Biomphalaria* embryonic cells (Bge cells; Yoshino & Laursen, 1995) leads to cercarial production (Ivanchenko *et al.* 1999; Cousteau & Yoshino, 2000) are promising in this respect. Experiments in which injections of or implants with (fractions of) extracts of free swimming cercariae were performed for studying the effects on the snail host should be considered very critically (see also Schallig, Sassen & De Jong-Brink, 1992). The products secreted by free swimming cercariae and/or schistosomula, on the other hand, are very important but only useful for studying effects on the definitive vertebrate host, especially its skin (Wilson, Coulson & Dixon, 1986; Ramaswamy *et al.* 1996; Rao & Ramaswamy, 2000).

Finally, the results presented in this review, showing that schistosomes apply multiple strategies to manipulate the regulatory systems in their intermediate snail host, are not only of interest from a scientific point of view. They also favour the option that scientists, too, should make use of multiple strategies to interrupt or inhibit the life cycle of parasites.

ACKNOWLEDGEMENTS

The authors are greatly indebted to Dr Elisabeth A. Meuleman for critical reading of the manuscript, to Mrs Carry Moorer-van Delft, Mrs Stèphany Buijs and Mr Anton Pieneman for technical help.

REFERENCES

ADEMA, C. M., HERTEL, L. A., MILLER, R. D. & LOKER, E. S. (1997). A family of fibrinogen-related proteins that precipitates parasite-derived molecules is produced by an invertebrate after infection. *Proceedings of the National Academy of Sciences, USA* **94**, 8691–8696.

AMEN, R. I., BAGGEN, J. M., BEZEMER, P. D. & DE JONG-BRINK, M. (1992). Modulation of the activity of the internal defence system of the pond snail *Lymnaea stagnalis* by the avian schistosome *Trichobilharzia ocellata*. *Parasitology* **104**, 33–40.

AMEN, R. I. & DE JONG-BRINK, M. (1992). *Trichobilharzia ocellata* in its snail host *Lymnaea stagnalis*: an *in vitro* study showing direct and indirect effects on the snail's internal defence system, via the host central nervous system. *Parasitology* **105**, 409–416.

AMEN, R. I. & MEULEMAN, E. A. (1992). Isolation of mother and daughter sporocysts of *Trichobilharzia ocellata* from *Lymnaea stagnalis*. *Parasitology Research* **78**, 265–266.

BAYNE, C. J., BUCKLEY, P. M. & DE WAN, P. C. (1980). Macrophagelike hemocytes of resistant *Biomphalaria glabrata* are cytotoxic for sporocysts of *Schistosoma mansoni in vitro*. *Journal of Parasitology* **66**, 413–419.

BECKAGE, N. E. (1997) New Insights: How parasites and pathogens alter the endocrine physiology and development of insect hosts. In *Parasites and Pathogens: Effects on Host Hormones and Behavior* (ed. Beckage, N. E.), pp. 3–36. New York: Chapman & Hall.

BECKAGE, N. E. & GELMAN, D. B. (2001). Parasitism of *Manduca sexta* by *Cotesia congregata*: A multitude of disruptive endocrine effects. In *Manipulating Hormonal Control* (ed. Edwards, J.) (in press).

BOER, H. H., GROOT, C., DE JONG-BRINK, M. & CORNELISSE, C. J. (1977). Polyploidy in the freshwater snail *Lymnaea stagnalis* (Gastropoda, Pulmonata). A cytophotometric analysis of the DNA in neurons and some other cell types. *Netherlands Journal of Zoology* 27, 245–252.

BOER, H. H. & MONTAGNE-WAJER, C. (1994). Functional morphology of the neuropeptidergic light yellow cell system in pulmonate snails. *Cell and Tissue Research* 277, 531–538.

BOER, H. H., MONTAGNE-WAJER, C., SMITH, F. G., PARISH, D. C., RAMKEMA, M. D., HOEK, R. M., VAN MINNEN, J. & BENJAMIN, P. R. (1994). Functional morphology of the light yellow cell and yellow cell (sodium influx-stimulating peptide) neuroendocrine systems of the pond snail *Lymnaea stagnalis*. *Cell and Tissue Research* 275, 361–368.

BOER, H. H., SCHOT, L. P., ROUBOS, E. W., TER MAAT, A., LODDER, J. C., REICHELT, D. & SWAAB, D. F. (1979). ACTH-like immunoreactivity in two electronically coupled giant neurons in the pond snail *Lymnaea stagnalis*. *Cell and Tissue Research* 202, 231–240.

BOGERD, J., GERAERTS, W. P., VAN HEERIKHUIZEN, H., KERKHOVEN, R. M. & JOOSSE, J. (1991). Characterization and evolutionary aspects of a transcript encoding a neuropeptide precursor of *Lymnaea* neurons, VD1 and RPD2. *Brain Research: Molecular Brain Research* 11, 47–54.

CONNORS, V. A. & YOSHINO, T. P. (1990). *In vitro* effect of larval *Schistosoma mansoni* excretory-secretory products on phagocytosis-stimulated superoxide production in hemocytes from *Biomphalaria glabrata*. *Journal of Parasitology* 76, 895–902.

COUSTEAU, C. & YOSHINO, T. P. (2000). Flukes without snails: advances in the *in vitro* cultivation of intramolluscan stages of trematodes. *Experimental Parasitology* 94, 62–66.

DE JONG-BRINK, M. (1992). Neuroendocrine mechanisms underlying the effects of schistosome parasites on their intermediate snail host. *Journal of Invertebrate Reproduction and Development* 22, 127–138.

DE JONG-BRINK, M. (1995). How schistosomes profit from the stress responses they elicit in their hosts. *Advances in Parasitology* 35, 177–256.

DE JONG-BRINK, M., HOEK, R. M., LAGEWEG, W. & SMIT, A. B. (1997) Schistosome parasites induce physiological changes in their snail hosts by interfering with two regulatory systems, the internal defense system and the neuroendocrine system. In *Parasites and Pathogens: Effects on Host Hormones and Behavior* (ed. Beckage, N. E.), pp. 57–75. New York: Chapman & Hall.

DE JONG-BRINK, M., REID, C. N., TENSEN, C. P. & TER MAAT, A. (1999a). Parasites flicking the NPY gene on the host's switchboard: why NPY? *FASEB Journal* 13, 1972–1984.

DE JONG-BRINK, M., VAN DER WAL, B., OP DE LAAK, X. E. & BERGAMIN-SASSEN, M. (1999b). Inhibition of the development of the reproductive tract in parasitized snails. *Invertebrate Reproduction and Development* 36, 223–227.

DE LANGE, R. P. J., MOORER-VAN DELFT, C. M., DE BOER, P. A. C. M., VAN MINNEN, J. & DE JONG-BRINK, M. (2001). Target-dependent differentiation and development of molluscan neurons and neuroendocrine cells, using parasitisation as a tool. *Neuroscience* 103, 289–299.

DUVAUX-MIRET, O., DISSOUS, C., GAUTRON, J. P., PATTOU, E., KORDON, C. & CAPRON, A. (1990). The helminth *Schistosoma mansoni* expresses a peptide similar to human beta-endorphin and possesses a proopiomelanocortin-related gene. *New Biologist* 2, 93–99.

DUVAUX-MIRET, O., STEFANO, G. B., SMITH, E. M., DISSOUS, C. & CAPRON, A. (1992). Immunosuppression in the definitive and intermediate hosts of the human parasite *Schistosoma mansoni* by release of immunoreactive neuropeptides. *Proceedings of the National Academy of Sciences, USA* 89, 778–781.

ELITSUR, Y., LUK, G. D., COLBERG, M., GESELL, M. S., DOSESCU, J. & MOSHIER, J. A. (1994). Neuropeptide Y (NPY) enhances proliferation of human colonic lamina propria lymphocytes. *Neuropeptides* 26, 289–295.

FRANKISH, H. M., DRYDEN, S., HOPKINS, D., WANG, Q. & WILLIAMS, G. (1995). Neuropeptide Y, the hypothalamus, and diabetes: insights into the central control of metabolism. *Peptides* 16, 757–771.

GARTNER, L. P. & HIATT, J. L. (1997). *Color Textbook of Histology*. Philadelphia, USA, W. B. Saunders Company. A Division of Harcourt Brace & Company, The Curtis Center.

GERAERTS, W. P. M. & JOOSSE, J. (1975). The control of vitellogenesis and of growth of female accessory sex organs by the dorsal body hormone (DBH) in the hermaphrodite freshwater snail *Lymanea stagnalis*. *General and Comparative Endocrinology* 27, 450–467.

HERMANN, P. M., VAN KESTEREN, R. E., WILDERING, W. C., PAINTER, S. D., RENO, J. M., SMITH, J. S., KUMAR, S. B., GERAERTS, W. P. M., ERICSSON, L. J., SMIT, A. B., BULLOCH, A. G. M. & NAGLE, G. T. (2000). Neurotrophic actions of a novel molluscan epidermal growth factor. *Journal of Neuroscience* 200, 6355–6364.

HOEK, R. M., LI, K. W., VAN MINNEN, J. & GERAERTS, W. P. (1992). Chemical characterization of a novel peptide from the neuroendocrine light yellow cells of *Lymnaea stagnalis*. *Molecular Brain Research* 16, 71–74.

HOEK, R. M., SMIT, A. B., FRINGS, H., VINK, J. M., DE JONG-BRINK, M. & GERAERTS, W. P. (1996). A new Ig-superfamily member, molluscan defence molecule (MDM) from *Lymnaea stagnalis*, is down-regulated during parasitosis. *European Journal of Immunology* 26, 939–944.

HOEK, R. M., VAN KESTEREN, R. E., SMIT, A. B., DE JONG-BRINK, M. & GERAERTS, W. P. (1997). Altered gene expression in the host brain caused by a trematode parasite: Neuropeptide genes are preferentially affected during parasitosis. *Proceedings of the National Academy of Sciences, USA* **94**, 14072–14076.

HORDIJK, P. L., DE JONG-BRINK, M., TER MAAT, A., PIENEMAN, A. W., LODDER, J. C. & KITS, K. S. (1992). The neuropeptide schistosomin and haemolymph from parasitized snails induce similar changes in excitability in neuroendocrine cells controlling reproduction and growth in a freshwater snail. *Neuroscience Letters* **136**, 193–197.

HURD, H. (1990). Physiological and behavioural interactions between parasites and invertebrate hosts. *Advances in Parasitology* **29**, 271–318.

IVANCHENKO, M. G., LERNER, J. P., MCCORMICK, R. S., TOUMADJE, A., ALLEN, B., FISCHER, K., HEDSTROM, O., HELMRICH, A., BARNES, D. W. & BAYNE, C. J. (1999). Continuous *in vitro* propagation and differentiation of cultures of the intramolluscan stages of the human parasite *Schistosoma mansoni*. *Proceedings of the National Academy of Sciences, USA* **96**, 4965–4970.

JOOSSE, J. & VAN ELK, R. (1986). *Trichobilharzia ocellata*: physiological characterization of giant growth, glycogen depletion, and absence of reproductive activity in the intermediate snail host, *Lymnaea stagnalis*. *Experimental Parasitology* **62**, 1–13.

KOOIJMAN, S. A. L. M. (2000). *Dynamic Energy and Mass Budgets in Biological Systems*, 2nd Edition. Cambridge, Cambridge University Press.

LAFFERTY, K. D. (1993*a*). Effects of parasitic castration on growth, reproduction and population dynamics of the marine snail *Cerithidea californica*. *Marine Ecology Progress Series* **96**, 229–237.

LAFFERTY, K. D. (1993*b*). The marine snail, *Cerithidea californica*, matures at smaller sizes where parasitism is high. *Oikos* **68**, 3–11.

LE GALL, S. (1981). Étude experimentale du facteur morphogenetique controlant la differenciation du tractus genetal male externe chez *Crepidula fornicata* L. (Mollusque hermaphrodite protandre). *General and Comparative Endocrinology* **43**, 51–62.

LIPTON, J. M. & CATANIA, A. (1997). Anti-inflammatory actions of the neuroimmunomodulator alpha-MSH. *Immunology Today* **18**, 140–145.

LODES, M. J., CONNORS, V. A. & YOSHINO, T. P. (1991). Isolation and functional characterization of snail hemocyte-modulating polypeptide from primary sporocysts of *Schistosoma mansoni*. *Molecular and Biochemical Parasitology* **49**, 1–10.

LODES, M. J. & YOSHINO, T. P. (1990). The effect of schistosome excretory-secretory on *Biomphalaria glabrata* haemocyte motility. *Journal of Invertebrate Pathology* **56**, 75–85.

MANGER, P., LI, J., CHRISTENSEN, B. M. & YOSHINO, T. P. (1996). Biogenic monoamines in the freshwater snail, *Biomphalaria glabrata*: influence of infection by the human blood fluke, *Schistosoma mansoni*. *Comparative Biochemistry and Physiology. Part A Physiology* **114**, 227–234.

MINCHELLA, D. J. & LOVERDE, P. T. (1981). A cost of increased early reproductive effort in the snail *Biomphalaria glabrata*. *American Naturalist* **118**, 876–881.

NÚÑEZ, P. E., ADEMA, C. M. & DE JONG-BRINK, M. (1994). Modulation of the bacterial clearance activity of haemocytes from the freshwater mollusc, *Lymnaea*

stagnalis, by the avian schistosome, *Triochobilharzia ocellata*. *Parasitology* **109**, 299–310.

NÚÑEZ, P. E. & DE JONG-BRINK, M. (1997). The suppressive excretory-secetory product of *Trichobilharzia ocellata*: a possible factor for determining compatibility in parasite-host interactions. *Parasitology* **115**, 193–203.

NÚÑEZ, P. E., MOLENAAR, M., LAGEWEG, W., LI, K. W. & DE JONG-BRINK, M. (1997). Excretory-secretory products of *Trichobilharzia ocellata* and their modulating effects on the internal defence system of *Lymnaea stagnalis*. *Parasitology* **114**, 135–144.

PATTERSON, F. H. (1992). Neuron-target interaction. In: *Introduction to Molecular Neurobiology* (ed. Hall, Z. W.). Sunderland, MA, Sinauer.

RAMASWAMY, K., SALAFSKY, B., POTLURI, S., HE, Y. X., LI, J.-W. & SHIBUYA, T. (1996). Secretion of an anti-inflammatory, immunomodulatory factor by schistosomula of *Schistosoma mansoni*. *Journal of Inflammation* **46**, 13–22.

RAO, K. V. & RAMASWAMY, K. (2000). Cloning and expression of a gene encoding Sm16, an anti-inflammatory protein from *Schistosoma mansoni*. *Molecular and Biochemical Parasitology* **108**, 101–108.

RIDGEWAY, R. L., SYED, N. I., LUKOWIAK, K. & BULLOCH, A. G. M. (1991). Nerve growth factor (NGF) induces sprouting of specific neurons of the snail, *Lymnaea stagnalis*. *Journal of Neurobiology* **22**, 377–390.

ROUBOS, E. W. & VAN DE VEN, A. M. H. (1987). Morphology of neurosecretory cells in Basommatophoran snails homologous with egg-laying and growth-hormone producing cells of *Lymnaea stagnalis*. *General and Comparative Endocrinology* **67**, 7–23.

SCHALLIG, H. D. F. H., SASSEN, M. J. & DE JONG-BRINK, M. (1992). *In vitro* release of the anti-gonadotropic hormone, schistosomin, from the central nervous system of *Lymnaea stagnalis* is induced with a methanolic extract of cercariae of *Trichobilharzia ocellata*. *Parasitology* **104**, 309–314.

SCHALLIG, H. D. F. H., SASSEN, M. J., HORDIJK, P. L. & DE JONG-BRINK, M. (1991). *Trichobilharzia ocellata*: influence of infection on the fecundity of its intermediate snail host *Lymnaea stagnalis* and cercarial induction of the release of schistosomin, a snail neuropeptide antagonizing female gonadotropic hormones. *Parasitology* **102**, 85–91.

SCHALLIG, H. D. F. H., SCHUT, A., VAN DER KNAAP, W. P. W. & DE JONG-BRINK, M. (1990). A simplified medium for the *in vitro* culture of mother sporocysts of the schistosome *Trichobilharzia ocellata*. *Parasitology Research* **76**, 278–279.

SLUITERS, J. F. (1981). Development of *Trichobilharzia ocellata* in *Lymnaea stagnalis* and the effects of infection on the reproductive system of the host. *Zeitschrift für Parasitenkunde* **64**, 303–319.

SLUITERS, J. F., ROUBOS, E. W. & JOOSSE, J. (1984). Increased activity of the female gonadotropic hormone producing dorsal bodies in *Lymnaea stagnalis* infected with *Trichobilharzia ocellata*. *Zeitschrift für Parasitenkunde* **70**, 67–72.

SMINIA, T. S. (1972). Structure and function of blood and connective tissue cells of the freshwater pulmonate *Lymnaea stagnalis* studied by electron microscopy and enzyme histochemistry. *Zeitschrift für Zellforschung* **130**, 497–526.

SMIT, A. B., HOEK, R. M. & GERAERTS, W. P. (1993). The isolation of a cDNA encoding a neuropeptide prohormone from the light yellow cells of *Lymnaea stagnalis. Cell Molecular Neurobiology* **13**, 263–270.

SMIT, A. B., VREUGDENHIL, E., EBBERINK, R. H., GERAERTS, W. P., KLOOTWIJK, J. & JOOSSE, J. (1988). Growth-controlling molluscan neurons produce the precursor of an insulin-related peptide. *Nature* **331**, 535–538.

SONETTI, D., MOLA, L. & STEFANO, G. B. (1999). Morphine signaling in invertebrate tissues. In *Recent Developments in Comparative Endocrinology and Neurobiology* (ed. Roubos, E. W., Wendelaar-Bonga, S. E., Vaudry, H. & De Loof, A.), pp. 346–348. Nijmegen: Shaker Publishing B. V.

SONETTI, D., OTTAVIANI, E., BIANCHI, F., RODRIGUEZ, M., STEFANO, M. L., SCHARRER, B. & STEFANO, G. B. (1994). Microglia in invertebrate ganglia. *Proceedings of the National Academy of Sciences, USA* **91**, 9180–9184.

STEARNS, S. C. (1989). Trade-offs in life-history evolution. *Functional Ecology* **3**, 259–268.

STEFANO, G. B., SALZET, B. & FRICCHIONE, G. L. (1998). Enkelytin and opioid peptide association in invertebrates and vertebrates: immune activation and pain. *Immunology Today* **19**, 265–268.

STEFANO, G. B., SCHARRER, B., SMITH, E. M., HUGHES, T. K. JR., MAGAZINE, H. I., BILFINGER, T. V., HARTMAN, A. R., FRICCHIONE, G. L., LIU, Y. & MAKMAN, M. H. (1996). Opioid and opiate immunoregulatory processes. *Critical Review in Immunology* **16**, 109–144.

SUN, S. C., LINDSTROM, I., BOMAN, H. G., FAYE, I. & SCHMIDT, O. (1990). Hemolin: an insect-immune protein belonging to the immunoglobulin superfamily. *Science* **250**, 1729–1732.

THORNHILL, J. A., JONES, J. T. & KUSEL, J. R. (1986). Increased oviposition and growth in immature *Biomphalaria glabrata* after exposure to *Schistosoma mansoni. Parasitology* **93**, 443–450.

VAN DER KNAAP, W. P. W. & LOKER, E. S. (1990). Immune mechanisms in trematode-snail interactions. *Parasitology Today* **6**, 175–182.

WHITE, B. D. & MARTIN, R. J. (1997). Evidence for a central mechanism of obesity in the zucker rat: role of neuropeptide Y and leptin. *Proceedings of the Society Experimental Biology and Medicine* **214**, 222–232.

WILSON, R. A., COULSON, P. S. & DIXON, B. (1986). Migration of the schistosomula of *Schistosoma mansoni* in mice vaccinated with radiation attenuated cercariae, and normal mice: an attempt to identify the timing and site of parasite death. *Parasitology* **92**, 101.

YOSHINO, T. P. & LAURSEN, J. R. (1995). Production of *Schistosoma mansoni* daughter sporocysts from mother sporocysts maintained in synxenic culture with *Biomphalaria glabrata* embryonic (Bge) cells. *Journal of Parasitology* **81**, 714–722.

Receptor–ligand interactions and cellular signalling at the host–parasite interface

T. P. YOSHINO*, J. P. BOYLE *and* J. E. HUMPHRIES

Department of Pathobiological Sciences, School of Veterinary Medicine, University of Wisconsin, Madison, WI 53706 USA

SUMMARY

Although the effects of trematode infection on snail host physiology or host responses on parasite development have been well described in the literature, very little is known regarding the underlying mechanisms and specific molecules responsible for mediating those effects. It is presumed that many host–parasite interactions are communicated through receptor-mediated events, in particular those involving haemocytic immune responses to invading parasites, larval motility and migration through host tissues, and larval acquisition of host molecules either as nutrients or critical developmental factors. The intent of this chapter is to review current knowledge of molecules (both receptors and their ligands or counter-receptors) involved in molecular communication at the interface between larval trematodes, especially the mother or primary sporocyst stage, and host cells/tissues in intimate proximity to developing larvae. Information to date suggests that the molecular exchange at this interface is a highly complex and dynamic process, and appears to be regulated in specific cases. Topics discussed will focus on snail cell receptor interactions with the sporocyst tegument and its secretions, host cell–cell and cell–substrate adhesion receptors and their related signal transduction pathways, and sporocyst tegumental surface receptors and ligands involved in the binding of soluble host molecules.

Key words: Digenea, flukes, schistosome, Mollusca, gastropod, immunity, haemocyte, receptor, encapsulation, signal transduction, *Biomphalaria glabrata* embryonic (Bge) cell line, adhesion, sporocyst, tegument, transporter, serotonin, review.

INTRODUCTION

Initiation and subsequent establishment of trematode infections within their molluscan intermediate hosts is a complex process involving a multitude of molecular interactions beginning with miracidial contact and entry into the host, its transformation to the primary or mother sporocyst and development through successive sporocyst generation(s), and finally, production of cercariae. The 'playing field' where molecular signals are exchanged during the course of intramolluscan development is referred to as the host–parasite interface (Read, Rothman & Simmons, 1963), and comprises that region between and including the host cell membrane (or other non-cellular host tissues) and that of the fluke's syncytial tegument. This interface represents a chemically complex and dynamic *milieu* in which host nutrients and other factors required for parasite growth and development are bound and taken up by intra-molluscan stages, where parasite excretory-secretory products are released, and where host cells and plasma interact with parasite molecules triggering immunological or pathological responses in the host.

From a biological perspective, the sum total of the

molecular exchange at the host–parasite interface ultimately dictates the success or failure of the parasite to establish an infection within its host. Dr C. A. Wright had a broad interest in the determinants of host–parasite compatibility, and in his book, *Flukes and Snails* (1973), he begins to create an integrated view of 'host specificity' by defining such terms as susceptibility, nonsusceptibility (= physiological unsuitability, Read *et al.* 1963; Loker & Bayne, 1986), and innate and acquired resistance in the context of trematode–snail interactions. In so doing, he also evokes such concepts as immune evasion (antigenic masking, antigen synthesis, immune suppression), age resistance and infectivity factors as underlying mechanisms responsible, at least in part, for determining infection success in any given encounter (Wright, 1973).

Although the presumed critical importance of molecular interactions at the snail-trematode interface has long been recognized, to date only limited information is available about the precise nature of the molecules being 'exchanged' (receptors and ligands), or the functional significance of these 'interface' molecules in host responses or trematode survival. The goal of the present paper is to review what is currently known regarding the molecules of host and parasite origin believed to be involved in the physiological interplay between larval digeneans and cells/tissues of the snail host. The scope of this paper will focus mainly on schistosomes and their pulmonate snails, although we are well aware that

* Corresponding author: Timothy P. Yoshino, Department of Pathobiological Sciences, School of Veterinary Medicine, University of Wisconsin, 2115 Observatory Drive (Biotron), Madison, WI 53706-1087 USA. Fax: 608–265–8122 E-mail: yoshinot@svm.vetmed.wisc.edu

Parasitology (2001), **123**, S143–S157.   © 2001 Cambridge University Press
DOI: 10.1017/S0031182001007685

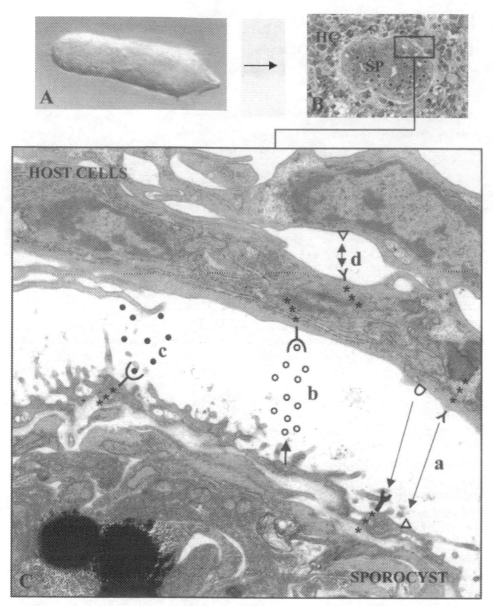

Fig. 1. Photographic depiction of the host–parasite interface. (A) Miracidium of *Schistosoma mansoni*; molluscan infective stage. (B) Early developing primary or mother sporocyst (SP) surrounded by *Biomphalaria glabrata* tissue (HC). 'Boxed' area represents the host–parasite interface presented in C. (C) Electronmicrographic representation of a generalized snail host–trematode interface illustrating various avenues of molecular communication. a. Direct receptor–ligand interactions between host cells and the sporocyst tegumental surface. Receptors on either parasite or host can result in the transduction of signals (*) and appropriate responses. b. Larval release of secretory ES products (open circles) bind to host cell receptors, triggering cellular responses. c. Receptors at the tegumental surface of developing sporocysts bind soluble plasma factors (closed circles), stimulating either transport into the tegumental syncytium or signal transduction events involved in host to parasite communication (*). d. Receptors and ligands associated with host cell–cell and cell–substrate interactions are required for directing normal processes, as well as those in response to larval infection.

one must be cautious about extrapolating broad generalizations based on a single parasite group, especially the blood flukes which, 'in many respects have aberrant characteristics' (Wright, 1973, Introduction, p. 14). Regardless, with that caveat, we have divided this presentation into several topics, each representing different, but interacting, components of the parasite–host interface: (1) receptor involvement in host cell interactions with larval excretory–secretory proteins (ESP) and sporocyst tegument-associated molecules, (2) adhesion receptors and their related signaling pathways in snail cells, and (3) tegumental membrane receptors for soluble snail factors. Emphasis in this review will be placed on host molecular interactions at the mother sporocyst interface (Fig. 1), since an understanding of the critical factors determining success or failure of this larval stage should reveal basic mechanisms underlying snail host compatibility in any given snail-trematode system.

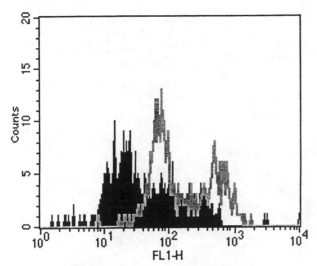

Fig. 2. Distribution of *Biomphalaria glabrata*
haemocytes pretreated with fucoidan (black histogram)
or untreated (gray histogram), followed by exposure to
fluorescent-labelled excretory–secretory proteins (fESP).
In this example, fucoidan pretreatment significantly
reduced the binding of fESP in this haemocyte
population.

PARASITE–HOST CELL INTERACTIONS

It is assumed that the receptor–counterreceptor
systems functioning at the host–parasite interface
are critical to two-way communication between the
mother sporocyst and snail cells. Upon miracidial
entry into the snail, a complex array of chemical
signals is broadcast to the host in the form of
secretions released by invading larvae, and as
membrane-associated molecules presented at the
sporocyst tegumental surface. Haemocytes, primary
effector cells involved in host immune defence,
monitor these signals, presumably through direct
binding of larval molecules to cellular receptors or,
indirectly, through detection of molecules released
from other host tissues in response to the parasite's
presence (see review by De Jong-Brink, 1995).
Haemocyte reactivity in response to larval secretions
or tegumental surface determinants are thought,
ultimately, to play an essential role in determining
the success or failure of trematodes to establish
infections within the snail host. However, what do
we know about the molecules that comprise the
'signalling' system at the haemocyte–sporocyst
interface?

*Host cell interactions with larval excretory–secretory
proteins*

As alluded to above, following parasite entry into the
snail, major sources of larval molecules introduced to
the host are in the form of secretions released during
transformation of the miracidium to mother sporocyst, and as a consequence of normal metabolism
during subsequent sporocyst development. These

secretions, commonly referred to as excretory–
secretory products/proteins (ESP) (Lodes &
Yoshino, 1989; Loker, Cimino & Hertel, 1992) are
known to have modulatory effects on haemocyte
function although we are still far from fully understanding the precise mechanisms by which these
secretory factors may be determining host susceptibility to trematode infection. Perhaps the
clearest case of trematode ESP-mediated haemocyte
modulation involves snail infection by larval echinostomatids. Following up on earlier observations by
Lie, Heyneman & Jeong (1976) that echinostome
sporocysts exert a strong immunosuppressive
influence on the snail host *Biomphalaria glabrata*
(termed 'interference'; Lie, 1982), Loker and
colleagues showed that natural *Echinostoma paraensei*
infection is associated with impairment of haemocyte
adhesion/spreading and phagocytosis *in vitro* (Noda
& Loker, 1989*a*, *b*), and that this influence is
mediated through ESP released by the developing
mother sporocyst stage (Loker *et al.* 1992; Adema *et
al.* 1994). Recently they also have presented evidence
for direct echinostome ESP binding to the surface of
B. glabrata haemocytes, and a corresponding ESP-
mediated induction of intracellular calcium fluxes, as
well as 'rounding', in spread haemocytes (Hertel,
Stricker & Loker, 2000). Although these results
strongly implicate ESP receptors at the haemocyte
membrane surface mediating transduction of specific
signals for cellular adhesion, to date the precise
molecular nature of these ESP receptors is unknown.
One potential candidate may be a group of soluble
fibrinogen-related proteins (FREPs) that are synthesized by *B. glabrata* haemocytes and capable of
binding echinostome ESP in precipitation reactions
(Adema *et al.* 1997). In this case, it is possible
that haemocyte membrane-associated FREPs could
be serving as ESP-binding receptors.

   In other trematode–snail systems, fluke-derived
ESP also have been shown to possess both
haemocyte-disruptive and -stimulatory activities,
although these effects generally are more subtle than
those exerted by echinostomes. For example there is
strong evidence that *Schistosoma mansoni* sporocysts
produce and release adrenocorticotropic hormone
(ACTH), which in turn is processed to α-melanocyte
stimulating hormone (αMSH) by a *B. glabrata*
haemocyte neural endopeptidase (Duvaux-Miret *et
al.* 1992). MSH subsequently was shown to inhibit
haemocyte adhesion/motility, suggesting that peptide hormone-like substances, released as component
of ESP, can bind to appropriate haemocyte receptors
and modulate an important immune function. Similarly, haemocytes from *Lymnaea stagnalis*, infected
by the avian schistosome *Trichobilharzia ocellata*,
initially are stimulated in their bacterial clearing
activity compared to uninfected controls, but then
exhibit decreased activity > 24 hour post-infection
(Nuñez, Adema & De Jong-Brink, 1994). These

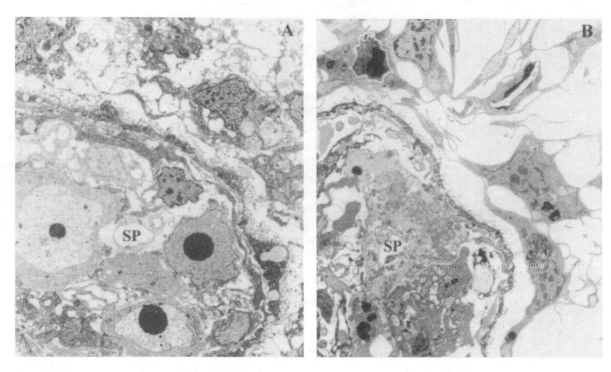

Fig. 3. Electronmicrographs demonstrating *in vitro* 'encapsulations' of mother sporocysts (SP) by either cells of the *Biomphalaria glabrata* embryonic cell line (A) or haemocytes of the susceptible NMRI strain of *B. glabrata* (B). Note the similar loose cellular network at the host–parasite interface created by both cell-types during capsule formation.

temporally-opposing effects appear to be due to two different fractions of *T. ocellata* ESP; one a 40 kDa fraction probably of sporocyst origin, mediating haemocyte suppression and another, an early-released 2 kDa fraction involved in cell activation (Nuñez & De Jong-Brink, 1997). Likewise, a dual effect of ESP from larval *S. mansoni* on snail haemocyte function also has been documented. *In vitro* exposure of haemocytes to early larval ESP strongly stimulates cellular protein biosynthesis (Yoshino & Lodes, 1988), while also inhibiting *in vitro* cell motility, phagocytosis and production of superoxide anions (Lodes & Yoshino, 1990; Connors & Yoshino, 1990). Taken together, these results are consistent with the involvement of a haemocyte multi-receptor system interacting with multiple ESP components.

However, with the exception of αMSH-mediated haemocyte suppression, which presumably is acting through MSH-binding receptors (Duvaux-Miret *et al.* 1992), only scant information is available concerning the nature of cellular ESP receptors or their specific ESP ligands. In a recent flow cytometric study of ESP–haemocyte interactions, fluorescent-labeled *S. mansoni* ESP (fESP) was used to investigate the binding of secretory glycoproteins to the surface of *B. glabrata* haemocytes. In carbohydrate inhibition experiments it was found that complex sulfated polysaccharides (e.g. fucoidan and dextran sulfate) and several glycoproteins significantly reduced fESP binding to haemocytes (Fig. 2),

suggesting the involvement of haemocyte membrane carbohydrate-binding receptors (CBRs) in ESP-adhesive reactions (Johnston & Yoshino, 2001). This finding, although not surprising in view of previous reports of CBRs associated with haemocytes of *B. glabrata* (Fryer, Hull & Bayne, 1989; Hahn, Bender & Bayne, 2000) and other gastropod species (Richards & Renwrantz, 1991), strongly suggests that interactions between snail haemocytes and larval secretions may be mediated through receptors recognizing specific glycan moieties associated with ESP. Diversity in haemocyte receptor specificity, and in the composition/sequence structure of parasite oligosaccharide ligands released during early sporocyst development could begin to explain the previously reported array of different biological effects of ESP on haemocyte function. A membrane FREP, binding ESP through its lectin-like domain (Adema *et al.* 1997) would be consistent with such a model of ESP–haemocyte adherence. As more information regarding the molecular composition of ESP and their cellular receptors is gained, a clearer understanding of the specific role of ESP–haemocyte interactions in determining snail host–trematode compatibility will undoubtedly emerge.

Host cell interactions with the sporocyst surface tegument

Direct contact and communication between the outer tegumental surface of developing sporocysts and the various host tissues which it encounters is critical to the parasite's survival within the snail host. For the parasite, information concerning suitability of location, nutrient availability, presence of developmental cues or potentially damaging immune factors, among others, is conveyed at this intimate interface. For the host, the ability of haemocytes to adhere to the sporocyst tegument and become cytotoxically activated, likewise, depends on the recognition of molecular cues presented at the host–parasite interface. Ultimately, the compatibility/incompatibility of any given combination of snail–trematode species or strains will be determined by the relative balance between these host-parasite interactions at the tegumental surface (Fig. 1). However, to date, very little is known regarding the specific molecules serving as receptors or ligands in mediating host cell binding to developing larval stages.

In attempts to begin identifying potentially important molecules involved in host cell adhesion to early developing trematode larvae, we have incorporated the *B. glabrata* embryonic (Bge) cell line, originally isolated and established by Hansen (1976*a*), as an *in vitro* model for circulating haemocyte adhesion to mother sporocysts of *S. mansoni* (Yoshino *et al.* 1999). As described previously (Hansen, 1976*b*; Yoshino & Laursen, 1995), Bge cells, when placed in coculture with *S. mansoni* sporocysts, adhere to and form cellular capsules around parasites in a reaction morphologically reminiscent of *in vitro* encapsulation by haemocytes of susceptible *B. glabrata* strains (Bayne, Buckley & DeWan, 1980) (Fig. 3). Also consistent with the proposed Bge-haemocyte adhesion model is the observation that both cell types are reduced in their capacity to adhere to *Echinostoma* spp. sporocysts under *in vitro* conditions (Loker, Boston, & Bayne, 1989; Loker & Adema, 1995; Ataev, Fournier & Coustau, 1998).

Due to the shared sporocyst-adhesive properties of *B. glabrata* haemocytes and Bge cells, recent investigations in our laboratory have focused on the mechanisms of Bge cell adherence to *S. mansoni* mother sporocysts. Using an *in vitro* cell adhesion assay to quantify Bge cell binding to the surface of living *S. mansoni* sporocysts, we have found that various carbohydrates, including fucoidan, dextran sulfate, heparin, mannose and mannose-6-phosphate significantly inhibited Bge cell-sporocyst adherence (unpublished data). Since inhibition patterns were selective for specific carbohydrates or glyco-conjugates, these results suggest that Bge cell lectin-like receptors may be involved, at least in part, in mediating cell–sporocyst binding interactions. This conclusion is supported by the recent cloning of a C-type lectin cDNA from *B. glabrata* tissue and Bge cells (Duclermortier *et al.* 1999), which upon further analysis of the predicted amino acid sequence

appears to possess significant homology to mammalian mannose-binding proteins (unpublished data). Furthermore, the finding of a *S. mansoni* tegument-binding receptor on Bge cells that is fucoidan-inhibitable is consistent with our demonstration of a functionally similar fucoidan-reactive receptor(s) on *B. glabrata* haemocytes that binds soluble ESP. Since previous studies have shown that glycopeptides associated with the tegumental surface and ESP of *S. mansoni* sporocysts share carbohydrate epitopes (Dunn & Yoshino, 1988), it is quite possible that the haemocyte and Bge cell 'fucoidan receptor(s)' may be structurally-related. Efforts currently are underway to isolate and characterize, both structurally and functionally, the Bge cell fucoidan receptor for eventual comparison with the haemocyte receptor. Because of the difficulty in obtaining sufficient numbers of haemocytes for detailed biochemical or molecular analyses, it is envisioned that immunological and molecular probes developed from and made available by Bge cell investigations will make comparative studies of similar molecules in snail haemocytes much more tractable.

ADHESION RECEPTORS AND SIGNALLING PATHWAYS IN SNAIL CELLS

As mentioned earlier, recognition and encapsulation of invading trematode larvae by circulating snail haemocytes require mechanisms of adherence and activation for effective elimination of parasite infections. However, with the exception of earlier work demonstrating the presence of carbohydrate-reactive lectins associated with the surface of molluscan haemocytes (Vasta *et al.* 1982; Renwrantz & Stahmer, 1983; Fryer *et al.* 1989; Richards & Renwrantz, 1991; Hahn, *et al.* 2000), relatively little is known about other, non-lectin haemocyte receptors that may be mediating cell–substrate or cell–cell adhesive behaviour, what factors regulate their expression, or what ligands/counter-receptors are recognized by these adhesion molecules. Recently, we have begun to explore the molecular basis for snail cell adhesion and the potential pathways used for cell signalling by again exploiting the Bge cell line as a model system for haemocyte function (Yoshino *et al.* 1999).

Cellular adhesion receptors

In addition to its sporocyst-binding capabilities, Bge cells also share with haemocytes several other functional characteristics including adherence and spreading on glass/plastic substrates, motility through filopodial extensions and phagocytosis (Yoshino *et al.* 1999). Davids & Yoshino (1998), while examining *B. glabrata* haemocyte adhesion and spreading, discovered that haemocyte spreading could be inhibited by the tetrapeptide, arg-gly-asp-ser (RGDS). Subsequently, haemocyte spreading in response to a panel of extracellular matrix (ECM) proteins was also examined. These studies demonstrated haemocyte clumping, or the formation of cell aggregates on ECM-coated glass slides, which similarly could be inhibited by RGDS. Together, these results imply the involvement of an integrin-like receptor in haemocyte spreading as it is frequently an RGD motif within ECM proteins that integrins recognize. In follow-up studies, Davids, Wu & Yoshino (1999) subsequently demonstrated that Bge cell spreading, like in haemocytes, was inhibited in the presence of RGDS, but not a glutamic-acid substituted control peptide (RGES), also suggesting the participation of an integrin-like receptor(s) in the Bge cell adhesion/spreading response. A β_1-like integrin subunit was cloned and characterized from Bge cell cDNA which demonstrated high similarity (45·6–52·7%) with human β integrin sequences (Davids *et al.* 1999). In follow-up studies, a very similar haemocyte cDNA fragment also was cloned and sequenced, suggesting that haemocytes express a β_1 integrin receptor subunit almost identical to that of Bge cells (Yoshino *et al.* 1999).

Integrin receptors are heterodimers comprising an α and a β subunit, and to date, approximately 18 different α and 8 different β subunits have been characterized in vertebrates (Hynes, 1999). These subunits can associate in a variety of pairings to create multiple integrin receptors. Specific combinations of α/β subunits act as receptors for different ECM proteins (De Fillipi *et al.* 1997), thus a cellular response to or preference for a particular ECM protein might indicate the presence of a distinct integrin α/β heterodimer. To further identify potential adhesion receptors, the spreading response of Bge cells to a panel of ECM proteins was investigated. Results showed that Bge cells were able to spread on all tested ECM-coated surfaces, although cells demonstrated a significant increase in spreading only on urea-denatured vitronectin when compared to a saline control (unpublished data). Furthermore, Bge cell spreading on all surfaces was RGD inhibitable. Subsequently, the migratory response of Bge cells towards the same ECM proteins was also investigated, and again Bge cells displayed a significant preference (chemotaxis) for vitronectin, both in its native and denatured conformations (unpublished data). Taken together, the Bge cell spreading and migration data demonstrate a strong affinity for vitronectin, suggesting the presence of a vitronectin receptor on Bge cells. Furthermore, RGD inhibition of spreading implies the proposed vitronectin receptor may be an integrin-like molecule, and since the majority of vitronectin receptors comprise an α_v subunit, it might be speculated that

Bge cells also possess an α_v-like orthologue. However, to date, no α subunits have been found in molluscan cells.

The finding that snail cell adhesive behaviour was selective for mammalian vitronectin was of particular interest since follow-up Western blot analyses using a polyclonal anti-human vitronectin antibody revealed the presence of a vitronectin-like protein in *B. glabrata* plasma (unpublished data). This approximately 70 kDa plasma protein may represent a native ligand for snail cell vitronectin receptors. Also, as mentioned earlier, haemocytes formed aggregations upon exposure to some ECM proteins, including vitronectin (Davids & Yoshino, 1998), suggesting the ECM proteins may be triggering cell–cell interactions similar to haemocytes during capsule formation. One may postulate that native molluscan ECMs (such as the plasma vitronectin-like protein), play a role in the haemocytic immune response against the parasite, perhaps acting at the host-parasite interface, stimulating haemocyte migration (chemotaxis) and binding (opsonization) to the parasite (Perry *et al.* 1997).

Signal transduction molecules regulating snail cell spreading

Haemocyte adhesion and motility are critical components of the *B. glabrata* immune response against *S. mansoni*. In resistant snails, haemocytes migrate towards the sporocyst, attach and spread over its surface tegument, thus producing a cytotoxic encapsulation reaction. This defence response depends upon the successful communication of signals from the haemocyte extracellular environment into the cells interior to elicit appropriate responses. However the molecules involved in cellular adhesion and spreading behaviour remain largely unknown in gastropod molluscs. A recent study addressing this area examined the effects of a variety of treatments upon the calcium fluxes and cell rounding *B. glabrata* haemocytes (Hertel *et al.* 2000). Haemocytes demonstrated rounding and calcium transients on exposure to *Echinostoma paraensei* sporocysts, their secretory-excretory products and rediae, while phorbol myristate acetate caused only calcium transients. Interestingly, haemocytes did not display similar responses to *S. mansoni* sporocysts or bacterial lipopolysaccharides. This study clearly demonstrates the ability of parasite-derived molecules to induce a signalling response in host immune cells, although we still know very little about the specific signal transduction pathways used by haemocytes or other molluscan cells to regulate adhesion and motility.

The recent identification of a receptor for activated protein kinase C (RACK) from *B. glabrata* and its immunolocalization in both Bge cells and haemocytes suggests that protein kinase C (PKC) is present

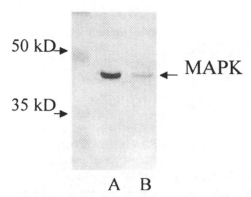

Fig. 4. Western blot showing detection of MAPK in *Biomphalaria glabrata* haemocytes (A) and embryonic (Bge) cells (B). MAPK appears to be more abundant in haemocytes than Bge cells, as each lane represents approximately 18 000 hemocytes and 25 000 Bge cells.

in these cells (Lardans *et al.* 1998). PKC has been reported to regulate cell spreading in a wide variety of vertebrate cell types, ranging from colon carcinoma cells (Rigot *et al.* 1998) to murine muscle cells (Disatnik & Rando, 1999), thus the question was posed whether Bge cells and haemocytes might also spread via a PKC-regulated mechanism. To address this question, the effect of inhibitors to PKC (calphostin C, chelerythrine chloride), Ras (Ras inhibitory peptide, FTase Inhibitor I) and mitogen-activated protein kinase kinase (MAPKK or Mek, inhibited by PD 98059), on Bge cell and haemocyte spreading was examined in an *in vitro* cell spreading assay. Results showed that spreading of both cell types was sensitive to all inhibitors, demonstrating a significant dose-dependent reduction in spreading (unpublished data). However, haemocytes typically required 10- to 100-fold higher inhibitor concentrations than Bge cells to inhibit spreading, possibly due to trace levels of plasma in haemocyte preparations. Low levels of contaminating plasma may contain several different ligands capable of stimulating haemocyte adhesion and spreading (such as the vitronectin-like plasma protein mentioned previously), reducing the effects of inhibitors. Moreover, as haemocytes are more differentiated than Bge cells, they might possess more surface receptors for the transduction of spreading signals compared to Bge cells. Regardless of these quantitative differences in the effects of the inhibitory drugs, the inhibition profile for both haemocytes and Bge cells suggests spreading behaviour is mediated through similar mechanisms under the regulation of PKC, Ras and Mek.

Extensive analyses of integrin-mediated signalling in vertebrates demonstrate that pathways vary depending on integrin subunit composition and cell type, but frequently the PKC–MAPK pathway has been reported as one pathway utilized to transduce integrin-mediated signals. To illustrate, fibroblast cells mediate integrin-induced signals through PKC

activating of MAPK, as well as Grb_2, Ras, Raf and Mek (Miranti, Ohno & Brugge, 1999). Focal adhesion kinase (FAK) is also often cited as one of the principal signal transduction molecules regulating cell spreading, such that PKC-dependent FAK phosphorylation has been demonstrated upon platelet spreading (Haimovich, Kaneshiki & Ji, 1996). However, the literature encompassing this area is conflicting and other researchers have found that FAK does not play a major role (Lin *et al.* 1997). Immunocytochemical analysis of haemocyte and Bge cells revealed crossreactivity with a polyclonal anti-FAK antibody, suggesting they possess a FAK-like protein which may participate in integrin-mediated snail cell spreading (unpublished data). FAK plays an important role in the formation of focal adhesion plaques, a complex of kinases and cytoskeletal proteins, regulating numerous signaling cascades, as well as cell structure, adhesion, and motility.

Downstream signalling molecules–MAPK

In addition to FAK, results of the Ras inhibitory peptide also suggests the presence of other homologues of vertebrate signal transduction molecules in *B. glabrata*. This inhibitor blocks the binding of Grb_2 to SOS, a complex which would otherwise activate Ras (Li *et al.* 1993). Therefore inhibition of Bge cell and haemocyte spreading implies Grb_2- and SOS-like molecules function in snail cells. Results acquired through the Bge cell and haemocyte spreading assays suggested PKC regulation of snail cell spreading. Often, PKC lies upstream of, and regulates activity of mitogen-activated protein kinase (MAPK) and has prompted an investigation of a possible MAPK homologue. Western blot analysis of Bge cells demonstrated phosphorylation of MAPK upon stimulation with the phorbol ester, PMA (unpublished data), which was detected using a monoclonal antibody that specifically recognizes the diphosphorylated (thr[183] and tyr[185]) activated loop of MAPK only.

Similar results were seen in *B. glabrata* haemocytes indicating that they too possess an immunoreactive MAPK-like molecule. Samples were prepared from unstimulated haemocytes and Bge cells, and this time MAPK was detected using a polyclonal antibody which recognizes total MAPK, irrespective of phosphorylation status (Fig. 4). On comparing band density and cell number, it appears that MAPK is substantially more abundant in haemocytes than Bge cells, perhaps because haemocytes are fully differentiated cells while Bge cells are of embryonic origin. Similar homologues of the MAPK family have been reported from only a few invertebrate species to date, including *Drosophila* (Biggs & Zipursky, 1992), *Cerratitis capitata* (Foukas *et al.*, 1998), and notably, another mollusc, *Hermissenda* (Crow *et al.* 1998). The functions of MAPK in

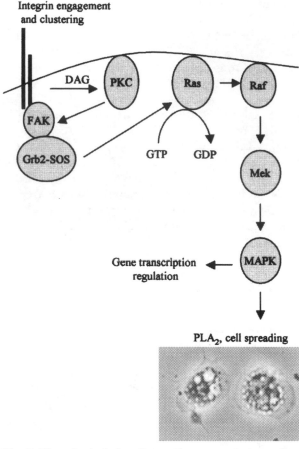

Fig. 5. Hypothetical signaling pathway regulating cell spreading behaviour in *Biomphalaria glabrata* haemocytes and in the embryonic (Bge) cell line. Integrin ligation and clustering causes an increase in diacylglycerol (DAG) levels, which stimulates PKC. Following, PKC activates FAK which upon phosphorylation binds Grb_2 by its SH2 domain. Grb_2, an adaptor protein, then forms a complex with SOS, a guanine nucleotide exchange factor, transporting SOS to Ras. SOS then activates Ras through the exchange of guanine triphosphate (GTP) for guanine diphosphate (GDP). Ras, upon activation is able to recruit Raf to the plasma membrane where it is phosphorylated by specific kinases. Raf may then activate/phosphorylate Mek, which in turn phosphorylates MAPK on threonine[183] and tyrosine[185] residues. Finally, MAPK may act upon either cytoplasmic or nucleic targets as mentioned previously.

vertebrate systems have been extensively studied but their equivalent substrates in invertebrates have yet to be discovered (Dickson & Hafen, 1994). In vertebrates, activated MAPK can translocate to the nucleus to regulate gene transcription, or alternatively act upon cytoplasmic targets such as phospholipase A_2 (PLA_2) (Durstin *et al.* 1994) and myosin light chain kinase (MLCK) regulating cell spreading and motility (Klemke *et al.* 1997). MAP kinases have been highly conserved at the molecular level throughout evolution, and presumably at the functional level also. Thus, *Biomphalaria* MAPK might transduce signals from the host–parasite interface,

which might in turn be involved in regulating gene expression, and/or cell spreading and migration in Bge cells and haemocytes.

The cumulative data acquired from these studies indicate that PKC, Ras, Mek and MAPK occur in both cell types, and specifically that PKC, Ras and Mek regulate cell spreading. An individual haemocyte, in the complex *milieu* of the host–parasite interface, undoubtedly is exposed to a wide array of signals, including those from the host tissues, other haemocytes and the sporocyst. Haemocytes require a variety of surface receptors to detect these different signals and an intricate signaling system with which to respond appropriately, possibly through upregulation of membrane receptors, modulation of spreading or motility, and communication with other haemocytes. Altogether, the data suggest that a number of homologues to vertebrate signal transduction molecules are indeed present in *B. glabrata*, and it is anticipated that numerous others will be discovered. Results also demonstrate that these molecules regulate snail cell spreading, an integral factor of the host immune response against the parasite. Although the exact signalling pathway regulating snail cell spreading cannot be determined from the available data, a hypothetical pathway is presented (Fig. 5). Snail cell spreading is speculated to be under the control of an intricate signalling network rather than a single pathway, utilizing various signalling routes depending on the specific receptor-ligand type being activated. Further detailed research is required to delineate the other signal transduction molecules which may be involved and how they interact with each other.

SPOROCYST TEGUMENTAL RECEPTORS FOR HOST MOLECULES

Following snail host entry, developing sporocysts find themselves in the same complex medium as host haemocytes and therefore subject to a multitude of environmental signals. As stressed by R. D. Lumsden 25 years ago in his review of parasitic helminth surface structure, the larval tegument is a complex, multifunctional organ: 'Tegument physiology [among digenetic trematodes] is ... diverse. This is perhaps most dramatically evidenced by larval trematodes, wherein the body surface may be specialized for locomotion, the elaboration of cysts, feeding, and sensory perception'. (Lumsden, 1975, p. 290). The tegument is constantly exposed to molecules present in snail plasma, including nutrients (glucose, amino acids), potentially harmful immune-related factors, as well as hormones and growth factors (see Fig. 1). However, given the recognized critical importance of this larval 'organ' for parasite survival, little is known of the membrane components or receptors responsible for mediating the complex trafficking of both parasite and host molecules across this interface barrier, nor of its role in transducing host-derived signals essential for larval development.

Larval surface carbohydrates and interactions with host plasma

The carbohydrate composition of the parasite surface is of considerable interest, given the potential role of lectin-like receptors on host haemocytes in mediating immune recognition and cellular encapsulation responses (Fryer, Hull & Bayne, 1989; Bayne & Fryer, 1994). A large majority of sporocyst surface proteins are thought to be glycosylated. This is illustrated by direct tegumental binding of a number of exogenous lectins, as well as studies demonstrating that specific snail plasma components bind to sporocyst surface glycoproteins (Yoshino, Cheng & Renwrantz, 1977; Stein & Basch, 1979; Spray & Granath, 1990; Uchikawa & Loker, 1991; Harris, Preston & Southgate, 1993; Johnston & Yoshino, 1996). Moreover, the importance of lectin-tegument interaction is emphasized by studies showing that the complement of lectin-reactive carbohydrates changes during larval development. Miracidia of *S. margrebowiei* bind a different complement of lectins than do mother sporocysts (Daniel, Preston & Southgate, 1992), and *S. mansoni* larvae lose the ability to bind two different lectins (eel serum agglutinin and *Dolichos* seed extract) after transformation from miracidium to sporocyst (Yoshino *et al.* 1977). In the avian schistosome, *T. szidati*, miracidial intercellular ridges bind a lectin from *Lotus tetragonolobus*, but the resulting sporocyst surface does not (Horak, 1995). While the exact implications of these results are unknown, developmental changes in surface lectin reactivity could serve a number of functions. It may be necessary for immune evasion, whether to prevent host opsonization of developing sporocysts (Fryer & Bayne, 1989) or recognition by circulating phagocytes. It may also reflect the need for each life stage to express a specific complement of surface receptor molecules to enable survival in different host microenvironments, with surface glycoproteins capable of transporting specific molecules or serving as receptors for essential growth factors.

In addition to these developmental changes in lectin reactivity, the surface of the developing sporocyst is capable of altering its characteristics in response to the binding of specific ligands. Sporocysts incubated in snail plasma rapidly acquire plasma proteins and, after wash-out, the antigens are rapidly (within 3 hours) cleared from the tegumental surface. In contrast, rabbit-anti-sporocyst antibody remained on the surface for 48 hours suggesting that the sporocyst response was specific to snail plasma protein binding (Bayne, Loker & Yui, 1986). A follow-up study by Dunn & Yoshino (1991) showed

that secondary or tertiary crosslinking of bound lectin or snail plasma enhances the clearance of surface proteins. Overall these studies demonstrate that the tegument is capable of modulating its surface chemistry in response to specific external stimuli, and suggest that carbohydrate-containing receptors capable of transducing signals from the external *milieu* are present at the host-parasite interface.

Despite as abundance of evidence for tegumental surface carbohydrates very few investigations have been directed toward identifying the specific structure of oligosaccharides expressed by intramolluscan trematode stages. In a recent collaboration with Drs. Richard Commins and Kwame Nyame (University of Oklahoma Health Sciences Center, Oklahoma City, OK), we employed several oligosaccharide-specific monoclonal antibodies that recognize Lewis X (LeX), LacdiNAc (GalNAcβ1-4GlcNAc) and fucosylated-LacdiNAc (GalNAcβ1-4[Fucα1-3]GlcNAc) to determine whether the mother and/or daughter sporocysts of *S. mansoni* expressed these oligosaccharide epitopes. Both sporocyst stages strongly expressed the LacdiNAc and Fuc-lacdiNAc epitopes at their tegumental surfaces and, under *in vivo* conditions, appeared to secrete epitope-bearing glycoconjugates into surrounding host tissues (unpublished data). These findings complement earlier reports of the occurrence of fucosyllactose (Mansour *et al.* 1995), LacdiNAc and Fuc-LacdiNAc (van Remoortere *et al.* 2000) on schistosome miracidia and cercariae, and underscore the possibility that the repertoire of saccharide structures expressed at the larval surface may actually be quite limited. Perhaps as we begin to learn more about the oligosaccharide composition/structure of these tegumental determinants, this information can then be used as a valuable tool for finding host receptor molecules capable of recognizing them.

Tegument-mediated transport of host molecules

The tegument is not only the primary site of the binding of host lectins, but also a critical organ for the acquisition of an array of other host molecules. A number of studies have addressed the nature of carrier-mediated uptake mechanisms (either facilitated diffusion or active transport) present in trematode larvae, since many of these molecules would be unlikely to enter the tegument in significant amounts via simple diffusion. Daughter sporocysts of *S. mansoni* (DiConza & Basch, 1974) and rediae of *Parorchis acanthus* (McDaniel & Dixon, 1967) transport glucose and thymidine by carrier-mediated mechanisms. Different life stages of *Proterometra macrostoma* demonstrated distinct forms of carrier-mediated glucose uptake, with rediae using an active transport mechanism (Na$^+$-dependent) and cercariae utilizing facilitated diffusion (Uglem, 1980). These studies highlight the diversity of transporter mechanisms that may be present on the sporocyst tegumental surface, and are consistent with studies on adult *S. mansoni* that demonstrate carrier-mediated transport capacity for over 75 different low molecular weight organic molecules, including sugars, amino acids, purines, pyrimidines, and fatty acids (Pappas & Read, 1975). Such a large number of potential uptake mechanisms raises a question regarding the molecular diversity of tegumental transporters. Does each substrate, or particular class of substrate, have a corresponding transporter or have tegumental transporters evolved to have a more broad substrate specificity in trematodes as an adaptation to their exposure to a complex host *milieu* containing a bevy of 'desirable' molecules? Recent molecular work has helped to address this question.

Shoemaker and colleagues recently cloned two facilitated diffusion glucose transporters (SGTP1 and SGTP4) and an amino acid permease light chain (SPRM1) from adult stages of *S. mansoni* (Skelly *et al.* 1994, 1999). Antibody-probe studies have shown that both SGTP1 and SPRM1 are expressed in miracidia, daughter sporocysts, and cercariae (Skelly & Shoemaker, 1996; Skelly *et al.* 1999). While the localization of SGTP1 in sporocysts is unknown, it localizes primarily to the basal syncytial membrane in adults, and may indeed be expressed at the surface of the sporocyst (Skelly & Shoemaker, 1996; Zong *et al.* 1995). SPRM1 is expressed on the tegumental surface of daughter sporocysts and intrasporocystic cercariae (Skelly *et al.* 1999). While the importance of these two transporters in the intramolluscan development of *S. mansoni* is not known, they provide the first molecular evidence that transport proteins are present in trematodes and are expressed at the tegumental surface where they have constant access to molecules in the host *milieu*. Studies carried out to determine substrate specificity of these gene products also shed light on the nature of transporter diversity in parasitic flatworms. Both SGTP1 and SPRM1 have a restricted substrate specificity (for D-glucose and amino acids, respectively), implying that a large number of other transporters may be present in the tegument (Mastroberardino *et al.* 1998, Skelly *et al.* 1994). However, certain characteristics of the amino acid transporter SPRM1 that differ from homologous proteins found in mammals are worthy of note. When expressed in frog oocytes, SPRM1 is most similar to the *system L* family of amino acid transporters, proteins responsible for the sodium-independent uptake of large, neutral amino acids. However, SPRM1 was also partially sodium-dependent, and a *system L*-specific substrate analogue failed to inhibit its activity; these characteristics are unlike any other *system L* transporter studied to date. Strikingly, SPRM1 was also shown to take up cationic amino acids, a group of substrates normally attributed to *system y+* or *y+L* transporters

(Mastroberardino *et al.* 1998). Overall, these results led the authors to conclude that SPRM1 possesses 'mixed properties', and suggest that schistosome tegumental transporters may have developed comparatively broad substrate specificities, perhaps to allow more efficient transport of a variety of molecules from the host environment. As more tegumental transporters are cloned and their substrate specificity characterized, this question may be addressed in more detail. In addition, these cloned cDNAs represent valuable tools for the elucidation of specific nutritional requirements of developing fluke larvae, as the dynamics and location of their expression can now be determined.

Serotonin transport in mother sporocysts of S. mansoni

Host glucose and other cell building blocks are essential during parasite development and work described above provides strong evidence for the presence of specific carriers for those molecules in the tegument. However, recent studies in our laboratory have examined larval transporter systems for molecules other than host nutrients. Specifically, we are interested in the role of serotonin (5-hydroxytryptamine; 5-HT) in the intramolluscan development of *S. mansoni*. Earlier results showed that snails infected with *S. mansoni* contain dramatically reduced levels of brain dopamine and 5-HT. Interestingly, treatment of infected snails with exogenous 5-HT reversed the classic effects of infection on host fecundity, a phenomenon known as parasitic castration (Manger *et al.* 1996). While exogenous 5-HT may be modulating any number of snail processes, including metabolic regulation and the release of ovulation-hormone from neurons in the central nervous system, these data raised the intriguing possibility that parasite utilization and/or depletion of host 5-HT played a role in parasite manipulation of host physiology and that 5-HT may be an important factor in parasite development, as has been suggested for adult stages of *S. mansoni* (Halton, Maule & Shaw, 1997).

Initial experiments showed that treatment of *in vitro*-transformed mother sporocysts with exogenous 5-HT, or the 5-HT analogue tryptamine, caused an increase in larval motility (Boyle, Zaide & Yoshino, 2000). Drug inhibition studies provided strong evidence that 5-HT exerted its action by binding to muscle 5-HT receptors with pharmacological similarities to those described previously in adult worms (Day, Bennett & Pax, 1994; Boyle *et al.* 2000). For this to occur, exogenous 5-HT must make its way through the two lipid bilayers bordering the tegumental cytoplasm before having access to the subtending muscle layer. As shown in Fig. 6A, however, the response to exogenous 5-HT occurs

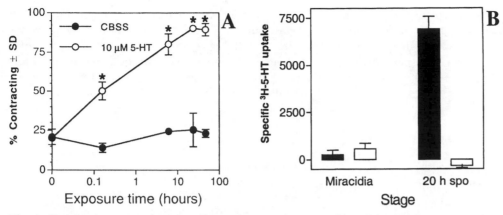

Fig. 6. Graphs demonstrating the effects of serotonin on motility of mother sporocysts (A) and the 5-HT uptake capacity of miracidia and mother sporocysts of *Schistosoma mansoni*. (A) Sporocysts were exposed to 10μM 5-HT (open circles) or normal saline (closed cirlces) and examined at different times post-treatment for overall motility (% of sporocysts exhibiting muscle contractions). Note that treated sporocysts were significantly more motile than controls within 10 min of 5-HT application. (B) Miracidia or 20 h old *in vitro*-transformed sporocysts were incubated for 10 min in 500 nM [^{3}H]5-HT, and in the absence (white bars) or presence (black bars) of 10μM fluoxetine. Miracidia demonstrated negligible specific 5-HT uptake, while 20 h sporocysts demonstrated robust [^{3}H]5-HT uptake that was highly sensitive to fluoxetine. (Fig. (A) reprinted with permission from Boyle, Zaide & Yoshino (2000), *Experimental Parasitology* **94**, 217–226.)

within 10 min of application of the agonist. Therefore, in order to explain this rapid passage of 5-HT through the tegument, we hypothesized that schistosome larvae possessed serotonin transporter (SERT) molecules on their tegumental surface.

To test this hypothesis, freshly hatched miracidia or *in vitro* transformed mother sporocysts (20 hours old) were incubated for 10 min in the presence of 500 nM [^{3}H] 5-HT and in the presence or absence of the specific serotonin reuptake inhibitor fluoxetine (10 μM), washed, and subjected to liquid scintillation counting. Non-specific controls were incubated on ice in the presence of excess (200 μM) unlabelled 5-HT. As shown in Fig. 6B, 20 hour sporocysts that have recently shed their ciliary plates and synthesized their tegument take up exogenous 5-HT. The specific component of this uptake was completely blocked in the presence of 10 μM fluoxetine. In contrast, free-swimming miracidia demonstrated no significant 5-HT uptake. These data suggest that mother sporocysts take up exogenous 5-HT *in vitro* and that this uptake capacity is developmentally regulated. Also, the site of expression of the transporter is most likely in the newly formed tegument, since miracidia, which lack a tegument do not demonstrate high SERT activity. Current studies are being carried out to determine the pharmacology, ion dependence and molecular basis for larval tegumental serotonin transporters. In addition, we are addressing broader questions regarding the role of host 5-HT in the development of larval *S. mansoni*. Overall these studies provide further support for high transporter diversity in sporocysts and to the hypothesis that serotonin uptake may be a crucial element in the development of *S. mansoni*.

Sporocyst regulatory processes: a role for the tegument?

It is difficult to imagine a group of parasites with more complex lifestyles than the trematodes. The establishment and maintenance of such a lifestyle is a complex and highly regulated process, involving parasite adaptation *in vivo* to changes in the host environment (Wilson & Draskau, 1976). Over the course of an infection, sporocysts must suffer assaults from the snail immune system, compete for nutrients without destroying the host and successfully migrate to suitable microhabitats. Evidence is also accumulating that *inter*larval communication is involved in spatially and energetically regulating snail infections (Théron, Moné & Gerard, 1992; Reddy & Fried, 1996; Théron, Pages & Rognon, 1997). These phenomena imply that the larval surface is capable of reading signals in the host environment (from haemocytes, plasma and other larvae) and transmitting the information to regulatory centers. Indeed, work in adults has provided pharmacological and molecular evidence that the tegument possesses receptors for acetylcholine, serotonin, tumour necrosis factor and transforming growth factor (Bennett & Bueding, 1973; Amiri *et al.* 1992; Camacho & Agnew, 1995; Davies, Shoemaker & Pearce, 1998). In that these are all receptors for neurotransmitters, hormones and other host 'signalling' molecules, they are intriguing candidates for mediating direct host-to-parasite communication. As more molecular information concerning tegumental receptor proteins accumulates, the overall importance of these interactions will hopefully be better understood. This information could then provide valuable tools for the development of therapeutics and methods for the control of the proliferation of schistosomes at the level of the intermediate host. In addition, these studies will provide valuable insights into the evolution of mechanisms for complex parasitic lifestyles.

CONCLUDING REMARKS

Successful transition of the free-living miracidium to the mother sporocyst stage and, in turn, its subsequent development within the snail host, represents a critical step in the eventual establishment of molluscan infections. Molecular communication between the sporocyst and host, the result of receptor–ligand interactions and exchange of chemical signals, is believed to be central to this developmental process, although we still know very little about the specific molecules represented at the host–parasite interface. In resistant snails, haemocytes infiltrate the tissues surrounding young sporocysts and form multi-layered cellular encapsulations that result in larval destruction within 24–48 hours. Specific adhesion receptors on host cells, some of which appear to be lectin-like carbohydrate-binding receptors (CBRs), may be involved in mediating such heamocytic encapsulations of sporocysts. This process also may be facilitated or inhibited by carbohydrate-reactive plasma factors. By contrast, in susceptible hosts, haemocytes are neither attracted to nor respond against the sporocyst. One hypothesis suggests that sporocysts may escape haemocyte reactivity by elaborating secretory glycoproteins (ESP) that either directly bind to and interfere with haemocyte function, or provide a disguise of host-like molecules to prevent larval recognition by hemocytes. CBRs also have been implicated in ESP–haemocyte binding interactions.

Still other cell receptors appear to be responsible for haemocyte motility during tissue migration (cell-substrate interactions) and/or cell-to-cell binding during encapsulation reactions. Integrin-like membrane receptors and peptide hormone-like receptors appear to be involved *in vitro* snail cell spreading and motility, although their specific role in regulating larval fluke infections remains to be investigated. Finally, molecules expressed at the tegumental surface of sporocysts are thought to assume multiple roles; (1) as oligosaccharide ligands for lectin-like host cell adhesion receptors, (2) for binding and transport of plasma macromolecules (e.g. haemoglobin, haemocyanin), presumably destined for digestion/absorption, (3) uptake of simple nutritive compounds such as sugars or amino acids, and (4) for transport of host neuropeptides or neurotransmitters (e.g. 5-HT) possibly serving a 'signalling' function related to larval behaviour or development.

Taken together, the host–parasite interface represents a complex and dynamic molecular environment that must be precisely balanced in order to ensure sporocyst survival. By expanding the diversity of 'model' snail–trematode systems being studied and by employing novel experimental approaches, such as the use of the Bge cell line, a full understanding of the mechanisms underlying compatibility between larval flukes and their snail intermediate hosts is clearly an attainable goal.

REFERENCES

ADEMA, C. M., ARGUELLO II, D. F., STRICKER, S. A. & LOKER, E. S. (1994). A time-lapse study of interactions between *Echinostoma paraensei* intramolluscan larval stages and adherent hemocytes from *Biomphalaria glabrata* and *Helix aspersa*. *Journal of Parasitology* **80**, 719–727.

ADEMA, C. M., HERTEL, L. A., MILLER, R. D. & LOKER, E. S. (1997). A family of fibrinogen-related proteins that precipitates parasite-derived molecules is produced by an invertebrate after infection. *Proceedings of the National Academy of Sciences, USA* **94**, 8691–8696.

ADEMA, C. M. & LOKER, E. S. (1997) Specificity and immunobiology of larval digenean–snail associations. In *Advances in Trematode Biology* (eds. Fried, B. & Graczyk, T. K.), pp. 229–263. Florida, CRC Press.

AMIRI, P., LOCKSLEY, R. M., PARSLOW, T. G., SADICK, M., RECTOR, E., RITTER, D. & MCKERROW, J. H. (1992). Tumour necrosis factor alpha restores granulomas and induces parasite egg-laying in schistosome-infected SCID mice. *Nature* **356**, 604–607.

ATAEV, G. L., FOURNIER, A. & COUSTAU, C. (1998). Comparison of *Echinostoma caproni* mother sporocyst development *in vivo* and *in vitro* using *Biomphalaria glabrata* snails and a *B. glabrata* embryonic cell line. *Journal of Parasitology* **84**, 227–235.

BAYNE, C. J., BUCKLEY, P. M. & DE WAN, P. C. (1980). Macrophagelike hemocytes of resistant *Biomphalaria glabrata* are cytotoxic for sporocysts of *Schistosoma mansoni in vitro*. *Journal of Parasitology* **66**, 413–419.

BAYNE, C. J. & FRYER, S. E. (1994). Phagocytosis and invertebrate opsonins in relation to parasitism. In *Primordial Immunity: Foundations for the Vertebrate Immune System* (eds. Beck, G., Cooper, E. L., Habicht, G. S. & Marchalonis, J. J.), pp. 162–177. New York, Annals of the New York Academy of Sciences.

BAYNE, C. J., LOKER, E. S. & YUI, M. A. (1986). Interactions between the plasma proteins of *Biomphalaria glabrata* (Gastropoda) and the sporocyst tegument of *Schistosoma mansoni* (Trematoda). *Parasitology* **92**, 653–664.

BENNETT, J. L. & BUEDING, E. (1973). Uptake of 5-hydroxytryptamine by *Schistosoma mansoni*. *Molecular Pharmacology* **9**, 311–319.

BIGGS, W. H. & ZIPURSKY, S. L. (1992). Primary structure, expression, and signal-dependent tyrosine phosphorylation of a *Drosophila* homolog of extracellular signal-regulated kinase. *Proceedings of the National Academy of Sciences, USA.* **89**, 6295–6299.

BOYLE, J. P., ZAIDE, J. V. & YOSHINO, T. P. (2000). *Schistosoma mansoni*: effects of serotonin and serotonin receptor antagonists on motility and length of primary sporocysts *in vitro*. *Experimental Parasitology* **94**, 217–226.

CAMACHO, M. & AGNEW, A. (1995). *Schistosoma*: rate of glucose import is altered by acetylcholine interaction with tegumental acetylcholine receptors and acetylcholinesterase. *Experimental Parasitology* **81**, 584–591.

CONNORS, V. A. & YOSHINO, T. P. (1990). *In vitro* effect of larval *Schistosoma mansoni* excretory-secretory products on phagocytosis-stimulated superoxide production in hemocytes of *Biomphalaria glabrata*. *Journal of Parasitology* **76**, 895–902.

CROW, T., XUE-BIAN, J. J., SIDIQI, V., KANG, Y. & NEARY, J. T. (1998). Phosphorylation of mitogen-activated protein kinase by one-trial and multi-trial classical conditioning. *Journal of Neuroscience* **18**, 3480–3487.

DANIEL, B. E., PRESTON, T. M. & SOUTHGATE, V. R. (1992). The *in vitro* transformation of the miracidium to the mother sporocyst of *Schistosoma margrebowlei*; changes in the parasite surface and implications for interactions with snail plasma factors. *Parasitology* **104**, 41–49.

DAVIDS, B. J., WU, X.-J. & YOSHINO, T. P. (1999). Cloning of a β integrin subunit cDNA from an embryonic cell line derived from the freshwater mollusc, *Biomphalaria glabrata*. *Gene* **228**, 213–223.

DAVIDS, B. J. & YOSHINO, T. P. (1998). Integrin-like RGD-dependent binding mechanism involved in the spreading response of circulating molluscan phagocytes. *Developmental and Comparative Immunology* **22**, 39–53.

DAVIES, S. J., SHOEMAKER, C. B. & PEARCE, E. J. (1998). A divergent member of the transforming growth factor beta receptor family from *Schistosoma mansoni* is expressed on the parasite surface membrane. *Journal of Biological Chemistry* **273**, 11234–11240.

DAY, T. A., BENNETT, J. L. & PAX, R. A. (1994). Serotonin and its requirement for maintenance of contractility in muscle fibres isolated from *Schistosoma mansoni*. *Parasitology* **108**, 425–432.

DEFILIPPI, P., GISMONDI, A., SANTONI, A., TARONE, G. (1997). *Signal Transduction by Integrins*, 188 p. Texas, USA, Landes Bioscience.

DE JONG-BRINK, M. (1995). How schistosomes profit from the stress responses they elicit in their hosts. *Advances in Parasitology* **35**, 177–256.

DICKSON, B. & HAFEN, E. (1994). Incorporation of ^{3}H-thymidine and ^{14}C-glucose by *Schistosoma mansoni* daughter sporocysts *in vitro*. *Journal of Parasitology* **60**, 1045–1046.

DISATNIK, M. H. & RANDO, T. A. (1999). Integrin-mediated muscle cell spreading. The role of protein kinase C in outside-in and inside-out signaling and evidence of integrin cross-talk. *Journal of Biological Chemistry* **274**, 32486–32492.

DUCLERMORTIER, P., LARDANS, V., SERRA, E., TROTTIEN, F. & DISSOUS, C. (1999). *Biomphalaria glabrata* embryonic cells express a protein with a domain homologous to the lectin domain of mammalian selectins. *Parasitology Research* **85**, 481–486.

DUNN, T. S. & YOSHINO, T. P. (1988). *Schistosoma mansoni*: Origin and expression of a tegumental surface antigen on the miracidium and primary sporocyst. *Experimental Parasitology* **67**, 167–181.

DUNN, T. S. & YOSHINO, T. P. (1991). Tegumental surface modulation in *Schistosoma mansoni* primary sporocysts

in response to ligand binding. *Parasite Immunology* **13**, 121–135.

DURSTIN, M., DURSTIN, S., MOLSKI, T. F. P., BECKER, E. L. & SHA'AFI, R. I. (1994). Cytoplasmic phospholipase A2 translocates to membrane fraction in human neutrophils activated by stimuli that phosphorylate mitogen-activated protein kinase. *Proceedings of the National Academy of Sciences, USA* **91**, 3142–3146.

DUVAUX-MIRET, O., STEFANO, G. B., SMITH, E. M., DISSOUS, C. & CAPRON, A. (1992). Immunosuppression in the definitive and intermediate hosts of the human parasite *Schistosoma mansoni* by release of immunoactive neuropeptides. *Proceedings of the National Academy of Sciences, USA* **89**, 778–781.

FOUKAS, L. C., HARALABOS, L. K., PARASKEVOPOULOU, N., METHENITI, A., LAMBROPOULOU, M. & MARMARAS, V. J. (1998). Phagocytosis of *Escherichia coli* by insect hemocytes requires both activation of the Ras/mitogen-activated protein kinase signal transduction pathway for attachment and β3 integrin for internalization. *Journal of Biological Chemistry* **272**, 14813–14818.

FRYER, S. E. & BAYNE, C. J. (1989). Opsonization of yeast by the plasma of *Biomphalaria glabrata* (Gastropoda): a strain-specific, time-dependent process. *Parasite Immunology* **11**, 269–278.

FRYER, S. E., HULL, C. J. & BAYNE, C. J. (1989). Phagocytosis of yeast by *Biomphalaria glabrata*: carbohydrate specificity of hemocyte receptors and a plasma opsonin. *Developmental and Comparative Immunology* **13**, 9–16.

HAHN, U. K., BENDER, R. C. & BAYNE, C. J. (2000). Production of reactive oxygen species by hemocytes of *Biomphalaria glabrata*: carbohydrate-specific stimulation. *Developmental and comparative Immunology* **24**, 531–541.

HAIMOVICH, B., KANESHIKI, N. & JI, P. (1996). Protein kinase C regulates tyrosine phosphorylation of pp125FAK in platelets adherent to fibrinogen. *Blood* **87**, 152–161.

HALTON, D. W., MAULE, A. G. & SHAW, C. (1997). Trematode Neurobiology. In *Advances in Trematode Biology* (eds. Fried, B., Graczyk, T. E.), pp. 345–382. New York, CRC Press.

HANSEN, E. (1976a). A cell line from embryos of *Biomphalaria glabrata* (Pulmonata): establishment and characteristics. In *Invertebrate Tissue Culture: Applications in Medicine, Biology, and Agriculture* (ed. Maramorosch, K.), pp. 75–97. New York, Academic Press.

HANSEN, E., (1976b). Application of tissue culture of a pulmonate snail to culture of larval *Schistosoma mansoni*. In *Invertebrate Tissue Culture: Applications in Medicine, Biology, and Agriculture* (eds. Kursak, E. & Maramorosch, K.), pp. 87–97. New York, Academic Press.

HARRIS, R. A., PRESTON, T. M. & SOUTHGATE, V. R. (1993). Purification of an agglutinin from the haemolymph of the snail *Bulinus nasutus* and demonstration of related proteins in other *Bulinus* spp. *Parasitology* **106**, 127–35.

HERTEL, L. A., STRICKER, S. A. & LOKER, E. S. (2000). Calcium dynamics of hemocytes of the gastropod

Biomphalaria glabrata: effects of digenetic trematodes and selected bioactive compounds. *Invertebrate Biology* **119**, 27–37.

HORAK, P. (1995). Developmentally regulated expression of surface carbohydrate residues on larval stages of the avian schistosome *Trichobilharzia szidati*. *Folia Parasitologica* **42**, 255–265.

HYNES, R. O. (1999). Cell adhesion: old and new questions. *Trends in Cell Biology* **9**, 33–37.

JOHNSTON, L. A. & YOSHINO, T. P. (1996). Analysis of lectin- and snail plasma-binding glycopeptides associated with the tegumental surface of the primary sporocysts of *Schistosoma mansoni*. *Parasitology* **112**, 469–479.

JOHNSTON, L. A. & YOSHINO, T. P. (2001). Larval *Schistosoma mansoni* excretory-secretory glycoproteins (ESP) bind to hemocytes of *Biomphalaria glabrata* (Gastropoda) via surface carbohydrate-binding receptors. *Journal of Parasitology* **87**, 786–793.

KLEMKE, R. L., CAI, S., GIANNINI, A. L., GALLAGHER, P. J., DE LANEROLLE, P. & CHERESH, D. A. (1997). Regulation of cell motility by mitogen-activated protein kinase. *Journal of Cell Biology* **137**, 481–492

LARDANS, V., SERRA, E., CAPRON, A. & DISSOUS, C. (1998). Characterization of an intracellular receptor for activated protein kinase C (RACK) from the mollusc *Biomphalaria glabrata*, the intermediate host for *Schistosoma mansoni*. *Experimental Parasitology* **88**, 194–199.

LI, N., BATZER, A., DALY, R., YAJNIK, V., SKOLNIK, E., CHARDIN, P., BAR-SAGI, D., MARGOLIS, B. & SCHLESSINGER, J. (1993). Guanine-nucleotide-releasing factor hSOS1 binds to GRB2 and links receptor tyrosine kinases to Ras signalling. *Nature* **363**, 85–88.

LIE, K. J. (1982). Survival of *Schistosoma mansoni* and other trematode larvae in the snail *Biomphalaria glabrata*. A discussion of the interference hypothesis. *Tropical and Geographic Medicine* **34**, 111–122.

LIE, K. J., HEYNEMAN, D. & JEONG, K. H. (1976). Studies on resistance in snails. 7. Evidence of interference with the defense reaction in *Biomphalaria glabrata* by trematode larvae. *Journal of Parasitology* **62**, 608–615.

LIN, T. H., APLIN, A. E., SHEN, Y., CHEN, Q., SCHALLER, M., ROMER, L., AUKHIL, I. & JULIANO, R. L. (1997). Integrin-mediated activation of MAP kinase is independent of FAK: evidence for dual integrin signaling pathways in fibroblasts. *Journal of Cell Biology* **136**, 1385–1395.

LODES, M. J. & YOSHINO, T. P. (1989). Characterization of excretory-secretory proteins synthesized *in vitro* by *Schistosoma mansoni* primary sporocysts. *Journal of Parasitology* **75**, 853–862.

LODES, M. J. & YOSHINO, T. P. (1990). The effect of schistosome excretory-secretory products on *Biomphalaria glabrata* hemocyte motility. *Journal of Invertebrate Pathology* **56**, 75–85.

LOKER, E. S & ADEMA, C. M. (1995). Schistosomes, echinostomes and snails: comparative immunobiology. *Parasitology Today* **11**, 120–124.

LOKER, E. S. & BAYNE, C. J. (1986). Immunity to trematode larvae in the snail *Biomphalaria*. In *Immune Mechanisms in Invertebrate Vectors* (ed. Lackie, A. M.), pp. 199–220. Symposia of the Zoological Society of London. Oxford, Clarendon Press.

LOKER, E. S, BOSTON, M. E. & BAYNE, C. J. (1989). Differential adherence of M line *Biomphalaria glabrata* hemocytes to *Schistosoma mansoni* and *Echinostoma paraensei* larvae, and experimental manipulation of hemocyte binding. *Journal of Invertebrate Pathology* **54**, 260–268.

LOKER, E. S., CIMINO, D. F. & HERTEL, L. A. (1992). Excretory-secretory products of *Echinostoma paraensei* sporocysts mediate interference with *Biomphalaria glabrata* hemocyte functions. *Journal of Parasitology* **78**, 104–115.

LUMSDEN, R. D. (1975). Surface ultrastucture and cytochemistry of parasitic helminths. *Experimental Parasitology* **37**, 267–339.

MANGER, P., LI, J., CHRISTENSEN, B. M. & YOSHINO, T. P. (1996). Biogenic monoamines in the freshwater snail, *Biomphalaria glabrata*: influence of infection by the human blood fluke, *Schistosoma mansoni*. *Comparative Biochemistry and Physiology* **114A**, 227–234.

MANSOUR, M. H., NAGM, H. I., SAAD, A. H. & TAALAB, N. I. (1995). Characterization of *Biomphalaria alexandrina*-derived lectins recognizing a fucosyllactose-related determinant on schistosomes. *Molecular and Biochemical Parasitology* **69**, 173–184.

MASTROBERARDINO, L., SPINDLER, B., PFEIFFER, R., SKELLY, P. J., LOFFING, J., SHOEMAKER, C. B. & VERREY, F. (1998). Amino-acid transport by heterodimers of 4F2hc/CD98 and members of a permease family. *Nature* **395**, 288–291.

MCDANIEL, J. S. & DIXON, K. E. (1967). Utilization of exogenous glucose by the rediae of *Parorchis acanthus* (Digenea: philophthalmidae) and *Cryptocotyle lingua* (Digenea: heterophyidae). *Biological Bulletin* **133**, 591–599.

MIRANTI, C. K., OHNO, S. & BRUGGE, J. S. (1999). Protein kinase C regulates integrin-induced activation of the extracellular regulated kinase pathway upstream of shc. *Journal of Biological Chemistry* **274**, 10571–10581.

NODA, S. & LOKER, E. S. (1989*a*). Effects of infection with *Echinostoma paraensei* on the circulating haemocyte population of the snail host *Biomphalaria glabrata*. *Parasitology* **98**, 35–41.

NODA, S. & LOKER, E. S. (1989*b*). Phagocytic activity of hemocytes of M-line *Biomphalaria glabrata* snails: effect of exposure to the trematode *Echinostoma paraensei*. *Journal of Parasitology* **75**, 261–269.

NUÑEZ, P. E., ADEMA, C. M. & DE JONG-BRINK, M. (1994). Modulation of the bacterial clearance activity of haemocytes from the freshwater mollusc, *Lymnaea stagnalis*, by the avian schistosome, *Trichobilharzia ocellata*. *Parasitology* **109**, 299–310.

NUÑEZ, P. E. & DE JONG-BRINK, M. (1997). The suppressive excretory-secretory product of *Trichobilharzia ocellata*: a possible factor for determining compatibility in parasite-host interactions. *Parasitology* **115**, 193–203.

PAPPAS, P. W. & READ, C. P. (1975). Membrane transport in helminth parasites: a review. *Experimental Parasitology* **37**, 469–530.

PERRY, D. G., WISNIOWSKI, P., DAUGHERTY, J. D. & MARTIN, II, W. J. (1997). Nonimmune phagocytosis of liposomes by rat alveolar macrophages is enhanced by vitronectin and is vitronectin-receptor mediated.

American Journal of Respiratory Cell and Molecular Biology **17**, 462–470.

READ, C. P., ROTHMAN, A. & SIMMONS, J. JR. (1963). Studies on membrane transport, with special reference to host parasite integration. *Annals of the New York Academy of Sciences* **113**, 154–205.

REDDY, A. & FRIED, B. (1996). *In vitro* studies on intraspecific and interspecific chemical attraction in daughter rediae of *Echinostoma trivolvis* and *E. caproni*. *International Journal for Parasitology* **26**, 1081–1085.

RENWRANTZ, L. & STAHMER, A. (1983). Opsonizing properties of an isolated hemolymph agglutinin and demonstration of lectin-like recognition molecules at the surface of hemocytes from *Mytilus edulis*. *Journal of Comparative Physiology* **149**, 535–546.

RICHARDS, E.H. & RENWRANTZ, L. R. (1991). Two lectins on the surface of *Helix pomatia* haemocytes: a Ca²⁺-dependent, GalNac-specific lectin and a Ca²⁺-independent manose 6-phosphate-specific lectin which recognises activated homologous opsonins. *Journal of Comparative Physiology* **161**, 43–54.

RIGOT, V., LEHMANN, M., ANDRE, F., DAEMI, N., MARVALDI, J. & LUIS, J. (1998). Integrin ligation and PKC activation are required for migration of colon carcinoma cells. *Journal of Cell Science* **111**, 3119–3127.

SKELLY, P. J., KIM, J. W., CUNNINGHAM, J. & SHOEMAKER, C. B. (1994). Cloning, characterization, and functional expression of cDNAs encoding glucose transporter proteins from the human parasite *Schistosoma mansoni*. *Journal of Biological Chemisty* **269**, 4247–4253.

SKELLY, P. J., PFEIFFER, R., VERREY, F. & SHOEMAKER, C. B. (1999). SPRMllc, a heterodimeric amino acid permease light chain of the human parasitic platyhelminth, *Schistosoma mansoni*. *Parasitology* **119**, 569–576.

SKELLY, P. J. & SHOEMAKER, C. B. (1996). Rapid appearance and asymmetric distribution of glucose transporter SGTP4 at the apical surface of intramammalian-stage *Schistosoma mansoni*. *Proceedings of the National Academy of Sciences, USA* **93**, 3642–3646.

SPRAY, F. J. & GRANATH JR., W. O. (1990). Differential binding of hemolymph proteins from schistosome-resistant and -susceptible *Biomphalaria glabrata* to *Schistosoma mansoni* sporocysts. *Journal of Parasitology* **76**, 225–229.

STEIN, P. C. & BASCH, P. F. (1979). Purification and binding properties of hemagglutinin from *Biomphalaria glabrata*. *Journal of Invertebrate Pathology* **33**, 10–18.

THÉRON, A., MONÉ, H. & GERARD, C. (1992). Spatial and energy compromise between host and parasite: the *Biomphalaria glabrata–Schistosoma mansoni* system. *International Journal for Parasitology* **22**, 91–94.

THÉRON, A., PAGES, J. R. & ROGNON, A. (1997). *Schistosoma mansoni*: distribution patterns of miracidia among

Biomphalaria glabrata snails as related to host susceptibility and sporocyst regulatory processes. *Experimental Parasitology* **85**, 1–9.

UCHIKAWA, R. & LOKER, E. S. (1991). Lectin-binding properties of the surfaces of *in vitro*-transformed *Schistosoma mansoni* and *Echinostoma paraensei* sporocysts. *Journal of Parasitology* **77**, 742–748.

UGLEM, G. L. (1980). Sugar transport by larval and adult *Proterometra macrostoma* (Digenea) in relation to environmental factors. *Journal of Parasitology* **66**, 748–758.

VAN REMOORTERE, A., HOKKE, C. H., VAN DAM, G. J., VAN DIE, I., DEELDER, A. M. & VAN DEN EIJNDEN, D. H. (2000). Various stages of *Schistosoma* express Lewisx, LacdiNAc, GalNAcβ1-4 (Fucl-3) GlcNAc and GalNAcβ1-4 (Fucl-2Fucl-3) GlcNAc carbohydrate epitopes: detection with monoclonal antibodies that are characterized by enzymatically synthesized neoglycoproteins. *Glycobiology* **10**, 601–609.

VASTA, G. R., SULLIVAN, J. T., CHENG, T. C., MARCHALONIS, J. J. & WARR, G. W. (1982). A cell membrane-associated lectin of the oyster hemocyte. *Journal of Invertebrate Pathology* **40**, 367–377.

WILSON, R. A. & DRASKAU, T. (1976). The stimulation of daughter redia production during the larval development of *Fasciola hepatica*. *Parasitology* **722**, 245–257.

WRIGHT, C. A. (1973). *Flukes and Snails*. New York, Hafner Press.

YOSHINO, T. P., CHENG, T. C. & RENWRANTZ, L. R. (1977). Lectin and human blood group determinants of *Schistosoma mansoni*: alteration following *in vitro* transformation of miracidium to mother sporocyst. *Journal of Parasitology* **63**, 818–824.

YOSHINO, T. P., COUSTAU, C., MODAT, S. & CASTILLO, M. (1999). The *Biomphalaria glabrata* embryonic molluscan cell line: establishment of an *in vitro* cellular model for the study of snail host–parasite interactions. *Malacologia* **41**, 331–343.

YOSHINO, T. P. & LAURSEN, J. R. (1995). Production of *Schistosoma mansoni* daughter sporocysts from mother sporocysts maintained in synxenic culture with *Biomphalaria glabrata* embryonic (Bge) cells. *Journal of Parasitology* **81**, 714–722.

YOSHINO, T. P. & LODES, M. J. (1988). Secretory protein biosynthesis in snail hemocytes: *in vitro* modulation by larval schistosome excretory-secretory products. *Journal of Parasitology* **74**, 538–547.

YOSHINO, T. P. & VASTA, G. R. (1996). Parasite–invertebrate host immune interactions. *Advances in Comparative and Environmental Physiology* **24**, 125–167.

ZHONG, C., SKELLY, P. J., LEAFFER, D., COHN, R. G., CAULFIELD, J. P. & SHOEMAKER, C. B. (1995). Immunolocalization of a *Schistosoma mansoni* facilitated diffusion glucose transporter to the basal, but not the apical, membranes of the surface syncytium. *Parasitology* **110**, 383–394.

Mechanisms of molluscan host resistance and of parasite strategies for survival

C. J. BAYNE*, U. K. HAHN *and* R. C. BENDER

Department of Zoology, Oregon State University, Corvallis, Oregon 97331, USA

SUMMARY

In parallel with massive research efforts in human schistosomiasis over the past 30 years, persistent efforts have been made to understand the basis for compatibility and incompatibility in molluscan schistosomiasis. Snail plasma contains molecules that are toxic to trematodes, but these seem to kill only species that never parasitize the mollusc used as the source of plasma. A sporocyst will be killed actively by haemocytes alone if they are from a snail that is resistant to the trematode. Oxygen-dependent killing mechanisms play a major role. Enzymes such as NADPH oxidase, superoxide dismutase, myeloperoxidase and nitric oxide synthase are critical components of the putative killing pathways. Metabolic intermediates such as hydrogen peroxide and nitric oxide appear to be more important against trematodes than the shorter-lived intermediates that are more important in anti-microbial defences. Products secreted by trematode larvae influence the physiology of snail haemocytes, implying active counter-defences mounted by the parasite, but these remain largely unexplored. A possible molecular basis for the susceptibility/resistance dichotomy in molluscan schistosomiasis is suggested to be deficient forms of enzymes in the respiratory burst pathway, and a selective disadvantage for schistosome resistance is an integral component of this model.

Key words: *Schistosoma*, *Biomphalaria*, compatibility, susceptibility, resistance, oxygen-dependent killing, haemocyte, cytotoxicity, myeloperoxidase, hydrogen peroxide, nitric oxide.

INTRODUCTION

What an exciting time it is to be re-examining aspects of the relationships between flukes and snails. In the thirty years since Wright's completion of his book (*Flukes and Snails*, Wright, 1971*a*), a great deal has been learned about trematodes and their intermediate hosts. Yet, with local exceptions, progress in reducing the numbers of infected people and livestock has not been impressive. New anti-trematode drugs have become available, but remain unaffordable to affected populations and do not provide protection from re-infection. For many individuals in endemic areas, vaccine development remains the leading hope; and that (hope) is the key word.

Many of those who contributed to this volume surely owe debts of gratitude to Wright. One of us (C. J. B.) began work in this field in 1971, and one's perspective on things then was constructively influenced by Wright's views on, for example, 'Susceptibility–Infectivity–Compatibility'.

Notwithstanding the success of Wright's book in provoking constructive and productive mind-sets and encouraging further research, the limits of knowledge at the time forced him to be satisfied to pose probing questions and suggest likely answers. At a time when the information was simply not there to proceed otherwise, Wright's way of dealing with immunity, for example, was to posit realistic

scenarios, such as paraphrased here: 'That the evasion of the innate host response is an active process on the part of the parasites is emphasized by the fact that immediately after they die they are attacked and encapsulated prior to removal.'

HOST RESISTANCE

The heading above really is an oxymoron and makes sense only in the context of the kind of host–parasite relationships exemplified by snails and trematodes. In general, when a trematode enters the body of a mollusc, it enters a hostile world. Unless the genetic profiles of the mollusc and the trematode are precisely concordant in some critical attributes, the parasite is killed. Thus one thinks of *individuals* in a population of the host species as being susceptible or resistant to *individuals* of the parasite species (Basch, 1975). Death, which awaits miracidia/sporocysts entering any non-susceptible individual, may be mediated by either humoral or cell components of the snail, or both. There is, indeed, a multiplicity of factors that contribute to the outcome of each host-parasite encounter, and this is well illustrated by the notion of encounter filters and compatibility filters, eloquently developed by Combes (1991).

Humoral and cellular components of host defences

It is likely that plasma molecules comprise one type of barrier. However, this may be operative only in cases of snail species that are resistant to all individuals of a given trematode species. This, of

* Corresponding author.

Parasitology (2001), **123**, S159–S167. © 2001 Cambridge University Press
DOI: 10.1017/S0031182001008137

course, includes the majority of potential host-parasite pairs. This conjecture is based on recent data (Sapp & Loker, 2000 a) implicating snail plasma factors as the agents of rapid (minutes to a few hours) killing of trematodes *in vitro*. Such toxic effects of plasma appear limited to trematodes that penetrate snails that are only distantly related to their normal host species. Thus flukes that parasitize planorbids are killed in plasma from lymnaeid snails, and *vice versa*. Earlier work (Bayne, Buckley & DeWan, 1980; Granath & Yoshino, 1984) had suggested that the presence of plasma from resistant planorbids enables susceptible haemocytes to kill *Schistosoma mansoni* sporocysts *in vitro*. However, this was not a potent effect and required long culture times. It later became apparent that the extra-cellular haemoglobin in *Biomphalaria glabrata* plasma will become oxidized and consequently generate hydrogen peroxide when exposed to normoxic conditions (Bender *et al.* 2001). Such events probably account for some, though perhaps not all, of the toxicity observed in the earlier studies.

While plasma toxicity appears not to be a problem for flukes entering individuals of their host or closely related species (Sapp & Loker, 2000 a), such snails are fully immunocompetent and deal normally with other immune challenges. Susceptibility is specific to a narrow parasite genotype. How, then, is survival achieved when a trematode is confronted with the impressive internal defensive armoury of an individual compatible snail (Bayne & Yoshino, 1989; van der Knaap & Loker, 1990; Loker, 1994)? Recent studies (Hahn, Bender & Bayne, 2001 a, b) have made it possible to construct an appealing model of how compatible and incompatible phenotypes may come about (see below), and they have potentially brought us to the brink of grasping the subtleties of at least some successful trematode–snail parasitisms.

The dominant effectors of sporocyst killing in resistant individuals of a host species are the haemocytes. In general, pulmonates seem to have two types of haemocyte, granulocytes and hyalinocytes. Definitive experiments have not been done to exclude the possibility of a cytotoxic role for the hyalinocytes, but ultrastructural evidence garnered in several laboratories implicates the granulocytes as the effector cells. When sporocysts and haemolymph are allowed to interact *in vivo* or *in vitro*, cells contacting the parasite appear to be granulocytes (Bayne *et al.* 1980; Loker *et al.* 1982). These data are consistent with those obtained in different host-parasite associations. Even when sporocysts of *S. mansoni* come into contact *in vitro* with haemocytes from resistant strains of *B. glabrata* in the absence of snail plasma (it is replaced with a culture medium), the parasites are killed. This killing is effected as fast as in whole haemolymph, and (unpublished observations) requires only sufficient cells to adhere to less

than half of the sporocyst's surface. As complete envelopment of the parasite in haemocyte encapsulations is not needed for a kill, it is clear that killing is an active process and not due simply to suffocation or starvation.

Does this leave any room for a plasma role in parasite killing in host–parasite pairs in which some individuals are compatible? Evidence from *in vivo* experiments is equivocal. Injections of plasma from resistant strains were reported to decrease infection frequencies in normally susceptible individual *B. glabrata* (Granath & Yoshino, 1984). Furthermore, when susceptible individuals were implanted with haematopoietic tissue from resistant individuals, in which the implants survived for weeks, the implanted individuals were less susceptible to miracidial challenge (Sullivan & Spence, 1994; 1999). In such snails, implanted cells may be able to synthesize and secrete plasma components, but the implanted tissue almost certainly gives rise to haemocytes with the resistant phenotype. The favoured interpretation is that resistance has been adoptively transferred and is mediated by the new cells.

Regardless, in completely plasma-free conditions, haemocytes of the 13-16-R1 strain of *B. glabrata* kill sporocysts of the PR1 strain of *S. mansoni* without reduced efficacy (Hahn *et al.* 2001 a).

The defensive armoury of the molluscan haemocyte

In the 1950s, 1960s and 1970s, both light and electron microscopy had been used to learn much about the organelles within molluscan haemocytes, and to observe the structures of cells responding to parasites and other foreign agents (the work of M. R. Tripp, T. C. Cheng and others, reviewed by Harris, 1975, and Bayne, 1983). Most of what was learned was, in retrospect, unsurprising. Haemocytes that circulate as free cells *in vivo* are typical phagocytic leukocytes: they will aggregate in response to trauma, phagocytose small foreign particles (bacteria, yeast), and encapsulate large ones (trematode larvae). By means of cytochemical stains, it was found that these cells contain lysosomes and these fuse with phagosomes releasing their digestive enzymes into the phagolysosomal vacuole whose membrane's proton pumps ensure a transient decrease in pH (Kroschinski & Renwrantz, 1988). Into the plasma, haemocytes may release enzymes. In a reciprocal manner, haemocyte behaviours are facilitated by constitutive plasma molecules, such as opsonizing lectins that bind to yeast thus targeting them for phagocytosis (Fryer, Hull & Bayne, 1989). Evidence of chemotactic behaviour proved elusive, but was eventually found (Kumazawa & Shimoji, 1991).

A little of what was learned in those early studies was perplexing and remains unexplained. For

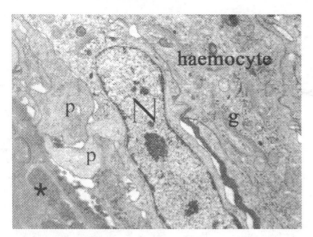

Fig. 1. Encapsulated and killed *Schistosoma mansoni* sporocyst (dark region of image, bottom left, *) within a group of haemocytes from the 10-R2 strain of *Biomphalaria glabrata*. The encapsulation occurred in a juvenile snail, and was fixed 24 h after it was exposed to miracidia. The cytoplasm of the haemocyte to the right of the nucleus (N) contains numerous organelles, including a Golgi apparatus (g). Close to the sporocyst surface, two pseudopods (p) display a finely granular cytoplasm that is free of organelles. (E. S. Loker & C. J. Bayne, unpublished.)

example, a glycocalyx on the haemocyte surface was found to be redistributed to the 'far' side of the *Bulinus guernei* haemocyte when the cells came close to a *S. haematobium* sporocyst (Krupa, Lewis & Del Vecchio, 1977). Intracellularly, the cytoplasm of *B. glabrata* haemocytes in contact with *S. mansoni* sporocysts was found to be finely granular and organelles were absent (Fig. 1; see also Loker *et al.* 1982). The meaning of these structural observations remains unexplored. Among remaining mysteries is this: *in vivo* encapsulations, naturally, grow to an effective size that is appropriate to the task; but the nature of the graded signal that terminates further cell accumulation remains unknown. And in multi-layered capsules, any differences in the functional roles of cells in actual contact with the target and those without contact remain unclear.

Building on earlier reports of the existence of lectins that serve as cell surface receptors (pattern recognition molecules) for carbohydrate structures (pathogen-associated molecular patterns, Janeway, 1989), evidence has recently been obtained for physiological functions of surface lectins on snail haemocytes (Hahn, Bender & Bayne, 2000). Development of this evidence exploited the respiratory burst potential of snail haemocytes (see below). Eight sugars, six of which are known to occur on the *S. mansoni* sporocyst surface, were tested for their ability to stimulate this response. No free sugars elicited the respiratory burst. However, *B. glabrata* haemocytes responded by producing reactive oxygen species when stimulated with BSA-galactose, BSA-mannose and BSA-fucose. In contrast, BSA conju-

gated with *N*-acetyl glucosamine, *N*-acetyl-galactosamine, lactose, glucose or melibiose was without effect. The responses were shown to be dependent on NADPH-oxidase. Since they were similar in cells from a susceptible and a resistant strain, it appears that the cells did not differ with respect to their lectin-type surface receptors. These results (1) imply that haemocytes from both strains are equally capable of producing reactive oxygen species, (2) imply that receptor activation requires a patterned presentation of multiple ligands, and (3) suggest that the susceptible-resistant dichotomy is likely due to something other than different recognition capabilities, though the possibility of strain-specific discrimination of more complex glycan structures has not been excluded.

The more molecular work that has typified recent research has revealed evidence for antimicrobial peptides in molluscan haemocytes (Charlet *et al.* 1996; Mitta, vandenBulckle & Roch, 2000; Mitta *et al.* 2000), though no evidence has been reported for toxicity of these peptides to metazoan parasites.

THE HOST–PARASITE MODEL

We have taken advantage of strains of *B. glabrata* that were developed by C. S. Richards (Richards & Merritt, 1972), and of the PR1 strain of *S. mansoni*. Both the snail and the trematode have been maintained in the laboratory for decades. Such models have attracted the criticism that what occurs in them may no longer represent events in the real world. Even though these inbred strains may not represent the full array of natural scenarios, they clearly lead to normally infective parasites: cercariae shed from susceptible individuals are fully infective for mammalian hosts. Furthermore, findings obtained through studies on inbred laboratory strains serve to do no more than direct ideas as to what factors may operate in the real world.

What has been repeatedly confirmed by subsequent research is that the haemocytes from individual snails that are resistant to PR1 *S. mansoni* are effective sporocyst killers on their own (Bayne *et al.* 1980; Hahn *et al.* 2001 *a*). A potential mechanism of this cell-mediated killing became apparent in studies involving the production of reactive oxygen species (ROS) by haemocytes. Dikkeboom and co-workers first demonstrated that gastropod haemocytes produce ROS (Dikkeboom *et al.* 1988 *a*), and that they do so in response to encounters with trematodes (Dikkeboom *et al.* 1988 *b*). Simultaneously, Shozawa, Suto & Kumada (1989) reported that *B. glabrata* haemocytes produce ROS. A direct involvement of ROS in haemocyte-mediated killing of trematodes was first demonstrated by Adema *et al.* (1994 *b*), using sporocysts of the avian schistosome *Trichobilharzia ocellata* and haemocytes

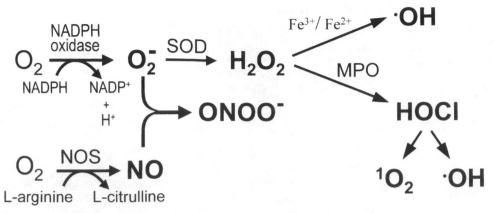

Fig. 2. Simplified schematic of the pathways that generate the major reactive oxygen and nitrogen species in phagocytic cells. The membrane-bound NADPH oxidase complex converts molecular oxygen (O_2) into superoxide (O_2^-), which can spontaneously dismutate to hydrogen peroxide (H_2O_2). However, this process is greatly accelerated by the enzyme superoxide dismutase (SOD). In the presence of chloride ion, H_2O_2 can be converted to hypochlorous acid (HOCl) by the enzyme myeloperoxidase (MPO). H_2O_2 can also react with iron (Fe^{3+}) to yield hydroxyl radical ($\cdot$OH; Fenton reaction). HOCl can potentially produce both $\cdot$OH and singlet oxygen ($^1O_2$). The enzyme nitric oxide synthase (NOS) utilizes O_2 and arginine to produce nitric oxide (NO), which can react with O_2^- to form the highly reactive peroxynitrite (ONOO$^-$).

from its intermediate host, *Lymnaea stagnalis*, as well as *S. mansoni* and *B. glabrata*. This study showed that when the generation of ROS by haemocytes was prevented, the haemocytes' ability to kill the parasite was compromised. Together, these studies paved the way for a more detailed examination of the contribution of ROS to parasite killing.

The burst of oxygen consumption (i.e. respiratory burst) by phagocytic cells and the consequent production of ROS is the result of activation and assembly of an NADPH oxidase complex in the phagocyte plasma membrane (reviewed by Babior, 1999). This enzyme complex catalyzes the production of superoxide (O_2^-), which serves as the first radical in the chain of production of additional reactive oxidants: hydrogen peroxide (H_2O_2), hydroxyl radical ($\cdot$OH), hypochlorous acid (HOCl), and singlet oxygen ($^1O_2$) (Fig. 2; reviewed by Hampton, Kettle & Winterbourn, 1998). Additionally, most phagocytic cell types possess an inducible nitric oxide synthase (iNOS), which generates nitric oxide (NO) from molecular oxygen and arginine. Although NO has cytotoxic activity on its own, it rapidly reacts with O_2^- to form the much more reactive oxidant, peroxynitrite (ONOO$^-$). In mammalian neutrophils, most of the O_2^- produced is converted to H_2O_2 (Makino *et al.* 1986) and most of the H_2O_2 is consumed by myeloperoxidase (MPO), an enzyme that uses H_2O_2 and chloride ion to produce HOCl, the most potent antibiotic produced by defence cells. Like their mammalian counterparts, molluscan haemocytes generate HOCl (Schlenk, Martinez & Livingstone, 1991; Torreilles, Guerin & Roch, 1997) and NO (Conte & Ottaviani *et al.* 1995; Torreilles & Guerin, 1999; Arumugam *et al.* 2000). Thus, haemocytes have the potential to produce both highly reactive (O_2^-, $\cdot$OH, $^1O_2$, HOCl and ONOO$^-$) and

relatively long-lived (NO and H_2O_2) oxidants. Each oxidant differs in its reactive properties (e.g. ability to cause DNA strand breaks, lipid peroxidation, enzyme inactivation, etc.). The precise roles of particular oxidants in pathogen killing remain unclear. Studies with MPO-deficient mice indicate the relative cytotoxic roles of specific oxidants are dependent on the bacteria or fungi involved (Aratani *et al.* 2000). Thus, it seems likely that pathogens employ a variety of anti-oxidant strategies and effective killing requires the production of the right oxidant at the right time and place. In light of these facts, we sought to determine the specific reactive oxidant pathways that allow resistant *B. glabrata* haemocytes to kill sporocysts of *S. mansoni*.

Utilizing an *in vitro* assay, killing of *S. mansoni* (PR 1) sporocysts by haemocytes from resistant (13-16-R1) *B. glabrata* was examined over a 48 h time-course. To determine which ROS is involved, specific oxidant scavengers or enzyme inhibitors were used (Hahn *et al.* 2001 *a*, *b*). Addition of superoxide dismutase (SOD), which scavenges O_2^-, and addition of hypotaurine, which scavenges hypochlorous acid (HOCl) and hydroxyl radicals ($\cdot$OH), did not protect sporocysts from being killed. In addition to eliminating HOCl and $\cdot$OH as direct mediators of sporocyst killing, these results exclude a cytotoxic role for singlet oxygen ($^1O_2$), since it is a product of HOCl. Addition of catalase, which converts H_2O_2 to water and molecular oxygen, did protect sporocysts significantly, indicating that H_2O_2 is the reactive oxygen species mainly responsible for sporocyst killing. These results are in contrast with results from mammalian studies which show that most of the H_2O_2 generated by neutrophils is consumed by the rapidly acting MPO and converted into HOCl (Kettle & Winterbourn, 1997). Although

HOCl is considered to be a more potent bactericidal agent than H_2O_2 (Klebanoff, 1968), *S. mansoni* sporocysts are less susceptible to HOCl than to H_2O_2: in an *in vitro* toxicity assay, 300 μM HOCl was required to achieve 100% sporocyst mortality compared to 150 μM H_2O_2 (Hahn *et al.* 2001 *a*). The possible involvement of NO in haemocyte-mediated killing of sporocysts was also examined. Addition of N_ω-nitro-L-arginine methylester (L-NAME), which inhibits both inducible and constitutive NOS activity, significantly reduced sporocyst killing. However, the presence of uric acid, which selectively scavenges ONOO$^-$, but does not react with NO, did not affect sporocyst killing.

In summary, these results indicate that H_2O_2 and NO are directly involved in haemocyte-mediated killing of *S. mansoni* sporocysts. Furthermore, these findings also suggest that HOCl, O_2^-, $^1O_2$, and ONOO$^-$ do not play a direct role in haemocyte-mediated killing. Since H_2O_2 and NO could be considered intermediates of the more toxic oxidants, HOCl, $\cdot$OH and ONOO$^-$, these results came as somewhat of a surprise. It is possible that the highly reactive nature of HOCl, $\cdot$OH and ONOO$^-$ limits their activity to the sporocyst tegument, which may be able to sustain substantial damage and recover due to its rapid turnover or its repair mechanisms. The less reactive nature of H_2O_2 and NO, coupled with their ability to cross membranes, may allow them to penetrate the tegument and act on targets within the sporocyst.

MECHANISTIC MODEL OF RESISTANCE/SUSCEPTIBILITY

The knowledge gained from these mechanistic studies provides the means to formulate hypotheses that address specific biochemical differences that may determine resistance/susceptibility. Since the biochemical pathways that produce H_2O_2 and NO appear to be directly involved in sporocyst killing, several hypotheses suggest themselves. Each addresses the possibility that susceptibility/resistance is mediated by a difference in one of these pathways.

Fig. 3 (A–C) illustrates potential 'enzyme deficiencies' in *B. glabrata* haemocytes that could promote susceptibility to *S. mansoni*. Since the production of H_2O_2 by haemocytes is essentially determined by the generation of O_2^- and its subsequent dismutation to H_2O_2, any anomalies involving either the NADPH oxidase complex or superoxide dismutase could potentially result in a decreased killing capacity (i.e. susceptibility). Similarly, anomalies involving NOS expression or activity could also lessen haemocyte-mediated killing. Regarding the first two scenarios, the evidence suggests that resistant and susceptible haemocytes

do not differ in H_2O_2 production when stimulated with zymosan (Dikkeboom *et al.* 1988 *a*) or with glycosylated albumin (Hahn *et al.* 2000). Attempts to compare H_2O_2 production by resistant and susceptible haemocytes after stimulation by live sporocysts have been inconclusive, due to high background as a result of sporocyst products interacting with the fluorescent probe. Although the possibility of different responses of resistant and susceptible haemocytes to *S. mansoni* sporocysts cannot be eliminated, all available evidence suggests that, in terms of H_2O_2 generation, the haemocytes from both strains are equally competent. The same case cannot be made for NO generation (Fig. 3 C). It remains to be determined if resistant and susceptible haemocytes differ in their ability to produce NO.

Fig. 3 D illustrates a potential enzyme deficiency in *B. glabrata* haemocytes that could promote resistance to *S. mansoni*. Two of our findings provide the basis for this scenario: haemocytes utilize H_2O_2 (and not HOCl) to kill sporocysts, and HOCl is significantly less toxic to *S. mansoni* sporocysts than H_2O_2. Thus, in terms of toxicity to *S. mansoni* sporocysts, haemocyte MPO would essentially function as a detoxifying enzyme. This presents the possibility that MPO activity prevents sporocyst killing in susceptible haemocytes by lowering the net quantity of H_2O_2 available for cytotoxic activity. A deficiency in this enzyme would allow accumulation of higher levels of H_2O_2, and thus promote resistance. This scenario is intriguing for two reasons. In human populations, MPO deficiency is the most common congenital neutrophil defect (Nauseef, 1990), with an estimated incidence of 1 in 2000–4000 in the United States (Parry *et al.* 1981). Even more interesting is the clinical significance of MPO deficiency. Most individuals with MPO deficiency are healthy and do not suffer from recurrent bacterial infections. However, increased susceptibility to fungal infections has been reported (Parry *et al.* 1981), and studies with MPO-deficient mice support the notion that the role of MPO in oxidative killing varies from critical to inconsequential depending on the pathogen (Aratani *et al.* 1999, 2000). Thus, in our hypothetical case of MPO-deficient snails, an increased ability to kill *S. mansoni* may come at the cost of an increased susceptibility to other pathogens (particularly fungus). The idea of 'resistance at a cost' in regard to schistosomes is not new. It stems from the perplexing observations that the prevalence of schistosome-susceptible snails is often quite high in both natural and laboratory populations of *Biomphalaria* sp. (Upatham, 1972; Fransden, 1979; Minchella & LoVerde, 1983). Wright suggested that resistant snails are not predominant because resistance is associated with a disadvantageous character or physiological defect (Wright, 1971 *b*). We concur with Wright and postulate that this defect comes in the form of MPO deficiency; the dis-

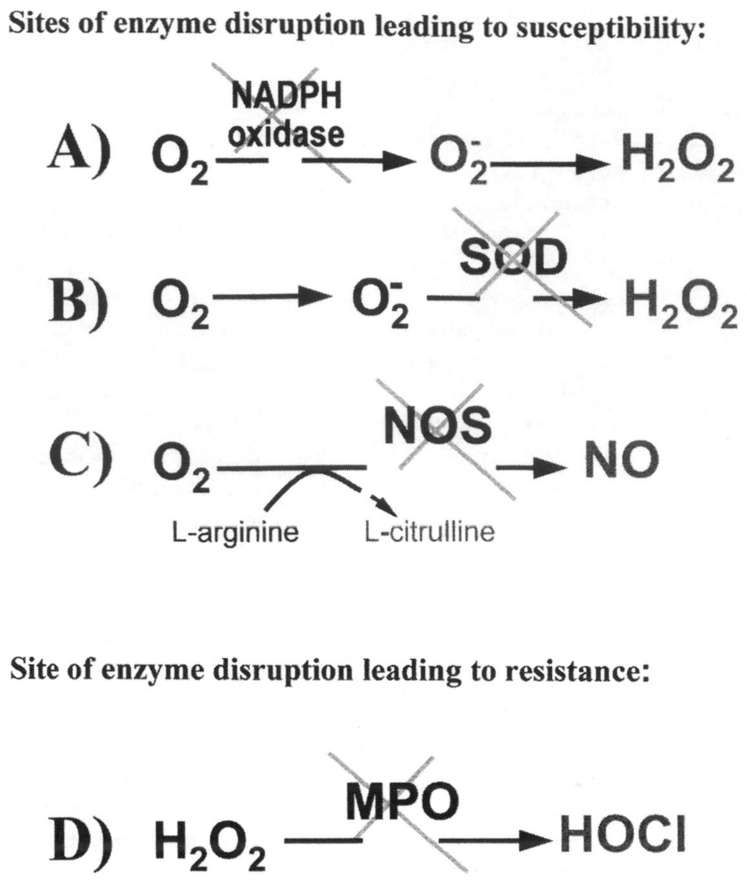

Sites of enzyme disruption leading to susceptibility:

A) $O_2 \xrightarrow{\text{NADPH oxidase}} O_2^- \longrightarrow H_2O_2$

B) $O_2 \longrightarrow O_2^- \xrightarrow{\text{SOD}} H_2O_2$

C) $O_2 \xrightarrow{\substack{\text{NOS} \\ \text{L-arginine} \quad \text{L-citrulline}}} NO$

Site of enzyme disruption leading to resistance:

D) $H_2O_2 \xrightarrow{\text{MPO}} HOCl$

Fig. 3. Hypothetical alterations in biochemical pathways known to be involved in *Schistosoma mansoni* killing by *Biomphalaria glabrata* haemocytes. Defects that would promote susceptibility to *S. mansoni* include decreased expression or reduced enzyme activity of: (A) NADPH oxidase, (B) superoxide dismutase (SOD), or (C) nitric oxide synthase (NOS). (D) Defects in the activity or expression of myeloperoxidase (MPO) would potentially promote resistance to *S. mansoni* by increasing the net production of hydrogen peroxide (H_2O_2).

advantage lies in an increased susceptibility to certain pathogens. If so, then this constitutes a mirror image of the parasite-driven retention of sickle cell genes in human populations that are sympatric with *Plasmodium* spp. In both cases, a mutation that would be detrimental in the absence of a parasite is retained because it provides survival value in parasitized individuals.

WHAT IS THE ROLE OF THE PARASITE IN AVOIDING DESTRUCTION?

First, flukes that parasitize planorbids must be presumed to be insensitive to whatever molecules kill lymnaeid parasites (Sapp & Loker, 2000*a*). However, until such a snail molecule is isolated it will remain unclear what strategy or mechanism is used by the flukes.

A question remains: does a sporocyst play an active role in avoiding destruction by the oxygen-dependent cytotoxic armoury of a snail? Possibilities include strategies that (a) avoid eliciting the respiratory burst of the host's haemocytes, (b) inhibit the host's systems for generating ROS, (c) rapidly

scavenge and detoxify products of the respiratory burst, or even (d) drive away the host haemocytes. Any or all of these may be used, and empirical data suggest that each may play a role. For example, relative to (a) avoidance, sporocysts mimic antigens expressed on host cells (Bayne & Stevens, 1983; Yoshino & Bayne, 1983) and acquire host plasma antigens (Bayne, Loker & Yui, 1986). Both of these strategies may reduce the apparent 'foreignness' of the parasite's surface (Damian, 1987), making it more difficult for the host to recognize the parasite. Relative to (b) inhibition, products secreted by schistosomes as they transform from miracidia into sporocysts (their excretory-secretory products; Lodes, Connors & Yoshino, 1991) alter the metabolism of host haemocytes (Loker, Bayne & Yui, 1986; Loker, Cimino & Hertel, 1992; Coustau & Yoshino, 1994), reducing both motility (Lodes & Yoshino, 1990) and phagocytosis (Connors & Yoshino, 1990), altering the profile of secreted molecules (Yoshino & Lodes, 1988) and modulating their production of ROS (Connors, Lodes & Yoshino, 1991). Relative to (c) scavenging and detoxifying, schistosomes express two enzymes capable of scavenging ROS – superoxide dismutase and glutathione peroxidase (Mei & LoVerde, 1997), and release proteases that degrade snail proteins (Yoshino *et al.* 1993). Finally, relative to (d) haemocyte repulsion, haemocytes that are attached and spread out on glass will round up if an echinostome sporocyst comes to reside nearby (Adema *et al.* 1994*a*). This ability appears non-functional for *S. mansoni* in *B. glabrata*, however, since no such effect on haemocytes was evident when these were evaluated *in vitro* (Sapp & Loker, 2000*b*). Furthermore, when sporocysts are brought into close contact with susceptible strain haemocytes *in vitro*, the parasite can become enveloped in a mass of haemocytes yet suffers no harm (Boehmler, Fryer & Bayne, 1996; Hahn *et al.* 2001*a*). It is clear that sporocysts smothered in haemocytes will survive well so long as the haemocytes come from susceptible snails. One is forced, then, to consider survival strategies other than repulsion of haemocytes.

One further strategy that parasites could use to protect themselves from host aggression exploits putative inhibitory receptors on host leukocytes. The notion draws on recent evidence that the failure of certain cytotoxic cells to respond aggressively to host (self) cells is due the expression, on the host cells, of ligands for receptors that switch off the aggressive machinery of the killer cell (Ljunggren & Karre, 1990). If such mechanisms exist at this level of phylogeny (as they surely must), then perhaps host antigens mimicked by a parasite may include such ligands. This idea has been examined in the *S. mansoni – B. glabrata* parasitism (Loker & Bayne, 2001), but no supporting evidence was found, perhaps because of reliance on reagents (antibodies) developed for use with mammalian systems.

CONCLUDING COMMENTS

Practically a decade into the genomic era, concrete hypotheses are emerging to account for the outcome when a fluke larva enters the haemocoel of a snail. Compared to 1971 when Wright published *Flukes and Snails*, knowledge of the mechanisms used by snails to attack parasites has expanded particularly in the area of oxygen-dependent mechanisms. Furthermore, knowledge of mechanisms used by parasites to survive in immunologically hostile body compartments has been extended to include enzymes that scavenge reactive oxygen species, and the extent of the cross-talk between the two species is now more fully appreciated. The road to understanding continues to reward the persistent investigator. In view of the extent of human suffering that results from flukes and their kin, it can be confidently predicted that progress in this area will continue. Perhaps the understanding will be complete before yet another generation of parasitologists complete their research careers focused on flukes and snails.

ACKNOWLEDGEMENTS

In addition to the benefits gained from the mind-setting ideas expressed by C. J. Wright (1971*a*), additional major influences on our work were the many papers of K. J. Lie and Donald Heyneman in San Francisco, Paul Basch in Stanford, and Chick Richards at the US NIH: no historical account of this subject would be worthy without paying attention to their publications. J. R. Allen, D. Barnes, C. A. Boswell, P. M. Buckley, J. B. Burch, R. Dikkeboom, S. E. Fryer, E. S. Loker, E. A. Meuleman, W. E. Noonan, A. Owczarzak, L. Renwrantz, H. H. Stibbs, W. P. W. van der Knaap, M. A. Yui and U. E. Zelck: these individuals, in addition to others mentioned in the text, have contributed significantly to the development of ideas explored in this paper, and we thank them for this. Financial support has been from NIH (AI-16137), The John D. and Catherine T. MacArthur Foundation, and the World Health Organization.

REFERENCES

ADEMA, C. M., ARGUELLO, D. F., STRICKER, S. A. & LOKER, E. S. (1994*a*). A time-lapse study of interactions between *Echinostoma paraensei* intramolluscan larval stages and adherent hemocytes from *Biomphalaria glabrata* and *Helix pomatia*. *Journal of Parasitology* **80**, 719–727.

ADEMA, C. M., DEUTOKOM-MULDER, E. C. V., VAN DER KNAAP, W. P. W. & SMINIA, T. (1994*b*). Schistosomicidal activities of *Lymnaea stagnalis* haemocytes: the role of oxygen radicals. *Parasitology* **109**, 479–485.

ARATANI, Y., KOYAMA, H., NYUI, S.-I., SUZUKI, K., KURA, F. & MAEDA, N. (1999). Severe impairment in early host defense against *Candida albicans* in mice deficient in myeloperoxidase. *Infection and Immunity* **67**, 1828–1836.

ARATANI, Y., KURA, F., WATANABE, H., AKAGAWA, H., TAKANO, Y., SUZUKI, K., MAEDA, N. & KOYAMA, H. (2000). Differential host susceptibility to pulmonary infections with bacteria and fungi in mice deficient in myeloperoxidase. *Journal of Infectious Diseases* **182**, 1276–1279.

ARUMUGAM, M., ROMESTAND, B., TORREILLES, J. & ROCH, P. (2000). *In vitro* production of superoxide and nitric oxide (as nitrite and nitrate) by *Mytilus galloprovincialis* haemocytes upon incubation with PMA or laminarin or during yeast phagocytosis. *European Journal of Cell Biology* **79**, 513–519.

BABIOR, B. M. (1999). NADPH oxidase: an update. *Blood* **93**, 1464–1476.

BASCH, P. F. (1975). An interpretation of snail-trematode infection rates: specificity based on concordance of compatible phenotypes. *International Journal for Parasitology* **5**, 449–452.

BAYNE, C. J. (1983). Molluscan Immunobiology. In *The Mollusca. Vol. 5. Physiology. Pt. 2.* (ed. Saleuddin A. S. M. & Wilbur, K. M.). pp. 407–486. San Diego: Academic Press.

BAYNE, C. J., BUCKLEY, P. M. & DEWAN, P. C. (1980). Macrophage-like hemocytes of resistant *Biomphalaria glabrata* are cytotoxic for sporocysts of *Schistosoma mansoni in vitro*. *Journal of Parasitology* **66**, 413–419.

BAYNE, C. J., LOKER, E. S. & YUI, M. A. (1986). Interactions between the plasma proteins of *Biomphalaria glabrata* (Gastropoda) and the sporocyst tegument of

Schistosoma mansoni (Trematoda). *Parasitology* **92**, 653–664.

BAYNE, C. J. & STEVENS, J. A. (1983). *Schistosoma mansoni* and *Biomphalaria glabrata* share epitopes: antibodies to sporocysts bind host snail hemocytes. *Journal of Invertebrate Pathology* **42**, 221–223.

BAYNE, C. J. & YOSHINO, T. P. (1989). Determinants of compatibility in mollusc–trematode parasitism. *American Zoologist* **29**, 399–407.

BENDER, R. C., BIXLER, L. M., LERNER, J. P. & BAYNE, C. J. (2001). Oxidative stress on *Schistosoma mansoni* sporocysts in culture: hemoglobin as a toxic component of host plasma *in vitro*. *Journal of Parasitology*. (In press).

BOEHMLER, A., FRYER, S. E. & BAYNE, C. J. (1996). Killing kinetics of *Schistosoma mansoni* sporocysts by *Biomphalaria glabrata* hemolymph *in vitro*: alteration of hemocyte behavior after poly-L-lysine treatment of plastic and the kinetics of killing by different host strains. *Journal of Parasitology* **82**, 332–335.

CHARLET, M., CHERNYSH, S., PHILIPPE, H., HETRU, C., HOFFMANN, J. A. & BULET, P. (1996). Innate immunity. Isolation of several cysteine-rich antimicrobial peptides from the blood of a mollusc, *Mytilus edulis*. *Journal of Biological Chemistry* **6**, 21808–21813.

COMBES, C. (1991). Evolution of parasite life cycles. In *Parasite–Host Associations. Coexistence or Conflict?* (ed. Toft, C. A., Aeschlimann, A. & Bolis, L.), pp. 62–82. Oxford: Oxford University Press.

CONNORS, V. A., LODES, M. J. & YOSHINO, T. P. (1991). Identification of a *Schistosoma mansoni* sporocyst excretory-secretory antioxidant molecule and its effect on superoxide production by *Biomphalaria glabrata* hemocytes. *Journal of Invertebrate Pathology* **58**, 387–395.

CONNORS, V. A. & YOSHINO, T. P. (1990). *In vitro* effect of larval *Schistosoma mansoni* excretory-secretory products on phagocytosis-stimulated superoxide production in hemocytes of *Biomphalaria glabrata*. *Journal of Parasitology* **76**, 895–902.

CONTE, A. & OTTAVIANI, E. (1995). Nitric oxide synthase activity in molluscan hemocytes. *FEBS Letters* **365**, 120–124.

COUSTAU, C. & YOSHINO, T. P. (1994). *Schistosoma mansoni*: modulation of haemocyte surface polypeptides detected in individual snails, *Biomphalaria glabrata*, following larval exposure. *Experimental Parasitology* **79**, 1–10.

DAMIAN, R. T. (1987). Molecular mimicry revisited. *Parasitology Today* **3**, 263–266.

DIKKEBOOM, R., BAYNE, C. J., VAN DER KNAAP, W. P. W. & TIJNAGEL, J. M. (1988b). Possible role of reactive forms of oxygen in *in vitro* killing of *Schistosoma mansoni* sporocysts by haemocytes of *Lymnaea stagnalis*. *Parasitology Research* **75**, 148–154.

DIKKEBOOM, R, VAN DER KNAAP, W. P. W., VAN DER BOVENKAMP, W. V. D., TIGNAGEL, J. M. G. H. & BAYNE, C. J. (1988a). The production of toxic oxygen metabolites by hemocytes of different snail species. *Developmental and Comparative Immunology* **12**, 509–520.

FRANSDEN, F. (1979). A study of the relationship between *Schistosoma* and their intermediate hosts. III. The

genus *Biomphalaria* and *Schistosoma mansoni* from Egypt, Kenya, Sudan, West Indies (St. Lucia) and Zaire (two different strains: Katanga and Kinshasa). *Journal of Helminthology* **53**, 321–348.

FRYER, S. E., HULL, C. J. & BAYNE, C. J. (1989). Phagocytosis of yeast by *Biomphalaria glabrata*: carbohydrate specificity of hemocyte receptors and a plasma opsonin. *Developmental and Comparative Immunology* **13**, 9–16.

GRANATH, W. O. & YOSHINO, T. P. (1984). *Schistosoma mansoni*: passive transfer of resistance by serum in the vector snail, *Biomphalaria glabrata*. *Experimental Parasitology* **58**, 188–193.

HAHN, U. K., BENDER, R. C. & BAYNE, C. J. (2000). Production of reactive oxygen species by hemocytes of *Biomphalaria glabrata*: carbohydrate-specific stimulation. *Developmental and Comparative Immunology* **24**, 531–541.

HAHN, U. K., BENDER, R. C. & BAYNE, C. J. (2001a). Killing of *Schistosoma mansoni* sporocysts by hemocytes from resistant *Biomphalaria glabrata*: role of reactive oxygen species. *Journal of Parasitology* **87**, 292–299.

HAHN, U. K., BENDER, R. C. & BAYNE, C. J. (2001b). Role of nitric oxide in killing of *Schistosoma mansoni* sporocysts by hemocytes from resistant *Biomphalaria glabrata*. *Journal of Parasitology*. (In press).

HAMPTON, M. B., KETTLE, A. J. & WINTERBOURN, C. C. (1998). Inside the neutrophil: oxidants, myeloperoxidase, and bacterial killing. *Blood* **92**, 3007–3017.

HARRIS, K. R. (1975). The fine structure of encapsulation in *Biomphalaria glabrata*. *Annals of the New York Academy of Sciences* **266**, 446–464.

JANEWAY, C. (1989). Approaching the asymptote? Evolution and revolution in immunology. *Cold Spring Harbor Symposia in Quantitative Biology* **54** Pt 1, 1–13.

KETTLE, A. J. & WINTERBOURN, C. C. (1997). Myeloperoxidase: a key regulator of neutrophil oxidant production. *Redox Reports* **3**, 3–9.

KLEBANOFF, S. J. (1968). Myeloperoxidase-halide-hydrogen peroxide antibacterial system. *Journal of Bacteriology* **95**, 2131–2138.

KROSCHINSKI, J. & RENWRANTZ, L. (1988). Determination of pH values inside the digestive vacuoles of hemocytes from *Mytilus edulis*. *Journal of Invertebrate Pathology* **51**, 73–79.

KRUPA, P. L., LEWIS, L. M. & DEL VECCHIO, P. (1977). *Schistosoma haematobium* in *Bulinus guernei*: electron microscopy of hemocyte-sporocyst interactions. *Journal of Invertebrate Pathology* **30**, 35–45.

KUMAZAWA, N. H. & SHIMOJI, Y. (1991). Plasma-dependent chemotactic activity of hemocytes derived from a juvenile estuarine gastropod mollusc, *Clithon retropictus*, to *Vibrio parahaemolyticus* and *Escherichia coli* strains. *Journal of Veterinary Medical Science* **53**, 883–887.

LODES, M. J., CONNORS, V. A. & YOSHINO, T. P. (1991). Isolation and functional characterization of snail hemocyte-modulating polypeptide from primary sporocysts of *Schistosoma mansoni*. *Molecular and Biochemical Parasitology* **49**, 1–10.

LODES, M. J. & YOSHINO, T. P. (1990). The effect of

schistosome excretory-secretory products on *Biomphalaria glabrata* hemocyte motility. *Journal of Invertebrate Pathology* **56**, 75–85.

LOKER, E. S. (1994). On being a parasite in an invertebrate host: a short survival course. *Journal of Parasitology* **80**, 728–747.

LOKER, E. S. & BAYNE, C. J. (2001). Molecular studies of the molluscan response to digenean infection. In *Phylogenetic Perspectives on the Vertebrate Immune System* (ed. Beck, G, Sugumaran, M. & Cooper, E. L.), pp. 209–222. New York, Kluwer Academic/Plenum.

LOKER, E. S., BAYNE, C. J., BUCKLEY, P. M. & KRUSE, K. T. (1982). Ultrastructure of encapsulation of *Schistosoma mansoni* mother sporocysts by hemocytes of juveniles of the 10-R2 strain of *Biomphalaria glabrata*. *Journal of Parasitology* **68**, 84–94.

LOKER, E. S., BAYNE, C. J. & YUI, M. A. (1986). *Echinostoma paraensei*: hemocytes of *Biomphalaria glabrata* as targets of echinostome-mediated interference with host snail resistance to *Schistosoma mansoni*. *Experimental Parasitology* **63**, 149–154.

LOKER, E. S., CIMINO, D. F. & HERTEL, L. A. (1992). Excretory-secretory products of *Echinostoma paraensei* sporocysts mediate interference with *Biomphalaria glabrata* hemocyte functions. *Journal of Parasitology* **78**, 104–115.

LJUNGGREN, H. G. & KARRE, K. (1990). In search of the 'missing self': MHC molecules and NK cell recognition. *Immunology Today* **7**, 237–244.

MAKINO, R., TANAKA, T., IIZUKA, T., ISHIMURA, Y. & KANEGASAKI, S. (1986). Stoichiometric conversion of oxygen to superoxide anion during the respiratory burst in neutrophils. Direct evidence by a new method for measurement of superoxide anion with diacetyldeuteroheme-substituted horseradish peroxidase. *Journal of Biological Chemistry* **261**, 11444–11447.

MEI, H. & LOVERDE, P. T. (1997). *Schistosoma mansoni*: the developmental regulation and immunolocalization of antioxidant enzymes. *Experimental Parasitology* **86**, 69–78.

MITTA, G., HUBERT, F., DYRYNDA, E. A., BOUDRY, P. & ROCH, P. (2000). Mytilin B and MGD2, two antimicrobial peptides of marine mussels: gene structure and expression analysis. *Developmental and Comparative Immunology* **24**, 381–393.

MITTA, G., VANDENBULCKLE, F. & ROCH, P. (2000). Original involvement of antimicrobial peptides in mussel innate immunity. *FEBS Letters* **486**, 185–190.

MINCHELLA, D. J. & LOVERDE, P. T. (1983). Laboratory comparison of the relative success of *Biomphalaria glabrata* stocks which are susceptible and insusceptible to infection with *Schistosoma mansoni*. *Parasitology* **86**, 335–344.

NAUSEEF, W. M. (1990). Myeloperoxidase deficiency. *Hematology and Pathology* **4**, 165–178.

PARRY, M. F., ROOT, R. K., METCALF, J. A., DELANEY, K. K., KAPLOW, L. S. & RICHAR, W. J. (1981). Myeloperoxidase deficiency: prevalence and clinical significance. *Annals of Internal Medicine* **95**, 293–301.

RICHARDS, C. S. & MERRITT, J. W. (1972). Genetic factors in the susceptibility of juvenile *Biomphalaria glabrata* to *Schistosoma mansoni* infection. *American Journal of Tropical Medicine and Hygiene* **21**, 425–434.

SAPP, K. K. & LOKER, E. S. (2000a). Mechanisms underlying digenean-snail specificity: role of miracidial attachment and host plasma factors. *Journal of Parasitology* **86**, 1012–1019.

SAPP, K. K. & LOKER, E. S. (2000b). A comparative study of mechanisms underlying digenean-snail specificity: *in vitro* interactions between hemocytes and digenean larvae. *Journal of Parasitology* **86**, 1020–1029.

SCHLENK, D., MARTINEZ, P. G. & LIVINGSTONE, D. R. (1991). Studies on myeloperoxidase activity in the common mussel, *Mytilus edulis* L. *Comparative Biochemistry and Physiology* **99C**, 63–68.

SHOZAWA, A., SUTO, C. & KUMADA, N. (1989). Superoxide production by the hemocytes of the freshwater snail, *Biomphalaria glabrata*, stimulated by miracidia of *Schistosoma mansoni*. *Zoological Science* **6**, 1019–1022.

SULLIVAN, J. T. & SPENCE, J. V. (1994). Transfer of resistance to *Schistosoma mansoni* in *Biomphalaria glabrata* by allografts of amoebocyte-producing organ. *Journal of Parasitology* **80**, 449–453.

SULLIVAN, J. T. & SPENCE, J. V. (1999). Factors affecting adoptive transfer of resistance to *Schistosoma mansoni* in the snail intermediate host *Biomphalaria glabrata*. *Journal of Parasitology* **85**, 1065–1071.

TORREILLES, J. & GUERIN, M. C. (1999). Production of peroxynitrite by zymosan stimulation of *Mytilus galloprovincialis* haemocytes *in vitro*. *Fish and Shellfish Immunology* **9**, 509–518.

TORREILLES, J., GUERIN, M. C. & ROCH, P. (1997). Peroxidase-release associated with phagocytosis in *Mytilus galloprovincialis* haemocytes. *Developmental and Comparative Immunology* **21**, 267–75.

UPATHAM, E. S. (1972). Exposure of caged *Biomphalaria glabrata* (Say) to investigate dispersion of miracidial of *Schistosoma mansoni* Sambon in outdoor habitats in St. Lucia. *Journal of Helminthology* **46**, 297–306.

VAN DER KNAAP, W. P. W. & LOKER, E. S. (1990). Immune mechanisms in trematode-snail interactions. *Parasitology Today* **6**, 175–182.

WRIGHT, C. A. (1971a). *Flukes and Snails*. New York: Macmillan.

WRIGHT, C. A. (1971b). Review of 'Genetics of a molluscan vector of schistosomiasis' by C. S. Richards. *Tropical Disease Bulletin* **68**, 333–335.

YOSHINO, T. P. & BAYNE, C. J. (1983). Mimicry of snail host antigens by miracidia and primary sporocysts of *Schistosoma mansoni*. *Parasite Immunology* **5**, 317–328.

YOSHINO, T. P. & LODES, M. J. (1988). Secretory protein biosynthesis in snail hemocytes: *in vitro* modulation by larval schistosome excretory-secretory products. *Journal of Parasitology* **74**, 538–547.

YOSHINO, T. P., LODES, M. J., REGE, A. A. & CHAPPELL, C. L. (1993). Proteinase activity in miracidia, transformation excretory secretory products, and primary sporocysts of *Schistosoma mansoni*. *Journal of Parasitology* **79**, 23–31.

The relationship between *Schistosoma mansoni* and *Biomphalaria glabrata*: genetic and molecular approaches

F. A. LEWIS, C. N. PATTERSON, M. KNIGHT *and* C. S. RICHARDS

Biomedical Research Institute, 12111 Parklawn Drive, Rockville, MD, 20852, USA

SUMMARY

Biomphalaria glabrata is a major intermediate host for the helminth parasite *Schistosoma mansoni*. Beginning in the mid-20th century, studies were carried out with this snail species to identify the immunological and genetic components that might be involved in controlling schistosome development. A number of genetically well-defined snail stocks were derived as a direct result of these studies and have since played major roles in helping investigators to identify important cellular and humoral components in the snail/schistosome relationship. This review will explore the historical development of these stocks and describe some of the major advances in several areas of medical malacology that have been made possible by their use.

Key words: *Biomphalaria glabrata*, *Schistosoma mansoni*, schistosomiasis.

INTRODUCTION

Of the numerous molluscs which serve as schistosome hosts, perhaps the most studied is the planorbid *Biomphalaria glabrata*. An important host for *Schistosoma mansoni*, it is the snail species most closely associated with human schistosomiasis in the Western hemisphere. Several biological characteristics of this species have been exploited to reveal basic features of schistosome interactions with their intermediate hosts. The snail has further proven to be a good model for studying planorbid biology in general. Compared to several other schistosome snail hosts, *B. glabrata* can be maintained in the laboratory with relative ease. This alone has probably contributed to a larger research literature for *S. mansoni* than for either *S. haematobium* or *S. japonicum*, whose snail hosts are considerably more fastidious (Liang, Bruce & Boyd, 1987). *B. glabrata* can also serve as host for certain non-schistosome digeneans, in particular some of the echinostomes, thus providing the opportunity to study the development of different trematode species in a common snail host (Huffman & Fried, 1990). In this review we will focus on the genetics of *B. glabrata* with particular emphasis on using various snail stocks to investigate the snail's relationship with the parasite. We show how the development of genetically-defined snail stocks has been important in a variety of research efforts and reveal areas where the genetics of this species may be useful in initiating novel research in this host/parasite relationship, as well as in other medically and economically important molluscs.

HISTORICAL DEVELOPMENT OF GENETICALLY-DEFINED *B. GLABRATA* STOCKS

The studies of *B. glabrata* by W. L. Newton in the early 1950s laid the groundwork for much of what we know today about its genetics. His fundamental methods of snail crossing remain practical for current studies (Newton 1953, 1954, 1955). The fact that albinism in *B. glabrata* is a simple Mendelian recessive trait allowed Newton to easily verify that crossing had occurred between a non-susceptible albino snail from Brazil and a susceptible pigmented snail of Puerto Rican origin. By obtaining large numbers of known hybrids from the albino parent and exposing them to *S. mansoni* miracidia, he found that several of the F1 progeny were susceptible, whereas all snails from the parental albino stock were non-susceptible, leading to the clear conclusion that susceptibility to *S. mansoni* is a heritable trait. Furthermore, after self-fertilization by the albino's F1 progeny, there was a continuous distribution of infection rates among F2 populations, suggesting multiple gene involvement. The heritable nature of susceptibility was supported by his finding that F3 snails derived from the F2 populations with high infection rates were more likely to be susceptible than F3 snails derived from the less susceptible F2 populations. Although age played some part in the susceptibility patterns, it was not until some time later that age factors were more thoroughly investigated (Richards & Merritt, 1972; Richards, 1984).

These early studies by Newton led to the possibility of looking more closely into genetically determined resistance in an invertebrate host, a relatively unexplored area of research in the mid-20th century. As in many other biological systems, detailed investigations of this genetic system required populations with more extensively defined phenotypes. A great deal of that need was met by the pioneering work of C. S. Richards and co-workers, who developed variously pigmented populations with well defined susceptibility, behavioural and morphological characters (see review of Richards & Shade, 1987). Table 1 lists many of these stocks that

Parasitology (2001), **123**, S169–S179. © 2001 Cambridge University Press
DOI: 10.1017/S0031182001007831

Table 1. Different susceptibility types of *Biomphalaria glabrata* stocks
(based on susceptibility to PR-1 strain of *S. mansoni*)*

	Origin	Reference
Type I		
10-R2	Lab derived	Richards, 1975*c*
13-16-R1	Lab derived	Richards, 1975*c*
BS-90	Brazilian isolate	Paraense and Correa, 1963
Type II		
442132	Lab derived	Richards, 1975*c*
2423432	Lab derived	Richards, 1975*c*
Type III		
M-line	Lab derived	Newton, 1955
NMRI	Lab derived	Lewis *et al.* 1986
13141	Lab derived	Richards, 1984
13142	Lab derived	Richards, 1984
PR-77	Puerto Rican isolate	Richards, 1984
PR 79	Pucrto Rican isolate	Richaɪds, 1984
831	Lab derived	Richards, 1984
832	Lab derived	Richards, 1984
Type IV		
NIH-Bg-D-1	Lab derived	Richards, 1977
93375	Lab derived	Sullivan & Richards, 1981
PR-78	Puerto Rican isolate	Richards, 1984

*Susceptibility types: Type I, Nonsusceptible at any age; Type II, Juvenile Susceptible/Adult Nonsusceptible; Type III, Susceptible at any age; Type IV, Juvenile Susceptible/Adult Variable.

have been useful for defining the different susceptibility phenotypes that are known to exist for *B. glabrata*, and a reference where further information on them may be found. We also include information on their derivation – whether as progeny of field isolates, or from laboratory crosses of other stocks. To understand the development and importance of these snail stocks we must first examine more closely the biology of the snail, its pigmentation patterns, and what one can expect to see in progeny populations after a cross.

Several biological aspects of this snail allow one to readily design crosses for exploring traits of interest. One of the more prominent features of these hermaphroditic snails is that preferential out-crossing, rather than self-fertilization, is regularly observed when mates are available. For *B. glabrata*, allosperm fertilize mature ova before the ova move into the hermaphroditic duct. On the other hand autosperm mature in the seminal vesicle and therefore fertilize only those ova that are well into the hermaphroditic duct (Paraense, 1976). With this in mind it is possible to alternate self-fertilization with crossing throughout the lifetime of a snail. Perhaps the most useful biological feature, however, is that *Biomphalaria* spp. possess two genes that determine pigmentation. They are not linked and their respective alleles are stable. The gene that determines mantle pigmentation has three alleles: the wild-type allele produces distinct spots, a codominant allele produces 'diffuse' unpatterned pigment in the

mantle, and the third allele, recessive to the others, produces no mantle pigment (Richards, 1985). A second pigment gene determines the distribution of pigment in the headfoot and mantle collar of the snail. There are four alleles involved with this trait, each dominant to the next as listed here: wildtype, pale-tentacle wildtype (Richards & Patterson, in press), blackeye (Richards, 1967*a*) and albino (Newton, 1954). It is clear that the various pigmentation phenotypes allow great flexibility in performing complex crosses. To date, pigment markers have been used most often in genetic studies of *B. glabrata* with respect to resistance/susceptibility to *S. mansoni* infection.

Richards has done much of the work on the genetics of morphological variation in *B. glabrata* and has used various morphological markers to study snail/schistosome compatibility. After finding extensive morphological variation among snail populations in Puerto Rico, he began selective mating experiments to determine the extent to which this variation was genetically determined (Richards, 1962). The set-up of these experiments was much like the earlier ones of Newton with known F1 hybrid, pigmented progeny obtained from albinos that were crossed with wild-type snails. Isolating F1 hybrids for self-fertilization and characterizing the recombinant F2s allowed Richards to determine if a unique character were heritable as a single locus character (as in Newton's study of albinism) or as a multigenic character. Moreover, rare recessive alleles

Table 2. *Heritable morphological characters isolated in laboratory snail stocks*

	Reference
Single locus characters	
Headfoot pigment	Richards, 1967*a*, 1973*d*, 1985
Mantle pigment	Richards, 1969*a*, 1973*d*
Edema horn (enlarged haemolymph sinuses)	Richards, 1980*b*
Pearl formation	Richards, 1970, 1972
Antler tentacles	Richards, 1973*d*, 1974*a*
Hepatopancreas location	Richards, unpublished
Abnormal pseudobranch	Richards, unpublished
Seminal vesicle degeneration with haemocyte involvement	Richards, unpublished
Formation of spherical aggregations of haemocytes	Richards, unpublished
Pericardial pigment	Richards, unpublished
Multiple loci characters	
Tentacle and eye abnormalities	Richards, 1969*b*, 1973*d*, 1974*a, b*
Pulmonary cavity growths	Richards, 1973*a*
Everted preputium and swollen tentacles	Richards, 1973*d*, 1974*b*
Biphallicism	Richards & Minchella 1986
Spired shell	Richards, 1971
Size at maturity	Richards & Merritt, 1975
Amoebocyte (haemocyte) accumulations	Richards, 1975*a*, 1976*a*, 1980*a*; LoVerde *et al.* 1982
Abnormal development of hepatopancreas	Richards, 1980*c, d*
Apertural lamellae formation and aestivation	Richards, 1963, 1964, 1967*b*, 1968
Variation in course of intestine	Richards, 1973*b*
Bulbous head growths	Richards, 1973*c*
'Shell mats' in Microsporidia-infected snails	Richards, 1974*c*
Dendritic crystals in egg capsules	Richards, unpublished
Egg clutch abnormalities	Richards, unpublished
Lethal embryonic malformations	Richards, unpublished
Longitudinal shell rings	Richards, unpublished

and single mutation events, often unexpressed and lost by genetic drift in populations, were propagated in self-fertilizing snail stocks. Many of the single and multiple gene traits that have been found are listed in Table 2. Most of the laboratory stocks displaying these traits can ultimately be traced back to the Brazilian albino used by Newton (1955) in his derivation of the susceptible M-line snail stock.

As mentioned previously, much of the work with the genetics of *B. glabrata* has been directed towards obtaining information relating to its interaction with *S. mansoni*. However, these single and multi-gene traits are each of singular interest. Gene expressions that lead to these morphological differences, in addition to being of value for studying snail/schistosome interactions, could be applied to many questions in comparative and developmental biology.

IDENTIFICATION OF SINGLE GENE CHARACTERS

For the discussions to follow it will be useful to give an example of the experimental steps that lead to defining inheritance patterns for characters of interest in the snail. One good illustration involves the recent discovery in our laboratory of a new single locus character. Haemocytes are known to play a role in resistance phenotypes, a genetically-determined

trait. It is to be expected that certain haemocyte functions are also under genetic control independent of their response to schistosomes. Several studies of haemocyte accumulation in *B. glabrata*, not associated with schistosome infection, have been reported (Richards, 1976*a*; 1980*a*; LoVerde, Gherson & Richards, 1982). The case we describe here is one of a haemocyte-rich, granuloma-like formation of spherical nodules that initially appear only in the headfoot of the snails. Several nodules were first observed in a selfing snail stock that was derived from a cross in an unrelated study. The fact that no nodules were found in parental stocks or their precross progeny suggested a mutation occurred in one of the snails used in the cross. Microscopic examination of fresh and fixed tissue clearly show that the nodules are composed of haemocytes (Fig. 1). A variety of experiments were performed to determine the possible involvement of an infectious agent in this phenomenon. None has yet been found. Adult snails displaying the haemocyte nodules were crossed with normal adults. In each cross, no hybrid offspring from either parent produced nodules. F1 progeny were allowed to self-fertilize and F2 offspring were observed. Every cross yielded some F2 progeny that produced haemocyte nodules. Examining many F2 snails, the ratio of F2 progeny without nodules to those with nodules was consistent with the expected 3:1 segregation pattern for a single

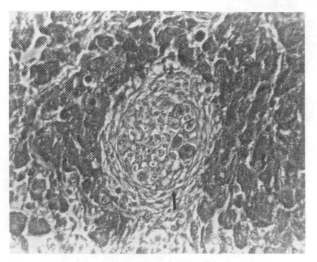

Figure 1. Haemocyte-rich nodule in headfoot of *B. glabrata*. Flattened haemocytes (arrow) surround central core of looser haemocyte aggregation. Pedal mucous gland cells are those darker stained cells surrounding the nodule. (Haematoxylin-eosin stain. 400×).

gene trait. Moreover, the F2 offspring with nodules, when allowed to self and produce subsequent generations, yielded only progeny that produced haemocyte nodules. The combined results of the snail crosses strongly suggest a single gene controls this haemocyte-rich nodule formation. Although no infectious agent has been found associated with this character, we cannot rule one out. It is possible that snails possessing the homozygous recessive condition for this gene are sensitive to an exogenous agent in the laboratory environment. By simple selection of F2 progeny from the various crosses, we derived albino, blackeye and wild-type snail stocks that produce haemocyte nodules, demonstrating no linkage of pigment with the locus controlling this trait.

The breeding experiment described above illustrates the simplicity of demonstrating that a single locus controls a character: hybrid F1 offspring from the recessively pigmented parent are recognized by the dominant pigment phenotype, F1s are uniform for the dominant phenotype of the character of interest, F2 progeny of dominant and recessive phenotypes are produced, and F2s with the recessive phenotype breed true in subsequent generations. If the trait of interest is dominant, F2 snails displaying the dominant phenotype are selected as breeding stock when their production of uniformly dominant F3s reveals their homozygosity for the trait. When back-crossed to a snail of the recessive parent stock, F1s with a dominant phenotype will produce approximately 50% dominant and 50% recessive offspring; homozygous dominant F2s will produce 100% dominant phenotype offspring.

B. glabrata has a haploid chromosome number of 18. None of the single locus characters that have been studied has shown linkage to another known single gene character. In addition, testing for linkage between every known morphological marker and adult snail resistance/susceptibility to schistosome infection has revealed no indication of any degree of linkage. Of course, juvenile snail susceptibility to schistosome infection is known to be influenced by more than a single genetic factor (Newton, 1953) and clear linkage of this to another character would not be expected. Recent evidence from our laboratory that molecular markers segregate with the adult resistance phenotype will be discussed later.

IDENTIFICATION OF MULTIGENE CHARACTERS

Traits influenced by more than one genetic locus are recognized in *B. glabrata* simply by the fact that they do not conform to the patterns of single gene inheritance as described above. Thus, F1 hybrids will not be uniform in expression of the trait, and populations of F2 progeny produced by selfing hybrid snails will show a wide distribution of the percentage of individuals displaying the trait. Furthermore, it often takes many generations of selection to achieve a true breeding stock, if a true breeding stock can in fact be obtained.

Juvenile snail resistance to schistosome infection is the multi-locus character that has drawn the most research interest. Since Newton's recognition of the complex character of the inheritance of this trait, Richards & Merritt (1972), in an exhaustive series of crosses, back-crosses and serial crosses between refractory and susceptible snails, proved the multi-locus nature of juvenile susceptibility. Evidence supporting their conclusion that juvenile susceptibility is controlled by at least four interacting genetic factors, possibly each with multiple alleles, is diagrammatically revealed in Figs 1 & 7 of their publication.

COMPLICATIONS IN DEFINING GENETIC CHARACTERS

When categorizing single and multi-locus characters that are associated with schistosome infection in the snail, several complicating factors must be kept in mind. Among these are (1) size and/or age of the snail, (2) miracidial dose on exposure, and (3) parasite strain. Following the initial report by Newton (1953), others have explored further the relationship between snail size and susceptibility (Anderson *et al.* 1982; Niemann & Lewis, 1990). For *B. glabrata*, and as is commonly seen among many molluscan species, size is not a good indicator of the physiological age of the snail. It is more of an indicator of the suitability of the environment in which they are maintained. The onset of egg laying in *B. glabrata*, indicating maturity, is very important in susceptibility, but is extremely variable and may occur when the snail is as small as 5 mm in diameter, or as large as 16 mm in diameter (Richards & Merritt, 1972). The seminal vesicle, in live snails the first visible indicator of development of the reproductive system, may be first seen in snails as small as 3 mm in diameter, or in others at 10 mm in diameter or larger. In general, in a population of snails of mixed susceptibility phenotypes, smaller snails are more likely to become patent upon exposure to miracidia than are larger snails. Indeed, if neonates (<2 mm in diameter) of one of the most refractory of stocks, BS90, are exposed to a large number of miracidia (NMRI strain) they will sometimes develop primary and secondary sporocysts if they survive the tissue damage caused by miracidial penetration. One BS-90 snail thus exposed survived and shed infective cercariae.

The role parasite genes play in the snail host/parasite relationship has received scant attention. In the past, the topic of schistosome strain differences has most often been discussed either in terms of differences in pathology in the definitive host or differences in drug sensitivity of certain isolates (Anderson & Cheever, 1972; Araujo *et al.* 1980; Tsai *et al.* 2000). We believe that the mollusc, rather than the definitive host, has likely been a more important force driving schistosome strain differentiation. A given geographic isolate of *B. glabrata* may or may not be a suitable host for a given strain of *S. mansoni*. A suitable vertebrate host, however, will (if exposed) be infected by *S. mansoni* regardless of the parasite's geographic origin. In field situations the snail can be considered the genetic bottleneck that determines which *S. mansoni* genotypes will reproduce. There can be dramatic strain variations in *S. mansoni* infectivity for a given laboratory snail stock of *B. glabrata*, even among schistosomes obtained from the same geographic area (Basch, 1976; Richards, 1976*b*; Lie, Heyneman & Richards, 1979; Richards & Shade, 1987). From schistosome crossing experiments and miracidial exposures of different snail stocks, it has been shown that the difference in infectivity for the snail is genetically determined and involves sex linkage (Richards, 1975*b*). Results emerging from the ongoing schistosome genome project (Tanaka *et al.* 1997) will enable us to look more closely at this subject – a relatively untapped research area but one that should provide novel and interesting information on trematode biology.

GENERAL HYPOTHESES ON RESISTANCE TO SCHISTOSOMES IN *B. GLABRATA*

The multi-locus nature of the juvenile snail susceptibility phenotype and its variation with snail maturation and miracidial dose are hallmarks of a threshold trait. In terms of quantitative genetics, the genotype of a given juvenile snail representing the additive genetic value of the alleles of multiple loci will place that snail at a point on a continuum of 'liability' for schistosome infection (Fig. 2). The relative position of a susceptibility threshold for a particular genotype will be influenced at least by the snail's state of maturation, compatibility with the particular strain of parasite and miracidial dose.

From the large number of crosses that have now been performed, some general hypotheses about resistance to *S. mansoni* in *B. glabrata* can be made. In addition to the four snail genes each with multiple alleles that influence juvenile susceptibility, there are at least two overriding resistance genes, whose effects can be characterized by the following. First, there is the Mr gene (Richards, 1984), which is not present in all *B. glabrata* populations. In some laboratory lines some snails may possess it, while others may not. Mr is only expressed in the adult snail and acts to suppress the juvenile susceptibility complex of genes, effectively obliterating the liability threshold and rendering the snails resistant. The mechanism of action of this gene and its product(s) are not known. When the Mr gene is expressed, the phenotype is a single, dominant character conferring resistance.

Strong evidence also exists for a separate resistance gene (R gene) that can be expressed only in non-susceptible snails, juvenile or adult (Richards, 1975*c*). The hallmark of this gene is the development of the haemocytic accumulation around the developing sporocyst. In order for this cellular response to the parasite to occur, the snails must be non-susceptible. The hallmark of the Mr gene is that in populations where the dominant Mr allele is fixed, juvenile snails may be susceptible but adult snails never are (Richards, 1975*a*). The Mr and R genes are known to be separate genes. Some snails may have one but not the other, and some may have both. When both are present, expression of Mr allows the expression of the R gene in adult snails. For juveniles, only those with genotypes conferring non-susceptibility for a given threshold will express R. Referring to Table 1, we believe the distribution of these genes among snails in the various susceptibility types to be the following. Type I snails express the R gene. Although snails in Type II possess the Mr gene, the R gene can be either present or absent. Type III snails do not possess Mr, but could possess R, although the latter would never be expressed. The presence of these genes in Type IV snails is less clear. Although Mr is likely present in type IV snails, and R can be either present or absent, confounding factors probably influence the ultimate susceptibility pattern for these snails.

Taking into account the various alleles of the four (or more) juvenile susceptibility genes and each one's interaction with the others, variation among recombinants (whether by selfing individuals or interbreeding populations) will be very large. It is not surprising that populations of field collected snails, once taken to the laboratory and exposed to a single strain of *S. mansoni*, often exhibit extremely

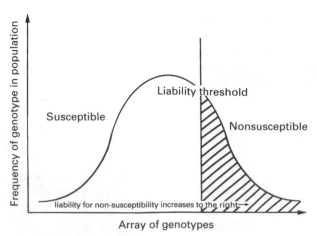

Figure 2. Distribution of population genotypes for the quantitative character of juvenile susceptibility. The shaded area represents snails having 'liability' above a critical threshold, rendering them nonsusceptible. For convenience a bell-shaped curve is shown in the example above, although the shape of the curve will depend on the particular susceptibility allele frequencies in the population. The position of the liability threshold can vary, depending on factors such as number of miracidia, strain of parasite, and maturity of the snails. For example, in the absence of the single, dominant Mr allele, adult susceptibility genotypes will show the same distribution, although the threshold may move to the left as snail size increases.

varied susceptibility levels. These genes confer great flexibility to a population and allow the snails to adapt to a considerable degree of parasite pressure.

THE SEARCH FOR MECHANISMS OF PARASITE DESTRUCTION

Although the mechanism of resistance to trematodes will be discussed by others in this supplement, something should be mentioned here about the search for soluble factors and cell types that play a role in resistance and the importance of defined snail stocks to these efforts. Several investigators have exploited inbred stocks of *B. glabrata* to explore the role of various humoral factors in the defence system of the snail. In addition, considerable work on putative soluble factors in trematode resistance has also been carried out using snail species other than *B. glabrata*. Thus, a relatively large literature has emerged on the lectins in *Helix* spp., *Lymnaea* spp. and *Achatina* spp. to name a few (see review by Adema & Loker, 1977).

Perhaps greater use of the genetically-defined *B. glabrata* stocks has been made for examining cellular contributions to parasite recognition and destruction. As mentioned before, histological studies on the fate of miracidia which penetrate resistant snails reveal that a haemocyte response occurs. This cellular response in the genetically resistant snails gives a visual benchmark to the parasite's demise. The classic studies of Lie and colleagues (see review

by Lie, Jeong & Heyneman, 1987) and subsequent studies of haemocytes provide an important chapter on invertebrate immunity in general and molluscan haemocytes in particular (Sminia, 1981; Granath & Yoshino, 1983; Adema, Harris & Van Deutekom-Mulder, 1992; Matricon-Gondran & Letocart, 1999).

The novel tissue transplantation studies by Sullivan & Spence (1994, 1999) give good recent examples of how the use of several of the genetically-defined *B. glabrata* stocks have been important in efforts to more clearly define the cellular components contributing to resistance. When the amoebocyte-producing organs (APO) of *S. mansoni*-resistant snails were transplanted to snails of a highly susceptible stock, some of the recipient snails acquired resistance to the parasite (Sullivan & Spence, 1994). Further experiments narrowed the anatomical regions of the snail in which the resistance factors or cells are expressed upon parasite challenge (Sullivan & Spence, 1999). The mechanism that transfers resistance however is not clearly understood. A combination of factors may contribute. From the reported data it is unlikely that all the cells that surrounded the invading miracidium in the recipient originated from the transplanted tissue.

As an extension of the studies on immunobiology/parasite compatibility of various snail stocks, an embryonic cell line from *B. glabrata*, termed Bge (Hansen, 1976), is in current use to experimentally determine soluble and cellular factors involved in parasite recognition and killing. To our knowledge the Bge cell line remains the only established molluscan cell line in existence. Characteristics of this line suggest it is fibroblastic in origin, based on morphology and ability to synthesize collagen-like molecules (see review of Yoshino *et al.* 1999). Similar to haemocytes, Bge cells form a granulomatous-type encapsulation around primary sporocysts *in vitro*, although killing of the sporocysts by the cells is not typically seen. Some intriguing aspects of co-cultivation of these cells with *S. mansoni* larvae are opening new lines of investigation on schistosome/molluscan interactions (Yoshino & Laursen, 1995; Coustau & Yoshino, 2000). Recently, co-cultivation of Bge cells with sporocysts, involving changing conditions during various growth stages, led to the generation of cercariae from the original miracidial culture, although infectivity of the cercariae was not verified to successfully complete the invertebrate phase of the life cycle (Ivanchenko *et al.* 1999). Nevertheless, it is clear that Bge cells provide a supportive environment in which parasite differentiation can proceed in the absence of the intact snail. Further use of this system will undoubtedly produce valuable information in a number of areas, for example: (1) the activity of various cell surface receptors in cellular adhesion; (2) gene expression during formation of the cellular sheath around the sporocyst and investigations into the various cytokine-like factors which may be of importance in cell trafficking; (3) regulation of enzymes in response to the parasite; and (4) multi-generational molecular genetic studies of *S. mansoni* by propagating clonal stocks of the parasite.

MOLECULAR STUDIES OF *BIOMPHALARIA* SPECIES

Molecular analysis of *B. glabrata*, including searches for linkage to resistance, began in the early 1980s with isozyme studies of numerous snail stocks (Fletcher, LoVerde & Richards, 1981; Mulvey & Vrijenhoek, 1981; Mulvey & Woodruff, 1985). Although no linkage could be shown between resistance and specific isozymes, these studies did emphasize the considerable degree of intra-strain polymorphism that existed in some of these snail stocks. In more recent years molecular approaches have yielded exciting clues to a variety of questions in basic snail biology and snail/schistosome interactions. Most of the molecular approaches to date have been designed with one of the following three goals in mind: (1) search for resistance genes or their markers; (2) species identification; (3) determining the extent of genetic polymorphism in field isolates. The use of genetically-defined snail stocks has been more important perhaps for the first of these three goals, although techniques developed during these studies have enhanced the progress in the other two areas.

Modern molecular methods applied to studying *B. glabrata* began with analyzing DNA polymorphisms between and among different snail stocks (Knight *et al.* 1991). The initial efforts used a restriction fragment length polymorphism (RFLP) approach in which heterologous *S. mansoni* rRNA probes were used. Snail specific probes were not available. An added benefit of this work was that differences in the structure of the rRNA gene were also found to be useful for distinguishing different species of *Biomphalaria*. Because of technical disadvantages of the RFLP-based approach, especially for field applications, more practical methods using RAPD–PCR were developed and have subsequently been used more frequently. Among other things, RAPD–PCR has been used to demonstrate genetic variability within laboratory stocks (Larson *et al.* 1996) and among snails in different schistosome transmission regions (Vidigal *et al.* 1994; and see also review by Carvalho *et al.*, this supplement). RAPD–PCR has also been useful for differentiating species of both *Biomphalaria* (Vidigal *et al.* 1998) and of *Bulinus* (review of Rollinson *et al.* 1998), the latter being the intermediate hosts for *S. haematobium*. The recent characterization of microsatellite loci in *B. glabrata* (Jones *et al.* 1999) will prove valuable in studies of snail population structure in the field.

Numerous recent advances have been made using molecular technology for identifying and studying genes involved in the snail/parasite relationship (reviewed in Knight, Ongele & Lewis, 2000). An accompanying review by Jones *et al.* (this supplement) goes into greater detail than will be presented here. In our laboratory, the use of RAPD–PCR has led to the identification of two heritable DNA markers (1·0 kb and 1·2 kb) associated with parasite resistance (Knight *et al.* 1999). Specifically, these markers segregated with the adult resistant phenotype in *B. glabrata*. Further characterizations of these markers among F1 and F2 juvenile-susceptible and -resistant snails are being conducted. The association of these markers with the R and/or Mr genes has not been demonstrated. Although both markers contain highly repeated sequences, non-repetitive sequences within these markers have now been identified. Using the non-repetitive sequences as probes should allow us to identify additional DNA sequence associated with the R genotype.

Since many of the earlier technical problems that plagued investigators working with molluscan DNA have now been solved, more laboratories are directing their efforts to gene identification in *B. glabrata*. One can point to the recent successes for characterizing the genes that encode for a novel β-integrin (Davids, Wu & Yoshino, 1999), a fibrinogen-like molecule (Adema *et al.* 1997), a receptor for protein kinase C (Lardans *et al.* 1998), and a snail homologue of human selectin, an adhesion receptor that may be involved in the binding of carbohydrate structures on the surface of developing sporocysts (Duclermortier *et al.* 1999). It is hoped that the use of Bge cells will rapidly improve and advance our knowledge of the genes involved in molluscan defence. Not only do Bge cells have several characteristics of haemocytes, but from a very practical standpoint, can be more easily obtained in large numbers than one can obtain circulating haemocytes.

In our laboratory, we have taken advantage of the techniques of subtractive hybridization and expressed sequence tags (EST) to search for genes that may be involved in snail/parasite interactions. Although the subtractive hybridization approach enabled us to identify a gene that was preferentially expressed in a resistant snail (Miller *et al.* 1996), the EST method was found to be more efficient in identifying genes from various snail tissues (Knight *et al.* 1998), including circulating haemocytes (Raghavan, personal communication). From these efforts, the public databases now contain approximately 1400 *B. glabrata* ESTs. A large majority of these ESTs are novel, showing no significant similarity to other sequences in the public databases.

The differential display technique (Liang & Pardee, 1992) is another gene discovery tool that has been useful for studying the snail/parasite relationship. Using this method Lockyer *et al.* (2000) reported the gene for cytochrome p450, along with 4 other genes identified in the snail during the intramolluscan stages of parasite development. In our laboratory, the differential display technique was used to identify a homologue of β-integrin interactor in *B. glabrata* (Knight *et al.* 2000), along with several transcripts from haemocytes that were differentially regulated by parasite exposure (Miller, personal communication). One novel finding from these efforts has been that, using both EST and differential display methodologies, several snail cDNA sequences have been identified that show significant sequence homology (71%) to reverse transcriptase (RT) of the invertebrate *Ciona intestinalis* (Knight *et al.* 2000). The role of RT in the snail as it relates to parasite exposure remains unknown, but it is intriguing that functional RT activity can be detected in *B. glabrata* and shown to vary in a time-dependent manner upon parasite exposure (Raghavan, personal communication). Findings such as this RT activity and a LINE-1 transposon in *B. glabrata* (Knight *et al.* 1992) challenge us with more questions than answers at this point. Although there is at present no clear evidence that RT activity or transposon-related sequences in the snail genome play any role in DNA modification or rearrangement in this organism, recent RAPD analysis of adult and juvenile F1 progeny snails (with the same primer) showed differences in the DNA profiles in the snails at different ages (Knight, unpublished). Whether such genetic variations are caused by mobile elements in the snail genome is speculative at the moment, but these observations, along with the frequent mutations detected among laboratory-reared *B. glabrata* populations (Richards, unpublished) suggest that a closer look at this subject may be warranted.

In this review we have summarized how studies of snail crossing in *B. glabrata*, starting in the 1950s and continuing to this day, have led to the rapid expansion of knowledge of specific gene expression in these snails, especially in association with *S. mansoni* infection. This knowledge may ultimately have substantial relevance for human schistosomiasis, through stimulating the development of novel methods for controlling transmission at the level of the intermediate host. On a different but also important level, this knowledge should increase our appreciation of how these molluscs have evolved to ensure their ecological niche in the face of continuous assault from the large number of trematode and non-trematode species that can infect them.

ACKNOWLEDGMENTS

Work in our laboratory was supported by NIH Grant AI-27777. We are grateful to Dr. Nithya Raghavan, Dr. Allen Cheever, and Andre Miller for helpful discussions and critical review of the manuscript.

REFERENCES

ADEMA, C. M., HARRIS, R. A. & VAN DEUTEKOM-MULDER, E. C. (1992). A comparative study of hemocytes from six different snails: morphology and functional aspects. *Journal of Invertebrate Pathology* **59**, 24–32.

ADEMA, C. M., HERTEL, L. A., MILLER, R. D. & LOKER, E. S. (1997). A family of fibrinogen-related proteins that precipitates parasite-derived molecules is produced by an invertebrate after infection. *Proceedings of the National Academy of Sciences, USA* **94**, 8691–8696.

ADEMA, C. M. & LOKER, E. S. (1997). Specificity and immunobiology of larval digenean-snail associations. In *Advances in Trematode Biology* (ed. Fried, B. & Graczyk, T. K.), pp. 229–263. Boca Raton & New York, CRC Press.

ANDERSON, L. A. & CHEEVER, A. W. (1972). Comparison of geographical strains of *Schistosoma mansoni* in the mouse. *Bulletin of the World Health Organization* **46**, 233–242.

ANDERSON, R. M., MERCER, J. G., WILSON, R. A. & CARTER, N. P. (1982). Transmission of *Schistosoma mansoni* from man to snail: Experimental studies of miracidial survival and infectivity in relation to larval age, water temperature, host size and host age. *Parasitology* **85**, 339–360.

ARAUJO, N., KATZ, N., DIAS, E. P. & DE SOUZA, C. P. (1980). Susceptibility to chemotherapeutic agents of strains of *Schistosoma mansoni* isolated from treated and untreated patients. *The American Journal of Tropical Medicine and Hygiene* **29**, 890–894.

BASCH, P. F. (1976). Intermediate host specificity to infection with *Schistosoma mansoni*. *Experimental Parasitology* **39**, 150–169.

COUSTAU, C. & YOSHINO, T. P. (2000). Flukes without snails: advances in the *in vitro* cultivation of intramolluscan stages of trematodes. *Experimental Parasitology* **94**, 62–66.

DAVIDS, B. J., WU, X. J. & YOSHINO, T. P. (1999). Cloning of a beta-integrin subunit cDNA from an embryonic cell line derived from the freshwater mollusk *Biomphalaria glabrata*. *Gene* **228**, 213–223.

DUCLERMORTIER, P., LARDANS, V., SERRA, E., TROTTEIN, F. & DISSOUS, C. (1999). *Biomphalaria glabrata* embryonic cells express a protein with a domain homologous to the lectin domain of mammalian selectins. *Parasitology Research* **85**, 481–486.

FLETCHER, M., LOVERDE, P. T. & RICHARDS, C. S. (1981). *Schistosoma mansoni*: electrophoretic characterization of strains selected for different levels of infectivity to snails. *Experimental Parasitology* **52**, 362–370.

GRANATH, W. O. & YOSHINO, T. P. (1983). Characterization of molluscan phagocyte subpopulations based on lysosomal enzyme markers. *Journal of Experimental Zoology* **226**, 205–210.

HANSEN, E. L. (1976). A cell line from embryos of *Biomphalaria glabrata* (Pulmonata): Establishment and characteristics. In *Invertebrate Tissue Culture* (ed. Maramorosch, K.), pp. 75–99. New York, Academic Press.

HUFFMAN, J. E. & FRIED, B. (1990). *Echinostoma* and echinostomiasis. *Advances in Parasitology* **29**, 215–269.

IVANCHENKO, M. G., LERNER, J. P., MCCORMICK, R. S., TOUMADJE, A., ALLEN, B., FISCHER, K., HEDSTROM, O.,

HELMRICH, A., BARNES, D. W. & BAYNE, C. J. (1999). Continuous *in vitro* propagation and differentiation of cultures of the intramolluscan stages of the human parasite *Schistosoma mansoni*. *Proceedings of the National Academy of Sciences, USA* **96**, 4965–4970.

JONES, C. S., LOCKYER, A. E., ROLLINSON, D., PIERTNEY, S. B. & NOBLE, L. R. (1999). Isolation and characterization of microsatellite loci in the freshwater gastropod, *Biomphalaria glabrata*, an intermediate host for *Schistosoma mansoni*. *Molecular Ecology* **8**, 2149–2151.

KNIGHT, M., BRINDLEY, P. J., RICHARDS, C. S. & LEWIS, F. A. (1991). *Schistosoma mansoni*: use of a cloned ribosomal RNA gene probe to detect restriction fragment length polymorphisms in the intermediate host *Biomphalaria glabrata*. *Experimental Parasitology* **73**, 285–294.

KNIGHT, M., MILLER, A. N., GEOGHAGEN, N. S. M., LEWIS, F. A. & KERLAVAGE, A. R. (1998). Expressed sequence tags (ESTs) of *Biomphalaria glabrata*, an intermediate snail host of *Schistosoma mansoni*: use in the identification of RFLP markers. *Malacologia* **39**, 175–182.

KNIGHT, M., MILLER, A. N., PATTERSON, C. A., ROWE, C. G., MICHAELS, G., CARR, D., RICHARDS, C. S. & LEWIS, F. A. (1999). The identification of markers segregating with resistance to *Schistosoma mansoni* infection in the snail *Biomphalaria glabrata*. *Proceedings of the National Academy of Sciences, USA* **96**, 1510–1515.

KNIGHT, M., MILLER, A., RAGHAVAN, N., RICHARDS, C. & LEWIS, F. A. (1992). Identification of a repetitive element in the snail *Biomphalaria glabrata*: relationship to the reverse transcriptase-encoding sequence in LINE-1 transposons. *Gene* **118**, 181–187.

KNIGHT, M., ONGELE, E. & LEWIS, F. A. (2000). Molecular studies of *Biomphalaria glabrata*, an intermediate host of *Schistosoma mansoni*. *International Journal for Parasitology* **30**, 535–541.

LARDANS, V., SERRA, E., CAPRON, A. & DISSOUS, C. (1998). Characterization of an intracellular receptor for activated protein kinase C (RACK) from the mollusc *Biomphalaria glabrata*, the intermediate host for *Schistosoma mansoni*. *Experimental Parasitology* **88**, 194–199.

LARSON, S. E., ANDERSON, P. L., MILLER, A. N., COUSIN, C. E., RICHARDS, C. S., LEWIS, F. A. & KNIGHT, M. (1996). Use of RAPD–PCR to differentiate genetically defined lines of an intermediate host of *Schistosoma mansoni*, *Biomphalaria glabrata*. *Journal of Parasitology* **82**, 237–244.

LEWIS, F. A., STIREWALT, M. A., SOUZA, C. P. & GAZZINELLI, G. (1986). Large-scale laboratory maintenance of *Schistosoma mansoni*, with observations on three schistosome/snail host combinations. *Journal of Parasitology* **72**, 813–829.

LIANG, Y.-S., BRUCE, J. I. & BOYD, D. A. (1987). Laboratory cultivation of schistosome vector snails and maintenance of schistosome life cycles. *Proceedings of the First Sino-American Symposium* **1**, 34–48.

LIANG, P. & PARDEE, A. B. (1992). Differential display of eukaryotic messenger RNA by means of the polymerase chain reaction. *Science* **257**, 967–971.

LIE, K. J., HEYNEMAN, D. & RICHARDS, C. S. (1979). Specificity of natural resistance to trematode infections in *Biomphalaria glabrata*. *International Journal for Parasitology* **9**, 529–531.

LIE, K. J., JEONG, K. H. & HEYNEMAN, D. (1987). Molluscan host reactions to helminthic infection. In *Immune Responses in Parasitic Infections* (ed. Soulsby, E. J. L.), pp. 211–270. Boca Raton & New York, CRC Press.

LOCKYER, A. E., JONES, C. S., NOBLE, L. R. & ROLLINSON, D. (2000). Use of differential display to detect changes in gene expression in the intermediate snail host *Biomphalaria glabrata* upon infection with *Schistosoma mansoni*. *Parasitology* **120**, 399–407.

LOVERDE, P. T., GHERSON, J. & RICHARDS, C. S. (1982). Amebocytic accumulations in *Biomphalaria glabrata*: fine structure. *Developmental and Comparative Immunology* **6**, 441–449.

MATRICON-GONDRAN, M. & LETOCART, M. (1999). Internal defenses of the snail *Biomphalaria glabrata*. I. Characterization of hemocytes and fixed phagocytes. *Journal of Invertebrate Pathology* **74**, 224–234.

MILLER, A. N., OFORI, K., LEWIS, F. & KNIGHT, M. (1996). *Schistosoma mansoni*: use of a subtractive cloning strategy to search for RFLPs in parasite-resistant *Biomphalaria glabrata*. *Experimental Parasitology* **84**, 420–428.

MULVEY, M. & VRIJENHOEK, R. C. (1981). Genetic variation among laboratory strains of the planorbid snail *Biomphalaria glabrata*. *Biochemical Genetics* **19**, 1169–1182.

MULVEY, M. & WOODRUFF, D. S. (1985). Genetics of *Biomphalaria glabrata*: linkage analysis of genes for pigmentation, enzymes, and resistance to *Schistosoma mansoni*. *Biochemical Genetics* **22**, 877–889.

NEWTON, W. L. (1953). The inheritance of susceptibility to infection with *Schistosoma mansoni* in *Australorbis glabratus*. *Experimental Parasitology* **2**, 242–257.

NEWTON, W. L. (1954). Albinism in *Australorbis glabratus*. *Proceedings of the Helminthological Society of Washington* **21**, 72–74.

NEWTON, W. L. (1955). The establishment of a strain of *Australorbis glabratus* which combines albinism and high susceptibility to infection with *Schistosoma mansoni*. *Journal of Parasitology* **29**, 539–544.

NIEMANN, G. M. & LEWIS, F. A. (1990). *Schistosoma mansoni*: Influence of *Biomphalaria glabrata* size on susceptibility to infection and resultant cercarial production. *Experimental Parasitology* **70**, 286–292.

PARAENSE, W. L. (1976). The sites of cross- and self-fertilization in planorbid snails. *Revista Brasileira Biologia* **36**, 535–539.

PARAENSE, W. L. & CORREA, L. R. (1963). Variation in susceptibility of populations of *Australorbis glabratus* to a strain of *Schistosoma mansoni*. *Revista do Instituto de Medicina Tropical de Sao Paulo* **5**, 15–22.

RICHARDS, C. S. (1962). Genetic crossing of pigmented Caribbean strains with an albino Venezuelan strain of *Australorbis glabratus*. *The American Journal of Tropical Medicine and Hygiene* **11**, 216–219.

RICHARDS, C. S. (1963). Apertural lamellae, epiphragms, and aestivation of planorbid mollusks. *The American Journal of Tropical Medicine and Hygiene* **12**, 254–263.

RICHARDS, C. S. (1964). Apertural lamellae as supporting structures in *Australorbis glabratus*. *Nautilis* **78**, 57–60.

RICHARDS, C. S. (1967a). Genetic studies on *Biomphalaria glabrata* (Basommatophora: Planorbidae), a third pigmentation allele. *Malacologia* **5**, 335–340.

RICHARDS, C. S. (1967b). Estivation of *Biomphalaria glabrata* (Basommatophora: Planorbidae). Associated characteristics and relation to infection with *Schistosoma mansoni*. *The American Journal of Tropical Medicine and Hygiene* **16**, 797–802.

RICHARDS, C. S. (1968). Aestivation of *Biomphalaria glabrata* (Basommatophora; Planorbidae) genetic studies. *Malacologia* **7**, 109–116.

RICHARDS, C. S. (1969a). Genetic studies on *Biomphalaria glabrata*: mantle pigmentation. *Malacologia* **9**, 339–348.

RICHARDS, C. S. (1969b). Genetic studies on *Biomphalaria glabrata*: tentacle and eye variations. *Malacologia* **9**, 327–338.

RICHARDS, C. S. (1970). Pearl formation by *Biomphalaria glabrata*. *Journal of Invertebrate Pathology* **15**, 459–460.

RICHARDS, C. S. (1971). *Biomphalaria glabrata* genetics: spire formation as a sublethal character. *Journal of Invertebrate Pathology* **17**, 53–58.

RICHARDS, C. S. (1972). *Biomphalaria glabrata* genetics: pearl formation. *Journal of Invertebrate Pathology* **20**, 37–40.

RICHARDS, C. S. (1973a). Tumors in the pulmonary cavity of *Biomphalaria glabrata*: genetic studies. *Journal of Invertebrate Pathology* **22**, 283–289.

RICHARDS, C. S. (1973b). Inherited variability in the course of the intestine in *Biomphalaria glabrata*. *Malacological Review* **6**, 133–138.

RICHARDS, C. S. (1973c). Bulbous head growths of *Biomphalaria glabrata*: genetic studies. *Journal of Invertebrate Pathology* **22**, 278–282.

RICHARDS, C. S. (1973d). Genetics of *Biomphalaria glabrata* (Gastropoda: Planorbidae). *Malacological Review* **6**, 199–202.

RICHARDS, C. S. (1974a). Antler tentacles of *Biomphalaria glabrata*: genetic studies. *Journal of Invertebrate Pathology* **24**, 49–54.

RICHARDS, C. S. (1974b). Everted preputium and swollen tentacles in *Biomphalaria glabrata*: genetic studies. *Journal of Invertebrate Pathology* **24**, 159–164.

RICHARDS, C. S. (1974c). Shell mats associated with Microsporida in *Biomphalaria glabrata*: genetic studies. *Journal of Invertebrate Pathology* **24**, 337–343.

RICHARDS, C. S. (1975a). Genetic studies of pathologic conditions and susceptibility to infection in *Biomphalaria glabrata*. *Annals of the New York Academy of Sciences* **266**, 394–410.

RICHARDS, C. S. (1975b). Genetic studies on variation in infectivity of *Schistosoma mansoni*. *Journal of Parasitology* **61**, 233–236.

RICHARDS, C. S. (1975c). Genetic factors in susceptibility of *Biomphalaria glabrata* for different strains of *Schistosoma mansoni*. *Parasitology* **70**, 231–241.

RICHARDS, C. S. (1976a). Dermal lesions and amebocytic accumulations in snails. *Marine Fisheries Review* **38**, 44–45.

RICHARDS, C. S. (1976b). Variations in infectivity for *Biomphalaria glabrata* in strains of *Schistosoma mansoni* from the same geographical area. *Bulletin of the World Health Organization* **54**, 706–707.

RICHARDS, C. S. (1977). *Schistosoma mansoni*: susceptibility reversal with age in the snail host *Biomphalaria glabrata*. *Experimental Parasitology* **42**, 165–168.

RICHARDS, C. S. (1980a). Genetic studies on amebocytic accumulations in *Biomphalaria glabrata*. *Journal of Invertebrate Pathology* **35**, 49–52.

RICHARDS, C. S. (1980b). Edema-horn, an abnormal mutant of *Biomphalaria glabrata*. *Journal of Invertebrate Pathology* **35**, 53–57.

RICHARDS, C. S. (1980c). Abnormal growths from the digestive gland of *Biomphalaria glabrata*. *Journal of Invertebrate Pathology* **35**, 318–319.

RICHARDS, C. S. (1980d). Abnormal development of the digestive gland in *Biomphalaria glabrata*. *Journal of Invertebrate Pathology* **36**, 121–122.

RICHARDS, C. S. (1984). Influence of snail age on genetic variations in susceptibility of *Biomphalaria glabrata* for infection with *Schistosoma mansoni*. *Malacologia* **25**, 493–502.

RICHARDS, C. S. (1985). A new pigmentation mutant in *Biomphalaria glabrata*. *Malacologia* **26**, 145–151.

RICHARDS, C. S. & MERRITT, J. W. JR. (1972). Genetic factors in the susceptibility of juvenile *Biomphalaria glabrata* to *Schistosoma mansoni* infection. *The American Journal of Tropical Medicine and Hygiene* **21**, 425–434.

RICHARDS, C. S. & MERRITT, J. W. (1975). Variation in size of *Biomphalaria glabrata* at maturity. *The Veliger* **17**, 393–395.

RICHARDS, C. S. & MINCHELLA, D. J. (1986). Genetic studies of biphallic *Biomphalaria glabrata*. *Malacologia* **27**, 243–247.

RICHARDS, C. S. & PATTERSON, C. (In press). Tentacle pigment variation in *Biomphalaria glabrata*. *Journal of Medical and Applied Malacology* (in press).

RICHARDS, C. S. & SHADE, P. C. (1987). The genetic variation of compatibility in *Biomphalaria glabrata* and *Schistosoma mansoni*. *Journal of Parasitology* **73**, 1146–1151.

ROLLINSON, D., STOTHARD, J. R., JONES, C. S., LOCKYER, A. E., DE SOUZA, C. P. & NOBLE, L. R. (1998). Molecular characterization of intermediate snail hosts and the search for resistance genes. *Memorias do Instituto Oswaldo Cruz* **93** (Suppl. 1), 111–116.

SMINIA, T. (1981). Phagocytic cells in mollusks. In *Aspects of Developmental and Comparative Immunology* (ed. Solomon, J. B.), pp. 125–132. Oxford, Pergamon.

SULLIVAN, J. T. & RICHARDS, C. S. (1981). *Schistosoma mansoni*, NIH-SM-PR-2 strain, in susceptible and nonsusceptible stocks of *Biomphalaria glabrata*: comparative histology. *Journal of Parasitology* **67**, 702–708.

SULLIVAN, J. T. & SPENCE, J. V. (1994). Transfer of resistance to *Schistosoma mansoni* in *Biomphalaria glabrata* by allografts of amoebocyte-producing organ. *Journal of Parasitology* **80**, 449–453.

SULLIVAN, J. T. & SPENCE, J. V. (1999). Factors affecting adoptive transfer of resistance to *Schistosoma mansoni* in the snail intermediate host, *Biomphalaria glabrata*. *Journal of Parasitology* **85**, 1065–1071.

TANAKA, M., TANAKA, T., INAZAWA, J., NAGAFUCHI, S.,

MUTSUI, Y., KAUKAS, A., JOHNSTON, D. A. & ROLLINSON, D. (1997). Proceedings of the schistosome genome project. *Memorias do Instituto Oswaldo Cruz* **92**, 829–834.

TSAI, M., MARX, K. A., ISMAIL, M. M. & TAO, L. (2000). Randomly amplified polymorphic DNA (RAPD) polymerase chain reaction assay for identification of *Schistosoma mansoni* strains sensitive or tolerant to anti-schistosomal drugs. *Journal of Parasitology* **86**, 146–149.

VIDIGAL, T. H. D. A., NETO, E. D., CARVALHO, O. D. S. & SIMPSON, A. J. G. (1994). *Biomphalaria glabrata*: extensive genetic variation in Brazilian isolates revealed by random amplified polymorphic DNA analysis. *Experimental Parasitology* **79**, 187–194.

VIDIGAL, T. H. D. A., SPATZ, L., NUNES, D. N., SIMPSON, A. J. G., CARVALHO, O. S. & NETO, E. D. (1998). *Biomphalaria* spp: identification of the intermediate snail hosts of *Schistosoma mansoni* by polymerase chain reaction amplification and restriction enzyme digestion of the ribosomal RNA gene intergenic spacer. *Experimental Parasitology* **89**, 180–187.

YOSHINO, T. P. & LAURSEN, J. R. (1995). Production of *Schistosoma mansoni* daughter sporocysts from mother sporocysts maintained in synxenic culture with *Biomphalaria glabrata* embryonic (Bge) cells. *Journal of Parasitology* **81**, 714–722.

YOSHINO, T. P., COUSTAU, C., MODAT, S. & CASTILLO, M. G. (1999). The *Biomphalaria glabrata* embryonic (BGE) molluscan cell line: establishment of an *in vitro* cellular model for the study of snail host-parastie interactions. *Malacologia* **41**, 331–343.

Molecular approaches in the study of *Biomphalaria glabrata* – *Schistosoma mansoni* interactions: linkage analysis and gene expression profiling

C. S. JONES[1], A. E. LOCKYER[1,2], D. ROLLINSON[2] *and* L. R. NOBLE[1]

[1] *Zoology Department, Aberdeen University, Tillydrone Avenue, Aberdeen, AB24 2TZ, UK*
[2] *Wolfson Wellcome Biomedical Laboratories, Zoology Department, Natural History Museum, Cromwell Road, London SW7 5BD, UK*

SUMMARY

Gene mapping and the generation of linkage groups are fundamental to an understanding of the organization and relationships of genes and marker sequences, providing a framework with which to investigate their association with traits of interest. The abundance of techniques available for generating polymorphic molecular markers, and recent advances in high throughput screening, have allowed the extension of map analysis to the tropical freshwater snail *Biomphalaria glabrata*, an important intermediate host for *Schistosoma mansoni*. Direct comparison of gene expression by differential display screening, without prior identification of candidate genes, can be combined with mapping to quantify the involvement of specific sequences in the schistosome resistance response, and other important host–parasite interactions. Here we discuss the application of current and emergent technologies to gene characterization and linkage analysis in snail–schistosome interactions. Preliminary results from the analysis of comparative gene expression in resistant and susceptible snails are also presented.

Key words: *Biomphalaria glabrata*, *Schistosoma mansoni*, linkage analyses, comparative gene expression.

INTRODUCTION

Current schistosomiasis control measures are based largely on public hygiene awareness programmes, chemotherapy to kill adult worms, and broad-based approaches to destroy the intermediate host snails. Snail control by molluscicide application or indirectly by habitat modification, may incur considerable collateral damage to the ecosystem (World Health Organization, 1993, 1997). Such measures can be locally effective, but require sustained input, generally in the form of both economic support and ongoing education of local inhabitants. Access to the requisite resources is often sporadic or unavailable in many areas where schistosomiasis constitutes a major public health problem. As yet an effective vaccine remains elusive (Bergquist & Colley, 1998; Gryseels, 2000) and there is a case for alternative sustainable snail-mediated control strategies. Research in this area is therefore timely, especially in view of the growing evidence for praziquantel resistance in schistosomes, promoting concerns that the parasite may become resistant to the most effective available drug (Fallon *et al.* 1996; Bennett *et al.* 1997). An alternative intermediate host-based strategy might involve intervention aimed at reducing or eradicating cercarial transmission (Lardans & Dissous, 1998). Snails incompatible with schistosome development/transmission (for a discussion of the forms in-

Author for correspondence: Dr. Catherine S. Jones.
Tel: 01224 272403. Fax: 01224 272396.
E-mail: c.s.jones@abdn.ac.uk

compatibility may take see Bayne, 1991; De Souza *et al.* 1997) have long been advocated as a simple control, either replacing susceptible snails or acting as decoys which block cercarial development (Hubendick, 1958). However, to be maximally effective such approaches require a thorough knowledge of the genetics of the host–parasite relationship.

Snail stocks incompatible to infection by schistosomes provide an opportunity to investigate the genetic basis of their resistance to this parasite. In the long term, identification and characterization of genomic regions associated with resistance to schistosomes may provide insights into the mechanism of snail defence responses. In the short term, the production of reliable molecular markers for identification of incompatible snails may be achieved. These would facilitate estimation of the selective pressures acting on incompatible and compatible genotypes, permitting predictive models of their spread in natural populations to be tested. Such studies would contribute to a general understanding of host–parasite interactions at the population level, ultimately benefiting the design of sustainable control measures.

To identify markers associated with resistance, molecular analysis of snail–schistosome systems demands laboratory maintenance and carefully controlled breeding programmes of snail and parasites, using isolates and strains of known pedigree and provenance. Many intermediate host species present difficulties for continued laboratory culture, however *Biomphalaria glabrata*, an intermediate host of

Parasitology (2001), **123**, S181–S196. © 2001 Cambridge University Press
DOI: 10.1017/S0031182001008174

Schistosoma mansoni, is the exception to this rule. This species has had a long history of laboratory culture, existing as a series of inbred lines which can be easily and reliably maintained under laboratory conditions (Richards, 1970; also see Lewis *et al.* this supplement). As such, *B. glabrata* provides a useful experimental system for investigating parasite–snail interactions.

Although genetic mapping is a powerful tool for identifying and ultimately determining the functional relationships of genes involved in resistance to parasites and modes of host and parasite interaction, no linkage or physical maps for *B. glabrata* have been published to date. Several experiments are currently underway in our laboratory to identify regions of the snail genome that confer resistance or susceptibility to schistosomes. Part of these experiments involve a systematic scan of the entire genome using large numbers of widely dispersed hypervariable DNA markers to genotype individuals within families. Here, we will outline the importance of linkage maps, offer our observations on the various methodologies and markers available for linkage mapping, and assess their utility in elucidation of the *B. glabrata* genome.

While mapping promises to contribute to a better understanding of the molecular genetic basis and evolution of resistance to schistosomes, examination of gene expression provides a more direct approach to identifying markers associated with resistance and ultimately defining the molecular basis of snail–trematode interactions. The intricacies of parasitic infection, the snail response to the parasite, particularly if it is resistant to infection, the schistosome response to the snail and the effect these have on parasite development, all require investigation. During recent years there have been considerable advances in the analysis of gene expression and in the technologies available for its study in freshwater snails, some of which we will discuss here. With the appropriate molecular tools and snail crosses, the search for markers associated with resistance to schistosomes has begun in earnest (Lockyer *et al.* 1998; Rollinson *et al.* 1998; Knight *et al.* 1999; Knight, Ongele & Lewis, 2000).

THE WHYS AND WHEREFORES OF MAPPING

Linkage mapping

A linkage map is constructed by segregation analyses; statistical analyses of the frequency of recombination between markers in the progeny of heterozygous parents. Linkage mapping is based on the observation of the genetic recombination events which occur as a result of crossing over between homologous chromosomes during meioses. Genetic markers located on different chromosomes, or far apart on the same chromosome, in most cases effectively show independent segregation, whereas physically linked markers exhibit co-segregation; the closer the spacing between markers on a chromosome, the lower the frequency of recombination. A linkage map contains information on the relative order of the markers along the chromosome as well as the map distance between markers; map distances are measured in centiMorgans (cM), with one cM being the distance over which recombination occurs once in a hundred meioses, which, in a genome the size of humans, relates to approximately 1 Mb.

Hence, in organisms that undergo sexual recombination and are amenable to laboratory breeding, genetic mapping can be used effectively to localise sequences that contribute to heritable phenotypes. A primary genetic linkage map, consisting of easily scored, polymorphic, marker loci spaced throughout a genome is a very a valuable framework for detailed molecular genetic analysis in any organism. Although the methodologies used to construct linkage maps can lead to errors, they remain an important standard to which physical maps are referenced and have the advantage, in tractable organisms, that they can be constructed rapidly, often combining a variety of molecular and phenotypic markers which need not be characterized in detail.

The requirements for a linkage mapping population are few; confirmed outcrossing between two individuals with fixed (homozygous) differences at a variety of scorable loci to produce a heterozygous F1 generation. Although *B. glabrata* is a facultatively self-fertile hermaphrodite (possessing the ability to both self- and cross-fertilize, when paired with a suitable compatible snail), it will preferentially outcross (Newton, 1953). Nevertheless, it is essential to confirm that progeny from paired snails are the product of outcrossing using molecular markers such as Random Amplified Polymorphic DNAs (RAPDs) (Vernon, Jones & Noble, 1995) or microsatellites (Jones *et al.* 1999), combined with visible markers such as body pigmentation (Richards, 1970). The main criteria for choosing parents for a laboratory cross is that they differ in phenotype for the trait of interest (e.g. resistance to *S. mansoni*). The F1s are allowed to cross *inter se* or, in the case of *B. glabrata*, to self-fertilize; this produces the same result as F1 individuals from inbred lines as each snail possesses identical unrecombined parental homologous chromosomes. Following recombination between homologues, an F2 segregating (mapping) population will contain a number of individuals that are recombinant between a variety of loci. Many individuals from three generations are scored for numerous polymorphic loci to determine the pattern of inheritance for each allele and to calculate recombination frequencies between marker pairs. The association between an inherited allele and resistant phenotype can be used to identify chromosomal regions that contain genes affecting resistance. This genetic approach, which uses genotype–phenotype associations to uncover markers linked to resistance, requires no *a priori* knowledge of biochemical events involved in the host–parasite interaction, and negates the need for prior identification of candidate genes.

If a trait is highly heritable, and a single gene controls most of the phenotypic variation in the progeny, then polymorphic markers tightly linked to the genomic region carrying that gene ought to co-segregate exactly with the phenotype. This will localize a crossover-defined interval that contains the gene controlling the inheritance of the trait. However, other factors dictate the ease with which trait-conferring loci can be mapped, including the number of genes involved and the magnitude of their effects, as well any the effect of 'environment' on expression. Map distances generated by linkage analysis are based on recombination frequencies in the F2 and, hence, require the unambiguous segregation of individual markers, unobscured by the effects of other genes or environmental variation. While molecular markers usually conform to these expectations, the genetic basis of some phenotypic traits are confounded by both genetic and environmental effects. This is particularly true of quantitative traits, including disease susceptibility, many of which are determined by the effects and interactions of multiple genes and so are quantitative; any loci which can be shown to have a significant effect are termed quantitative trait loci (QTLs). Mapping QTLs calls for an entirely different experimental approach and statistically more demanding analyses.

Experiments on selected lines of *B. glabrata* have established that schistosome resistance has a genetic basis and adult resistance (as defined by Richards, Knight & Lewis, 1992) is a Mendelian trait governed by a single locus, with resistance dominant (Richards, 1970; for a review see Richards, Knight & Lewis, 1992 and this supplement). To identify molecular markers associated with adult resistance, Knight and colleagues (1999) performed genetic crosses between a resistant and susceptible line. The F1 showed the resistant phenotype only and F2 progeny, when exposed to *S. mansoni*, displayed an approximate 3:1 segregation ratio for resistant versus susceptible phenotype, again confirming that this trait is controlled by a single dominant gene. However, in our laboratory, F2 progeny of crosses between the same resistant line and a different susceptible line, exposed to a different *S. mansoni* strain, did not segregate according to Mendelian expectations of resistance being determined by a single dominant gene (an approximate ratio of 1:1·5 segregation ratio of resistant:susceptible). Instead resistance seemed to approximate to a continuous variable, according to cercarial shedding counts, with truly resistant snails the minority class. The dominance hierarchy of the resistance gene, when

crossed onto a novel genetic background, may have altered, in addition to other genes exerting a marked influence in response to a novel parasite. This suggests that, in other snail–parasite combinations at least, adult resistance may actually be under the influence of a number of other independently assorting genes, possibly with major effects, each contributing to the resistance phenotype. Hence, the resistance trait observed in our crosses may be a quantitative, and not a single gene effect. Similarly, Richards *et al.* (1992) suggest that juvenile resistance is a polygenic trait (for further details see Lewis *et al.* this supplement). However, snail–trematode compatibility may result from the ability of the parasite to interfere with host internal defences (Bayne & Yoshino, 1989; Duvaux-Miret *et al.* 1992; Dissous, Grzych & Capron, 1986; Dissous & Capron, 1995). Such studies suggest that compatibility is a very complex phenomenon, resulting from many variables relating to the genetic diversity of both the intermediate host and the trematode. Despite this complex scenario, resolution of polygenic traits (including juvenile susceptibility) into discrete quantitative trait loci (QTL) should be possible through the development of high resolution linkage maps in conjunction with molecular approaches.

QTL mapping

If resistance to schistosomes in *B. glabrata* is truly a quantitative trait, it is more appropriate to use a strategy which takes account of this to map resistance genes. The statistical detection of resistance QTL depends on examining the patterns of inheritance of and association between molecular markers and quantitative traits in the F2 generation. Once detected, the QTL are localized on the genetic map and their contribution to the observed phenotypic variation is determined. The linkage between a genetic marker and importance/effect of its QTL can be assessed in a number of ways including multiple regression analysis (for binary data), general linear modelling, interval mapping and Bayesian approaches (Lynch & Walsh, 1998). These methods enable the estimation of all the genetic parameters associated with a particular resistance phenotype: position of QTLs on the chromosome(s), amount of genetic variance contributed by each QTL and any QTL interactions (Lynch & Walsh, 1998; Hoeschele *et al.* 1997).

Perhaps the approach with the greatest utility is that of interval mapping (Lander & Botstein, 1989). As each QTL may have a considerably different effect on the magnitude of the trait, an adaptation of the logarithm of odds (LOD) score of genetic analysis is used to estimate the statistical likelihood of linkage between the trait and the marker in question. A threshold value, determined by the number of loci examined and the estimated size of the genome, is calculated and where along the genetic map (effectively, along the chromosome) the LOD score exceeds that value a QTL is likely to exist. Hence, interval mapping can indicate which markers are located closest to QTLs, and suggest which regions of the genome would profit from the generation of additional markers.

To apply interval mapping to the *B. glabrata*–schistosome system to assess QTLs for the resistance phenotype involves accurately determining the comparative resistance of as many F2 individuals as possible to exposure to a strain of schistosome. The more precisely the resistance phenotype of the individual snails can be determined, the more confident the placing of QTL(s) on the linkage map. This approach requires neither a single large family nor that existing markers be closely linked to the trait of interest, and data from experimental replicates is additive.

MARKERS FOR MAPPING

A prerequisite for the development of a good linkage map is the availability of a large number of genetic markers. Ideally these should be sufficiently polymorphic, selectively neutral, co-dominant, abundant, evenly distributed throughout the genome, easy to assay (with high reproducibility) and readily exchangeable between laboratories. It is difficult to find a molecular marker that meets all the above criteria, although most of those used fulfil at least some of the required merits (Table 1). It is often the technical problems associated with each approach, or conversely their ease of use, which dictates the eventual choice. The optimal approach we have adopted is a mixed strategy, using high-density random marker maps anchored with a smaller set of reliable, highly polymorphic microsatellite loci and several expressed sequence markers, to allow comparison of random marker maps generated in different laboratories. In future, additional markers from other laboratories may be integrated into the existing linkage groupings to improve map resolution and accuracy.

Our intention is not to list exhaustively all possible marker systems but to discuss only those of which we have most experience and are using to elucidate the *B. glabrata* genome. A more comprehensive and practical set of reviews is given in Dear (1997), including HAPPY mapping, an approach not outlined here.

Microsatellites or simple sequence repeats

Microsatellites (Simple Sequence Repeats – SSR) (Tautz, 1989) are the preferred marker for genetic mapping because of their abundance, high heterozygosity, hypervariability, co-dominance and ease of PCR typing, providing reliable and conveniently scored polymorphisms from minute quantities of DNA (for reviews see Chambers & Macavoy, 2000; Goldstein & Schlotterer, 1999). Chromosomal locations of many microsatellites have been established in other organisms (Dietrich *et al.* 1996); their distribution is uniform across most intergenic regions (except telomeres) and repeats are rare (and deleterious) in coding regions. The basis for microsatellite polymorphisms in di-, tri-, tetra- or penta-nucleotide tandem repeat sequences is size difference between alleles. This variation is caused by different numbers of tandem repeats revealed when PCR products are amplified using primers corresponding to unique sequence flanking regions, analysed by high resolution polyacrylamide electrophoresis.

The choice of which type of SSR to use in mapping studies relates mostly to technical issues concerned with their detection/scoring. For instance, di-nucleotide repeats often show one or more 'stutter' or shadow bands (multiple PCR products that are one to two repeats shorter than the genuine allele, and usually less intense than the full-length product). Such artefacts mean that care and experience are required to interpret gels accurately, leading to a preference for loci containing tri- or tetra-nucleotide repeat motifs. Conversely, other factors often mitigate against the use of tetra-nucleotide repeats; di-nucleotides occur more frequently within the genome than tri- or tetra-nucleotide arrays, an important factor when, of necessity, most mapping studies require the maximum practical number of scorable markers. Other practical requirements include having sufficient loci with distinct, non-overlapping alleles, which can be amplified under a standard set of PCR conditions. Multiplexing of PCRs simplifies the design of large screening experiments required for linkage analyses of extended pedigrees. Although typing of microsatellite markers is slow relative to techniques using random primers, the application of automated genetic linkage analyses using fluorescent microsatellite markers will improve the efficiency and accuracy of genotyping (Mansfield *et al.* 1994; Wang, Kafatos & Zheng, 1999).

The major drawback to the use of microsatellites is that, unless dealing with a organism for which much genomic sequence data is publicly available, every marker has to be isolated *de novo* from a genomic library of the species of interest. In the majority of species, development of microsatellite markers is technically demanding, costly and time consuming, although a number of simple protocol guides are available (Bruford *et al.* 1996; Schlotterer, 1998). In many cases, microsatellites originally developed from a single focal species can be adapted for use in closely related taxa (Primmer & Ellegren, 1998). Many microsatellite markers have been isolated from *B. glabrata* (Jones *et al.* 1999; Mavarez *et al.* 2000). Recently, we have shown that the Neotropical

Molecular study of B. glabrata–S. mansoni *interactions*

Table 1. Properties of marker systems for mapping studies

	Marker system				
	RAPDs	AFLPs	Microsatellites	Expressed sequences	SNPs
Rationale	DNA amplification with short arbitrary PCR primers; multilocus fragment profiles/genome scanning	Use of specific PCR primers complimentary to adaptor sequences ligated to restriction fragments; multilocus fragment profiles/genome scanning	PCR of simple sequence repeats; single locus profiles	Restriction digestion, Southern blotting and hybridization or direct PCR or restriction digestion of PCR products; single locus profiles	PCR of single nucleotide polymorphisms; single locus profiles
Polymorphism type	Single base substitutions (in priming site) Insertions/Deletions (within & between sites)	Single base substitutions (in restriction sites) Insertions/Deletions (within & between sites)	Changes in lengths of repeats	Single base substitutions Insertions/Deletions	Single base substitutions
Reproducibility	Low	High	High	High	High
Abundance in genome	High	High	Moderate	Moderate	Very high
Amount of polymorphism	Moderate/High (primer dependent)	Moderate	High	Low/Moderate	Low/Moderate
Dominance	Dominant	Dominant	Co-dominant	Co-dominant	Co-dominant
Quantity of DNA needed for assays	10–25 ng	100–500 ng	10–25 ng	10 ng–10 µg[1]	10–25 ng
Necessity for prior sequence information	No	No	Yes	Yes	Yes
Development and set-up costs	Low	Low	High	Moderate/High	High[2]

[1] Amount of DNA required depends on the method of genotyping; large amounts are required for restriction digestion of genomic DNA, followed by Southern blotting and hybridization, wheras small quantities are sufficient for PCR-based approaches.
[2] Costs high unless database sequences exist.

B. glabrata has an African affinity (Campbell *et al.* 2000) which is further supported by the observation that microsatellite loci of African *B. pfeifferi* are amplified more readily than those of any other Neotropical taxon using primers designed specifically from *B. glabrata* (Jones *et al.* 1999). Hence, microsatellites isolated from the closely related African species, *B. pfeifferi* (C. S. Jones, unpublished data; Charbonnel *et al.* 2000) are also being incorporated into our *B. glabrata* linkage analyses.

Random markers

Many linkage maps have been constructed using random marker approaches such as RAPDs (Welsh & Mcclelland, 1990; Williams *et al.* 1990), or Amplified Fragment Length Polymorphisms (AFLPs) (Vos *et al.* 1995). RAPDs use single, short oligonucleotide primers of arbitrary sequence to initiate DNA strand synthesis under low stringency at a number of complementary binding sites scattered randomly throughout the genome. RAPD fragment profiles visualized on ethidium bromide-stained agarose gels are technically simple and straightforward. However, the main disadvantages of RAPDs for gene mapping include poor reliability and fidelity, and sensitivity to experimental conditions, hence, strict standardization of all reagents and protocols is required (Grosberg, Leviton & Cameron, 1996).

AFLPs combine PCR and RFLPs (restriction fragment length polymorphisms) to assay anonymous genomic sequences. Genomic DNA is digested with two restriction enzymes simultaneously (a rare and common cutter respectively), followed by ligation of double-stranded adapters to the digested fragments, which are then amplified using two PCR primers, containing the common sequences of the adapters and 1–3 arbitrary nucleotides as selective sequences at their 3′ end. Only that population of restriction fragments in which the bases flanking the restriction site match the selective nucleotides will be amplified. As with RAPDs, AFLPs require no prior knowledge of template DNA sequences, potentially detect variation over the entire genome but they have the major advantage of being robust and reliable because amplification is performed under more stringent reaction conditions. However, the drawbacks of AFLPs compared with RAPDs include the relative complexity of the procedure and visualization methods, and the requirement of larger quantities of genomic DNA. The latter precludes their protracted use on small organisms, such as freshwater snails, because of the large number of markers which must be screened per individual for effective linkage analysis.

One of the greatest advantages of both AFLP and RAPD approaches is the simultaneous amplification of large numbers of polymorphic DNA fragments providing information for many putative loci in a single assay. Hence, for rapid construction of dense genetic maps suitable for QTL analyses, random markers are by far the fastest and most inexpensive route. Rapid analyses of random markers however come at a price. In our laboratory, some genotyping errors, such as non-Mendelian segregation of a few fragments, have been encountered although these are limited to only a few primers/primer combinations with AFLP and RAPD markers. Although both RAPDs and AFLPs are typically dominant markers which lack the utility and resolving power of co-dominant systems for linkage and pedigree analyses, they are still of considerable use. Given an almost inexhaustible supply of arbitrary primers there is virtually no limit to the numbers of these markers in a genome. Yet, random markers are much less transferable among laboratories, crosses and species, hence alignment of maps produced from different crosses may be impossible unless markers are cloned and sequenced or amplified products are digested with restriction enzymes to establish their homology (Rieseberg, 1996).

Expressed sequence tags (ESTs) and other expressed sequences

Expressed sequence tags (ESTs) are short (< 750 bp), single pass sequences generated from randomly selected cDNA library clones (Adams *et al.* 1991). The EST approach is useful not only for identification of new genes but also for genome mapping. The availability of human ESTs enabled the early mapping of over 16 000 genes in the human genome (Schuler *et al.* 1996). EST markers are advantageous for mapping because they unambiguously represent genes, which may serve as useful anchor points for comparative mapping, should useful polymorphisms be found in either the EST or the gene's introns. Also, development of EST markers is fast, straightforward and technically easy.

Using whole body or cerebral ganglia *B. glabrata* ESTs as probes, Knight and colleagues (1998) have demonstrated the occurrence of RFLPs between parasite resistant and susceptible snail stocks, which may prove useful for genetic linkage mapping. Single pass sequencing analysis produces sufficient data for PCR primer design from the 5′ or 3′ ends of the cDNAs. Direct PCR analysis of genomic DNA of stocks differing in phenotype of the trait of interest (e.g. resistance to *S. mansoni*) from which F2 reference families are produced would indicate whether these markers reveal length polymorphisms useful for linkage analyses.

Future of mapping studies

Technological advances have led to Single Nucleotide Polymorphisms (SNPs) being identified as the new marker of choice for genetic mapping (Kruglyak, 1997) for organisms with large amounts of sequence data available. SNPs are single-base differences in the DNA sequence observed between individuals in a population (Zhao *et al.* 1998; Brookes, 1999). These markers are highly abundant, stable, and the most common form of genomic variation. They account for 90% of sequence variation in humans and are estimated to occur once every 1000 base pairs between any two individuals (Wang *et al.* 1998; Cargill *et al.* 1999), giving roughly 3 million in the human genome. These markers are relatively easy to detect and are amenable to high throughput, low cost, automated typing. Each SNP has only two alleles, compared with microsatellites which tend to be hypervariable, yet the abundance and density of SNPs makes them potentially much more informative (Wang *et al.* 1998). Additionally, genotyping of this type of marker requires only a presence/absence assay rather than length determination.

However, in most organisms an efficient method for reliable identification of new SNPs remains a significant challenge (Buetow, Edmonson & Cassidy 1999), but there is presently intensive technological investment in this field and it is very likely that large-scale screenings of SNPs could be performed at low cost in the future. Matrix-Assisted Laser Desorption-Ionization-Time-Of-Flight (MALDI-TOF) mass spectrometry is just one promising tool for the high throughput screening of SNPs (Griffin & Smith, 2000). But, currently, owing to the lack of sequence data available for freshwater snails, a more cost effective and productive strategy is to identify and genotype SNPs in candidate genomic regions highlighted by genetic mapping studies.

READING THE GENETIC MAP AND THE SEARCH FOR RESISTANCE GENES

Once a genomic scan identifies a region containing one or more resistance related QTLs, map-based (positional) cloning is used to isolate the gene(s) responsible, without the need for biochemical or molecular data for the trait. Early positional cloning projects were expensive, time consuming and beyond the scope of resources available for analysis of most species. However, the availability of YAC or BAC libraries (Yeast/Bacterial Artificial Chromosomes) for cloning large fragments of genomic DNA, combined with fast, efficient techniques for screening many thousands of markers, has made positional cloning more feasible for at least some species of non-model organisms. The full potential of this approach will not be realized until large insert *B. glabrata* libraries and further markers are made more generally available.

Positional cloning aims to narrow the genetic window in which the gene of interest (the resistance gene) lies. This is achieved by linkage analysis and refined through physical mapping. Firstly, the chromosome or chromosomal segment containing the resistance gene is identified by classical linkage analysis and a saturated map of the region is constructed. To identify large fragments of DNA that are near to, or contain, the resistance gene, *B. glabrata* BAC or YAC libraries should be screened with markers suggested to be closest to the gene and flanking it on either side. It possible to move from the seed clones containing the marker tags to the clone in which the resistance gene must lie, through a series of overlapping clones by a combination of chromosome walking or jumping (Poustka *et al.* 1987) and additional linkage analysis. If the locus is saturated with markers, it is possible to go straight to the gene by chromosome landing (Tanksley, Ganal & Martin 1995). The defined genomic region is then sequenced and analysed to identify predicted candidate genes, which should be confirmed by expression studies and further characterized to determine the molecular basis of resistance.

Nevertheless, positional cloning of a QTL can be a formidable undertaking. The genotype at any given QTL locus cannot be inferred directly from the phenotype due to the interaction of other QTLs and environmental effects. There can be little doubt that alternative strategies for resistant gene isolation involving essentially non-map based methods such as mRNA expression analyses will become increasingly important in 'map poor' species such as *B. glabrata*.

COMPARATIVE GENE EXPRESSION ANALYSIS

A variety of strategies are available for comparative gene expression analysis including identification and investigation of candidate genes; subtractive hybridization techniques, sequencing based approaches such as ESTs, the Serial Analysis of Gene Expression (SAGE), PCR-based approaches such as Differential Display (DD) and cDNA Representational Difference Analysis (RDA) and microarray analysis. Table 2 compares various approaches, and each is discussed below with regard to actual or potential application to snail–parasite investigation. For a review, including more detailed descriptions of the techniques, see Kozian & Kirschbaum (1999).

Classical approaches

Approaches to investigate gene expression in snails have included the identification and analysis of candidate genes, already characterized in other organisms and then identified in snails on the basis of sequence homology. For example, β-integrin (Davids, Wu & Yoshino, 1999) and a receptor for activated protein kinase c (RACK) (Lardans *et al.*

Table 2. Properties of comparative gene expression systems

| | Comparative gene expression system | | | | | | |
	Candidate gene characterization	Subtractive hybridization	cDNA RDA	Differential display	Expressed Sequence Tags	SAGE	DNA microarrays	Proteomics
Amount of RNA /tissue required	High	High	Moderate	Low	High	Moderate	Low	Moderate (tissue)
Speed of analysis	Slow	Slow	Moderate	Quick	Slow	Quick	Moderate	Moderate
Necessity for prior sequence information	Yes	No	No	No	No	Yes	Either	Either
Level of investigation	mRNA	mRNA	mRNA	mRNA	mRNA	mRNA	mRNA	Protein
Comparison	n/a	Dual	Unidirectional	Multiple	Multiple	Multiple	Dual /multiple	Multiple
Output	Informative	Full/long length cDNAs	cDNA fragments	cDNA fragments	Many cDNA clones/long length sequences	Sequence tags	Gene or clone identified	Peptide mass fingerprint/Peptide sequence tag
Comments				Identification of false positives	Identifies SNPs Redundancy of sequences	High throughput Quantitative	Initial set up cost high	Low abundance proteins hard to identify

1998) have been identified from *B. glabrata*. This strategy has been a mainstay of genetic investigation. Target genes are those which, due to function, would be predicted to be involved in host–parasite interactions. Initial isolation of such candidate genes in snails may involve degenerate PCR based on sequence homology, or cross-species hybridization. The full length sequence may then be obtained directly, if identified from a high quality cDNA library, or indirectly by cloning the 5′ and 3′ ends separately by Rapid Amplification of cDNA Ends (RACE) (Frohman, Dush & Martin, 1988). Traditional methods for investigation of the expression of genes identified in this way include northern blotting (Alwine, Kemp & Stark, 1977) and RNA protection assays (Lee & Costlow, 1987). These techniques are indispensable for detailed characterization of the sequence and expression of genes, but are conducted on a gene by gene basis and require the initial identification of candidate genes from other organisms. What if previously unknown genes are involved?

Approaches requiring no candidate genes include subtractive hybridization techniques (reviewed by Sangerstrom, Sun & Sive, 1997), and this strategy has been used successfully to produce a resistant genotype-enriched *B. glabrata* cDNA library (Miller *et al.* 1996). cDNAs from a susceptible snail line and a resistant snail line were hybridized and the resulting subset of non-hybridized cDNA cloned. The library was screened and two transcripts identified which showed elevated expression in resistant snail lines; one was a novel, developmentally-regulated albumin gland transcript which also contained a useful RFLP. This approach requires much material for the generation of cDNA libraries, limiting its application. Further, the technique favours identification of abundant transcripts, and transcripts associated with the snail resistance may be rare. An advantage of cloning novel products from a cDNA library is that a substantial fragment or the full length sequence is accessible.

PCR-based approaches

Classical approaches to the analysis of mRNA expression such as northern blotting, the screening of cDNA libraries by plaque hybridization or subtractive cloning are time consuming and material expensive. The advent of PCR-based techniques, which are now widely used for comparative gene expression analysis, has led to an explosion of papers detailing differential gene expression. PCR techniques in general require much less starting material and are quicker and less expensive than traditional techniques, making them useful for systems where traditional methodologies could not be applied. Two such techniques for differential gene expression analysis are cDNA RDA and DD.

RDA is a cross between subtractive hybridization and PCR-based methodologies. Originally developed to examine the difference between genomes, using congenic strains (Lisitsyn, Lisitsyn & Wigler, 1993; Lisitsyn *et al.* 1994) RDA has since been adapted for cDNA analysis (Hubank & Schatz, 1994; O'Neill & Sinclair, 1997). Briefly, the technique involves the synthesis of cDNA from two mRNA populations (control and experimental). Following restriction enzyme digestion, linkers are ligated onto the experimental cDNA to allow PCR amplification. The two populations are then hybridized (with an excess of the control cDNA) and then reamplified, enriching the cDNAs found only in the experimental fraction. Although the use of PCR allows smaller quantities of starting material to be employed (Michiels *et al.* 1998) this is still a subtractive technique which only allows identification of the difference between two mRNA populations, not a comparison of many different tissue types, time periods or strains. It is also unidirectional, enriching the mRNAs found only in one population, and must be carried out reciprocally to identify those in the other population. Some of these problems have been overcome by gene expression profiling techniques.

Differential display

RNA profiling techniques such as differential display (Liang & Pardee, 1992) and RNA arbitrarily primed PCR (RAP-PCR – Welsh *et al.* 1992) provide for more detailed analysis of gene expression in snails. These techniques produce cDNA profiles with random primers using only small quantities of RNA, enabling the parallel investigation of a large number of tissues or time periods. The use of random primers means that no previous sequence knowledge is required and there is the potential for high throughput screening of many different samples with many primers. Using a differential display protocol adapted from Rothschild, Brewer & Bowden (1997), Lockyer *et al.* (2000) investigated changes in gene expression in *B. glabrata* when exposed to *S. mansoni*. This approach, which combined the advantages of both DD and RAP-PCR by synthesizing cDNA using an anchored oligo dT primer and amplifying this with 18mer random primers to produce stable profiles, was successful in identifying 6 fragments from profiles generated from ovotestis and mantle tissue. However, by semi-quantitative RT-PCR it was possible to confirm differential expression of only one of these fragments, which was down-regulated in the ovotestis of exposed snails (Lockyer *et al.* 2000). This partial transcript showed considerable similarity to genes in the cytochrome p450 superfamily. This experiment demonstrated the suitability of the technique for identifying differential expressed transcripts in

Primer 1   Primer 2
Resistant Susceptible Resistant Susceptible

Ladder   E   C   E   C   E   C   E   C   Ladder

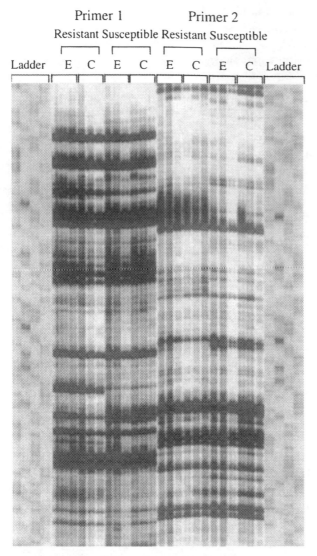

Fig. 1. A typical differential display profile obtained from resistant and susceptible snails, demonstrating that the profiles from the two snail lines are similar for two 18 mer RAP primers. Profiles are generated by reverse transcribing 1 µg, 500 ng and 250 ng of total RNA. A diluted aliquot of this is used for amplification with IRD700 labeled primers. The resulting fragments are visualized on a LICOR automated sequencer. For each resistant and susceptible snail sample 3 concentrations of RNA are shown for 2 labeled primers. E, exposed snails; C, control snails. Outside lanes: sequencing ladders as size standards.

snails; however it also confirmed that, due to the high number of false positives generated (Debouck, 1995), it is imperative to confirm differential expression using northern blotting or semi-quantitative RT-PCR (Yeates & Powis, 1997; Yoshikawa *et al.* 1998). The latter has been particularly useful for confirming changed expression from minute amounts of mRNA.

In a similar study, Knight *et al.* (2000) compared schistosome-exposed and unexposed *B. glabrata*, using differential display. They identified several fragments from DD profiles, whose expression

Resistant   Susceptible
Exposed Control Exposed Control

1 2 3   4 5 6 7 8 9 10 11 12

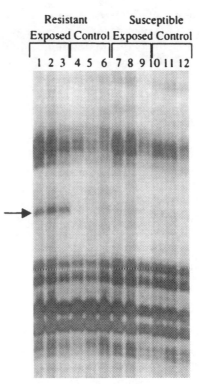

Fig. 2. Identification of differences between profiles from exposed and control resistant snails. For each sample 3 concentrations of RNA were analysed. Lanes 1–3: exposed resistant snails, lanes, 4–6: control resistant snails. Lanes 7–9: exposed susceptible snails, lanes 10–12: control susceptible snails. A 390 bp fragment found only in exposed resistant snails is arrowed.

appeared altered after miracidial exposure, although this was unconfirmed by either semi-quantitiative RT-PCR or northern blotting. Those sequence fragments which had homology to sequences already in GenBANK were a metalloprotease, a reverse transcriptase and $\beta 4$ integrin interactor.

These experiments demonstrated the usefulness of DD in investigating changes in mRNA expression in the snail on parasite exposure. Our previous experiment examined these changes using a strain of *B. glabrata* shown to be 70% resistance to *S. mansoni* in laboratory trials (Lockyer *et al.* 2000). However, it was not possible to distinguish between a specific resistant response and a more general snail response to infection. Therefore, in order to examine specifically the response from resistant snails, an adult 100% resistant and a susceptible snail line have been compared. The DD technique itself has been further modified by the use of a LI-COR automated sequencer to generate profiles from PCRs with fluorescent primers, speeding up the screening process and allowing the examination of more samples with additional primers. Preliminary results show that the cDNA profiles obtained from the two snail lines are similar (Fig. 1) and hence comparable. Initially, RNA extracted from mantle tissue, obtained from both snail strains 6 hours after parasite

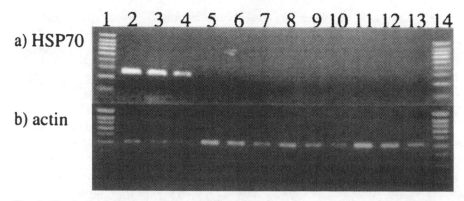

Fig. 3. Confirmation of differential expression: (a) Semi-quantitative RT-PCR using primers specific to the sequenced HSP70-like fragment and limiting the cycles in the PCR until the linear phase is found, shows this transcript to be upregulated in the exposed resistant snails confirming the DD profile. For each sample 3 concentrations of RNA (1 µg, 500 ng and 250 ng) were used to make cDNA. Lanes 2–4 exposed resistant snails, lanes 5–7 control resistant snails, lanes 8–10: exposed susceptible snails, lanes 11–13: control susceptible snails. Lanes 1 and 14: molecular weight marker (100 base pair ladder, Bioline Inc.). (b) Semi-quantitative RT-PCR using primers specific to actin (a non-regulated housekeeping gene), giving a 414 bp product, was used as a control to check that the samples were comparable and confirm that equivalent quantities of RNA has been used for the specific HSP70 primers. Lanes loaded are as for (a).

exposure was examined and compared to unexposed (control) snails. cDNA fragments which were present only in resistant snails that have been exposed to the parasite have been identified.

One example is a ~ 390 bp fragment present only in the exposed resistant snails but absent in control (unexposed) resistant snails and also absent in both control and exposed susceptible snails (Fig. 2). This fragment was successfully excised from the differential display gel, cloned and sequenced. Database searches showed the transcript to be almost identical (97 % identity) to the previously sequenced *B. glabrata* HSP70 (Laursen *et al.* 1997). The up-regulation of this transcript only in exposed resistant snails has been confirmed by semi-quantitative PCR (Fig. 3) using primers specifically designed from the fragment. Semi-quantitative RT-PCR in parallel with actin specific primers (a non-regulated house keeping gene, whose transcript levels are expected to be equal in equivalent mRNA quantities), demonstrated that the samples were comparable. The significance of up-regulation of HSP70, only in resistant snails, has yet to be elucidated.

Sequence-based approaches

Some progress in characterizing the *B. glabrata* genome has also been made by generating expressed sequence tags (ESTs) from cDNA libraries (Knight *et al.* 1998). This approach rapidly generates a database of expressed *B. glabrata* genes containing considerable sequence data and source clones, which will be a long-term valuable resource and may eventually provide much useful information. Many of the ESTs identified are novel, and so function cannot be ascribed by sequence homology to other genes characterized in different organisms. One

potential use for such a resource will be the comparison *in silico* of EST databases to identify differentially expressed transcripts (Lee *et al.* 1995; Vesmatzis *et al.* 1998). This comparative EST approach requires the recorded sequence data to be representative of the tissues from which the cDNA is derived for large numbers of tissues to be sampled; and for parasite-specific data from infected stages to be available, to allow elimination of parasite-derived sequences.

A second sequence-based approach which could potentially be applied to the investigation of snail mRNA expression is SAGE (Bertelsen & Velculescu, 1998; Polyak *et al.* 1997; Velculescu *et al.* 1995), which is most useful for the expression profiling of fully sequenced genes, allowing simultaneous analysis of sequences derived from different tissues. This approach requires the generation of unique sequence tags (10–14 bp) from each cDNA. These tags are then ligated into long concatemers and sequenced. The tags are long enough to identify each source gene from its sequence on the database. The greater the level of expression of a gene, the more common its tag in the dataset. Different cDNA populations can then be compared to discover absolute or relative differences in expression of the source genes. SAGE is a high throughput method but requires large amounts of mRNA (2·5–5·0 µg poly(A) RNA), although recent modifications have reduced this (Datson *et al.* 1999). The chief drawback to this approach for analysis of *B. glabrata* is that it allows investigation only of sequenced genes or ESTs, which currently makes it inappropriate for the analysis of snail gene expression. However, as the amount of sequence data for *B. glabrata* in the database increases this technique will become a real possibility.

Future prospects

New approaches which have potential for analysis of snail gene expression make use of microarray technologies, or DNA chips (for reviews see Gerhold, Rushmore & Caskey, 1999; Lennon, 2000). Conventional differential screening used filter arrays; cDNA clones arrayed on nylon membranes and screened by traditional hybridization techniques with different pools of mRNAs (Gress et al. 1992). DNA microarrays are miniature grids consisting of cDNA tags (clones or PCR-amplified clone inserts) deposited onto a glass slide by high speed printing methods (Schena et al. 1995) or oligonucleotide tags synthesized in situ (DNA chips – Pease et al. 1994). These microarrays have many advantages over traditional filter arrays, including increasing the number of sequences that can be screened, reducing the area for hybridization so that less material is required and allowing faster screening with non-radioactive techniques to generate information on simultaneous and relative expression. It is also possible to construct 'shot-gun' microarrays using random unknown genomic DNA sequences, as have been generated for the investigation of stage-specific gene expression in Plasmodium falciparum (Hayward et al. 2000). To screen any of these arrays, cDNA derived from two mRNA populations is labelled with different fluorescent dyes (e.g. red and green) and hybridized to DNA on the slide. The relative abundance of each transcript is measured by laser scanning of the array. Each spot will fluoresce red or green if cDNA from only one cDNA population is present, yellow if both have hybridized, or will remain black if neither contain that sequence. In this way the relative abundance of cDNAs and therefore mRNA expression between the two can be assessed. The principal advantages of this approach are the ability to analyse expression patterns in a parallel fashion and the immediate interpretation of hybridization results. The progression towards DNA arrays for snail expression analysis using either a random sequence (shot-gun) or cDNA fragment approach is a realistic possibility despite high initial costs, and the expansion of the EST database will enable the quick identification of full length sequences.

A complementary approach to mRNA studies is proteomics (for a reviews see Anderson, Matheson & Steiner, 2000; Pandey & Mann, 2000). A proteome is defined as an entire PROTein population expressed from a genOME (Wasinger et al. 1995). Proteomics describes protein analysis by two dimensional (2D) gel electrophoresis, followed by mass spectrometry and/or protein sequencing. This could be a useful approach with regard to snail gene expression analysis. Sophisticated software tools exist to compare protein spots on 2D gels between different cell types, tissues or disease states, to identify differences and highlight regulated proteins for further analysis. However, the number of proteins in a proteome can far exceed the number of genes present, as one mRNA can give rise to many protein products via alternative splicing or post-translational modifications. 2D gel electrophoresis is not a new technique (O'Farrell, 1975), but its utility has been improved by development of immobilised pH gradients to increase sensitivity and reproducibility, and image analysis systems to allow automated scanning followed by protein isolation and characterization. Hence this approach now allows the quantitative and qualitative expression analysis of thousands of proteins simultaneously. Protein spots of interest can be cut out of the gel for identification and characterization using a variety of mass spectrometry (MS)-based protocols (Andersen & Mann, 2000) including the MALDI-TOF analysis of peptides generated from enzymatic digestion of isolated proteins. These characteristic Peptide Mass Fingerprints (PMF), reveal the accurate masses of the tryptic peptides which can be used to screen known databases to identify the isolated protein. Should this not provide an unambiguous identification, perhaps because, like B. glabrata, full length sequences are not well represented in the databases, then 'peptide sequence tags' (Mann & Wilm, 1994) can be obtained by tandem mass spectrometry (MS/MS), coupled with electrospray ionization (ESI), with a collision cell for induced fragmentation (full method abbreviated to ESI-MS/MS). This combined approach generates additional partial sequence to complement the mass data and these 'protein sequence tags' may then be used to search protein and EST databases with very high specificity. Should this not provide an unambiguous identification, then the design of degenerate PCR primers based on the peptide sequence tag may be used to recover sequences from cDNA libraries or genomic DNA.

One disadvantage of looking directly at proteins is that there is no technique for amplification of peptides equivalent to PCR for nucleic acids and this results in difficulty in investigating low abundance proteins. However, the technique examines expressed proteins directly and allows investigation of the relative abundance of protein products, post-translational modifications, sub-cellular location, turnover, and interaction with other proteins, all of which are not reflected in mRNA analyses.

CONCLUSIONS

It is obvious that the efficient and rapid identification of snail and parasite genes involved in this complex host–parasite relationship would benefit enormously from a multi-disciplinary approach. Thus, our combined strategy using map-based and non-map-based approaches for the identification and characterization of schistosome resistance genes in the freshwater intermediate snail host, *B. glabrata*, is outlined.

Segregation/linkage analysis, genome-wide mapping and candidate gene studies will benefit from the application of the genomic and genetic resources discussed in this review to unravel the combinations of host and parasite genes influencing resistance to schistosomes in intermediate host freshwater snails. Studies using the candidate gene approach are likely to be generally limited by a lack of knowledge of their function in *B. glabrata*. However, as functional genomics of other invertebrates becomes more widely investigated with recent technological advances described in this review, such as microchip- and mass spectrometry-based detection methods for assaying known polymorphisms for mapping purposes, this should improve the extent and efficacy of candidate resistance gene analysis in freshwater snails.

Some progress has already been made in characterizing specific snail genes and identifying differential gene expression using RNA profiling. There is no doubt that, with the progression in techniques available for multiple gene expression analysis, in particular microarrays and proteomics, their application to the study of snail–parasite interactions is imminent.

The identification of genes involved in snail–schistosome interactions, and most especially those involved in the resistance of the snail to the parasite, will advance knowledge of the complex interaction between a host and its parasite. But only through international collaboration and pooling of resources, which has been the hallmark of several successful global genome projects such as that for the helminth parasites (Rollinson & Blackwell, 1999; Williams & Johnston, 1999), will progress in this field be significantly enhanced. Such advances could ultimately lead to marked improvements in public health in both developed and developing countries.

ACKNOWLEDGEMENTS

This work was supported by the BBSRC (Advanced Fellowship to CSJ, grant number 1/AF09056). We would also like to thank Dr Fred Lewis for the resistant *B. glabrata* strain.

REFERENCES

ADAMS, M. D., KELLEY, J. M., GOCAYNE, J. D. *et al.* (1991). Complementary DNA sequencing: expressed sequence tags and human genome project. *Science* **252**, 1651–1656.

ALWINE, J. C., KEMP, D. J. & STARK, G. R. (1977). Method for detection of specific RNAs in agarose gels by transfer to diazobenzyloxymethyl-paper and hybridization with DNA probes. *Proceedings of the National Academy of Sciences, USA* **74**, 5350–5354.

ANDERSEN, J. S. & MANN, M. (2000). Functional genomics by mass spectrometry. *FEBS Letters* **480**, 25–31.

ANDERSON, N. L., MATHESON, A. D. & STEINER, S. (2000). Proteomics: applications in basic and applied biology. *Current Opinion in Biotechnology* **11**, 408–412.

BAYNE, C. J. (1991). Invertebrate host immune mechanisms and parasite escapes. In *Parasite-Host Associations: Coexistence or Conflict?* (ed. Toft, C. A. Aeschlimann, A. & Bolis, L.), pp. 299–315. Oxford University Press, Oxford.

BAYNE, C. J. & YOSHINO, T. P. (1989). Determinants of compatibility in mollusc-trematode parasitism. *American Zoologist* **29**, 399–407.

BENNETT, J. L., DAY, T., FENG-TAO, L., ISMAIL, M. M. & FARGHALY, A. (1997). The development of resistance to antihelmintics: a perspective with an emphasis on the antischistosomal drug Praziquantel. *Experimental Parasitology* **87**, 260–267.

BERGQUIST, N. R. & COLLEY, D. G. (1998). Schistosome vaccines: research to development. *Parasitology Today* **14**, 99–104.

BERTELSEN, A. H. & VELCULESCU, V. E. (1998). High-throughput gene expression analysis using SAGE. *Drug Discovery Today* **3**, 152–159.

BROOKES, A. J. (1999). The essence of SNPs. *Gene* **234**, 177–186.

BRUFORD, M. W., CHEESMAN, D. J., COOTE, T., GREEN, H. A. A., HAINES, S. A., O'RYAN, C. & WILLIAMS, T. R. (1996). Microsatellites and their application to conservation. In *Molecular Genetic Approaches in Conservation* (ed. Smith, T. B. & Wayne, R. K.), pp. 278–297. New York, Oxford University Press.

BUETOW, K. H., EDMONSON, M. N. & CASSIDY, A. B. (1999). Reliable identification of large numbers of candidate SNPs from public EST data. *Nature Genetics* **21**, 323–325.

CAMPBELL, G., JONES, C. S., LOCKYER, A. E., HUGHES, S., BROWN, D., NOBLE, L. R. & ROLLINSON, D. (2000). Molecular evidence supports an African affinity of the Neotropical freshwater gastropod, *Biomphalaria glabrata* (Say, 1818), an intermediate host for *Schistosoma mansoni*. *Proceedings of the Royal Society of London, Series B* **267**, 2351–2358.

CARGILL, M., ALTSHULER, D., IRELAND, J. *et al.* (1999). Characterization of single nucleotide polymorphisms in coding regions of human genes. *Nature Genetics* **22**, 231–237.

CHAMBERS, G. K. & MACAVOY, E. S. (2000). Microsatellites: consensus and controversy. *Comparative Biochemistry and Physiology Part B* **126**, 455–476.

CHARBONNEL, N., ANGERS, B., RAZATAVONJIZAY, R., BREMOND, P. & JARNE, P. (2000). Microsatellite variation in the freshwater snail *Biomphalaria pfeifferi*. *Molecular Ecology* **9**, 1006–1007.

DATSON, N. A., VAN DER PERK-DE JONG, J., VAN DEN BERG, M. P., DE KLOET, E. R. & VREUGDENHIL, E. (1999). MicroSAGE: a modified procedure for serial analysis of gene expression in limited amounts of tissue. *Nucleic Acids Research* **27**, 1300–1307.

DAVIDS, B. J., WU, X.-J. & YOSHINO, T. P. (1999). Cloning of a *β* integrin subunit cDNA from an embryonic cell line derived from the freshwater mollusc, *Biomphalaria glabrata*. *Gene* **228**, 213–223.

DEAR, P. H. (1997). Ed. *Genome Mapping : A Practical Approach*. Oxford, IRL Press.

DEBOUCK, C. (1995). Differential display or differential dismay? *Current Opinion in Biotechnology* **6**, 597–599.

DE SOUZA, C. P., BORGES, C. C., SANTANA, A. G. & ANDRADE, Z. A. (1997). Comparative histopathology of *Biomphalaria glabrata*, *B. tenagophila* and *B. straminea* with variable degrees of resistance to *Schistosoma mansoni* miracidia. *Memorias do Instituto Oswaldo Cruz* **92**, 517–522.

DIETRICH, W. F., MILLER, J., STEEN, R., *et al.* (1996). A comprehensive genetic map of the mouse genome. *Nature* **308**, 149–152.

DISSOUS, C. & CAPRON, A. (1995). Convergent evolution of tropomyosin epitopes. *Parasitology Today* **11**, 45–46.

DISSOUS, C., GRZYCH, J. M. & CAPRON, A. (1986). *Schistosoma mansoni* shares a protective oligosacccharide epitope with freshwater and marine snails. *Nature* **323**, 443–445.

DUVAUX-MIRET, O., STEFANO, G. B., SMITH, E. M., DISSOUS, C. & CAPRON, A. (1992). Immunosuppression in the definitive and intermediate hosts of the human parasite *Schistosoma mansoni* by release of immunoactive neuropeptides. *Proceedings of the National Academy of Sciences, USA* **89**, 778–781.

FALLON, P. G., TAO, L. F., ISMAIL, M. M. & BENNETT, J. L. (1996). Schistosome resistance to Praziquantel: fact or artefact? *Parasitology Today* **12**, 316–320.

FROHMAN, M. A., DUSH, M. K. & MARTIN, G. R. (1988). Rapid production of full-length cDNAs from rare transcripts: amplification using a single gene-specific oligonucleotide primer. *Proceedings of the National Academy of Sciences, USA* **85**, 8998–9002.

GERHOLD, D., RUSHMORE, T. & CASKEY, C. T. (1999). DNA chips: promising toys have become powerful tools. *Trends in Biochemical Sciences* **24**, 168–173.

GOLDSTEIN, D. B. & SCHLOTTERER, C. (EDS.) (1999). *Microsatellites : Evolution and Applications*. Oxford, Oxford University Press.

GRESS, T. M., HOHEISEL, J. D., LENNON, G. G., ZEHETNER, G. & LEHRACH, H. (1992). Hybridization fingerprinting of high-density cDNA library arrays with cDNA pools derived from whole tissues. *Mammalian Genome* **3**, 609–619.

GRIFFIN, T. J. & SMITH, L. M. (2000). Single-nucleotide polymorphism analysis by MALDI-TOF mass spectrometry. *Trends in Biotechniques* **18**, 77–84.

GROSBERG, R. K., LEVITAN, D. R. & CAMERON, B. B. (1996). Characterization of genetic structure and genealogies using RAPD-PCR markers: a random primer for the novice and nervous. In *Molecular Zoology : Advances, Strategies and Protocols* (ed. Ferraris, J. D. & Palumbi, S. R.), pp. 67–100. New York, Wiley-Liss.

GRYSEELS, B. (2000). Schistosomiasis vaccines: a devils' advocate view. *Parasitology Today* **16**, 46–48.

HAYWARD, R. E., DERISI, J. L., ALFADHLI, S., KASLOW, D. C., BROWN, P. O. & RATHOD, P. K. (2000). Shotgun DNA microarrays and stage specific gene expression in *Plasmodium falciparum* malaria. *Molecular Microbiology* **35**, 6–14.

HOESCHELE, I., UIMARI, P., GRIGNOLA, F. E., ZHANG, Q. & GAGE, K. M. (1997). Advances in statistical methods to map quantitative trait loci in outbred populations. *Genetics* **147**, 1445–1457.

HUBANK, M. & SCHATZ, D. G. (1994). Identifying differences in mRNA expression by representational difference analysis of cDNA. *Nucleic Acids Research* **22**, 5640–5648.

HUBENDICK, B. (1958). A possible method for schistosome-vector control by competition between resistant and susceptible strains. *Bulletin of the World Health Organization* **18**, 1113–1116.

JONES, C. S., LOCKYER, A. E., PIERTNEY, S. B., ROLLINSON, D. & NOBLE, L. R. (1999). Isolation and characterization of microsatellite loci in the freshwater gastropod, *Biomphalaria glabrata*, an intermediate host for *Schistosoma mansoni*. *Molecular Ecology* **8**, 2149–2151.

KNIGHT, M., MILLER, A. N., GEOGHAGEN, N. S. M., LEWIS, F. A. & KERLAVAGE, A. R. (1998). Expressed Sequence Tags (ESTs) of *Biomphalaria glabrata*, an intermediate host of *Schistosoma mansoni*: use in the identification of RFLP markers. *Malacologia* **39**, 175–182.

KNIGHT, M., MILLER, A. N., PATTERSON, C. N., ROWE, C. G., MICHAELS, G., CARR, D., RICHARDS, C. S. & LEWIS, F. A. (1999). The identification of markers segregating with resistance to *Schistosoma mansoni* infection in the snail *Biomphalaria glabrata*. *Proceedings of the National Academy of Sciences, USA* **96**, 1510–1515.

KNIGHT, M., ONGELE, E. & LEWIS, F. A. (2000). Molecular studies of *Biomphalaria glabrata*, an intermediate host of *Schistosoma mansoni*. *International Journal for Parasitology* **30**, 535–541.

KOZIAN, D. H. & KIRSCHBAUM, B. J. (1999). Comparative gene-expression analysis. *Trends in Biotechnology* **17**, 73–78.

KRUGLYAK, L. (1997). The use of a genetic map of biallelic markers in linkage studies. *Nature Genetics* **17**, 21–24.

LANDER, E. S. & BOTSTEIN, D. (1989). Mapping Mendelian factors underlying quantitative traits using RFLP linkage maps. *Genetics* **121**, 185–199.

LARDANS, V. & DISSOUS, C. (1998). Snail control strategies for reduction of schistosomiasis transmission. *Parasitology Today* **14**, 413–417.

LARDANS, V., SERRA, E., CAPRON, A. & DISSOUS, C. (1998). Characterization of an intracellular receptor for activated protein kinase C (RACK) from the mollusc *Biomphalaria glabrata*, the intermediate host for *Schistosoma mansoni*. *Experimental Parasitology* **88**, 194–199.

LAURSEN, J. R., DI LIU, H., WU, X.-J. & YOSHINO, T. P. (1997). Heat-shock response in a molluscan cell line: Characterization of the response and cloning of an inducible HSP70 cDNA. *Journal of Invertebrate Pathology* **70**, 226–233.

LEE, J. J. & COSTLOW, N. A. (1987). A molecular titration assay to measure transcript prevalence levels. *Methods in Enzymology* **152**, 633–648.

LEE, N. H., WEINSTOCK, K. G., KIRKNESS, E. F., *et al.* (1995). Comparative expressed-sequence-tag analysis of differential gene expression profiles in PC-12 cells before and after nerve growth factor treatment. *Proceedings of the National Academy of Sciences, USA* **92**, 8303–8307.

LENNON, G. G. (2000). High-throughput gene expression analysis for drug discovery. *Drug Discovery Today* **5**, 59–66.

LIANG, P. & PARDEE, A. B. (1992). Differential display of eukaryotic messenger RNA by means of the polymerase chain reaction. *Science* **257**, 967–971.

LISITSYN, N. A., LISITSYN, N. M. & WIGLER, M. H. (1993). Cloning the difference between two complex genomes. *Science* **259**, 946–951.

LISITSYN, N. A., SEGRE, J. A., KUSUMI, K., LISITSYN, N. M., NADEAU, J. H., FRANKEL, M. N., WIGLER, M. H. & LANDER, E. S. (1994). Direct isolation of polymorphic markers linked to a trait by genetically directed representational difference analysis. *Nature Genetics* **6**, 57–63.

LOCKYER, A. E., JONES, C. S., ROLLINSON, D. & NOBLE, L. R. (1998). Identification and preliminary characterization of molecular markers linked to schistosome resistance in *Biomphalaria glabrata*. In *Proceedings of the Danish Bilharziasis Laboratory workshop, Zimbabwe, 1997* (ed. Madsen, H. Appleton, C. C. & Chimbari, M.), pp. 219–223. Charlottenlund, Danish Bilharziasis Laboratory.

LOCKYER, A. E., JONES, C. S., NOBLE, L. R. & ROLLINSON, D. (2000). Use of differential display to detect changes in gene expression in the intermediate snail host *Biomphalaria glabrata* upon infection with *Schistosoma mansoni*. *Parasitology* **120**, 399–407.

LYNCH, M. & WALSH, B. (1998). *Genetics and Analysis of Quantitative Traits*. Massachusetts, USA, Sinauer Associates, Inc.

MANN, M. & WILM, M. (1994). Error tolerant identification of peptides in sequence databases by peptide sequence tags. *Analytical Chemistry* **66**, 4390–4399.

MANSFIELD, D. C., BROWN, A. F., GREEN, D. K., CAROTHERS, A. D., MORRIS, S. W., EVANS, H. J. & WRIGHT, A. F. (1994). Automation of genetic linkage analyses using fluorescent microsatellite markers. *Genomics* **24**, 225–233.

MAVAREZ, J., AMARISTA, M., POINTIER, J.-P. & JARNE, P. (2000). Microsatellite variation in the freshwater schistosome-transmitting snail *Biomphalaria glabrata*. *Molecular Ecology* **9**, 1009–1011.

MICHIELS, L., VAN LEUVEN, F., VAN DEN OORD, J. J., DE WOLF-PEETERS, C. & DELABIE, J. (1998). Representational difference analysis using minute quantities of DNA. *Nucleic Acids Research* **26**, 3608–3610.

MILLER, A. N., OFORI, K., LEWIS, F. & KNIGHT, M. (1996). *Schistosoma mansoni*: use of a subtractive cloning strategy to search for RFLPs in parasite resistant *Biomphalaria glabrata*. *Experimental Parasitology* **84**, 420–428.

NEWTON, W. L. (1953). The inheritance of susceptibility to infection with *Schistosoma mansoni* in *Australorbis glabratus*. *Experimental Parasitology* **2**, 242–257.

O'FARRELL, P. H. (1975). High resolution two-dimensional electrophoresis of proteins. *Journal of Biological Chemistry* **250**, 4007–4021.

O'NEILL, M. J. & SINCLAIR, A. H. (1997). Isolation of rare transcripts by representational difference analysis. *Nucleic Acids Research* **25**, 2681–2682.

PANDEY, A. & MANN, M. (2000). Proteomics to study genes and genomes. *Nature* **405**, 837–846.

PEASE, A. C., SOLAS, D., SULLIVAN, E. J., CRONIN, M. T.,

HOLMES, C. P. & FODOR, S. P. A. (1994). Light-generated oligonucleotide arrays for rapid DNA sequence analysis. *Proceedings of the National Academy of Sciences, USA* **91**, 5022–5026.

POLYAK, K., XIA, Y., ZWEIER, J. L., KINZLER, K. W. & VOGELSTEIN, B. (1997). A model for p53-induced apoptosis. *Nature* **389**, 300–305.

POUSTKA, A., ROHL, T. M., BARLOW, D. P., FRISCHAUF, A. M. & LEHRACH, H. (1987). Construction and use of human chromosome jumping libraries from *Not*I-digested DNA. *Nature* **325**, 353–355.

PRIMMER, C. R. & ELLEGREN, H. (1998). Pattern of molecular evolution in avian microsatellites. *Molecular Biology and Evolution* **15**, 997–1008.

RICHARDS, C. S. (1970). Genetics of a molluscan vector of schistosomiasis. *Nature* **227**, 806–810.

RICHARDS, C. S., KNIGHT, M. & LEWIS, F. A. (1992). Genetics of *Biomphalaria glabrata* and its effect on the outcome of *Schistosoma mansoni* infection. *Parasitology Today* **8**, 171–174.

RIESEBERG, L. (1996). Homology among RAPD fragments in interspecific comparisons. *Molecular Ecology* **5**, 99–105.

ROLLINSON, D. & BLACKWELL, J. (1999). Exploring parasite genomes. *Parasitology* **118**, S1.

ROLLINSON, D., STOTHARD, J. R., JONES, C. S., LOCKYER, A. E., DE SOUZA, C. P. & NOBLE, L. R. (1998). Molecular characterisation of intermediate snail hosts and the search for resistance genes. *Memorias do Instituto Oswaldo Cruz* **93**, Suppl. 1, 111–116.

ROTHSCHILD, C. B., BREWER, C. S. & BOWDEN, D. W. (1997). DD/AP-PCR: combination of differential display and arbitrarily primed PCR of oligo(dT) cDNA. *Analytical Biochemistry* **245**, 48–54.

SANGERSTROM, C. G., SUN, B. I. & SIVE, H. L. (1997). Subtractive cloning: past, present and future. *Annual Review of Biochemistry* **66**, 751–783.

SCHENA, M., SHALON, D., DAVIS, R. W. & BROWN, P. O. (1995). Quantitative monitoring of gene expression patterns with a complementary DNA microarray. *Science* **270**, 467–470.

SCHLOTTERER, C. (1998). Microsatellites. In *Molecular Genetic Analysis of Populations: A Practical Approach* (ed. Hoelzel, A. R.), pp. 237–261. Oxford, IRL Press.

SCHULER, G. D., BOGUSKI, M. S., STEWART, E. A., *et al.* (1996). A gene map of the human genome. *Science* **274**, 540–546.

TANKSLEY, S. D., GANAL, M. W. & MARTIN, G. B. (1995). Chromosome landing: a paradigm for map-based gene cloning in plants with large genomes. *Trends in Genetics* **11**, 63–68.

TAUTZ, D. (1989). Hypervariability of simple sequences as a general source for polymorphic DNA markers. *Nucleic Acids Research* **17**, 6463–6471.

VELCULESCU, V. E., ZHANG, L., VOGELSTEIN, B. & KINZLER, K. W. (1995). Serial analysis of gene expression. *Science* **270**, 484–487.

VERNON, J. G., JONES, C. S. & NOBLE, L. R. (1995). Random amplified polymorphic DNA (RAPD) markers reveal cross-fertilisation in *Biomphalaria glabrata* from wild populations. *Journal of Molluscan Studies* **61**, 455–465.

VESMATZIS, G., ESSAND, M., BRINKMANN, U., LEE, B. & PASTAN, I. (1998). Discovery of three genes specifically expressed in human prostate by expressed sequence tag database analysis. *Proceedings of the National Academy of Sciences, USA* **95**, 300–304.

VOS, P., HOGERS, R., BLEEKER, M., REIJANS, M., VAN DE LEE, T., HORNES, M., FRIJTERS, A., POT, J., PELEMAN, J., KUIPER, M. & ZABEAU, M. (1995). AFLP: a new technique for DNA fingerprinting. *Nucleic Acids Research* **23**, 4407–4414.

WANG, D. G., FAN, J.-B., SIAO, C.-J., *et al.* (1998). Large-scale identification, mapping and genotyping of single-nucleotide polymorphisms in the human genome. *Science* **280**, 1077–1082.

WANG, R., KAFATOS, F. C. & ZHENG, L. (1999). Microsatellite markers and genotyping procedures for *Anopheles gambiae*. *Parasitology Today* **15**, 33–37.

WASINGER, V. C., CORDWELL, S. J., CERPA POLJAK, A., YAN, J. X., GOOLEY, A. A., WILKINS, M. R., DUNCAN, M. W., HARRIS, R., WILLIAMS, K.L. & HUMPHERY SMITH, I. (1995). Progress with gene-product mapping of the Mollicutes – *Mycoplasma genitalium*. *Electrophoresis* **16**, 1090–1094.

WELSH, J., CHADA, K., DALAL, S. S., CHENG, R., RALPH, D. & MCCLELLAND, M. (1992). Arbitrarily primed PCR fingerprinting of RNA. *Nucleic Acids Research* **20**, 4965–4970.

WELSH, J. & MCCLELLAND, M. (1990). Fingerprinting genomes using PCR with arbitrary primers. *Nucleic Acids Research* **18**, 7213–7218.

WILLIAMS, J. G. K., KUBELIK, A. R., LIVAK, K. J., RAFOLSKI, J. A. & TINGEY, S. V. (1990). DNA polymorphisms amplified by arbitrary primers are useful as genetic markers. *Nucleic Acids Research* **18**, 6531–6535.

WILLIAMS, S. A. & JOHNSTON, D. A. (1999). Helminth genome analysis: the current status of the filarial and schistosome genome projects. *Parasitology* **118**, S19–S38.

WORLD HEALTH ORGANIZATION (1993). The control of Schistosomiasis. *WHO Technical Report Series* No. 830. Geneva, WHO.

WORLD HEALTH ORGANIZATION (1997). Schistosomiasis. *Aide Mémoire* **115**, 1–4.

YEATES, L. C. & POWIS, G. (1997). The expression of the molecular chaperone calnexin is decreased in cancer cells grown as colonies compared to monolayer. *Biochemical and Biophysical Research Communications* **238**, 66–70.

YOSHIKAWA, Y., MUKAI, H., ASADA, K., HINO, F. & KATO, I. (1998). Differential display with Carboxy-X-rhodamine-labeled primers and the selection of differentially amplified cDNA fragments without cloning. *Analytical Biochemistry* **256**, 82–91.

ZHAO, L., ARAGAKI, C., HSU, L. & QUIAOIT, F. (1998). Mapping of complex traits by single nucleotide polymorphisms. *American Journal of Human Genetics* **63**, 225–240.

Genetic variability and molecular identification of Brazilian *Biomphalaria* species (Mollusca: Planorbidae)

O. S. CARVALHO[1]*, R. L. CALDEIRA[1], A. J. G. SIMPSON[2] *and* T. H. D. A. VIDIGAL[1,3]

[1] *Centro de Pesquisas René Rachou, Laboratório de Helmintoses Intestinais, FIOCRUZ, Belo Horizonte, MG, BRASIL*
[2] *Laboratório de Genética do Câncer, Instituto Ludwig de Pesquisa sobre o Câncer, São Paulo, SP, BRASIL*
[3] *Instituto de Ciências Biológicas, Depto. de Zoologia - Universidade Federal de Minas Gerais, Belo Horizonte, MG, BRASIL*

SUMMARY

Freshwater snails belonging to the genus *Biomphalaria* are intermediate hosts of the trematode *Schistosoma mansoni* in the Neotropical region and Africa. In Brazil, one subspecies and ten species of *Biomphalaria* have been identified: *B. glabrata*, *B. tenagophila*, *B. straminea*, *B. occidentalis*, *B. peregrina*, *B. kuhniana*, *B. schrammi*, *B. amazonica*, *B. oligoza*, *B. intermedia* and *B.t. guaibensis*. However, only the first three species are found naturally infected with *S. mansoni*. The classical identification of these planorbids is based on comparison of morphological characteristics of the shell and male and female reproductive organs, which is greatly complicated by the extensive intra-specific variation. Several molecular techniques have been used in studies on the identification, genetic structure as well as phylogenetic relationships between these groups of organisms. Using the randomly amplified polymorphic DNAs (RAPD) analysis we demonstrated that *B. glabrata* exhibits a remarkable degree of intra-specific polymorphism. Thus, the genetics of the snail host may be more important to the epidemiology of schistosomiasis than those of the parasite itself. Using the simple sequence repeat anchored polymerase chain reaction (SSR-PCR) in intra-populational and intra-specific studies we have demonstrated that snails belonging to the *B. straminea* complex (*B. straminea*, *B. kuhniana* and *B. intermedia*) clearly presented higher heterogeneity. Using the low stringency polymerase chain reaction (LS-PCR) technique we were able to separate *B. glabrata* from *B. tenagophila* and *B. tenagophila* from *B. occidentalis*. To separate all Brazilian *Biomphalaria* species we used the restriction fragment length polymorphism (PCR-RFLP) of the internal transcribed spacer region (ITS) of the DNA gene. The method also proved to be efficient for the specific identification of DNA extracted from snail eggs. Recently we have sequenced the ITS2 region for phylogenetic studies of all *Biomphalaria* snails from Brazil.

Key words: *Biomphalaria*, Brazil, polymerase chain reaction, genetic variability, identification, ribosomal RNA, internal transcribed spacer, phylogeny.

BIOMPHALARIA SNAILS: IMPORTANCE AND BIOLOGICAL CHARACTERIZATION

Freshwater snails, belonging to the genus *Biomphalaria* Preston, 1910 (Mollusca: Pulmonata, Planorbidae) are intermediate hosts for the parasitic trematode *Schistosoma mansoni* in the Neotropical region and Africa. According to Baker (1945), geological records of the Planorbidae prove its presence in Europe and United States since the Jurassic Period.

These molluscs can be found in a variety of natural and artificial (lakes, dams, river streams and irrigation ditches) environments. Thus, such snails readily adapt to different environmental conditions and are able to withstand a range of variation in the physical, chemical and biological characteristics in the environment in which they live (Paraense, 1972).

Biomphalaria snails are hermaphrodites but usually reproduce by cross-fertilization when paired. Such biological characteristics are evolutionarily important, providing the organism with the ability to establish colonies from a single individual while maintaining genetic variation through sexual reproduction (Paraense, 1955). Most populations are under significant environmental pressure (rain and drought) that can dramatically reduce population levels, requiring re-colonization from a low number of remaining snails, founder effect (Paraense, 1955). Paraense, Pereira & Pinto (1955) comment that the fixation of populations may occur in a very short period of time, due to the high reproductive potential of the species, enhanced by the absence of intra-specific competition. On the other hand, limited gene flow (a lack of interconnecting waterways) between different populations could enable the formation of local strains, resulting in inter-populational variability. The observation of several levels of reproductive isolation, detected in allopatric conspecific individuals, also indicated extensive genetic heterogeneity among populations of *Biomphalaria* snails from different localities (Paraense, 1959).

* Corresponding author. Centro de Pesquisas René Rachou - Laboratório de Helmintoses Intestinais, Avenida Augusto de Lima, 1715, Barro Preto, 30190.002, Belo Horizonte, MG - BRASIL. Tel: + 55 3132953566. Fax: +55 3132953115. E-mail: omar@cpqrr.fiocruz.br

BRAZILIAN *BIOMPHALARIA* SNAILS: SPECIES,
GEOGRAPHICAL DISTRIBUTION AND
SUSCEPTIBILITY

In Brazil, ten species and one subspecies of *Biomphalaria* are recognised: *B. glabrata* (Say, 1818), *B. tenagophila* (Orbigny, 1835), *B. straminea* (Dunker, 1848), *B. peregrina* (Orbigny, 1835), *B. schrammi* (Crosse, 1864), *B. kuhniana* (Clessin, 1883), *B. intermedia* Paraense & Deslandes, 1962, *B. amazonica* Paraense, 1966, *B. oligoza* Paraense, 1974, *B. occidentalis* Paraense, 1981 and *B. t. guaibensis* (Paraense, 1984). However, only the first three species are found naturally infected with *S. mansoni*. *Biomphalaria peregrina* and *B. amazonica* are considered potential hosts based on experimental infection (Paraense & Corrêa, 1973; Corrêa & Paraense, 1971).

Biomphalaria glabrata is considered the most important *S. mansoni* intermediate host in the Neotropical region, due to its wide distribution and high susceptibility to the trematode (Paraense & Corrêa, 1963). On the other hand, *B. straminea* plays an important role as a vector in the Northwest of Brazil (Paraense, 1986; Paraense & Corrêa, 1989) and *B. tenagophila* in some areas in the South of Brazil (Paraense & Corrêa, 1987).

Richards & Shade (1987) reported that *Biomphalaria* susceptibility to *S. mansoni* infection varies among geographic areas, populations in the same area, individuals within the same population and also among snails of different ages. In fact, Richards (1984) showed that in *B. glabrata* adults resistance to infection is controlled by a single gene, the inheritance of which follows simple Mendelian genetics, with dominance to resistance. In juvenile snails, however, resistance appears to be a polygenic trait (Richards & Merrit, 1972). Richards & Shade (1987) point out that the genetic variation in snail susceptibility and parasite infectivity result in a variety of host-parasite relations.

Thus, studies on the parasite and intermediate host genome remain important for the study of the epidemiology and control of schistosomiasis.

MOLECULAR STUDIES OF *BIOMPHALARIA*
SNAILS

Molecular studies to further understand the genetic variability and population structure of *Biomphalaria* snails are needed due to: (1) intraspecific variation being detected at both morphological (Paraense, 1975 *a*) and genetic levels (Knight *et al.* 1991; Vidigal *et al.* 1994), and (2) differences in susceptibility and compatibility among *Biomphalaria* snails to infection with *S. mansoni* (Paraense & Corrêa, 1963; Souza, Passos-Jannoti & Freitas, 1995).

Isoenzyme techniques have been shown to be useful in the identification of *Bulinus*, *Oncomelania*, *Lymnaea* and *Biomphalaria* as well as for the investigation of phylogenetic relationships and the genetic structure of these groups of organisms (Bandoni, Mulvey & Loker 1995; Mascara & Morgante, 1995; Zahab-Jabbour *et al.* 1997; Zhou & Kristensen, 1999; Mukaratirwa *et al.* 1998).

Knight *et al.* (1991) used the ribosomal RNA gene to detect restriction fragment length polymorphisms (RFLPs) and demonstrated the variability between resistant and susceptible *B. glabrata* snails to *S. mansoni*. The development of the polymerase chain reaction (PCR) has allowed considerable progress in the molecular study of *Biomphalaria*. The arbitrarily primed polymerase chain reaction (AP-PCR–Welsh & McClelland, 1990; Williams *et al.* 1990) has been used to study genomic variability in different groups of organisms (Gomes *et al.* 2000; De Sousa, Dutari & Gardenal, 1999). Vernon, Jones & Noble (1995) showed the potential of RAPDs as molecular markers for analysis of fertilisation in wild-type *B. glabrata*. In addition, promising results concerning resistance and susceptibility to *S. mansoni* were reported by Larson *et al.* (1996), Lewis, Knight & Richards (1997), Knight *et al.* (1999) and by Knight, Ongele & Lewis (2000). Abdel-Hamid *et al.* (1999) also employed RAPD analysis for the study of susceptibility and resistance to *S. mansoni* in *B. tenagophila*. These studies showed genetic variability between the susceptible and resistant strains. A high level of genetic variability among populations of *B. pfeifferi*, obtained along a 6 km stretch of the Zimbabwe river, was demonstrated by Hoffmam *et al.* (1998). On Guadalupe Island, with a limited area of 45 km², a correlation between genetic distance and geographic distance was reported by Langand *et al.* (1999), with a high level of genetic variability.

Microsatellites (MS) are useful genetic markers due to their frequencies within eukaryotic genomes, their ease of amplification by PCR and their degree of polymorphism. Microsatellites have been widely used for studies of genetic variability in many organisms (Donnelly *et al.* 1999; Ferdig & Su, 2000) including snails such as *Bulinus*, *Melania* and *Littorina* (Viard *et al.* 1997 *a*, *b*, *c*; Samadi *et al.* 1999; Tie, Boulding & Naish, 2000).

Recently, Jones *et al.* (1999) reported the identification and characterization of the first microsatellite *loci* in *B. glabrata* and also demonstrated divergence between resistant and susceptible stocks to *S. mansoni*. These authors indicated that microsatellite markers could contribute to the elucidation of evolutionary relationships among African and Neotropical *Biomphalaria* species. Indeed, *B. pfeifferi* was found to be more polymorphic than the neotropical species (*B. straminea*, *B. occidentalis* and *B. tenagophila*). Mavárez *et al.* (2000) characterized nine microsatellite *loci* in *B. glabrata* populations from Venezuela and also detected at least eight *loci*

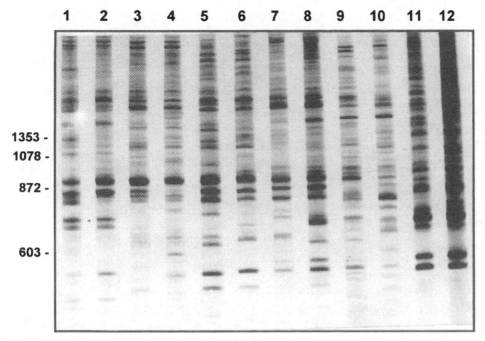

Fig. 1. Silver-stained polyacrylamide gel (4%) showing AP-PCR profiles obtained with primer 3307 and twelve Brazilian *Biomphalaria glabrata* specimens. Lanes 1 and 2: Specimens from Belém (Pará); lanes 3 and 4: from Cururupu (Maranhão); lanes 5 and 6: from Touros (Rio Grande do Norte); lanes 7 and 8: from Pontezinha (Pernambuco); lanes 9 and 10: from Aracaju (Sergipe); lanes 11 and 12: from Jacobina (Bahia). Molecular size markers are shown on the left of the gel. Vidigal *et al.* (1994).

suitable for studies of population structure, reproductive systems and resistance to *S. mansoni*.

GENETIC VARIATION IN BRAZILIAN *BIOMPHALARIA*

The available methodologies, based on molecular analysis, have permitted the generation of more consistent information concerning the population structure of *Biomphalaria*. Such studies have allowed inference of the origin, colonization processes and dispersion of populations and species of the *Biomphalaria* genus (Woodruff, Mulvey & Yipp, 1985; Mulvey, Newman & Woodruff, 1988; Woodruff & Mulvey, 1997).

By means of AP-PCR, an estimate of intra and interpopulation variability of *B. glabrata*, from different localities, in Brazil, was achieved (Vidigal *et al.* 1994). The comparison among specimens, from different field populations, presented a limited intrapopulation but high interpopulation heterogeneity. Such results suggest that individual localized populations are homogenous but that the global *B. glabrata* population is highly heterogeneous. Snails, which had been laboratory reared and maintained for different periods of time, were also analysed and showed no differences in intrapopulation variability as compared to field populations. This rules out the possibility that selective breeding in the laboratory produced inbred strains, regardless of the time period of maintenance in the

laboratory. The high level of genetic variability found in *B. glabrata* was confirmed by analysis of two randomly selected individuals from each of six populations, both from laboratory and field, using the AP-PCR primers 3307 (Fig. 1) and 3302. Less than 10% of the amplified fragments were present in all analysed samples. The average percentage of shared bands between each pair from the same locality was 74·5% and between all possible pairs, 43%. These results are in accordance with those obtained for *B. glabrata* from Puerto Rico and *B. prona* from Venezuela using isoenzymes (Mulvey & Vrijenhoek, 1982; Mulvey *et al.* 1988; Paraense *et al.* 1992), or for *Bulinus* snails by RAPD analysis (Langand *et al.* 1993). In summary, the studies using AP-PCR demonstrated that it is appropriate for the study of *Biomphalaria* at the genetic level, and that it allowed confirmation of the wide genetic variability of these snails, previously observed in isoenzyme and morphological studies.

Zietkiewicz, Rafalski & Labuda (1994) introduced the simple sequence repeat anchored PCR (SSR-PCR) technique to the study of several eukariotic species. This technique is based on the anchoring of PCR primers at the 3′ or 5′ ends of microsatellites. Its advantage lies in the reduction of the number of other possible targets for annealing and hence results are easier to interpret. Among the seven tested primers the $(CA)_8RY$ gave the best results. Oliveira *et al.* (1997) used SSR-PCR with the $(CA)_8RY$ primer to study the intraspecific variability of *Trypanosoma cruzi*, *Leishmania braziliensis* and

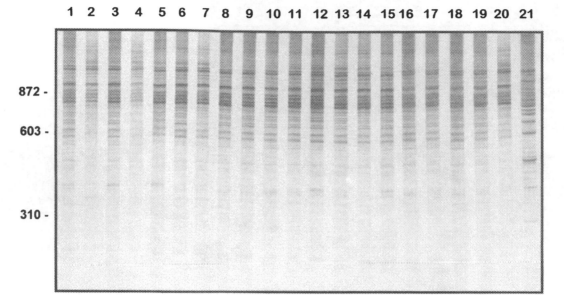

Fig. 2. Silver-stained polyacrylamide gel (6%) showing simple sequence repeat-anchored polymerase chain reaction amplification profiles obtained with the K7 primer and twenty Brazilian *Biomphalaria kuhniana* specimens. Lanes 1 to 20: from Tucuruí (Pará, Brazil) and one outgroup, *B. glabrata* specimen lane 21: from Esteio (Rio Grande do Sul, Brazil). Molecular size markers are shown on the left of the gel. Caldeira *et al.* (2001).

S. mansoni, showing that the patterns obtained were comparable to those resulting from AP-PCR. Due to its applicability and the quality of the results obtained, we have standardized the SSR-PCR technique using the primers (CA)$_8$RY and K7 for the study of genetic variability among Brazilian populations of *B. straminea*, *B. intermedia* and *B. kuhniana* (Caldeira *et al.*, in press). Fig. 2 illustrates the reproducibility of the profiles obtained with the SSR-PCR technique using 20 specimens of *B. kuhniana* and the primer K7. The trees obtained by UPGMA and neighbour-joining methods showed similar topologies. We observed greater heterogeneity between populations than within. The mean percentage of shared bands among all possible pairs from the same locality was 83%. This result suggests, in the three species studied, the existence of uniform populations. In contrast, *B. straminea* and *B. intermedia* displayed intraspecific genetic heterogeneity.

SPECIFIC IDENTIFICATION OF *BIOMPHALARIA* SNAILS

The morphological identification of medically important freshwater snails is greatly complicated by the extensive intraspecific variation in anatomical and morphological characteristics commonly used for classical identification (Paraense, 1975*a*, *b*, 1981, 1984, 1988). To overcome this problem, the use of molecular techniques, in conjunction with morphological characters, has been proposed for the identification of these snails. The remarkable degree of intraspecific polymorphism detected in *B.*

glabrata, through AP-PCR, suggests that this technique is not suitable for identification of this species (Vidigal *et al.* 1994). Testing the low stringency polymerase chain reaction technique (LS-PCR–Dias Neto *et al.* 1993), Vidigal *et al.* (1996) used primers (NS1 and ET1) designed to amplify a portion of the rDNA gene, which permitted the differentiation of *B. glabrata* and *B. tenagophila* (Fig. 3) from six localities in Brazil. The primers produced a complex pattern of bands with 4 species-specific fragments for *B. glabrata* and 3 for *B. tenagophila*. Pires *et al.* (1997), using the same technique, were able to differentiate *B. tenagophila* from *B. occidentalis*, which is not possible using morphological characters done (Paraense, 1981). Fig. 4 shows different profiles of *B. tenagophila* specimens with the primers NS1 and ET1. Note the presence of a double 500 bp band (T1) and two fragments of 400 and 310 bp (T2 and T3). *B. occidentalis* specimens were characterized by a band of 130 bp (O1). The profiles were polymorphic but were found to be reproducible in all specimens from each species.

Preliminary experiments with *B. straminea* populations, using the same methodology, suggest that this species has higher levels of interpopulation variability than *B. glabrata*, *B. tenagophila* or *B. occidentalis*. No fragments were shared among all studied specimens (Vidigal *et al.* 1996; Pires *et al.* 1997). Although LS-PCR was able to differentiate the species studied, polymorphic profiles were produced making the interpretation of results very difficult.

Vidigal *et al.* (1998) examined sequence polymorphism within the internal transcribed spacer region (ITS) of the rDNA which includes the 5·8S

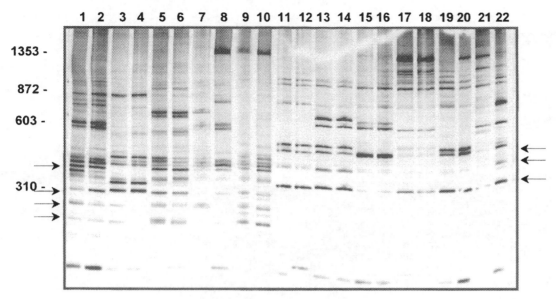

Fig. 3. Silver-stained polyacrylamide gel (4%) showing LS-PCR profiles obtained with primers NS1-ET1 and Brazilian *Biomphalaria* species. Lanes 1 and 2: *B. glabrata* from Belém (Pará); lanes 3 and 4: *B. glabrata* from Cururupu (Maranhão); lanes 5 and 6: *B. glabrata* from Touros (Rio Grande do Norte); lanes 7 and 8: *B. glabrata* from Pontezinha (Pernambuco); lanes 9 and 10: *B. glabrata* from Aracaju (Sergipe); lanes 11 and 12: *B. tenagophila* from Paracambi (Rio de Janeiro); lanes 13 and 14: *B. tenagophila* from Imbé (Rio Grande do Sul); lanes 15 and 16: *B. tenagophila* from Joinville (Santa Catarina); lanes 17 and 18: *B. tenagophila* from Araçatuba (São Paulo); lanes 19 and 20: *B. tenagophila* from Formosa (Goiás); lanes 21 and 22: *B. tenagophila* from Vila Velha (Espirito Santo). The species diagnostic bands are indicated by arrows, and the molecular size markers are shown on the left of the gel. Vidigal *et al.* (1996).

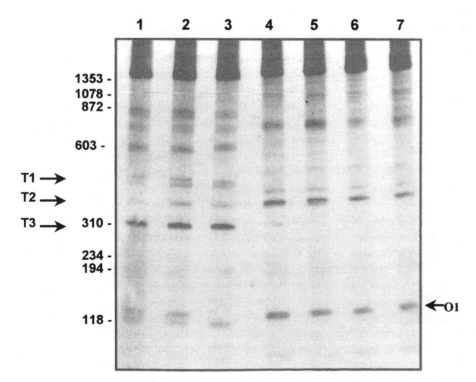

Fig. 4. Comparison of LSP of profiles of *Biomphalaria occidentalis* and *B. tenagophila* obtained with the primer pair NS1-ET1 and 1 ng of DNA template. Lanes 1 to 3: *B. tenagophila* specimens from Vespasiano, MG; Vitória, ES and Joinville, SC, respectively. Lanes 4 to 7: *B. occidentalis* specimens from Campo Grande, MS; Ladário, MS; Assis, SP and Dracema, SP, respectively. The LS-PCR amplification products were visualized in a 4% polyacrilamide gel stained with silver. The species diagnostic bands are indicated by arrows, and the molecular size markers are shown on the left of the gel. Pires *et al.* (1997).

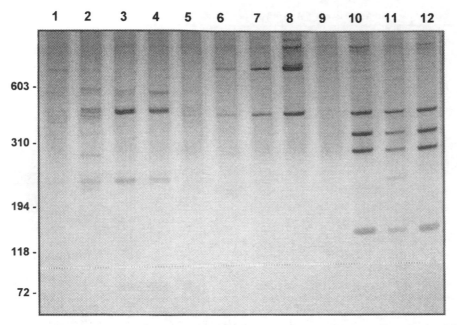

Fig. 5. Silver stained polycrylamide gel (6 %) showing the RFLP profiles obtained by digesting the DNA ITS with
*Dde*I. The DNA used was extracted from the eggs of *Biomphalaria glabrata*, *B. tenagophila* and *B. straminea* on
different days and compared with the profile produced with adult control DNA. Lane 1: 1 day eggs of *B. glabrata*;
lane 2: 2 day eggs of *B. glabrata*; lane 3: 3 day eggs; lane 4: adult *B. glabrata*; lane 5: 1 day eggs of *B. tenagophila*;
lane 6: 2 day eggs of *B. tenagophila*, lane 7: 3 day eggs of *B. tenagophila*; lane 8: adult *B. tenagophila*; lane 9: 1 day
eggs lane 10: 2 day eggs; lane 11: 3 days layed eggs; lane 12: adult *B. straminea*. Vidigal *et al.* (1998).

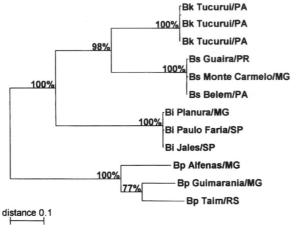

Fig. 6. Neighbour-joining tree unrooted of *Biomphalaria
kuhniana, B. straminea, B. intermedia* and *B. peregrina*
constructed using the PCR-RFLPs profiles produced
with all enzymes. The numbers are bootstrap percent
values based on 1000 replications. Caldeira *et al.* (1998).

rDNA gene together with the flanking ITS1 and
ITS2 spacers. They used PCR amplification fol-
lowed by digestion with several restriction enzymes
(RFLP). This method was employed in systematic
studies of snails within the genera *Oncomelania*,
Bulinus and *Stagnicola* (Hope & McManus, 1994;
Stothard, Hughes & Rollinson, 1996; Stothard &
Rollinson, 1997; Remigio & Blair, 1997). Firstly,
different *B. glabrata*, *B. tenagophila* and *B. straminea*
populations from Brazil were analysed. The entire
ITS region was amplified using the primers ETTS1
and ETTS2 anchored respectively in the conserved

extremities of the 18S and 28S ribosomal genes
(Stothard, Hughes & Rollinson, 1996; Kane &
Rollinson, 1994). The PCR-specific amplification of
the *Biomphalaria* ITS resulted in a product of
approximately 1·3 kb. The amplified ITS region,
from all species studied, contains sites for a variety of
restriction enzymes including *Alu*I, *Dde*I, *Hae*III,
*Mnl*I, *Msp*I, *Rsa*I and *Sau*3aI. The most promising
RFLP profiles were those produced with *Hae*III and
*Dde*I which included species-specific fragments for
the three species studied (data not shown).
Biomphalaria glabrata is characterized by the presence of three fragments of approximately 500, 220
and 80 bp. *Biomphalaria tenagophila* had two fragments of approximately 800 and 470 bp. The
majority of the *B. straminea* snails were characterized
by the presence of four fragments of approximately
470, 310, 280 and 120 bp, although two specimens
from Porto Alegre showed only two fragments (470
and 310 bp). The reproducibility of the *Dde*I–ITS
RFLP profiles was demonstrated by analysis of
several specimens obtained from distinct localities
within Brazil. The possibility of using eggs as a
source of DNA was reported by Knight *et al.* (1992)
and our data confirmed the usefulness of this
approach (Fig. 5). This possibility has important
practical implications when the number of collected
specimens is low, permitting the maintenance of live
adults in the laboratory for further studies.

Using the same methodology Caldeira *et al.* (1998)
studied morphologically similar species *B. straminea*,
B. intermedia, *B. kuhniana* and *B. peregrina*. Six

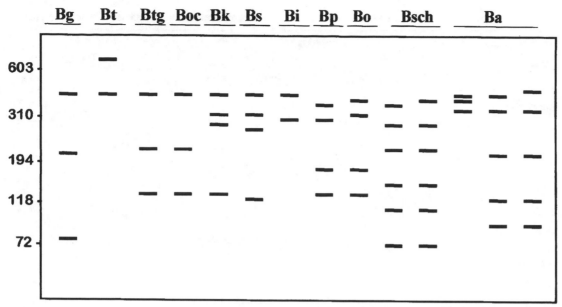

Fig. 7. Schematic representation of the ITS rDNA restriction patterns of 10 Brazilian *Biomphalaria* species and one subspecies produced with *Dde*I. The abbreviation for each species is: *B. glabrata* (Bg); *B. tenagophila* (Bt); *B. occidentalis* (Boc); *B. t. guaibensis* (Btg); *B. straminea* (Bs); *B. intermedia* (Bi); *B. kuhniana* (Bk); *B. peregrina* (Bp); *B. oligoza* (Bo); *B. schrammi* (Bsch) and *B. amazonica* (Ba). Molecular size markers are shown on the left of the gel. Vidigal *et al.* (2000).

enzymes were tested (*Dde*I, *Mnl*I, *Hae*III, *Rsa*I, *Hpa*II and *Alu*I) and the best profile for the differentiation of these species was obtained with *Dde*I. Two methods of clusters and two coefficients (NJ and UPGMA; distance coefficient of Nei and Li and similarity coefficient of Dice) were used. Trees with the same topology identifying three distinct groups were produced: (1) *B. straminea* and *B. kuhniana*; (2) *B. intermedia*; (3) *B. peregrina*. Fig. 6 shows the tree constructed with NJ and distance coefficient of Nei and Li from 79 bands generated by six enzymes. Branches reflect the shared percentages of bands among the 12 specimens of the four snail species. The mean percentage of bands shared among all the possible pairs was 44%. The three groups were supported by high bootstrap values (98% and 100%). We verified a high level of similarity between the first and second groups, which reinforces the data from Paraense (1988) who, based on the morphology characters, grouped *B. straminea*, *B. kuhniana* and *B. intermedia* in the *B. straminea* complex.

Using PCR-RFLP Spatz *et al.* (1999) studied *B. tenagophila*, *B. t. guaibensis* and *B. occidentalis* which are indistinguishable on the basis of shell morphology and are similar across most organs of the genital system. Snails from different localities in Brazil, Argentina and Uruguay were analysed using seven enzymes (*Hae*III, *Dde*I, *Alu*I, *Mnl*I, *Hpa*II, *Hfa*I, *Rsa*I). The *Alu*I profiles were the most informative and distinguish these species (data not shown). Profiles obtained with the other enzymes did not permit species identification as extensive intraspecific polymorphism was observed. Restric-

tion profiles obtained with all enzymes were used to calculate the percentage of shared bands between all possible snail pairs and these data were used for a cluster analysis (data not shown). A closer relationship was observed between *B. occidentalis* and *B. t. guaibensis* than between *B. tenagophila* and the subspecies *B. t. guaibensis*. Based on previous morphological and molecular data, it was proposed that *B. tenagophila*, *B. occidentalis* and *B. t. guaibensis* should be grouped into a complex named *B. tenagophila*.

PCR-RFLP was also used for the identification of ten Brazilian species and one subspecies of *Biomphalaria* with emphasis on the analysis of *B. oligoza*, *B. schrammi* and *B. amazonica* (Vidigal *et al.* 2000). The profiles obtained with *Dde*I (data not shown) allowed the ready separation of the majority of species studied, with the exception of *B. occidentalis* and *B. t. guaibensis*, and *B. oligoza* and *B. peregrina* which have very similar profiles. The correct identification of *B. oligoza* and *B. peregrina* species is also difficult at the morphological level. *Biomphalaria peregrina* is widely distributed in South America and is considered a potential intermediate host of *S. mansoni*. Thus, with the aim of obtaining a suitable molecular separation between *B. oligoza* and *B. peregrina* other enzymes were tested (*Hpa*II, *Mnl*I, *Alu*I and *Hae*III). Despite the similarity among these profiles, *Alu*I, *Hpa*II, *Hae*III permitted the molecular separation of these species. The profiles obtained with *Dde*I for the 10 species and one subspecies of the Brazilian *Biomphalaria* snails are all represented in the diagram in Fig. 7. We have been routinely using this schematic representation of

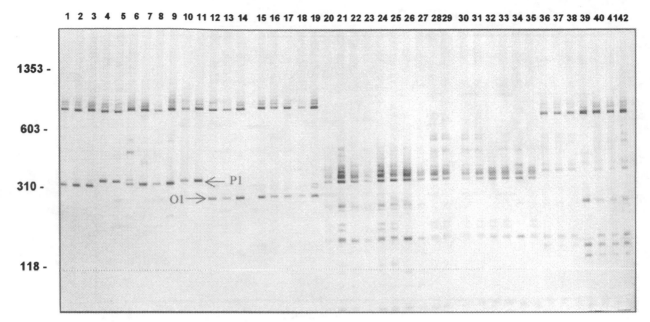

Fig. 8. Silver stained polyacrylamide gel (6 %) showing the polymerase chain reaction and restriction fragment length polymorphism profiles obtained following the digestion of the rDNA internal transcribed spacer with *Mva*I. Lane 1: *Biomphalaria peregrina* from Alfenas, Minas Gerais, (Brazil); lane 2: *B peregrina* from Bom Jesus da Penha, Minas Gerais (Brazil); lane 3: *B. peregrina* from Alfenas, Minas Gerais (Brazil); lanes 4–5: *B. peregrina* from Buenos Aires (Argentina); lanes 6–7: *B. peregrina* from Paso de los Toros (Uruguay); lanes 8–9: *B. peregrina* from Cholila (Argentina); lanes 10–11: *B. peregrina* from Córdoba Province (Argentina); lanes 12–13: *B. oligoza* from Florianópolis, Santa Catarina (Brazil); lanes: 14–15: *B. oligoza* from Eldorado do Sul, Rio Grande do Sul (Brazil); lanes 16–19: *B. oligoza* from Córdoba Province (Argentina); lanes 20–25: *B. orbignyi* from San Roque Corrientes Province (Argentina); lanes 26–27: *B. orbignyi* Termas, Arapey (Uruguay); lanes 28–31: *B. orbignyi* Sierra Quijadas, San Luis Province (Argentina); lanes 32–34: *B. orbignyi* La Carlota, Córdoba Province (Argentina); lanes 35–38: *B. orbignyi* from Chamical, La Rioja Province (Argentina); lanes 39–42: *B. orbignyi* from Patquia, La Rioja Province (Argentina). Molecular size markers are shown on the left of the gel. The arrows indicate species-specific fragments of *B. oligoza* (O1) and *B. peregrina* (P1). Spatz *et al.* (2000).

ITS restriction patterns in our laboratory as a model for comparative molecular identification of Brazilian *Biomphalaria* populations.

Using several enzymes Spatz *et al.* (2000) studied *B. oligoza* and *B. peregrina* from Brazil, Argentina and Uruguay. *Biomphalaria orbignyi* (Paraense, 1975*b*) was included due to its high morphological similarity to *B. peregrina*. This species was originally described in Argentina and has been shown to be refractory to *S. mansoni* infection (Paraense, 1975*b*). The enzyme *Mva*I (Fig. 8) produced two distinct profiles for *B. peregrina* and *B. oligoza* snails. These profiles exhibit (a) one fragment of approximately 700 bp, shared between *B. peregrina* and *B. oligoza*, (b) one fragment of approximately 300 bp (P1) specific for *B. peregrina* and (c) one fragment specific for *B. oligoza* of approximately 200 bp (O1). This enzyme produced polymorphic profiles for *B. orbignyi* specimens (profile 1: lanes 20 to 35, profile 2: lanes 36 to 42). Similarity and genetic distance analyses of the three species were performed using 146 bands obtained with six restriction enzymes. The phenetic trees, obtained using UPGMA and NJ analysis suggested that *B. peregrina* and *B. oligoza* are the most closely related species (data not shown).

The ITS region contains useful genetic markers

for the identification of Brazilian *Biomphalaria* snails. PCR-RFLP is a simple and rapid technique that will improve the precision of snail population surveys undertaken in South America. The efficiency and utility of this methodology was confirmed when snail populations from Venezuela, previously identified as *B. straminea*, were characterized as *B. kuhniana*, at the molecular level (Caldeira *et al.* 2000).

In recent years, the eukaryotic ribosomal RNA locus has been extensively utilised for phylogenetic reconstruction. Sequence comparisons of the two internal transcribed spacers (ITS1 and ITS2) have been used for phylogenetic reconstruction of closely related species from a wide range of organisms. In molluscs, ITS1 sequences have been used to study the relationship of different genera such as: *Isabellaria*, *Albinaria*, *Bulinus* and *Stagnicola* (Schilthuizen, Gittenberger & Gultyaev 1995; Stothard *et al.* 1996; Remigio & Blair, 1997).

Vidigal *et al.* (2000), generated data for 27 *Biomphalaria* snails from Brazil, two from Venezuela and two specimens of *Helisoma duryi* from Brazil in order to determine the molecular relationships among Brazilian *Biomphalaria* snails (Table 1). This study was performed by ITS2 sequence analysis,

Table 1. Species, localities, geographical coordinates, abbreviations and GenBank accession number of the rRNA ITS2 for the snail populations studied

Species	Specimens	Locality	Geographical coordinates	Abbreviation	GenBank accession number
Biomphalaria glabrata	01	Belém, PA, Brazil	01s27/48w30	B.glPA	AF198659
	02	Esteio, RS, Brazil	29s51/51w10	B.glRS-1	AF198661
				B.glRS-2	AF198660
	01	Portuguesa, Chabasquen, Venezuela		B.glVEN	AF198662
B. tenagophila	01	Vespasiano, MG, Brazil	19s41/43w55	B.tenMG	AF198654
	01	Imbé, RS, Brazil*	29s58/50w07	B.tenRS	AF198655
	01	Formosa, GO, Brazil*	15s32/47w20	B.tenGO	AF198656
B. straminea	01	Picos, PI, Brazil*	07s04/41w28	B.strPI	AF198672
	02	Porto Alegre, RS, Brazil*	29s58/51w17	B.strPA-1	AF198669
				B.strPA-2	AF198670
	01	Florianópolis, SC, Brazil	27s35/48w32	B.strSC	AF198668
	01	Belém, PA, Brazil	01s27/48w30	B.str PA	AF198671
B. intermedia	01	Jales, SP, Brazil	20s16/50w32	B.int SP-1	AF198674
	01	Pindorama, SP, Brazil	21s11/48w54	B.int SP-2	AF198675
	01	Itapagipe, MG, Brazil	19s54/49w22	B.int MG	AF198673
B. peregrina	01	Alfenas, MG, Brazil	21s25/45w56	B.per MG-1	AF198676
	01	Taim, RS, Brazil	32s29/52w34	B.per RS	AF198677
	01	Bom Jesus da Penha, MG, Brazil	21s01/46w31	B.perMG-2	AF198678
B. schrammi	02	Ilicinea, MG, Brazil	20s56/45w49	B.schMG-1	AF198681
				B.schMG-2	AF198682
B. kuhniana	01	Tucurui, PA, Brazil	03s46/49w40	B.kun PA	AF198666
	01	Aragua, Villa de Cura, Venezuela		B.kun VEN	AF198667
B. occidentalis	01	Capetinga, MG, Brazil	20s36/47w03	B. ocMG	AF198658
	01	Campo Grande, MS, Brazil*	20s26/54w38	B. ocMT	AF198657
B. oligoza	01	Eldorado do Sul , RS, Brazil	30s05/51w36	B.oliRS	AF198680
	01	Florianópolis, SC, Brazil	27s35/48w32	B.oli SC	AF198679
B. amazonica	02	Bejamin Constant, AM, Brazil	04s22/70w01	B.aAM-1	AF198663
				B.aAM-2	AF198664
	01	Barão de Melgaço, MT, Brazil*	16s13/55w58	B.aMT-1	AF198665
Helisoma duryi	02	Uberaba, MG, Brazil	19s45/47w56	H.duryi-3	AF267503
				H.duryi-5	AF267504

* Laboratory populations. From Vidigal *et al.* (2000).

which could be confidently aligned. The multiple sequence alignment revealed a substantial degree of variation between snails from different species with substitutions, insertions and deletions being present. However, there was minimal sequence variation between snails from the same species collected in different localities. The phylogenetic analysis of ITS2 sequences, performed with the three different methods of phylogenetic reconstruction (distance, maximum parsimony and maximum likelihood), generated very similar trees. However, only the parsimony and neighbour-joining algorithm (NJ) with the addition of *Helisoma* were showed (Fig. 9A and B). Parsimony (MP) analysis yielded 1274 equally parsimonious trees of length 281. Only one of the trees is shown, however any bifurcating branches that are not present in the strict consensus

of all trees are indicated in bold (Fig. 9A). All trees were found on a single island. The tree generated by maximum-likelihood (ML) has a log likelihood of −2199·87746. In the case of distance analyses, the trees were constructed using the neighbour-joining algorithm (NJ) with and without the additional taxa. In all cases, the trees were unrooted for analysis, however *B. glabrata* was specified as the outgroup followed the arguments of Bandoni *et al.* (1995) and Woodruff & Mullvey (1997). All four trees show the same major groupings: I–VI (Fig. 9A). The branch uniting the members of group VI and the branch uniting groups I–III are quite long and separate these groups well, however the branch uniting groups IV and V is shorter and less well supported by bootstrap analysis (63%). Some species show invariant relationships to each other in all trees

(A)

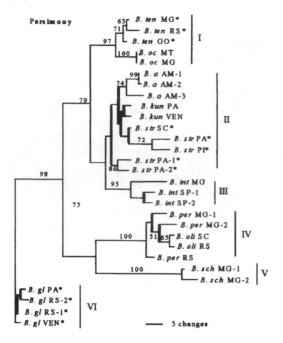

(B)

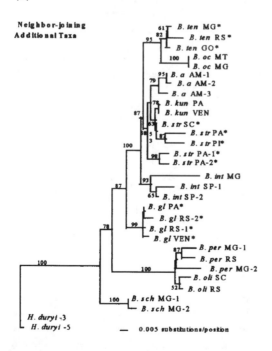

Fig. 9. Trees resulting from phylogenetic analysis of ITS2 sequences of the *Biomphalaria species*. Species abbreviations are defined in Table 1. Bootstrap values above 50 % are listed above branches. Scale bars are as indicated. Species names followed by an asterisk* are natural hosts for the parasite *Schistosoma mansoni*. Roman numerals (I–VI) shown in (A) define the major lineages found in both trees. (A) maximum parsimony analysis, bold vertical bars represent bifurcations not present on the strict consensus of 1274 equally parsimonious trees; (B) neighbour-joining analysis using additional taxa. Vidigal *et al.* (2000).

(*B. tenagophila, B. occidentalis, B. amazonica, B. intermedia* and *B. schrammi*) however, others are variable. The pairs of species that are most volatile in terms of their position on the trees are *B. straminea/B. kuhniana* and *B. peregrina/B. oligoza*. In both cases, the specimens from each of these species pairs are intermingled and do not form monophyletic groups by any of the methods used. The ITS2-based trees exhibit very similar overall topologies that are compatible with the traditional taxonomy based on morphology.

Of interest is the placement of *B. glabrata* in the interior of the tree when *H. duryi* is specified as the outgroup (Fig. 9B). This tree clearly places *B. schrammi* outside of *B. glabrata* and is basal to the remaining *Biomphalaria* species. Our inspection of the sequence alignment supports this finding. *Biomphalaria glabrata* is clearly distinct from the other species examined. This distinction has been made previously by Bandoni *et al.* (1995) and Woodruff & Mulvey (1997). The results obtained by the first group led them to suggest that *B. glabrata* was a sister taxon to the African species. They demonstrated African affinities of *B. glabrata* obtained from Brazil and the Dominican Republic. These results are in accordance with those of Woodruff & Mulvey (1997), which show greater affinity of the American *B. glabrata* with African *Biomphalaria* species than with other American species. The fact that the *B. glabrata* specimens cluster differently in the NJ tree than they do in the MP (Fig. 9A) and ML trees can be explained by genetic variability attributed to this species as mentioned above or by the marker used. However, our results suggested that *B. glabrata* was not the most distant *Biomphalaria* species in Brazil (Fig. 9B). The three species that are natural hosts of *S. mansoni* do not form a monophyletic group. There is at least one potential or natural intermediate host species for *S. mansoni* in four of the *Biomphalaria* groups. In group I: *B. tenagophila*; in group II: *B. straminea*; in group IV: *B. peregrina* and in group VI: *B. glabrata*. The latter shows the highest infection rates both naturally and experimentally and is almost always associated with the presence of schistosomiasis. Thus, the ability to act as an intermediate host for *S. mansoni* may be 'easily' acquired.

We believe that molecular-based methodologies may be very useful in furthering understanding of *Biomphalaria* species by means of systematic and population genetics studies. Indeed, Simpson *et al.* (1995), analyzing the *Schistosoma* and *B. glabrata* genomes, observed that while parasite genetic intraspecific polymorphism is limited, that of the mollusc is very extensive, suggesting that the genetics of the intermediate host play a more important role in the epidemiological control of schistosomiasis than those of the parasite itself.

ACKNOWLEDGEMENTS

This work was partially supported by grants from the CAPES and FAPEMIG.

REFERENCES

ABDEL-HAMID, A. H., DE MOLFETTA, J. B., FERNANDEZ, V. & RODRIGUES, V. (1999). Genetic variation between susceptible and non-suceptible snails to *Schistosoma* infection using random amplified polymorphic DNA analysis (RAPDs). *Revista do Instituto de Medicina Tropical de São Paulo* 5, 291–5.

BAKER, F. C. (1945). Subfamilies, Genera and Subgenera – Recent and Fossil. In *The Molluscan Family Planorbidae* (ed. Baker, F. C.), pp. 80–95. Urbana, University Illinois Press.

BANDONI, S. M., MULVEY, M. & LOKER, E. S. (1995). Phylogenetic analysis of eleven species of *Biomphalaria* Preston, 1910 (Gastropoda, Planorbidae) based comparisons of allozymes. *Biological Journal of the Linnean Society* 54, 1–27.

CALDEIRA, R. L., VIDIGAL, T. H. D. A., MATINELA, L., SIMPSON, A. J. G. & CARVALHO, O. S. (2000). Identification of planorbids from Venezuela by polymerase chain reaction amplification and restriction fragment length polymorphism (PCR-RFLP). *Memórias do Instituto Oswaldo Cruz* 95, 171–177.

CALDEIRA, R. L., VIDIGAL, T. H. D. A., PAULINELLI, S. T., SIMPSON, A. J. G. & CARVALHO, O. S. (1998). Molecular identification of similar species of the genus *Biomphalaria* (Mollusca, Planorbidae) determined by a PCR-RFLP. *Memórias do Instituto Oswaldo Cruz* 93, 219–225.

CALDEIRA, R. L., VIDIGAL, T. H. D. A., SIMPSON, A. J. G. & CARVALHO, O. S. (2001). Genetic variability in Brazilian populations of *Biomphalaria straminea* complex detected by simple sequence repeat anchored polymerase chain reaction amplification (SSR-PCR). *Memórias do Instituto Oswaldo Cruz* 96, 535–544.

CORRÊA, L. R. & PARAENSE, W. L. (1971). Susceptibility of *Biomphalaria amazonica* to infection with two strains of *Schistosoma mansoni*. *Revista do Instituto de Medicina Tropical de São Paulo* 13, 387–390.

DE SOUSA, G. B., DUTARI, G. P. & GARDENAL, C. N. (1999). Genetic structure of *Aedes albifasciatus* (Diptera, Culicidae) populations in central Argentina determined by random amplified polymorphic DNA-polymerase chain reaction markers. *Journal of Medical Entomology* 36, 400–404.

DIAS NETO, E., SANTOS, F. R., PENA, S. D. J. & SIMPSON, A. J. G. (1993). Sex determination by low stringency PCR (LS-PCR). *Nucleic Acids Research* 21, 763–764.

DONNELLY, M. J., CUAMBA, N., CHARIWOOD, J. D., COLLINS, F. H. & TOWNSON, H. (1999). Population structure in the malaria vector, *Anopheles arabiensis* Patton, in East Africa. *Heredity* 83, 408–417.

FERDIG, M. T & SU, X. Z. (2000). Microsatellite markers and genetic mapping in *Plasmodium falciparum*. *Parasitology Today* 16, 307–312.

GOMES, M. A., MELO, N. M, MACEDO, A. M, FURST, C. & SILVA, E. F. (2000). RAPD in the analysis of isolates of *Entamoeba histolytica*. *Acta Tropica* 75, 71–77.

HOFFMAN, J. I., WEBSTER, J. P., NDAMBA, J. & WOOLHOUSE, M. E. (1998). Extensive genetic variation revealed in adjacent populations of the schistosome intermediate host *Biomphalaria pfeifferi* from single river system. *Annals of Tropical Medicine and Parasitology* 92, 693–698.

HOPE, M. & MCMANUS, D. P. (1994). Genetic variations in geographically isolated populations and subspecies of *Oncomelania hupensis* determined by a PCR-based RFLP method. *Acta Tropica* 57, 75–82.

JARNE, P. & DELAY, B. (1991). Populations genetics of freshwater snails. *Tree* 6, 383–386.

JONES, C. S., LOCKYER, A. E., ROLLINSON, D., PIERTNEY, S. B. & NOBLE, L. R. (1999). Isolation and characterisation of microsatellite loci in the freshwater gastropod, *Biomphalaria glabrata*, an intermediate host for *Schistosoma mansoni*. *Molecular Ecology* 8, 2141–2152.

KANE, R. A. & ROLLISON, D. (1994). Repetitive sequences in the ribosomal DNA internal transcribed spacer of *Schistosoma haematobium, Schistosoma intercalatum* and *Schistosoma mattheii*. *Molecular and Biochemical Parasitology* 63, 153–156.

KNIGHT, M., BRINDLEY, P. J., RICHARDS, C. S. & LEWIS, F. A. (1991). *Schistosoma mansoni*, use of a cloned ribosomal RNA gene probe to detect restriction fragment length polymorphisms in the intermediate host *Biomphlaria glabrata*. *Experimental Parasitology* 73, 285–294.

KNIGHT, M., MILLER, A., RAGHAVAN, N., RICHARDS, C. & LEWIS, F. (1992). Identification of a repetitive element in the snail *Biomphalaria glabrata*: relationship to the reverse transcriptase-encoding sequence in *LINE-1* transposons. *Gene* 118, 181–187.

KNIGHT, M., MILLER, A. N., PATTERSON, C. N., ROWE, C. G., MICHAELS, G., CARR, D., RICHARDS, C. S. & LEWIS, F. A. (1999). The identification of markers segregating with resistance to *Schistosoma mansoni* infection in the snail *Biomphalaria glabrata*. *Proceedings of the National Academy of Science, USA* 96, 1510–1515.

KNIGHT, M., ONGELE, E. & LEWIS, F. A. (2000). Molecular studies of *Biomphalaria glabrata*, an intermediate host of *Schistosoma mansoni*. *International Journal for Parasitology* 14, 535–541.

LANGAND, J., BARRAL, V., DELAY, B. & JOURDANE, J. (1993). Detection of genetic diversity within snail intermediate hosts of the genus *Bulinus* by using random amplified polymorphic DNA markers (RAPDs). *Acta Tropica* 55, 205–215.

LANGAND, J., THERON, A., POINTIER, J. P., DELAY, B. & JOURDANE, J. (1999). Population structure of *Biomphalaria glabrata*, intermediate snail host of *Schistosoma mansoni* in Guadeloupe Island using RAPD markers. *Journal of Molluscan Studies* 65, 425–433.

LARSON, S. E., ANDERSEN, P. L., MILLER, N. A., COUSIN, C. E., RICHARDS, C. S., LEWIS, F. A. & KNIGHT, M. (1996). Use of RAPD-PCR to differentiate genetically defined lines of an intermediate host of *Schistosoma mansoni, Biomphalaria glabrata*. *Journal of Parasitology* 82, 237–244.

LEWIS, F. A., KNIGHT, M. & RICHARDS, C. S. (1997). A laboratory-based approach to biological control of snails. *Memórias do Instituto Oswaldo Cruz* **92**, 661–662.

MASCARA, D. & MORGANTE, J. S. (1995). Use of isozyme patterns in the identification of *Biomphalaria tenagophila* (D'Orbigny, 1835) and *B. occidentalis* (Paraense, (1981)) (Gastropoda, Planorbidae). *Memórias do Instituto Oswaldo Cruz* **90**, 359–366.

MAVAREZ, J., AMARISTA, M., POINTIER, J. P. & JARNE, P. (2000). Microsatellite variation in the freshwater schistosome-transmitting snail. *Biomphalaria glabrata*. *Molecular Ecology* **9**, 1009–1011.

MUKARATIRWA, S., KRISTENSEN, T. K., SIEGISMUND, H. R. & CHANDIWANA, S. K. (1998). Genetic and morphological variation of populations belonging to the *Bulinus truncatus/tropicus* complex (Gastropoda, Planorbidae) in south western Zimbabwe. *Journal of Molluscan Studies* **64**, 435–446.

MULVEY, M., NEWMAN, M. C. & WOODRUFF, D. S. (1988). Genetic differentiation among west indian populations of the schistosome-transmitting snail *Biomphalaria glabrata*. *Malacology* **29**, 309–317.

MULVEY, M. & VRIJENHOEK, R. C. (1982). Population structure in *Biomphalaria glabrata*: examination of hypothesis for the patchy distribuition of susceptibility to schistosomes. *American Journal of Tropical Medicine and Hygiene* **31**, 1195–1200.

OLIVEIRA, R. P., MACEDO, A. M., CHIARI, E. & PENA, S. D. J. (1997). An alternative approach to evaluating the intraspecific genetic variability of parasites. *Parasitology Today* **5**, 196–200.

PARAENSE, W. L. (1955). Self and cross-fertilization in *Australorbis glabratus*. *Memórias do Instituto Oswaldo Cruz* **53**, 285–291.

PARAENSE, W. L. (1959). One-sided reproductive isolation between geographically remote populations of a planorbid snail. *American Naturalist* **XCII**, 93–101.

PARAENSE, W. L. (1972). Fauna planorbídica do Brasil. In *Introdução à geografia médica do Brasil* (ed. Lacaz, C. S. Baruzzi, R. G. & Siqueira Junior, W.), pp. 213–239. São Paulo, Edgard Blücher, Editora Universidade de São Paulo.

PARAENSE, W. L. (1975a). Estado atual da sistemática dos planorbídeos brasileiros. *Arquivo do Museu Nacional do Rio de Janeiro* **55**, 105–128.

PARAENSE, W. L. (1975b). *Biomphalaria orbignyi* sp. n. From Argentina (Gastropoda, Basommatophora, Planorbidae). *Revista Brasileira de Biologia* **35**, 211–222.

PARAENSE, W. L. (1981). *Biomphalaria occidentalis* sp. n. from South America (Mollusca, Basommatophora, Pulmonata). *Memórias do Instituto Oswaldo Cruz* **76**, 199–211.

PARAENSE, W. L. (1984). *Biomphalaria tenagophila guaibensis* ssp. n. from southern Brazil and Uruguay (Pulmonata, Planorbidae). I. Morphology. *Memórias do Instituto Oswaldo Cruz* **79**, 465–469.

PARAENSE, W. L. (1986). Distribuição dos caramujos no Brasil. In *Modernos conhecimentos sobre a esquistossomose mansônica* (ed. Reis, F. A. Faria, I. & Katz, N.), pp. 117–128. Belo Horizonte, Brazil, Academia Mineira de Medicina.

PARAENSE, W. L. (1988). *Biomphalaria kuhniana* (Clessin, 1883), planorbid mollusc from South America. *Memórias do Instituto Oswaldo Cruz* **83**, 1–12.

PARAENSE, W. L. & CORRÊA, L. R. (1963). Variation in susceptibility of populations of *Australorbis glabratus* to a strain of *Schistosoma mansoni*. *Revista do Instituto de Medicina Tropical de São Paulo* **5**, 15–22.

PARAENSE, W. L. & CORRÊA, L. R. (1973). Susceptibility of *Biomphalaria peregrina* from Brazil and Ecuador to two strains of *Schistosoma mansoni*. *Revista do Instituto de Medicina Tropical de São Paulo* **15**, 127–130.

PARAENSE, W. L. & CORRÊA, L. R. (1987). Probable extension of schistosomiasis mansoni to southern most Brazil. *Memórias do Instituto Oswaldo Cruz* **82**, 577.

PARAENSE, W. L. & CORRÊA, L. R. (1989). A potential vector of *Schistosoma mansoni* in Uruguay. *Memórias do Instituto Oswaldo Cruz* **84**, 281–288.

PARAENSE, W. L., PEREIRA, O. & PINTO, D. B. (1955). Um aspecto da ecologia do "*Australorbis glabratus*" que favorece a reinfestação dos criadouros. *Revista do Serviço Especial de Saúde Pública* **7**, 573–581.

PARAENSE, W. L., POINTIER, J. P., DELAY, B., PERNOT, A. F., INCANI, R. N., BALZAN, C. & CHROSCIECHOWSKI, P. (1992). *Biomphalaria prona* (Gastropoda, Planorbidae), a morphological and biochemical study. *Memórias do Instituto Oswaldo Cruz* **87**, 171–179.

PIRES, E. R., VIDIGAL, T. H. D. A., TELES, H. M. S., SIMPSON, A. J. G. & CARVALHO, O. S. (1997). Specific Identification of *Biomphalaria tenagophila* and *Biomphalaria occidentalis* populations by the low stringency polymerase chain reaction. *Memórias do Instituto Oswaldo Cruz* **92**, 101–106.

REMIGIO, E. A. & BLAIR, D. (1997). Relationships among problematic North American stagnicoline snails (Pulmonata, Lymnaeidae) reinvestigated using nuclear ribosomal DNA internal transcribed spacer sequences. *Canadian Journal of Zoology* **75**, 1540–1545.

RICHARDS, C. S. (1984). Influence of snail age on genetic variations in susceptibility of *Biomphalaria glabrata* for infection with *Schistosoma mansoni*. *Malacologia* **25**, 493–502.

RICHARDS, C. S. & MERRITT, J. W. (1972). Genetic factors in susceptibility of juvenile *Biomphalaria glabrata* to *Schistosoma mansoni* infection. *American Journal of Tropical Medicine and Hygiene* **21**, 425.

RICHARDS, C. S. & SHADE, P. C. (1987). The genetic variation of compatibility in *Biomphalaria glabrata* and *Schistosoma mansoni*. *Journal of Parasitology* **73**, 1146–1151.

SAMADI, S., MALVAREZ, J., POINTIER, J. P. & JARNE, P. (1999). Microsatellite and morphological analysis of population structure in the parthenogenetic freshwater snail *Melanoides tuberculata*, insight into the creation of clonal variability. *Molecular Ecology* **8**, 1141–1153.

SCHILTHUIZEN, M., GITTENBERGER, E. & GULTYAEV, A. P. (1995). Phylogenetic relationships inferred from the sequence and secondary structure of ITS1 rRNA in *Albinaria* and putative *Isabellaria* species (Gastropoda, Pulmonata, Clausillidae). *Molecular Phylogenetics and Evolution* **4**, 457–462.

SIMPSON, A. J. G., DIAS NETO, E., VIDIGAL, T. H. D. A., PENA, H. B., CARVALHO, O. S. & PENA, S. D. J. (1995). DNA polymorphism of schistosomes and the snail hosts. *Memórias do Instituto Oswaldo Cruz* **90**, 211–213.

SOUZA, C. P., PASSOS-JANNOTTI, L. K. & FREITAS, J. R. (1995). Degree of host-parasite compatibility between *Schistosoma mansoni* and their intermediate molluscan hosts in Brazil. *Memórias do Instituto Oswaldo Cruz* **90**, 5–10.

SPATZ, L., VIDIGAL, T. H. D. A., CALDEIRA, R. L., DIAS NETO, E., CAPPA, S. M. G. & CARVALHO, O. S. (1999). Study of *Biomphalaria tenagophila, B. t. guaibensis* and *B. occidentalis* by polymerase chain reaction amplification and restriction enzyme digestion of the ribosomal RNA gene intergenic spacer. *Journal of Molluscan Studies* **65**, 143–149.

SPATZ, L., VIDIGAL, T. H. D. A., SILVA, M. C. A., GONZÁLEZ CAPPA, S. M. & CARVALHO, O. S. (2000). Characterization of *Biomphalaria orbignyi, B. peregrina* and *B. oligoza* by polymerase chain reaction and restriction enzyme digestion of the internal transcribed spacer region of the RNA ribosomal gene. *Memórias do Instituto Oswaldo Cruz* **95**, 807–814.

STOTHARD, J. R., HUGHES, S. & ROLLINSON, D. (1996). Variation within the internal transcribed spacer (ITS) of ribosomal DNA genes of intermediate snail hosts within the genus *Bulinus* (Gastropoda, Planorbidae). *Acta Tropica* **61**, 19–29.

STOTHARD, J. R. & ROLLINSON, D. (1997). Molecular characterization of *Bulinus globosus* and *B. nasutus* on Zanzibar, and an investigation of their roles in the epidemiology of *Schistosoma haematobium*. *Transactions of the Royal Society for Tropical Medicine and Hygiene* **91**, 353–357.

TIE, A. D., BOULDING, E. G. & NAISH, K. A. (2000). Polymorphic microsatellite DNA markers for the marine gastropod *Littorina subrotundata*. *Molecular Ecology* **9**, 108–110.

VERNON, J. G., JONES, C. & NOBLE, L. R. (1995). Random amplified polymorphic DNA (RAPD) markers reveal cross-fertilisation in *Biomphalaria glabrata* (Pulmonata, Basommatophora). *Journal of Molluscan Studies* **61**, 455–465.

VIARD, F., JUSTY, F. & JARNE, P. (1997*a*). Selfing, sexual polymorphism and microsatellites in the hermafroditic freshwater snail *Bulinus truncatus*. *Proceedings of the Royal Society of London* **264**, 39–44.

VIARD, F., JUSTY, F. & JARNE, P. (1997*b*). The influence of self-fertilization and populations dynamics on the genetic structure of subdivided populations, case study using microsatellite markers in the freshwater snail *Bulinus truncatus*. *Evolution* **51**, 1518–1528.

VIARD, F., JUSTY, F. & JARNE, P. (1997*c*). Population dynamics inferred from temporal variation at microsatellite loci in the selfing snail *Bulinus truncatus*. *Genetics* **3**, 973–982.

VIDIGAL, T. H. D. A., CALDEIRA, R. L., SIMPSON, A. J. G. & CARVALHO, O. S. (2000). Further studies on the molecular systematics of *Biomphalaria* snails from Brazil. *Memórias do Instituto Oswaldo Cruz* **95**, 57–66.

VIDIGAL, T. H. D. A., DIAS NETO, E., CARVALHO, O. S. & SIMPSON, A. J. G. (1994). *Biomphalaria glabrata*: extensive genetic variation in Brazilian isolates by random amplified polymorphic DNA analysis. *Experimental Parasitology* **79**, 187–194.

VIDIGAL, T. H. D. A., DIAS NETO, E., SIMPSON, A. J. G. & CARVALHO, O. S. (1996). A low stringency polymerase chain reaction approach to identification of *Biomphalaria glabrata* and *Biomphalaria tenagophila* intermediate snail hosts of *Schistosoma mansoni* in Brazil. *Memórias do Instituto Oswaldo Cruz* **91**, 739–744.

VIDIGAL, T. H. D. A., KISSINGER, J. C., CALDEIRA, R. L., PIRES, E. C. R., MONTEIRO, E., SIMPSON, A. J. G. & CARVALHO, O. S. (2000). Phylogenetic relationships among Brazilian *Biomphalaria* species (Mollusca, Planorbidae) based upon analysis of ribosomal ITS2 sequences. *Parasitology* **121**, 611–620.

VIDIGAL, T. H. D. A., SPATZ, L., NUNES, D. N., SIMPSON, A. J. G., CARVALHO, O. S. & DIAS NETO, E. (1998). *Biomphalaria* spp: identification of the intermediate snail hosts of *Schistosoma mansoni* by polymerase chain reaction amplification and restriction enzyme digestion of the ribosomal RNA gene Intergenic spacer. *Experimental Parasitology* **89**, 180–187.

WELSH, J. & MCCLELLAND, M. (1990). Fingerprinting genomes using PCR with arbitrary primers. *Nucleic Acids Research* **18**, 7213–7218.

WILLIAMS, J. G. K., KUBELICK, A. R., LIVAK, K. J., RAFALSKI, J. A. & TINGEY, S. V. (1990). DNA polymorphisms amplified by arbitrary primers are useful as genetic markers. *Nucleic Acids Research* **18**, 6531–6535.

WOODRUFF, D. S & MULVEY, M. (1997). Neotropical schistosomiasis: African affinites of the host snail *Biomphalaria glabrata* (Gastropoda, Planorbidae). *Biological Journal of the Linnean Society* **60**, 505–516.

WOODRUFF, D. S., MULVEY, M. & YIPP, M. W. (1985). Population genetics of *Biomphalaria straminea* in Hong Kong. *Journal of Heredity* **76**, 355–360.

ZAHAB-JABBOUR, R., POINTIER, J. P., JOURDANE, J., JARNE, P., OVIEDO, J. A., BARGUES, M. D., MAS-COMA, S., ANGLÉS, R., PERERA, G., BALZAN, C., KHALLAYOUNE, K. & RENAUD, F. (1997). Phylogeography and genetic divergence of some lymnaeid snails, intermediate hosts of human and animal fascioliasis with special reference to lymnaeids from the Bolivian Altiplano. *Acta Tropica* **64**, 191–203.

ZHOU, X. & KRISTENSEN, T. K. (1999). Genetic and morphological variations in populations of *Oncomelania* spp in China. *Southeast Asian Journal of Tropical Medicine and Public Health* **30**, 166–176.

ZIETKIEWICZ, E., RAFALSKI, A. & LABUDA, D. (1994). Genome fingerprinting by simple sequence repeat (SSR)-anchored polymerase chain reaction amplification. *Genomics* **20**, 176–193.

Schistosoma mansoni and Biomphalaria: past history and future trends

J. A. T. MORGAN[1], R. J. DEJONG[1], S. D. SNYDER[2], G. M. MKOJI[3] and E. S. LOKER[1]*

[1] Department of Biology, University of New Mexico, Albuquerque, NM 87131 USA
[2] Department of Biology & Microbiology, University of Wisconsin Oshkosh, 800 Algoma Boulevard, Oshkosh, WI 54901-8640 USA
[3] Center for Biotechnology Research and Development, Kenya Medical Research Institute, P.O. Box 54840, Nairobi, Kenya

SUMMARY

Schistosoma mansoni is one of the most abundant infectious agents of humankind. Its widespread distribution is permitted by the broad geographic range of susceptible species of the freshwater snail genus Biomphalaria that serve as obligatory hosts for its larval stages. Molecular phylogenetic studies suggest that Schistosoma originated in Asia, and that a pulmonate-transmitted progenitor colonized Africa and gave rise to both terminal-spined and lateral-spined egg species groups, the latter containing S. mansoni. Schistosoma mansoni likely appeared only after the trans-Atlantic dispersal of Biomphalaria from the Neotropics to Africa, an event that, based on the present African fossil record, occurred only 2–5 million years ago. This parasite became abundant in tropical Africa and then entered the New World with the slave trade. It prospered in the Neotropics because a remarkably susceptible and productive host, B. glabrata, was widely distributed there. Indeed, a snail similar to B. glabrata may have given rise to the African species of Biomphalaria. Schistosoma mansoni has since spread into other Neotropical Biomphalaria species and mammalian hosts. The distribution of S. mansoni is in a state of flux. In Egypt, S. mansoni has nearly completely replaced S. haematobium in the Nile Delta, and has spread to other regions of the country. A susceptible host snail, B. straminea, has been introduced into Asia and there is evidence of S. mansoni transmission in Nepal. Dam and barrage construction has lead to an epidemic of S. mansoni in Senegal, and the parasite continues its spread in Brazil. Because of competition with introduced aquatic species and environmental changes, B. glabrata and consequently S. mansoni have become less abundant on the Caribbean islands. Control of S. mansoni using praziquantel and oxamniquine has reduced global prevalence but control is difficult to sustain, and S. mansoni can develop tolerance/resistance to praziquantel, raising concerns about its future efficacy. Because of legitimate environmental concerns, snail control is unlikely to be an option in future control efforts. Global warming will impact the distribution of Biomphalaria and S. mansoni, but the magnitude and nature of the effects are poorly understood.

Key words: Schistosoma mansoni, Biomphalaria, evolution, distribution, control.

INTRODUCTION: S. MANSONI'S PRESENT ABUNDANCE, DISTRIBUTION AND HOST PREFERENCES

Schistosoma mansoni Sambon, 1907 has the distinction of being the most intensively studied of all members of the phylum Platyhelminthes. The number of papers published annually regarding this species is 3–4 times higher than for any other platyhelminth species, reflecting its importance as a laboratory model for studying schistosomiasis. It is likely that the first complete genome sequence obtained for a flatworm will be that of S. mansoni (Snyder et al. 2001). The scrutiny given this organism is well deserved for S. mansoni is the most widely distributed of all the schistosomes infecting humans, being found in sub-Saharan Africa where it is particularly abundant, in the valley and delta of the Nile, in parts of southwest Asia, in Brazil and other parts of northeastern South America, and in

isolated foci on some of the Caribbean islands (Fig. 1). A recent estimate places the number of people infected with S. mansoni at about 83 million (Crompton, 1999), in 54 countries (Chitsulo et al. 2000). In addition to infecting humans, S. mansoni is also found in rodents (Théron & Pointier, 1995; D'Andrea et al. 2000) and in wild primates, particularly baboons (Ghandour et al. 1995; Müller-Graf et al. 1997; Munene et al. 1998).

The geographic distribution of S. mansoni is closely tied to that of freshwater pulmonate snails that serve as its obligatory molluscan hosts, susceptible species of the planorbid genus Biomphalaria. Although there are reports of successful experimental S. mansoni infections in another discoidal planorbid species, Planorbarius metidjensis (Barbosa, Barbosa & Morais-Rêgo, 1959), the extent to which S. mansoni is a specialist on Biomphalaria snails is remarkable. In sub-Saharan Africa, S. mansoni is predominantly transmitted through Biomphalaria pfeifferi even though all of the 12 African species of Biomphalaria are susceptible to infection and can play some role in transmission in certain situations (Brown, 1994). Biomphalaria alexandrina is the most

* Corresponding author: Eric S. Loker, Department of Biology, University of New Mexico, Albuquerque, New Mexico 87131 USA. Tel: (505) 277 5508. Fax: (505) 277 0304. E-mail: esloker@unm.edu

Parasitology (2001), 123, S211–S228.   © 2001 Cambridge University Press
DOI: 10.1017/S0031182001007703

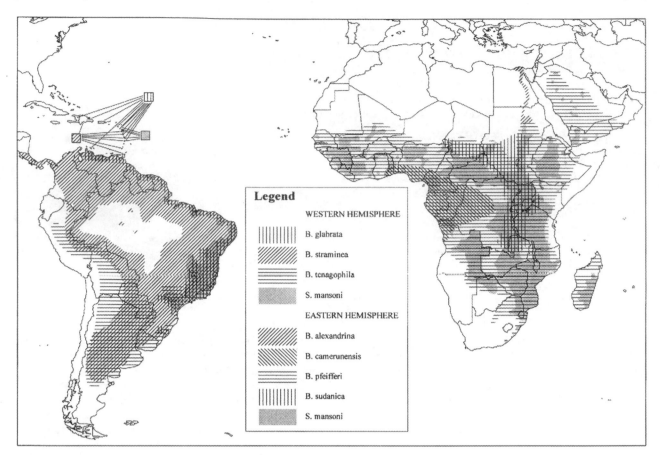

Fig. 1. Distributions of *Schistosoma mansoni* and *Biomphalaria* host species of major medical importance. After Brown (1994), de Souza & Lima (1997), Doumenge *et al.* (1987) and Malek (1985). Distributions are intended to indicate the main areas of occurrence; continuity of distribution is not implied and there may be significant discontinuities within these areas.

common host in Egypt. In South America and the Caribbean region, *B. glabrata* is the most important snail host, although *B. straminea* and *B. tenagophila* can also be found naturally infected. In South America there are six *Biomphalaria* species that have not been implicated in *S. mansoni* transmission but that are susceptible experimentally and eight species that are apparently refractory to infection (Malek, 1985).

The factors underlying compatibility between *S. mansoni* and *Biomphalaria* are complex and have been the subject of a number of studies that have been reviewed elsewhere (Loker & Bayne, 1986; Richards, Knight & Lewis, 1992; Yoshino & Vasta, 1996; Adema & Loker, 1997). Suffice it to say here that both the resistance status of the snail host and the infectivity of the parasite are genetically controlled. Also because both snail and parasite are variable entities (Vidigal *et al.* 1994; Mulvey & Bandoni, 1994; Curtis & Minchella, 2000), tremendous variation in the compatibility of local host and parasite strains has been noted (Basch, 1976; Frandsen, 1979*b*).

For many of us, *S. mansoni* exists as a laboratory abstraction, yet it is vitally important to try to understand its existence in the real world. With this

goal in mind, it is helpful to gain a perspective on where *S. mansoni* has been and where it is likely to go in coming years.

VIEWS ON WHERE *S. MANSONI* CAME FROM

Origin of the genus

Schistosomes have left no fossil record making it difficult to pinpoint where or when the genus originated. *Schistosoma mansoni* probably arose in Africa (Davis, 1980, 1992; Snyder & Loker, 2000) but how and when did its progenitors get there? Two theories are currently available to describe the origins of the genus.

Out of Africa theory. The African origin theory was proposed by George M. Davis and has been developed over a number of years (Davis, 1980, 1992). The theory proposes that the genus *Schistosoma* originated in Gondwanaland (the supercontinent consisting of Africa, Antarctica, India and South America) within both pulmonate and pomatiopsid snails and dispersed via continental drift and collisions, see Fig. 2A (Davis, 1980, 1992). Thus, the ancestor of all Asian *Schistosoma* rafted

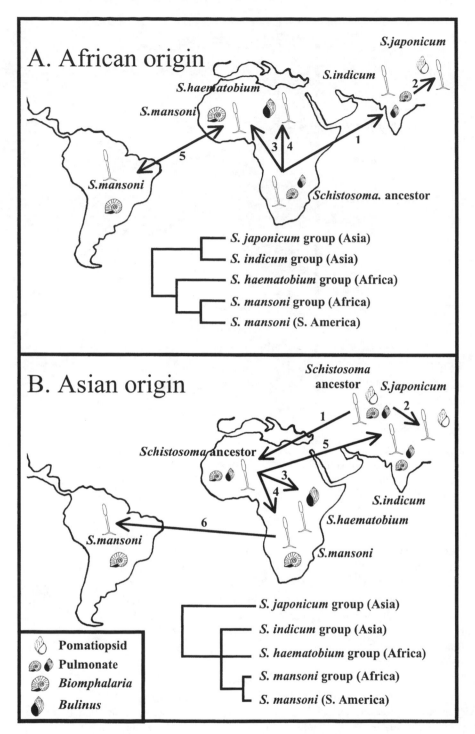

Fig. 2. Maps displaying the two hypotheses of origin of *S. mansoni*, A. the African origin and B. the Asian origin. Below each map is a tree depicting expected species group relationships given the theorized origin. (A) African origin (after Davis, 1980; 1992); 1 and 2 The ancestral African *Schistosoma* rafts to Asia on the Indian plate 70–148 MYA forming the *S. indicum* and *S. japonicum* groups. 3 and 4 The *Schistosoma* ancestor remaining in Africa diverges > 120 MYA to form the *S. mansoni* and *S. haematobium* groups. 5 *Schistosoma mansoni* disperses to South America 80–120 MYA, before continental drift splits Gondwanaland. (B) Asian origin (after Snyder & Loker, 2000 and other sources): 1 The ancestral Asian *Schistosoma* (likely a parasite of pulmonate snails) moves to Africa 12–19 MYA via widespread mammal migration. 2 The *Schistosoma* ancestor remaining in Asia becomes the *S. japonicum* group. 3 and 4 The African *Schistosoma* ancestor diverges 1–4 MYA to form the *S. mansoni* and *S. haematobium* groups. 5 An *S. indicum* ancestor also diverges from the African ancestor 1–4 MYA and migrates back to India, probably with early humans and their domestic animals (Barker & Blair, 1996). 6 *Schistosoma mansoni* disperses to South America 150–500 YA via the transport of African slaves.

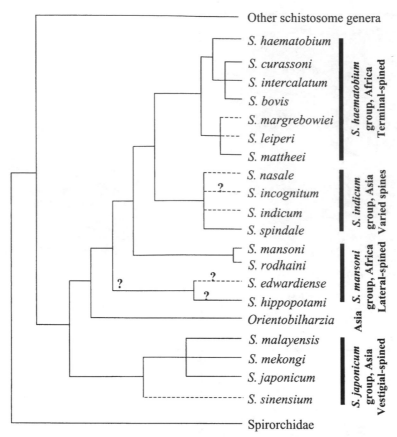

Fig. 3. Manually constructed schistosome tree summarising published phylogenies. Tree is a compilation of trees published by Snyder & Loker (2000), trees reviewed in Rollinson *et al.* (1997) and a super-tree generated by Morand & Müller-Graf (2000). Dotted lines mark predicted branch positions for species with no available data. The question marks (?) indicate branches that vary in different publications and are placed speculatively on this tree.

there from Africa on the Indian plate, which split from Africa 70–148 million years ago (MYA; Després *et al.* 1992). On reaching Asia the ancestors of *Schistosoma japonicum* radiated along with their pomatiopsid snail hosts. Schistosomes remaining in India retained their association with pulmonate snails and became the ancestors of the *S. indicum* group. The ancestral *Schistosoma* remaining in Africa, diverged over 120 MYA to form the ancestors to today's *S. mansoni* and *S. haemotobium* groups. *Schistosoma mansoni* had to be present in Gondwanaland at least 80 MYA, within *Biomphalaria*, to enable them both to access South America before the continents separated.

Out of Asia theory. A more recent theory on the origin of *Schistosoma* suggests that the genus arose in Asia and was introduced to Africa (Fig, 2B; Snyder & Loker, 2000). The Asian ancestral *Schistosoma* may have had either a pomatiopsid or a pulmonate snail host. Davis (1992) has linked the diversification of the *S. japonicum* group to the radiation of their pomatiopsid hosts in the mid-Miocene. Based on this link, Snyder & Loker (2000) reasoned that the colonization of Africa by *Schistosoma* occurred no earlier than this time (15 MYA), well after the fragmentation of Gondwanaland. On reaching

Africa, the descendants underwent extensive radiation becoming exclusive parasites of pulmonate snails in the family Planorbidae. According to this theory, *Schistosoma mansoni* was not in Gondwanaland when Africa separated from South America; thus the species must have utilized an alternative method for dispersal.

Molecular data support Asian theory

Taking a closer look at relationships among members of the genus *Schistosoma* at a molecular level may help us to understand the origins of *S. mansoni*. Predicted phylogenetic trees based on the two theories outlined above are drawn in Fig. 2.

A tree summarizing the genetic relationships among members of the genus *Schistosoma* is shown in Fig. 3. The tree has been compiled from a number of published phylogenies generated from both mitochondrial and nuclear genes (Snyder & Loker, 2000, trees reviewed in Rollinson *et al.* 1997 and a super tree generated by Morand & Müller-Graf, 2000). Predicted branches leading to species for which sequence is lacking are indicated with dotted lines. With the exception of *S. hippopotami*, the genetic phylogenetic trees produced agree with the four general groupings within the genus (*S. mansoni* group, *S. haematobium* group, *S. japonicum* group and *S. indicum* group), which were originally based on morphology and life history characteristics. To date only one representative of the *S. indicum* group has been sequenced. The level of genetic diversity among species within the *S. japonicum*, *S. haematobium* and *S. mansoni* groups appears to be very low compared to that measured between the species groups (reviewed in Rollinson *et al.* 1997).

The position of *S. hippopotami* is the one exception. The rDNA ITS2 from a single specimen has been sequenced and despite *S. hippopotami* displaying *S. mansoni* group characteristics (lateral-spined eggs and thought to infect *Biomphalaria*) it fails to cluster with the *S. mansoni* clade (Després *et al.* 1995; Rollinson *et al.* 1997). Thurston (1963) indicated that *S. hippopotami* resembles *S. incognitum* in egg size, egg morphology (subterminal-spine) and adult morphology. Efforts to complete the life cycle of *S. hippopotami* have been unsuccessful (Thurston, 1971) and to date its intermediate host remains unknown. *Schistosoma incognitum* infects lymnaeid snails and adults are common in pigs and dogs from India (Sinha & Srivastava, 1960) and rodents from Indonesia (Carney *et al.* 1977). Sequence data for *S. incognitum* are currently unavailable so its position within the *S. indicum* group (Fig. 3) is questionable. In the future it may in fact be recognized as separate from the *S. indicum* group, all other members of which infect the planorbid snail *Indoplanorbis exustus*, and show a closer affiliation to *S. hippopotami*. The ancestral African *Schistosoma* might have been a *S. incognitum*-like parasite, transmitted by a pulmonate (lymnaeid or planorbid?) snail, that made the passage from Asia in pigs, or their close relatives, hippos. The position of *S. edwardiense* is even less clear. This species infects hippos, has *S. margrebowiei*-shaped eggs and appears to infect *Biomphalaria*. It is intriguing that *S. edwardiense* also has cercariae with long tail-stems, like those of *S. incognitum*.

The Asian origin theory is based on molecular evidence. Snyder & Loker (2000) generated a tree from rDNA 28S sequences representing ten of the 13 genera belonging to the family Schistosomatidae. The tree depicts the genus *Schistosoma* as a paraphyletic assemblage with the Asian genus *Orientobilharzia* positioned as sister taxon to the African *Schistosoma* with *S. japonicum* more distant (Fig. 3). The African clade appears to be the most derived of the *Schistosoma* species, supporting the Asian origin theory. It is interesting to note that, like *S. incognitum*, *Lymnaea* snails are also intermediate hosts for *Orientobilharzia*.

Trees constructed from rRNA 18S (Johnston, Kane & Rollinson 1993) and 28S (Barker & Blair, 1996) sequences place *S. spindale*, a member of the *S. indicum* group, within the African species clade as a sister taxon to the *S. haematobium* group. Barker & Blair (1996) suggested that the position of *S. spindale*, and its lack of genetic divergence reflected a more recent introduction into Asia, probably with early humans and their domestic animals as they dispersed from Africa. Unfortunately sequence data are currently unavailable for comparisons with other members of the *S. indicum* group.

In favour of the Asian origin theory, the *S. japonicum* group displays the greatest divergence in all of the molecular studies to date. Species variation within the group (sequence data available for *S. japonicum*, *S. malayensis* and *S. mekongi*) is also higher than that observed among species within the *S. haematobium* and *S. mansoni* (excluding *S. hippopotami*) groups (Bowles, Blair & McManus, 1995; Rollinson *et al.* 1997).

Recent research has discovered changes in the mitochondrial gene order of Asian versus African schistosomes (Le *et al.* 2000). Their results add further support to an Asian origin for the genus. The Asian species, represented by *S. japonicum* and *S. mekongi*, display a gene order similar to that observed in other trematodes and cestodes while the African species (*S. mansoni*, *S. rodhaini*, *S. haematobium*, *S. bovis*, *S. curassoni*, *S. intercalatum*, *S. mattheei* and *S. margrebowiei*) have a rearranged mitochondrial genome. Although mitochondrial gene order cannot in itself be used as evidence for an Asian origin it does suggest that the African taxa may be more derived. The *Schistosoma* Genome Network has undertaken a collaborative study to sequence and analyse the entire mitochondrial genome of *S. mansoni* and *S. japonicum* (Franco *et al.* 2000).

Resolving the origins of *Schistosoma* may require additional sequences from those rarer species and genera missing from current studies: *S. sinensium*, *S. indicum*, *S. nasale*, *S. incognitum*, *S. hippopotami*, *S. edwardiense*, *S. leiperi*, *Griphobilharzia*, *Bivitellobilharzia* and *Macrobilharzia*.

Timing of events

The time frame placed on the divergence of the African and Asian *Schistosoma*, given an African origin, is 70–148 million years (Després *et al.* 1992). According to the Asian origin hypothesis, if the time of diversification of the *S. japonicum* group in the mid-Miocene is correct, then Africa was colonized no sooner than 15 MYA. The obvious question is: Is 15 million years enough time to account for the diversification of African *Schistosoma*? Barker & Blair (1996) completed a log likelihood ratio test on their rDNA 28S trees to determine if a molecular clock could be used to describe the evolution of the data set. Unfortunately, the molecular clock was rejected and they were unable to speculate about the age of the genus. Després *et al.* (1992) estimated divergence times by calibrating their nuclear rDNA ITS2 and mitochondrial 16S sequence based trees against rodents, which have a clear fossil record. They estimated a divergence for 16S of 1–2 % per million years and for ITS2, 0·3–0·8 % per million years. Based on these rates, the African and Asian schistosomes split 24–70 MYA. Separation of the two African groups (*S. mansoni* and *S. haematobium*) was dated 10–30 MYA and divergence within the African clades was 1–10 MYA for the *S. mansoni* group and 1–6 MYA for the *S. haematobium* group.

The mammalian fossil record shows an influx of Asian mammals into Africa 12–19 MYA during a series of collisions between Arabia and Turkey (Cox, 2000). Lowered sea-levels and a continuous tropical environment encouraged the transfer of carnivores, pigs, bovids and rodents from Asia into Africa (Cox, 2000). The conditions were probably well suited for the transfer of a *Schistosoma* ancestor within one of these hosts.

Origin of the species

The ancestral African *Schistosoma* species diverged to form the two African clades known today, the lateral-spined *S. mansoni* group and the terminal-spined *S. haematobium* group. The basal position of *S. hippopotami*, with respect to the African species, suggests that the African ancestor had a subterminal-spined egg. Note that the evidence to support the position of *S. hippopotami* is not strong, it relies upon a 300 base pair sequence from a single animal. Sequence data from *S. incognitum* may provide additional support for a subterminal-spined ancestor. A snail other than *Biomphalaria* must have been susceptible to the African ancestor as evidence presented below suggests *Biomphalaria* was not in Africa at this time.

The *S. mansoni* ancestor radiated into at least two, and possibly four, species and the *S. haematobium* ancestor radiated into seven. The human acquisition of African schistosomes is believed to be the result of independent lateral transfers from other animals (Combes, 1990). It is unlikely that primates were the ancestral host because no *Schistosoma* specific to non-human primates have been described (Combes, 1990). Ungulates are thought to be the original hosts of the *S. haematobium* group and Combes (1990) suggested that primates acquired *S. mansoni* from rodents. The lack of genetic differentiation detected between *S. mansoni* and *S. rodhiani* (0·56 % difference through ITS2) suggests a recent separation of these species, and divergence rates indicate it could be as little as 1 MYA (Després *et al.* 1992). Given that hominids have been present in Africa for considerably longer than this, it is possible that they acquired the ancestor of *S. mansoni* from other mammals and that the parasite has since undergone a lateral transfer into rodents. Further research is required to resolve the origin and diversification of the *S. mansoni* group.

Despite greater speciation within the *S. haematobium* group, it was *S. mansoni* that radiated geographically. Populations of *S. mansoni* now occur in the New World and have spread throughout the Neotropics. Analysis of RAPD data has shown that *S. mansoni* isolates display considerably less intra-specific variation than is observed within *S. intercalatum* and *S. haematobium* isolates despite greater geographic separation (Rollinson *et al.* 1997). The current distribution of *S. mansoni* is strongly correlated to that of *Biomphalaria*, which is outlined in the following section.

Biomphalaria *fossil record and phylogenetics*

To fully understand the past history and future trends of *S. mansoni*, it is important to place this parasite in the context of the origin and evolutionary history of *Biomphalaria*. The current distribution of *Biomphalaria*, primarily in South America and Africa, can be described as Gondwanian. The geographic distributions of the most important intermediate hosts for *S. mansoni* are shown in Fig. 1. Given this distribution, it is reasonable to consider the possibility that *Biomphalaria* originated in Gondwanaland more than 100 MYA, and as the African and South American continents split apart, two lineages of *Biomphalaria* were formed: one South American and the other African (Pilsbry, 1911; Davis, 1980, 1992; Meier-Brook, 1984). Davis (1980, 1992) proposed that schistosomes existed along with pomatiopsid, bulinid and biomphalarid snail hosts before the breakup of the two continents. However, it is perplexing why bulinids, if Gondwanian in origin, are neither extant nor have a fossil record in South America.

A Gondwanian origin for *Biomphalaria* is challenged by the fossil record as well. Although *Biomphalaria*-like shells are represented in South America as early as the Paleocene (55–65 MYA; Parodiz, 1969), there is no corresponding record in North Africa. The oldest known occurrence of *Biomphalaria* in Africa is only mid to late Pleistocene (1–2 MYA; Van Damme, 1984). In fact, the African fossil record is relatively devoid of planorbids until the Pleistocene, with the exception of *Planorbis*, which has possible fossils dating to the Lower Eocene (42–55 MYA; Van Damme, 1984).

The pre-Pleistocene fossil record of *Biomphalaria* is not restricted to South America, however, but appears to include North America and Eurasia. Due to the sub-tropical and tropical conditions that existed over much of North America extending as far north as southern Canada during the Eocene and Oligocene (23–55 MYA), fossil *Biomphalaria* are quite common in sediments of this age (Pierce, 1993). Similar conditions presumably existed in Eurasia where such fossils have been described from Eocene strata (34–55 MYA; McKenna, Robinson &

Taylor, 1962) These tropical conditions waned in the late Oligocene and early Miocene (16–29 MYA), and *Biomphalaria* fossils are less common, restricted to western Montana (Pierce, 1993). This fossil literature is less frequently considered by medical malacology, but there is a need for incorporating this extensive fossil evidence, only touched upon here, into the discussion of existing *Biomphalaria* species and distribution. Caution is certainly warranted when interpreting these records; not all specimens may be correctly identified. Clearly though, the past distributional and evolutionary history of *Biomphalaria* may be more complex than the current distribution would imply.

In 1997 some striking new possibilities regarding the continent of origin and subsequent diversification for *Biomphalaria* were proposed. Based on allozyme studies, Woodruff & Mulvey (1997) concluded that *Biomphalaria* was of South American (not Gondwanian) origin. According to their theory, *Biomphalaria* underwent a west-to-east trans-Atlantic colonization event, either by rafting or via aquatic birds, between 2·3 and 4·5 MYA, suggesting the 12 extant species of African *Biomphalaria* must have diverged since this colonization event. Woodruff & Mulvey (1997) suggest that the fossil record (as discussed above) is consistent with this theory. They also note that the morphological differences amongst African *Biomphalaria* species are small compared to those amongst South American species (Mello, 1972), suggestive of a more recent origin of the African species. In light of the extensive, albeit uncertain, North American fossils, the concept of a South American origin should perhaps be modified to a concept of an American origin, allowing for a North American or South American origin. We suggest that a longer existence in the Americas, than in Africa, is also supported by: (1) there are more recognized species in the Neotropics; (2) these species exhibit a greater range in size than do the African *Biomphalaria*; and (3) the Neotropical species show greater diversity with respect to *S. mansoni* susceptibility (Paraense, Ibañez & Miranda, 1964; Paraense & Corrêa, 1982; Malek, 1985).

A Pliocene (2–5 MYA) colonization of Africa by *Biomphalaria* is significant because it raises the possibility that *S. mansoni* did not differentiate from its immediate ancestor until this time. The ancestor, possibly a pulmonate-transmitted schistosome similar to *S. hippopotami* or *S. incognitum*, might have switched to *Biomphalaria* when it became available, and given rise to the *S. mansoni* species group.

Woodruff & Mulvey (1997) also provide some provocative suggestions regarding the most important host of *S. mansoni* in the Neotropics, *B. glabrata*. This snail occurs in Venezuela, French Guiana, Surinam, Guyana, eastern Brazil and on some of the Caribbean islands (Fig. 1). The analyses of

Woodruff & Mulvey (1997) supported a close relationship between this species and the 3 African species included in their study. In fact, their analyses indicated *B. glabrata* to be derived from African ancestors, prompting them to suggest that the presence of this species in the Neotropics is the result of colonization with the slave trade, with *B. glabrata* or proto-*B. glabrata* carried to the Caribbean and South America via water casks aboard slave ships (though they also allowed for an earlier east-to-west trans-Atlantic event). If *B. glabrata* is a recent colonist (< 500 years ago), one would expect that this species might still be extending its range, which has been documented (Paraense 1983; Paraense & Araujo, 1984). However, the presence of *B. glabrata*-like shells at two Upper Pleistocene sites in Brazil (Lima, 1984, 1987) suggests that *B. glabrata* existed in Brazil before the slave trade.

A particularly puzzling question that arises from the Woodruff & Mulvey (1997) hypothesis is: which particular African taxon might be the immediate ancestor of *B. glabrata*? There is no obvious answer based on morphology. If the colonization event is as recent as the slave trade, one expects there to be a reasonably similar snail somewhere in West Africa. Based on its relatively large size, and West African distribution (Fig. 1), *B. camerunensis* may be a candidate. However, *B. camerunensis* lacks a renal ridge, a defining character for *B. glabrata* (Malek, 1985; Paraense & Deslandes, 1959). It is also a rainforest species, unlike *B. glabrata*, and is less susceptible to *S. mansoni* infection (Frandsen, 1979 *a*, *b*: Greer *et al* 1990). Furthermore, if the slave trade commonly facilitated colonization of African *Biomphalaria* to the Neotropics, why wasn't *B. pfeifferi*, the most common species in Africa (including West Africa; Fig. 1), or *Bulinus* introduced? West Africa has been surveyed for *Biomphalaria* (Doumenge *et al.* 1987), so the existence of a yet undescribed species as conspicuous as *B. glabrata* seems unlikely.

Bandoni, Mulvey & Loker (1995) also found a close association between *B. glabrata* and the African species, yet the results of the two studies differ in an important way. Bandoni *et al.* (1995) found *B. glabrata* to be the sister group of the African species and therefore a proto-*B. glabrata* could be the possible ancestor of the African lineage they found to be monophyletic. This scenario requires only a single colonization event, that of a proto-*B. glabrata* colonizing Africa, with subsequent radiation giving rise to the African species. The success of *S. mansoni* in so many different locations in South America and on several Caribbean islands is more easily explained under this scenario, since a suitable endemic host in *B. glabrata* would be present before the arrival of the parasite. If *B. glabrata* is of African origin, and colonized South America via water casks, the establishment of *S. mansoni* in multiple locations would require the introduction and rapid success of the African snail in each location, which seems somewhat unlikely. However, we credit Woodruff & Mulvey (1997) for demonstrating the close association of *B. glabrata* with the African species and providing a thought-provoking suggestion regarding its origin. Both of the above studies were based on allozyme data, and were limited by low numbers of species (9 and 11), high frequencies of laboratory populations instead of field-derived populations, and poor representation of populations from West Africa.

Recent molecular studies, which have incorporated field-derived and West African taxa, are consistent with Bandoni *et al.* (1995). A study of 7 *Biomphalaria* species (Campbell *et al.* 2000) and data obtained from 23 species in our laboratory (DeJong *et al.*, unpublished observations) support an American origin for *Biomphalaria*, and a relatively recent colonization of Africa by proto-*B. glabrata*.

COLONIZATION OF THE NEW WORLD BY *S. MANSONI*

If the fossil record is correct in suggesting that *Biomphalaria* has been in Africa for less than 5 million years and that *S. mansoni* originated in Africa less than 4 MYA, the New World colonization by *S. mansoni* must have occurred since then. Files (1951) suggested a recent introduction of *S. mansoni* into the New World via the extensive 16th to 19th centuries slave trade. Enzyme electrophoresis studies by Fletcher, Loverde & Woodruff (1981) supported this theory as they detected very little variation between South American and African isolates. A comparison of nuclear and mitochondrial ribosomal gene sequences detected greater variation within South American isolates than could be measured between isolates originating from the two continents (Després *et al.* 1992). They concluded that the South American *S. mansoni* was a recent introduction and that the relatively high level of genetic diversity observed among South American isolates was a consequence of multiple transfer events originating from many parts of Africa. Further research, by Després and her co-workers, characterizing *S. mansoni* mitochondrial DNA, produced similar results providing additional support for the recent introduction theory (Després, Imbert-Establet & Monnerot, 1993).

Although intraspecific variation among *S. mansoni* isolates appears to be low compared to other *Schistosoma* species (Rollinson *et al.* 1997), localized strain differences have been recorded among New World isolates. An RFLP study focusing on the intergenic spacer and 18S rRNA was able to detect differences among isolates from Brazil (Vieira *et al.* 1991). Isolates from the southeast showed less variation than those from the northeast, suggesting

S. mansoni is radiating from the northeast of the country.

More sensitive markers are now being sought to detect *S. mansoni* population differences. A repeating element within the mitochondrial genome of *S. mansoni* is being targeted. The number of copies of this element can cause length polymorphisms among isolates that can vary up to 8400 bases (Minchella *et al.* 1994; Minchella, Sollenberger & Desouza, 1995; Johnston *et al.* 1993; Curtis & Minchella, 2000; Després *et al.* 1993). A set of microsatellite markers has also been identified for *S. mansoni* but they are yet to be applied to populations outside of Guadeloupe (Durand, Sire & Theron, 2000).

Definitive host capture appears to be more common in the New World with *S. mansoni* frequently recovered from rodents while more strictly primate transmitted isolates occur in East Africa (Combes, Léger & Golvan, 1975; Jourdane, 1978). Isolates of *S. mansoni* recovered from humans and rodents originating from Brazil could not be distinguished using mitochondrial DNA 16S sequences (Després *et al.* 1992). The authors suggested that, rather than co-evolving with a single host, *S. mansoni* is simply expanding its host range. A similar transfer into rodents has been observed in Guadeloupe (Bremond *et al.* 1993; Théron & Pointier, 1995). In Guadeloupe, differences between human and rodent transmitted isolates of *S. mansoni*, including egg morphology and patterns of cercarial shedding, are suggestive of incipient speciation (Théron & Pointier, 1995). If the origin of *S. rodhaini* in Africa was a result of a lateral transfer by *S. mansoni* from humans into rodents, then we may be seeing the start of a similar event in the New World.

CHANGES IN SNAIL DISTRIBUTIONS AND IMPLICATIONS FOR *S. MANSONI*

The transmission of *S. mansoni* is restricted to freshwater habitats in geographic regions where susceptible species of *Biomphalaria* are present and where the specific local ecological circumstances enable biomphalarids to exist. Not surprisingly, with increasing human demands placed on the world's supplies of freshwater, pervasive changes have occurred in tropical freshwater habitats that influence *Biomphalaria* and consequently *S. mansoni*. Each of the countries/geographic regions discussed below provides its own insights regarding changing patterns of snail distributions and the attendant impacts on *S. mansoni*.

S. mansoni *ascendant – the example of Egypt*

Egypt presents several intriguing observations and questions regarding *S. mansoni*. The first systematic studies of the abundance of *S. mansoni* in Egypt were by Scott (1937) who noted that this parasite was restricted to the northern and eastern part of the Nile Delta and was rare south of Cairo. By comparison, *S. haematobium* was not only much more abundant in the Delta, but also found in Upper Egypt. *Biomphalaria alexandrina*, the Egyptian snail host for *S. mansoni*, was apparently then confined to the Nile Delta. In 1955, the snail was reported for the first time south of Cairo, and by 1979 had been reported as far south as Aswan (Vrijenhoek & Graven, 1992). Studies of allozyme variation in *B. alexandrina* suggested that colonization of the upper Nile involved a series of stepwise founder events originating from the Delta, each marked by a loss of some allelic diversity (Vrijenhoek & Graven, 1992). Not surprisingly, focal transmission of *S. mansoni* from regions such as Fayoum south of Cairo began to occur (Abdel-Wahab *et al.* 2000). Judging from the increased abundance of *S. mansoni* in the Delta, *B. alexandrina* has also become relatively more abundant there as well. At the same time, the prevalence of *S. haematobium* throughout the Nile Delta has declined sharply, and reduced numbers of *Bulinus truncatus* have been noted from the region.

The underlying reasons for these twentieth century changes in snail abundance are not known. There is a general sense that construction of the Low and High dams at Aswan, and the resultant shift from inundative to perennial irrigation, has had the effect of creating more impounded water bodies with a sufficient degree of stability to favour the proliferation of *B. alexandrina* (Malone *et al.* 1997; El Khoby *et al.* 2000). The role of other factors such as extensive pollution of the irrigation canals and the introduction there of exotic species has not been assessed. Whatever the underlying causes, *S. mansoni* has now nearly completely replaced *S. haematobium* in the Nile Delta and appears likely to continue to increase in other areas as well. To make matters worse, the Neotropical *B. glabrata* has been introduced into Egyptian canals, possibly as early as 1981 (Pflüger, 1982), and there is evidence that it is hybridizing with *B. alexandrina* (Kristensen, Yousif & Raahauge, 1999). As *B. glabrata* is such an excellent host for *S. mansoni*, it seems its presence can only favour increased transmission.

It is also interesting to contemplate the history of *S. mansoni* in Egypt on a longer time-scale. Although studies of mummified remains and numerous references to hematuria in ancient writings clearly place *S. haematobium* in Egypt as early as the Middle Kingdom (1500 BC) (Ruffer, 1910; Contis & David, 1996) a definitive presence at that time for *S. mansoni* is less certain. *Schistosoma mansoni*'s symptomatology is less dramatic than for *S. haematobium* so its presence would have been understandably overlooked by the ancients. It is intriguing, however, that there are no clear demonstrations of *S. mansoni* eggs from Egyptian mummies.

A better understanding of the biogeography of *Biomphalaria* in Egypt would help to clarify the duration of the presence there of *S. mansoni*. In this regard, the relatively restricted, fragmented and disjunct geographic range of *B. alexandrina* is noteworthy. It is abundant in the Delta and as noted above, has been found south of Cairo only in recent years. Populations of *B. alexandrina* are also reported from northern Sudan, nearly 1000 km to the south (Fig. 1; Williams & Hunter, 1968). Also inhabiting the Nile drainage in the Sudan is *B. sudanica*, a species that according to our ongoing phylogenetic studies is a very close relative of *B. alexandrina*. For reasons that are not clear, perhaps because of annual flooding, the course of the Nile between Khartoum and Cairo was apparently not heavily colonized historically by *Biomphalaria* snails. One scenario is that *B. sudanica*-like snails from the Sudan managed to colonize the favorable habitat of the Nile Delta, perhaps by downstream or avian-mediated dispersal. The intervening stretch of the Nile may have served as a barrier sufficiently strong to enable this founder population to diversify into the endemic taxon recognized today as *B. alexandrina*. The scarcity of *Biomphalaria* in Upper Egypt could also have been a significant barrier that delayed the colonization of Egypt by *S. mansoni* from large endemic foci to the south. Thus, to a certain extent, the increase in *S. mansoni* noted in Egypt today, could represent part of a longer term colonization that has been underway potentially for thousands of years.

Epidemic schistosomaisis in Senegal '*If you build it, they will come*' (slightly modified from *Field of Dreams*, MCA Universal Films)

Senegal provides a striking modern example of the rapidity and extent to which large water development projects can favour colonization by *Biomphalaria* and transmission of *S. mansoni* (Southgate, 1997). In 1985, a large barrage was built about 40 km from the mouth of the Senegal River to prevent the intrusion of salt water into the river during times of low flow. Prior to construction of the barrage, *S. haematobium* existed at a low level along the lower stretches of the river. It was thought that the temperature was too high to allow *B. pfeifferi* to survive there and *S. mansoni* was not present. By as early as 1988, for the first time in the area, *S. mansoni* infections were being reported in the town of Richard Toll, 140 km upstream of the barrage (Talla *et al.* 1990). By 1989, 49·3% of patients examined were infected with *S. mansoni* and *B. pfeifferi* had become very common in the area. By 1994–5, mean prevalence in villages around Richard Toll had reached 72% (Picquet *et al.* 1996). Construction of the dam reduced salinity and increased irrigation as expected, but also had the unforeseen effect of increasing the pH of the river water (Southgate, 1997). All of these changes favoured colonization by both *Biomphalaria* and *Bulinus* snails. To make matters worse, compatibility between the local *B. pfeifferi* and *S. mansoni* was shown to be extraordinarily high (Tchuente *et al.* 1999).

The lower Senegal river basin is now considered to be one of the world's most intense foci of *S. mansoni* infection; mean values of 1793 ± 848 eggs/g of faeces have been reported from the area near Richard Toll (Picquet *et al.* 1996). Further compounding the problem, rates of treatment success with praziquantel in this area have been low. Several factors contribute to poor response to treatment (Southgate, 1997), and among them must be included the observation that the Senegalese isolate of *S. mansoni* is inherently less responsive to the drug (see below).

Ongoing conquest of a new world – S. mansoni *in Brazil*

Today we are witnessing a large-scale, ongoing colonization event by *S. mansoni* that began four to five hundred years ago when African slaves were brought to Brazil. This colonization is a complex process, one favoured by the built-in presence (Lima, 1987) of populations of a very susceptible snail host, *B. glabrata*, around the coastal areas originally inhabited by infected slaves. Since the time of the original introductions, massive human-imposed changes in the environment have favoured the spread of *B. glabrata* and other indigenous *Biomphalaria* species. For example, locally-acquired cases of *S. mansoni* have been reported for the first time from Brazil's most southern state, Rio Grande do Sul (Graeff-Teixeira *et al.* 1999), thus fulfilling a prediction made by Paraense & Corrêa (1987). The *S. mansoni* cases are associated with the recent first report of *B. glabrata* from Rio Grande do Sul. The sequence of events leading to the colonization of this area by *B. glabrata* is not known. Range extensions of *B. glabrata* have probably occurred elsewhere in Brazil (Paraense, 1983; Paraense & Araujo, 1984), creating new opportunities for *S. mansoni* transmission.

Another major factor contributing to the spread of *S. mansoni* has been the host capture of other species of *Biomphalaria* that initially were poorly susceptible to the parasite. Paraense & Corrêa (1963; 1987) noted that A. Lutz, during his pioneering work on schistosomiasis in Brazil in 1916, was unable to infect *B. tenagophila*. Since that time, the parasite has adapted to this snail species which is now an efficient host along the southeastern coastal region of the country. Similarly, to the northeast, the parasite has adapted to *B. straminea*. The adoption of different intermediate host species has been accompanied by the differentiation of isolates of *S.*

mansoni (Paraense & Corrêa, 1963, 1981). The parasite has also shown a remarkable ability to adapt to new mammalian hosts, most particularly the aquatic rodent, *Nectomys squamipes* (D'Andrea *et al.* 2000). One of the interesting questions for the future is whether *S. mansoni* will continue its Neotropical spread. *Biomphalaria straminea* is known from as far north as Costa Rica (Paraense, Zeledón & Rojas, 1981) and has disseminated widely throughout the Caribbean (Pointier, Paraense & Mazille, 1993). A *B. straminea*-like snail is known from Uruguay and is susceptible to *S. mansoni* (Paraense & Corrêa, 1989). *Biomphalaria peregrina* from Equador (Paraense & Corrêa, 1973) and *B. tenagophila* from Peru (Paraense, Ibañez & Miranda, 1964) have been shown to be susceptible to *S. mansoni*.

One of the factors that may work to reduce the prevalence of *S. mansoni* in Brazil and other parts of South America is the introduction of thiarid snails such as *Melanoides tuberculata* (e.g. Junior, 1999). Thiarids are now present in Minas Gerais and also in the Rio de Janiero area. As noted below, for reasons that are still not well understood, following their introduction thiarids often become staggeringly abundant, and displace native snail species, including *Biomphalaria*.

Biomphalaria, S. mansoni *and Asia – will long-range colonization events result in endemicity in tropical Asia?*

In addition to the introduction of *B. glabrata* into Egypt noted above, other potentially troublesome long-range introductions of *Biomphalaria* are known. The Neotropical snail *B. straminea* was first noticed in a small stream in Hong Kong in 1973 (Meier-Brook, 1974). This introduction was likely aided by the trade in aquarium plants and fishes. The exact source of the introduced specimens has not been determined. Since that time the snail has spread into several adjacent habitats and it has become locally abundant (Woodruff *et al.* 1985*a*; Yipp, 1990). Yipp (1990) noted that *B. straminea* is likely to colonize organically polluted sites where other freshwater snails have been eliminated. Woodruff *et al.* (1985*b*) provided allozyme evidence to indicate that a second, separate introduction of *B. straminea* occurred in Hong Kong in 1981–1982. They commented that there is every reason to expect additional introductions of *B. straminea* to occur, and once established in Asia, that it was likely to be spread secondarily, both to other Asian localities and elsewhere. They also indicated surprise that *B. glabrata* had not yet colonized Asia. Walker (1978) noted that *B. straminea* was found among fishes imported into Australia from Hong Kong, but fortunately the snails were intercepted, and there still is no indication of the presence of *Biomphalaria* in tropical Australia. Although there is no evidence for *B. straminea*-mediated transmission of *S. mansoni* in China, the snail is found naturally infected in Brazil and can maintain transmission there, so the potential exists for Asian transmission of *S. mansoni* by *B. straminea*.

Another more recent and troubling report concerns the discovery of eggs resembling those of *S. mansoni* from human stool samples in southern Nepal (Sherchand & O'Hara, 1997). A serological study found 18·1% of the 518 sera examined to be positive for antibodies to *S. mansoni* (Sherchand *et al.* 1999). Very little is presently known about this focus, but there has been extensive environmental change in the region, including deforestation and construction of irrigation schemes. It is conceivable that *Biomphalaria* has been introduced into these schemes and that this has been followed by the unfortunate introduction of *S. mansoni*, possibly with immigrant workers. An alternative possibility is that the eggs observed in Nepal are those of *Schistosoma sinensium*, a species with lateral-spined eggs known from southern China (Greer, Kitikoon & Lohachit, 1989). Eggs of this species may have been ingested and then passed by humans.

'*No worm is an island ...*' – *the rise and fall of* S. mansoni *in the Caribbean*

The relatively brief history of *S. mansoni* in the Caribbean islands, although poorly documented, is nonetheless instructive in several regards: *Biomphalaria glabrata* is either present or used to be present on 17 of 31 Caribbean islands, and *S. mansoni* at one time or another colonized 10 of the 17 islands (Bundy, 1984). Significant foci of infection occurred in St. Lucia, Guadeloupe, Martinique, Puerto Rico and the Dominican Republic, and limited transmission also occurred on St. Martin, St. Kitts, Vieques, Antiqua and Montserrat. Today, although the overall status of human schistosomiasis in the area is remarkably poorly known, it is likely that transmission to humans is marginal at best throughout the islands. Puerto Rico provides a good example of this trend. In 1945, the overall prevalence of *S. mansoni* in Puerto Rico was 13·5%. A limited survey carried out in previously endemic areas by Giboda, Malek & Correa (1997) revealed only 3 cases, all in older individuals. The cane-growing areas of the Dominican Republic may be the largest remaining foci of human infection on the Caribbean islands (Vargas, Malek & Perez, 1990). Thriving but highly focal areas of infection involving *Rattus rattus* as definitive host are known from the island of Guadeloupe (Théron & Pointier, 1995).

The failure of *S. mansoni* to persist in the region following colonization with the slave trade is due to several interacting factors. The first, and likely most important, is simply that these foci existed on islands. Populations on islands are inherently more vulnerable to extinction events, particularly if the islands are small. Islands like St. Martin and St. Kitts are tiny with few habitats to support snails and deforestation and water diversion projects altered these few habitats sufficiently to terminate transmission. Control programmes are more likely to succeed on islands because the target populations are small and confined to start with, and are less likely to be replenished from surrounding areas by immigration. Thus, a concerted schistosomiasis control programme, featuring the use of molluscicides and chemotherapy, succeeded in eliminating *S. mansoni* from the relatively small island of Vieques in 1962 (Ferguson, Palmer & Jobin, 1968). Human-induced changes in the islands have also contributed to the demise of *S. mansoni*. In Puerto Rico, the channeling of streams through enclosed cement viaducts, as part of a pervasive trend of urbanization, has played a role in reducing *B. glabrata* populations there. 'Economic development and well being' are described as the control strategy in Puerto Rico leading to eradication (Hillyer & deGalanes, 1999).

Natural catastrophic events also influence *S. mansoni* in the Caribbean. Hurricanes, such as Georges that devastated Puerto Rico in 1998, may interrupt local electric and water services, forcing people to wash clothing in streams. The attendant water contact would favour transmission (Hillyer & deGalanes, 1999). Hurricanes can eliminate freshwater snails by inundating habitats with saltwater. Another reality for the Caribbean is volcanism, dramatically demonstrated most recently in Monserrat. The volcano there, after being dormant for 400 years, erupted in 1995. As late as 1978, 14% of people in two local villages were serologically positive for *S. mansoni* (Tikasingh *et al.* 1982). Ash and lava flows may have altered or obliterated habitats occupied by *B. glabrata*, but little is known of the eruption's effects.

Of all the factors influencing the decline in prevalence of *S. mansoni* in the Caribbean, probably the most important has been the introduction, both accidental and deliberate, of exotic snails that compete with and/or prey upon *Biomphalaria* (Pointier & Giboda, 1999). *Marisa cornuarietis*, an ampullariid snail indigenous to the Orinoco drainage of Venezuela, was introduced in 1958 into 30 of Puerto Rico's water reservoirs and by 1976, only 5 of these reservoirs still harboured *B. glabrata* (Jobin *et al.* 1977). The ampullariid both competes for resources with *B. glabrata* and preys upon its egg masses and young. A focus of murine schistosomiasis in Grand Etang lake in Gaudeloupe was also eliminated by the combined actions of ampullariids *Pomacea glauca* and *M. cornuarietis* (Pointier *et al.* 1991).

The islands of the Caribbean have experienced a remarkable biological invasion that began in the 1940s, probably as a consequence of the trade in aquatic plants and fishes (Pointier & Giboda, 1999). In 1954, the Oriental snail *Thiara granifera* was first reported in Puerto Rico, and by 1968 had spread throughout the island (Chaniotis *et al.* 1980). Butler *et al.* (1980) documented the ability of *T. granifera* to displace *B. glabrata* from permanent streams in Puerto Rico. Similar results were obtained by Prentice (1983) in St. Lucia, and by Perez, Vargas & Malek (1991) in the Dominican Republic. Giboda, Malek & Correa (1997) concluded that the main reason for the decline of *S. mansoni* in Puerto Rico was the lack of *B. glabrata* and attributed this to the probable competitive impacts of thiarid snails. In St. Lucia, where *S. mansoni* prevalence was once as high as 57 % in some villages, *B. glabrata* is now scarce in former transmission sites. The related thiarid snail *Melanoides tuberculata*, an African species, was introduced in 1978, and only two years later had displaced *B. glabrata* where introduced (Prentice, 1983). Pointier (1993) found *B. glabrata* to be abundant in only two of 26 sites where it was formerly abundant. These two sites lacked *M. tuberculata*.

In Martinique, the introduction in 1983 of *M. tuberculata* into watercress beds, where *S. mansoni* was still transmitted, resulted in practically complete elimination of *B. glabrata* and *B. straminea* by 1990, and the reduction of *S. mansoni* transmission to a handful of cases (Pointier & Guyard, 1992). The thiarid has since colonized the entire island, and successive waves of colonization of the island by different morphs have been noted (Pointier *et al.* 1993). This species has since largely been replaced, although not eliminated, on Martinique by *T. granifera* (Pointier *et al.* 1998). *Melanoides tuberculata* did not effectively eliminate *B. glabrata* from the marshy forest transmission foci of Guadeloupe (Pointier, Theron & Borel, 1993) and, as noted above, *S. mansoni* is still actively transmitted there by rats.

Although *S. mansoni* is clearly on the wane in the Caribbean, it is premature to assume its eventual extinction. The situation on Guadeloupe indicates that in some locations thiarids are not able to displace *B. glabrata*, and that the parasite has exhibited a remarkable adaptability for infecting rats. Although thiarids are present to stay in the Caribbean region, it is conceivable that with time their abundance may diminish and that biomphalarids will come to coexist with them. In East Africa, where both *M. tuberculata* and *B. pfeifferi* are normal components of the snail fauna, the two species are known to occupy the same habitat for extended periods, without one species eliminating the other (Mkoji *et al.* 1992). Schistosomiasis thrives in regions of Africa where *M. tuberculata* is present (Brown, 1994). The same human behaviours that have introduced thiarids everywhere will also favour the colonization or re-colonization of some

Caribbean islands or habitas by either *B. glabrata*, or the more peripatetic species, *B. straminea*. Whether *S. mansoni* can continue to exist on such shifting biological terrain remains to be seen, but its ability to adapt to introduced populations of *B. straminea* may prove to be critical to its survival.

Finally, with respect to *S. mansoni* in the Caribbean, both Haiti and Cuba present interesting situations. Although *S. mansoni* is present in the Dominican Republic, and *B. glabrata* is known to be present in parts of Haiti (Raccurt *et al.* 1985), there is still no definitive evidence for the presence of schistosomiasis in Haiti. Thiarid snails may be increasingly limiting the distribution of *B. glabrata* in both countries, making it improbable that new foci of transmission will appear in Haiti. Cuba is of interest simply because *B. glabrata*, and hence *S. mansoni*, do not exist on the island. This raises the more general issue of what are the underlying determinants of *B. glabrata*'s distribution throughout the Caribbean region? It seems unlikely that islands as large as Cuba and Jamaica, which also lacks *B. glabrata*, were never colonized by this snail. Other species of *Biomphalaria* occur in Cuba, so it also seems unlikely that the hydrogeography of the island is unsuitable. Perhaps Cuba's more northernly latitude renders its climate too cool to support the tropical *B. glabrata*? Thiarids are also present in Cuba and so may prevent future colonization of the island by *B. glabrata*.

The impact of introduced aquatic species on snail intermediate hosts has been most dramatic with thiarids in the Caribbean region, but in Africa other introduced species are also likely to affect the distribution of *S. mansoni*. In Kenya, the North American crayfish *Procambarus clarkii* has become common in some drainage systems and where present, schistosome intermediate hosts are not found (Hofkin *et al.* 1991). The crayfish is a voracious predator of snails and other aquatic organisms. Where the crayfish establishes it has the potential to stop transmission of schistosomiasis (Mkoji *et al.* 1999). It is also present in the irrigation canals of the Nile Delta, and is known from the Sudan, Uganda, South Africa, Zimbabwe and Zambia (Hobbs, Jass & Huner, 1989). Unfortunately, the aggressive and omnivorous tendencies of this species pose a threat to the integrity of African freshwater ecosystems. Also, now widely present in Africa is another North American invader, the snail *Physa acuta*. This species has several attributes including high fecundity (Brackenbury & Appleton, 1991), effective defense against predators (Wilken & Appleton, 1991), upstream migratory tendencies (Appleton & Branch, 1989) and a high tolerance for polluted waters that seem to give it a competitive advantage over endemic species. These attributes may eventually favor the displacement of indigenous African *Biomphalaria* species.

CONTROL OF *S. MANSONI* AND SOME OF THE IMPLICATIONS

The most effective means to control *S. mansoni* has been the use of chemotherapy. In particular, use of praziquantel has been associated with an overall decline in the global prevalence of this parasite (Chitsulo *et al.* 2000). The experience in Egypt has shown, however, that control of *S. mansoni* based purely on chemotherapy is very hard to sustain, and prevalence rates have remained stubbornly high (El Khoby *et al.* 2000). Furthermore, evidence from a variety of sources suggests that resistance to praziquantel may be developing. In the laboratory, isolates of *S. mansoni* with a significant degree of resistance can be developed by simply exposing infected mice to subcurative doses of praziquantel. After seven generations of selection 93 % of the worms were unresponsive to high doses of praziquantel (Fallon & Doenhoff, 1994). In field situations, in both Senegal and Egypt, there is evidence for the presence of *S. mansoni* isolates that are relatively unresponsive to the drug, although the underlying basis for this seems to be different in each case. In Senegal, the poor response of *S. mansoni* in the new focus on the delta of the Senegal River has been partially explained by the presence of very high worm burdens and the lack of immunity in this newly-exposed human population (see discussion in Southgate, 1997). Nonetheless, laboratory studies (Fallon *et al.* 1995) suggest that the Senegalese isolate is intrinsically less responsive to praziquantel. The underlying reasons for the presence of a tolerant isolate in Senegal remain poorly understood. The presence in this area of individuals with relatively heavy infections of both *S. haematobium* and *S. mansoni* has lead to some unforeseen situations with respect to treatment. Egg counts for *S. haematobium* were shown to decline sharply in such individuals following treatment, whereas *S. mansoni* egg counts increased seven-fold. One explanation for this result was that *S. haematobium* males had paired with *S. mansoni* females, and following the praziquantel-induced demise of the *S. haematobium* males, the relatively drug tolerant *S. mansoni* females were then freed to pair with *S. mansoni* males (Ernould, Ba & Sellin, 1999).

Considerably more troubling is the situation in the Nile Delta of Egypt, where praziquantel has been used aggressively for more than 10 years. Here it is probable that drug selection has favoured the emergence of resistant genotypes. *Schistosoma mansoni* isolates derived from patients that continued to pass eggs following treatment were, when passaged through mice, less responsive to praziquantel (Ismail *et al.* 1996, 1999), indicating that the diminished responsiveness was not somehow due to patient-related factors. *Schistosoma mansoni* from the Lake Albert region of Uganda may also becoming

less responsive to praziquantel (Doenhoff, Kimani & Cioli, 2000). Fortunately, praziquantel resistance has not yet become pervasive and praziquantel-resistant worms are still susceptible to oxamniquine. Regarding the latter drug, resistance has also been produced experimentally and a recent study from Brazil suggests that *S. mansoni* isolates recovered from patients, that did not respond to oxamniquine, were less susceptible to the drug when in mice (Conceição, Argento & Corrêa, 2000).

One of the cornerstones of past schistosomiasis control programmes has been snail control, usually achieved with molluscicidal chemicals, although biological control and environmental control measures have also been employed in some contexts. One of the realities for schistosomiasis control programmes of the future will be an increasingly strong resistance to the use of measures to kill snails. About two-thirds of the 330 species of freshwater and brackish snails of Africa, are classified as "threatened" and programmes to control snails that serve as intermediate hosts for parasites of medical and veterinary importance can be perceived as a threat to this diversity (Kristensen & Brown, 1999).

Given that snail control is not economically feasible or environmentally acceptable, and that some cracks in the chemotherapy façade are showing, development of alternatives to praziquantel should be encouraged, and the much anticipated schistosomiasis vaccine would certainly be useful. Other effective methods of control including improved sanitation, provision of piped water, and health education all need to be encouraged as well.

Whither S. mansoni ?

Several interacting factors will influence the future prevalence of *S. mansoni*. The continued application of chemotherapeutic and other (new?) control measures and a hoped-for overall rise in the standard of living in the developing world can be expected to lower both intensity and prevalence of infection. Deliberate control efforts will probably be unwittingly abetted by increased industrial water pollution, urbanization and continued introductions of exotic aquatic competitor/predator species, all of which will limit the distribution of biomphalarid snails. In the future, species of *Biomphalaria* once considered pestiferous and whose eradication was actively sought, will be regarded as endangered and in need of protection.

On the other hand, the traditional hand maidens of parasitic disease – poverty, over-crowding and civil war in developing countries – will conspire to prevent access to clean water and proper sanitation and favour *S. mansoni*. Although the effects of global warming are presently difficult to ascertain with any real degree of accuracy, some models predict an expansion of geographic areas susceptible to schistosomiasis (Martens *et al.* 1995) whereas others predict a decrease in transmission (Martens, Jetten & Focks, 1997). By eliminating dense vegetation cover and opening up potential snail habitats to colonization, deforestation can favour the spread of schistoso-miasis (Walsh, Molyneux & Birley, 1993). The destruction of the forests of the central highlands of Madagascar followed by the encroachment of *S. mansoni* provides a bleak example (Ollivier, Brutus & Cot, 1999). Major new areas of endemicity may thus arise in rainforest areas like West Africa and Brazil. Construction of massive water development schemes favour transmission and continued application of drug pressure may favour the emergence of drug-resistant parasites. Human-mediated spread of snails such as *B. straminea* and *B. glabrata* will also favour *S. mansoni*. Much of the future success of this parasite may hinge on its ability to adapt to non-human hosts, as has happened in Guadeloupe and Brazil and in its ability to colonize invasive species of snails, particularly *B. straminea*. The coming years will prove to be fascinating in deciphering the net impact of these contrary trends.

ACKNOWLEDGEMENT

We would like to thank Wade Wilson, Nicole Rupple and Pascale Léonard for their editorial assistance in prep-aration of the manuscript. Thanks also go to Lynn Hertel for her map-womanship. We also wish to thank David Blair for his informative comments on the manuscript. This project was supported by NIH grant AI44913.

REFERENCES

ABDEL-WAHAB, M. F., ESMAT, G., RAMZY, I., NAROOZ, S., MEDHAT, E., IBRAHIM, M., ELBORAEY, Y. & STRICKLAND, G. T. (2000). The epidemiology of schistosomiasis in Egypt: Fayoum Governorate. *American Journal of Tropical Medicine and Hygiene* **62**, 55–64.

ADEMA, C. M. & LOKER, E. S. (1997). Specificity and immunobiology of larval digenean-snail interactions. In *Advances in Trematode Biology* (ed. Fried, B. & Graczyk, T. K.), pp. 229–264. Boca Raton, CRC Press.

APPLETON, C. C. & BRANCH, G. M. (1989). Upstream migration by the invasive snail, *Physa acuta*, in Cape town, South Africa. *South African Journal of Science* **85**, 189–190.

BANDONI, S. M., MULVEY, M. & LOKER, E. S. (1995). Phylogenetic analysis of eleven species of *Biomphalaria* Preston, 1910 (Gastropoda: Planorbidae) based on comparisons of allozymes. *Biological Journal of the Linnean Society* **54**, 1–27.

BARBOSA, F. S., BARBOSA, I. & MORAIS-RÊGO, A. (1959). Laboratory infection of the snail *Planorbarius metidjensis* (Forbes) from French Morocco with a Brazilian strain of *Schistosoma mansoni*. *Annals of Tropical Medicine and Parasitology* **53**, 314–315.

BARKER, S. C. & BLAIR, D. (1996). Molecular phylogeny of *Schistosoma* species supports traditional groupings within the genus. *Journal of Parasitology* **82**, 292–298.

BASCH, P. F. (1976). Intermediate host specificity in *Schistosoma mansoni*. *Experimental Parasitology* **39**, 150–169.

BOWLES, J., BLAIR, D. & MCMANUS, D. (1995). A molecular phylogeny of the human schistosomes. *Molecular Phylogenetics and Evolution* **4**, 103–109.

BRACKENBURY, T. D. & APPLETON, C. C. (1991). Morphology of the mature spermatozoon of *Physa acuta* (Draparnaud, 1801) (Gastropoda, Physidae). *Journal of Molluscan Studies* **57**, 211–218.

BREMOND, P., PASTEUR, N., COMBES, C., RENAUD, F. & THERON, A. (1993). Experimental host-induced selection in *Schistosoma mansoni* strains from Guadeloupe and comparison with natural observations. *Heredity* **70**, 33–37.

BROWN, D. S. (1994). *Freshwater Snails of Africa and their Medical Importance.* 2nd edn. London, Taylor and Francis.

BUNDY, D. A. P. (1984). Caribbean schistosomiasis. *Parasitology* **89**, 377–406.

BUTLER, J. M., FERGUSON, F. F., PALMER, J. R. & JOBIN, W. L. (1980). Displacement of a colony of *Biomphalaria glabrata* by an invading population of *Tarebia granifera* in a small stream in Puerto Rico. *Caribbean Journal of Science* **16**, 73–79.

CAMPBELL, G., JONES, C. S., LOCKYER, A. E., HUGHES, S., BROWN, D., NOBLE, L. R. & ROLLINSON, D. (2000). Molecular evidence supports an African affinity of the Neotropical freshwater gastropod, *Biomphalaria glabrata*, Say 1818, an intermediate host for *Schistosoma mansoni*. *Proceedings of the Royal Society of London Series B–Biological Sciences* **267**, 2351–2358.

CARNEY, W. P., PURNOMO, I. B., VAN PEENEN, P. F. D., BROWN, R. J. & SUDOMO, M. (1977). *Schistosoma incognitum* from mammals of central Sulawesi, Indonesia. *Proceedings of the Helminthological Society of Washington* **44**, 150–155.

CHANIOTIS, B. N., BUTLER, J. M., FERGUSON, F. F. & JOBIN, W. L. (1980). Thermal limits, dessication tolerance, and humidity reactions of *Thiara (Tarebia) granifera mauiensis* (Gastropoda: Thiaridae) host of the Asiatic lung fluke disease. *Caribbean Journal of Science* **16**, 91–93.

CHITSULO, L., ENGELS, D., MONTRESOR, A. & SAVIOLI, L. (2000). The global status of schistosomiasis and its control. *Acta Tropica* **77**, 41–51.

COMBES, C. (1990). Where do human Schistosomes come from? An evolutionary approach. *Trends in Ecology and Evolution* **5**, 334–337.

COMBES, C., LÉGER, N. & GOLVAN, Y. J. (1975). The role of the rat in the dynamics of endemic schistosomiasis in Guadeloupe. *Comptes Rendus Hebdomadaires des Séances de L'Académie des Sciences D: Sciences Naturelles* **281**, 1059–1061.

CONCEIÇÃO, M. J., ARGENTO, C. A. & CORRÊA, A. (2000). Study of *Schistosoma mansoni* isolates from patients with failure of treatment with oxamniquine. *Memorias do Instituto Oswaldo Cruz* **95**, 375–380.

CONTIS, G. & DAVID, A. R. (1996). The epidemiology of bilharzia in ancient Egypt: 5000 years of schistosomiasis. *Parasitology Today* **12**, 253–255.

COX, C. B. (2000). Plate tectonics, seaways and climate in the historical biogeography of mammals. *Memorias do Instituto Oswaldo Cruz* **95**, 509–516.

CROMPTON, D. W. T. (1999). How much human helminthiasis is there in the world? *Journal of Parasitology* **85**, 397–403.

CURTIS, J. & MINCHELLA, D. J. (2000). Schistosome population genetic structure: When clumping worms is not just splitting hairs. *Parasitology Today* **16**, 68–71.

D'ANDREA, P. S., MAROJA, L. S., GENTILE, R., CERQUEIRA, R., MALDONADO, A. & REY, L. (2000). The parasitism of *Schistosoma mansoni* (Digenea Trematoda) in a naturally infected population of water rats, *Nectomys squamipes* (Rodentia Sigmodontinae) in Brazil. *Parasitology* **120**, 573–582.

DAVIS G. M. (1980). Snail hosts of Asian *Schistosoma* infecting man: evolution and coevolution. In *The Mekong Schistosome.* (ed. Bruce, J. & Sornmani, S.), pp. 195–238. Michigan, USA. Malacological Review.

DAVIS, G. M. (1992). Evolution of Prosobranch snails transmitting Asian *Schistosoma*; coevolution with *Schistosoma*: a reveiw. *Progress in Clinical Parasitology* **3**, 145–204.

DE SOUZA, C. P. & LIMA, L. C. (1997). *Moluscos de Interesse Parasitológico do Brasil.* Belo Horizonte, Brazil. Serie de esquistossomase n. 1. FIOCRUZ/CPqRR.

DESPRÉS, L., IMBERT-ESTABLET, D., COMBES, C. & BONHOMME, F. (1992). Molecular evidence linking hominid evolution to recent radiation of schistosomes (Platyhelminthes: Trematoda). *Molecular Phylogenetics and Evolution* **1**, 295–304.

DESPRÉS, L., IMBERT-ESTABLET, D. & MONNEROT, M. (1993). Molecular characterization of mitochondrial DNA provides evidence for the recent introduction of *Schistosoma mansoni* into America. *Molecular and Biochemical Parasitology* **60**, 221–230.

DESPRÉS, L., KRUGER, F. J., IMBERT-ESTABLET, D. & ADAMSON, M. L. (1995). ITS2 ribosomal RNA indicates *Schistosoma hippopotami* is a distinct species. *International Journal for Parasitology* **25**, 1509–1514.

DOENHOFF, M. J., KIMANI, G. & CIOLI, D. (2000). Praziquantel and the control of schistosomiasis. *Parasitology Today* **16**, 364–366.

DOUMENGE, J., MOTT, K. E., CHEUNG, C., VILLENAVE, D., CHAPUIS, O., PERRIN M. F. & REAUD-THOMAS, G. (1987) *Centre d'Etude de Géographie Tropicale/WHO Atlas of the Global Distribution of Schistosomiasis.* Bordeaux, Presses Universitaires de Bordeaux.

DURAND, P., SIRE, C. & THERON, A. (2000). Isolation of microsatellite markers in the digenetic trematode *Schistosoma mansoni* from Guadeloupe Island. *Molecular Ecology* **9**, 997–998.

EL KHOBY, T., GALAL, N., FENWICK, A., BARAKAT, R., ELHAWEY, A., NOOMAN, Z., HABIB, M., ABDEL WAHAB, F., GABR, N. S., HAMMAN, H. M. & HUSSEIN, M. H. (2000). The epidemiology of schistosomiasis in Egypt: Summary findings in nine governorates. *American Journal of Tropical Medicine and Hygiene* **62**, 88–99.

ERNOULD, J. C., BA, K. & SELLIN, B. (1999). Increase of intestinal schistosomiasis after praziquantel treatment in a *Schistosoma haematobium* and *Schistosoma mansoni* mixed focus. *Acta Tropica* **73**, 143–152.

FALLON, P. G. & DOENHOFF, M. J. (1994). Drug-resistant schistosomiasis: Resistance to praziquantel and oxamniquine induced in *Schistosoma mansoni* in mice

is drug-specific. *American Journal of Tropical Medicine and Hygiene* **51**, 83–88.

FALLON, P. G., STURROCK, R. F., NIANG, C. M. & DOENHOFF, M. J. (1995). Diminished susceptibility to praziquantel in a Senegal isolate of *Schistosoma mansoni*. *American Journal of Tropical Medicine and Hygiene* **53**, 61–62.

FERGUSON, F. F., PALMER, J. R. & JOBIN, W. L. (1968). Control of schistosomiasis on Vielgues Island. *American Journal of Tropical Medicine and Hygiene* **17**, 858–863.

FILES, V. S. (1951). A study of the vector-parasite relationships in *Schistosoma mansoni*. *Parasitology* **41**, 264–269.

FLETCHER, M., LOVERDE, P. T. & WOODRUFF, D. S. (1981). Genetic variation in *Schistosoma mansoni*: Enzyme polymorphisms in populations from Africa, Southwest Asia, South America , and the West Indies. *American Journal of Tropical Medicine and Hygiene* **30**, 406–421.

FRANCO, G. R., VALADAO, A. F., AZEVEDO, V. & RABELO, E. M. L. (2000). The *Schistosoma* gene discovery program: state of the art. *International Journal for Parasitology* **30**, 453–463.

FRANDSEN, F. (1979a). Studies on the relationship between *Schistosoma* and their intermediate hosts. III. The genus *Biomphalaria* and *Schistosoma mansoni* from Egypt, Kenya, Sudan, Uganda, West Indies (St. Lucia) and Zaire (two different strains: Katanga and Kinshasha). *Journal of Helminthology* **53**, 433–452.

FRANDSEN, F. (1979b). Discussion of the relationships between *Schistosoma* and their intermediate hosts, assessment of the degree of host-parasite compatibility and the evaluation of schistosome taxonomy. *Zeitschrift für Parasitenkunde* **58**, 272–296.

GHANDOUR, A. M., ZAHID, N. Z., BANAJA, A. A., KAMAL, K. B. & BOUQ, A. I. (1995). Zoonotic intestinal parasites of Hamadryas baboons *Papio hamadryas* in the western and northern regions of Saudi Arabia. *Journal of Tropical Medicine and Hygiene* **98**, 431–439.

GIBODA, M., MALEK, E. A. & CORREA, R. (1997). Human schistosomiasis in Puerto Rico: Reduced prevalence rate and absence of *Biomphalaria glabrata*. *American Journal of Tropical Medicine and Hygiene* **57**, 564–568.

GRAEFF-TEIXEIRA, C., DOSANJOS, C. B., DEOLIVEIRA, V. C., VELLOSO, C. F. P., DAFONSECA, M. B. S., VALAR, C., MORAES, C., GARRIDO, C. T. & DOAMARAL, R. S. (1999). Identification of a transmission focus of *Schistosoma mansoni* in the southernmost Brazilian State, Rio Grande do Sul. *Memorias do Instituto Oswaldo Cruz* **94**, 9–10.

GREER, G. J., KITIKOON, V. & LOHACHIT, C. (1989). Morphology and life-cycle of *Schistosoma sinesium* Pao, 1959, from northwest Thailand. *Journal of Parasitology* **75**, 98–101.

GREER, G. J., MIMPFOUNDI, R., MALEK, E. A., JOKY, A., NGONSEU, E. & RATARD, R. C. (1990). Human schistosomiasis in Cameroon. II. Distribution of the snail hosts. *American Journal of Tropical Medicine and Hygiene* **42**, 573–580.

HILLYER, G. V. & DEGALANES, M. S. (1999). Seroepidemiology of schistosomiasis in Puerto Rico: Evidence for vanishing endemicity. *American Journal of Tropical Medicine and Hygiene* **60**, 827–830.

HOBBS, H. H., JASS, J. P. & HUNER, J. V. (1989). A review of global crayfish introductions with particular emphasis on 2 North-American species (Decapoda, Cambaridae). *Crustaceana* **56**, 299–316.

HOFKIN, B. V., MKOJI, G. M., KOECH, D. K. & LOKER, E. S. (1991). Control of schistosome transmitting snails in Kenya by the North American crayfish *Procambarus clarkii*. *American Journal of Tropical Medicine and Hygiene* **45**, 339–344.

ISMAIL, M., BOTROS, S., METWALLY, A., WILLIAM, S., FARGHALLY, A., TAO, L. F., DAY, T. A. & BENNETT, J. L. (1999). Resistance to praziquantel: Direct evidence from *Schistosoma mansoni* isolated from Egyptian villagers. *American Journal of Tropical Medicine and Hygiene* **60**, 932–935.

ISMAIL, M., METWALLY, A., FARGHALY, A., BRUCE, J., TAO, L. F. & BENNETT, J. L. (1996). Characterization of isolates of *Schistosoma mansoni* from Egyptian villagers that tolerate high doses of praziquantel. *American Journal of Tropical Medicine and Hygiene* **55**, 214 218.

JOBIN, W. R., BROWN, R. A., VELEZ, S. P. & FERGUSON, F. F. (1977). Biological control of *Biomphalaria glabrata* in major reservoirs of Puerto Rico. *American Journal of Tropical Medicine and Hygiene* **26**, 1018–1024.

JOHNSTON, D. A., KANE, R. A. & ROLLINSON, D. (1993). Small subunit (18S) ribosomal RNA gene divergence in the genus *Schistosoma*. *Parasitology* **107**, 147–156.

JOHNSTON, D. A., DIAS NETO, E., SIMPSON, A. J. G. & ROLLINSON, D. (1993). Opening the can of worms: molecular analysis of schistosome populations. *Parasitology Today* **92**, 86–291.

JOURDANE, J. (1978). In contrast to the laboratory rat, the rat (*Rattus rattus*) of Guadeloupe is a favorable host for the life cycle of *S. mansoni*. *Comptes Rendus de L'Academie des Sciences, Series D* **286**, 1001–1004.

JUNIOR, P. D. (1999). Invasion by the introduced aquatic snail *Melanoides tuberculata* (Muller, 1774) (Gastropoda: Prosobranchia: Thiaridae) of the Rio Doce State Park, Minas Gerais, Brazil. *Studies on Neotropical Fauna and Environment* **34**, 186–189.

KRISTENSEN, T. K. & BROWN, D. S. (1999). Control of intermediate host snails for parasitic diseases: a threat to biodiversity in African freshwaters? *Malacologia* **41**, 379–391.

KRISTENSEN, T. K., YOUSIF, F. & RAAHAUGE, P. (1999). Molecular characterization of *Biomphalaria* spp in Egypt. *Journal of Molluscan Studies* **65**, 133–136.

LE, T. H., BLAIR, D., AGATSUMA, T., HUMAIR, P. F., CAMPBELL, N. J. H., IWAGAMI, M., LITTLEWOOD, D. T. J., PEACOCK, B., JOHNSTON, D. A., BARTLEY, J., ROLLINSON, D., HERNIOU, E. A., ZARLENGA, D. S. & McMANUS, D. P. (2000). Phylogenies inferred from mitochondrial gene orders: a cautionary tale from the parasitic flatworms. *Molecular Biology and Evolution* **17**, 1123–1125.

LIMA, L. C. (1984). *Biomphalaria* aff. *glabrata* do Pleistoceno de Janaúba, Minas Gerais. *Memórias do Instituto Oswaldo Cruz* **79**, 55–58.

LIMA, L. C. (1987). Ocorrência de planorbideos Pleistocênicos no município de Jacobina, Bahia. *Memórias do Instituto Oswaldo Cruz* **82**, 71–72.

LOKER, E. S. & BAYNE, C. J. (1986). Immunity to trematode larvae in the snail *Biomphalaria*. *London Zoological Society Symposium* **56**, 199–220.

MALEK, E. A. (1985) *Snail Hosts of Schistosomiasis and Other Snail-Transmitted Diseases in Tropical America:*

A Manual. Washington, Pan American Health Organization Scientific Publication No. 478, PAHO.

MALONE, J. B., ABDELRAHMAN, M. S., ELBAHY, M. M., HUH, O. K., SHAFIK, M. & BAVIA, M. (1997). Geographic information systems and the distribution of *Schistosoma mansoni* in the Nile delta. *Parasitology Today* **13**, 112–119.

MARTENS, W. J. M., JETTEN, T. H., ROTMANS, J. & NIESSEN, L. W. (1995). Climate-change and vector-borne diseases: a global modeling perspective. *Global Environmental Change – Human and Policy Dimensions* **5**, 195–209.

MARTENS, W. J. M., JETTEN, T. H. & FOCKS, D. A. (1997). Sensitivity of malaria, schistosomiasis and dengue to global warming. *Climatic Change* **35**, 145–156.

MCKENNA, M. C., ROBINSON, P. & TAYLOR, D. W. (1962). Notes on Eocene Mammalia and Mollusca from Tabernacle Butte, Wyoming. *American Museum Novitates* **No. 2102**, 33 pp.

MEIER-BROOK, C. (1974). A snail intermediate host of *Schistosoma mansoni* introduced into Hong Kong. *Bulletin of the World Health Organization* **15**, 661.

MEIER-BROOK, C. (1984). A preliminary biogeography of freshwater pulmonate gastropods. In *World-wide Snails: Biogeographic Studies on Non-Marine Mollusca* (ed. Solem, A., & Van Bruggen, A. C.), pp. 23–37. Leiden, E. J. Brill & W. Backhuys.

MELLO, D. A. (1972). The comparative morphology of the genital system of some African species of *Biomphalaria* (Mollusca, Planorbidae). *Review of Brasilian Biology* **32**, 443–450.

MINCHELLA, D. J., LEWIS, F. A., SOLLENBERGER, K. M. & WILLIAMS, J. A. (1994). Genetic diversity of *Schistosoma mansoni*: quantifying strain heterogeneity using a polymorphic DNA element. *Molecular and Biochemical Parasitology* **68**, 307–313.

MINCHELLA, D. J., SOLLENBERGER, K. M. & DESOUZA, C. P. (1995). Distribution of schistosome genetic diversity within molluscan intermediate hosts. *Parasitology* **111**, 217–220.

MKOJI, G. M., HOFKIN, B. V., KURIS, A. M., STEWARTOATEN, A., MUNGAI, B. N., KIHARA, J. H., MUNGAI, F., YUNDU, J., MBUI, J., RASHID, J. R. & KARIUKI, C. H. (1999). Impact of the crayfish *Procambarus clarkii* on *Schistosoma haematobium* transmission in Kenya. *American Journal of Tropical Medicine and Hygiene* **61**, 751–759.

MKOJI, G. M., MUNGAI, B. N., KOECH, D. K., HOFKIN, B. V., LOKER, E. S., KIHARA, J. H. & KAGENI, F. M. (1992). Does the snail *Melanoides tuberculata* have a role in biological control of *Biomphalaria pfeifferi* and other medically important African pulmonates? *Annals of Tropical Medicine and Parasitology* **86**, 201–204.

MORAND, S. & MÜLLER-GRAF, C. D. M. (2000). Muscles or testes? Comparative evidence for sexual competition among dioecious blood parasites (Schistosomatidae) of vertebrates. *Parasitology* **120**, 45–56.

MÜLLER-GRAF, C. D. M., COLLINS, D. A., PACKER, C. & WOOLHOUSE, M. E. J. (1997). *Schistosoma mansoni* infection in a natural population of olive baboons (*Papio cynocephalus anubis*) in Gombe Stream National Park. Tanzania. *Parasitology* **115**, 621–627.

MULVEY, M. & BANDONI, S. M. (1994). Genetic variability in the M-line stock of *Biomphalaria glabrata*

(Mollusca, Planorbidae). *Journal of the Helminthological Society of Washington* **61**, 103–108.

MUNENE, E., OTSYULA, M., MBAABU, D. A. N., MUTAHI, W. T., MURIUKI, S. M. K. & MUCHEMI, G. M. (1998). Helminth and protozoan gastrointestinal tract parasites in captive and wild-trapped African non-human primates. *Veterinary Parasitology* **78**, 195–201.

OLLIVIER, G., BRUTUS, L. & COT, M. (1999). Schistosomiasis due to *Schistosoma mansoni* in Madagascar: Spread and focal patterns. *Bulletin de la Société de Pathologie Exotique* **92**, 99–103.

PARAENSE, W. L. (1983). A survey of planorbid molluscs in the Amazonian region of Brazil. *Memorias do Instituto Oswaldo Cruz* **78**, 343–361.

PARAENSE, W. L. & ARAUJO, M. V. (1984). *Biomphalaria glabrata* no estado do Piauf. *Memorias do Instituto Oswaldo Cruz* **79**, 385–387.

PARAENSE, W. L. & CORRÊA, L. R. (1963). Susceptibility of *Australorbis tenagophilus* to infection with *Schistosoma mansoni*. *Revista do Instituto de Medicina Tropical de São Paulo* **5**, 23–29.

PARAENSE, W. L. & CORRÊA, L. R. (1973). Susceptibility of *Biomphalaria peregrina* from Brazil and Ecuador to two strains of *Schistosoma mansoni*. *Revista do Instituto de Medicina Tropical de São Paulo* **15**, 127–130.

PARAENSE, W. L. & CORRÊA, L. R. (1981). Observations on two biological races of *Schistosoma mansoni*. *Memorias do Instituto Oswaldo Cruz* **76**, 287–291.

PARAENSE, W. L. & CORRÊA, L. R. (1982). Unsusceptibility of *Biomphalaria occidentalis* to infection with a strain of *Schistosoma mansoni*. *Memorias do Instituto Oswaldo Cruz* **77**, 55–58.

PARAENSE, W. L. & CORRÊA, L. R. (1987). Probable extension of schistosomiasis mansoni to southernmost Brazil. *Memorias do Instituto Oswaldo Cruz* **82**, 577.

PARAENSE, W. L. & CORRÊA, L. R. (1989). A potential vector of *Schistosoma mansoni* in Uruguay. *Memorias do Instituto Oswaldo Cruz* **84**, 281–288.

PARAENSE, W. L. & DESLANDES, N. (1959). The renal ridge as a reliable character for separating *Taphius glabatus* from *Taphius tenagophilus*. *American Journal of Tropical Medicine and Hygiene* **8**, 456–472.

PARAENSE, W. L., IBAÑEZ, H. N. & MIRANDA, C. H. (1964). *Australorbis tenagophilus* in Peru, and its susceptibility to *Schistosoma mansoni*. *American Journal of Tropical Medicine and Hygiene* **13**, 534–540.

PARAENSE, W. L., ZELEDÓN, R. & ROJAS, G. (1981). *Biomphalaria straminea* and other planorbid molluscs in Costa Rica. *Journal of Parasitology* **67**, 282–283.

PARODIZ, J. J. (1969). The Tertiary non-marine Mollusca of South America. *Annals of the Carnegie Museum* **40**, 1–242.

PEREZ, J. G., VARGAS, M. & MALEK, E. A. (1991). Displacement of *Biomphalaria glabrata* by *Thiara granifera* under natural conditions in the Dominican Republic. *Memorias do Instituto Oswaldo Cruz* **86**, 341–347.

PFLUGER, W. (1982). Introduction of *Biomphalaria glabrata* to Egypt and other African countries. *Transactions of the Royal Society of Tropical Medicine and Hygiene* **76**, 567–567.

PICQUET, M., ERNOULD, J. C., VERCRUYSSE, J., SOUTHGATE, V. R., MBAYE, A., SAMBOU, B., NIANG, M. & ROLLINSON,

D. (1996). The epidemiology of human schistosomiasis in the Senegal river basin. *Transactions of the Royal Society of Tropical Medicine and Hygiene* **90**, 340–346.

PIERCE, H. G. (1993). The nonmarine mollusks of the Late Oligocene–Early Miocene Cabbage Patch fauna of western Montana III. Aquatic mollusks and conclusions. *Journal of Paleontology* **67**, 980–993.

PILSBRY, H. A. (1911). Non-marine mollusca of Patagonia. In *Report of the Princeton University Expedition to Patagonia*, 1896–1899 (ed. Scott, W. B.), **3**: 513–633.

POINTIER, J. P. (1993). The introduction of *Melanoides tuberculata* (Mollusca, Thiaridae) to the island of Saint Lucia (West Indies) and its role in the decline of *Biomphalaria glabrata*, the snail intermediate host of *Schistosoma mansoni*. *Acta Tropica* **54**, 13–18.

POINTIER, J. P. & GIBODA, M. (1999). The case for biological control of snail intermediate hosts of *Schistosoma mansoni*. *Parasitology Today* **15**, 395–397.

POINTIER, J. P. & GUYARD, A. (1992). Biological-control of the snail intermediate hosts of *Schistosoma mansoni* in Martinique, French West Indies. *Tropical Medicine and Parasitology* **43**, 98–101.

POINTIER, J. P., PARAENSE, W. L. & MAZILLE, V. (1993). Introduction and spreading of *Biomphalaria straminea* (Dunker, 1848) (Mollusca, Pulmonata, Planorbidae) in Guadeloupe, French West Indies. *Memorias do Instituto Oswaldo Cruz* **88**, 449–455.

POINTIER, J. P., SAMADI, S., JARNE, P. & DELAY, B. (1998). Introduction and spread of *Thiara granifera* (Lamarck, 1822) in Martinique, French West Indies. *Biodiversity and Conservation* **7**, 1277–1290.

POINTIER, J. P., THALER, L., PERNOT, A. F. & DELAY, B. (1993). Invasion of the Martinique island by the parthenogenetic snail *Melanoides tuberculata* and the succession of morphs. *Acta Oecologica–International Journal of Ecology* **14**, 33–42.

POINTIER, J. P., THERON, A. & BOREL, G. (1993). Ecology of the introduced snail *Melanoides tuberculata* (Gastropoda, Thiaridae) in relation to *Biomphalaria glabrata* in the marshy forest zone of Guadeloupe, French West Indies. *Journal of Molluscan Studies* **59**, 421–428.

POINTIER, J. P., THERON, A., IMBERT-ESTABLET, D. & BOREL, G. (1991). Eradication of a sylvatic focus of *Schistosoma mansoni* using biological control by competitor snails. *Biological Control* **1**, 244–247.

PRENTICE, M. A. (1983). Displacement of *Biomphalaria glabrata* by the snail *Thiara granifera* in field habitats in St Lucia, West Indies. *Annals of Tropical Medicine and Parasitology* **77**, 51–59.

RACCURT, C. P., SODEMAN, W. A., RODRICK, G. L. & BOYD, W. P. (1985). *Biomphalaria glabrata* in Haiti. *Transactions of the Royal Society of Tropical Medicine and Hygiene* **79**, 455–457.

RICHARDS, C. S., KNIGHT, M. & LEWIS, F. A. (1992). Genetics of *Biomphalaria glabrata* and its effect on the outcome of *Schistosoma mansoni* infection. *Parasitology Today* **8**, 171–174.

ROLLINSON, D., KAUKAS, A., JOHNSTON, D. A., SIMPSON, A. J. G. & TANAKA, M. (1997). Some molecular insights into Schistosome evolution. *International Journal for Parasitology* **27**, 11–28.

RUFFER, M. A. (1910). Note on the presence of *Bilharzia haematobium* in Egyptian mummies of the twentieth dynasty (1250–1000 B.C.). *British Medical Journal* **1**, 16–23.

SCOTT, J. A. (1937). The incidence and distribution of human schistosomiasis in Egypt. *American Journal of Hygiene* **25**, 566–614.

SHERCHAND, J. B. & O'HARA, H. (1997). *Schistosoma mansoni*-like eggs detected in stool of inhabitants in southern Nepal. *Journal, Nepal Medical Association* **37**, 386–387.

SHERCHAND, J. B., OHARA, H., SHERCHAND, S. & MATSUDA, H. (1999). The suspected existence of *Schistosoma mansoni* in Dhanusha district, southern Nepal. *Annals of Tropical Medicine and Parasitology* **93**, 273–278.

SINHA, P. K. & SRIVASTAVA, H. D. (1960). Studies on *Schistosoma incognitum* Chandler, 1926, II. On the life history of the blood fluke. *Journal of Parasitology* **46**, 629–641.

SNYDER, S. D. & LOKER, E. S. (2000). Evolutionary relationships among the Schistosomatidae (Platyhelminthes: Digenea) and an Asian origin for *Schistosoma*. *Journal of Parasitology* **86**, 283–288.

SNYDER, S. D., LOKER, E. S., JOHNSTON, D. & ROLLINSON, D. (2001). The Schistosomatidae: Advances in phylogenetics and genomics. In *Interrelationships of the Platyhelminthes* (ed. Littlewood D. T. J. and Bray R. A.), pp. 194–200. London, Taylor and Francis.

SOUTHGATE, V. R. (1997). Schistosomiasis in the Senegal river basin: Before and after the construction of the dams at Diama, Senegal and Manantali, Mali and future prospects. *Journal of Helminthology* **71**, 125–132.

TALLA, I., KONGS, A., VERLE, P., BELOT, J., SARR, S. & COLL, A. M. (1990). Outbreak of intestinal schistosomiasis in the Senegal River basin. *Annales de la Societe Belge de Medecine Tropicale* **70**, 173–180.

TCHUENTE, L. A. T., SOUTHGATE, V. R., THERON, A., JOURDANE, J., LY, A. & GRYSEELS, B. (1999). Compatibility of *Schistosoma mansoni* and *Biomphalaria pfeifferi* in Northern Senegal. *Parasitology* **118**, 595–603.

THÉRON, A. & POINTIER, J. P. (1995). Teaching parasitology: ecology, dynamics, genetics and divergence of trematode populations in heterogeneous environments: The model of *Schistosoma mansoni* in the insular focus of Guadeloupe. *Research and Reviews in Parasitology* **55**, 49–64.

THURSTON, J. P. (1963). Schistosomes from *Hippopotamus amphibius* L., I. The morphology of *Schistosoma hippopotami* sp. nov. *Parasitology* **53**, 49–54.

THURSTON, J. P. (1971). Further studies on *Schistosoma hippopotami* and *Schistosoma edwardiense* in Uganda. *Revue de Zoologie et de Botanique Africaines* **84**, 145–152.

TIKASINGH, E. S., WOODING, C. D., LONG, E., LEE, C. P. & EDWARDS, C. (1982). The presence of *Schistosoma mansoni* in Montserrat leeward islands. *Journal of Tropical Medicine and Hygiene* **85**, 41–43.

VAN DAMME, D. (1984). *The Freshwater Mollusca of Northern Africa*. Dordrecht, Junk Publishers.

VARGAS, M., MALEK, E. A. & PEREZ, J. G. (1990). Schistosomiasis mansoni in the Dominican Republic: prevalence and intensity in various urban and rural communities, 1982–1987. *Tropical Medicine and Parasitology* **41**, 415–418.

VIDIGAL, T. H. D. A., DIAS NETO, E., CARVALHO, O. D. & SIMPSON, A. J. G. (1994). *Biomphalaria glabrata*: Extensive genetic-variation in Brazilian isolates revealed by random amplified polymorphic DNA analysis. *Experimental Parasitology* **79**, 187–194.

VIEIRA, L. Q., CORREA-OLIVEIRA, R., KATZ, N., DESOUZA, C. P., CARVALHO, O. S., ARAUJO, N., SHER, A. & BRINDLEY, P. J. (1991). Genomic variability in field populations of *Schistosoma mansoni* in Brazil as detected with a ribosomal gene probe. *American Journal of Tropical Medicine and Hygiene* **44**, 69–78.

VRIJENHOEK, R. C. & GRAVEN, M. A. (1992). Population genetics of Egyptian *Biomphalaria alexandrina* (Gastropoda, Planorbidae). *Journal of Heredity* **83**, 255–261.

WALKER, J. (1978). The finding of *Biomphalaria straminea* amongst fish imported into Australia. *WHO Document WHO/Schisto/78·46*, Geneva, WHO.

WALSH, J. F., MOLYNEUX, D. H. & BIRLEY, M. H. (1993). Deforestation: Effects on vector-borne disease. *Parasitology* **106**, S55–S75.

WILKEN, G. B. & APPLETON, C. C. (1991). Avoidance responses of some indigenous and exotic fresh-water pulmonate snails to leech predation in South Africa. *South African Journal of Zoology–Suid-Afrikaanse Tydskrif Vir Dierkunde* **26**, 6–10.

WILLIAMS, S. N. & HUNTER, P. J. (1968). The distribution of *Bulinus* and *Biomphalaria* in Khartoum and Blue Nile provinces, Sudan. *Bulletin of the World Health Organization* **39**, 949–954.

WOODRUFF, D. S., MULVEY, M. & YIPP, M. W. (1985a). Population genetics of *Biomphalaria straminea* in Hong Kong: a neotropical schistosome transmitting snail recently introduced into China. *Journal of Heredity* **76**, 355–360.

WOODRUFF, D. S., MULVEY, M. & YIPP, M. W. (1985b). The continued introduction of intermediate host snails of *Schistosoma mansoni* into Hong Kong. *Bulletin of The World Health Organization* **63**, 621–622.

WOODRUFF, D. S. & MULVEY, M. (1997). Neotropical schistosomiasis: African affinities of the snail *Biomphalaria glabrata* (Gastropoda: Planorbidae). *Biological Journal of the Linnean Society* **60**, 505–516.

YIPP, M. W. (1990). Distribution of the schistosome vector snail, *Biomphalaria straminea* (Pulmonata, Planorbiadae) in Hong Kong. *Journal of Molluscan Studies* **56**, 47–55.

YOSHINO, T. P. & VASTA, G. R. (1996). Parasite–invertebrate host immune interactions. *Advances in Comparative and Environmental Physiology* **24**, 125–167.

Evolutionary relationships between trematodes and snails emphasizing schistosomes and paragonimids

D. BLAIR[1]*, G. M. DAVIS[2] and B. WU[3]

[1] School of Tropical Biology, James Cook University, Townsville, Qld 4811, Australia
[2] Department of Microbiology and Tropical Medicine, The George Washington University Medical Center, Washington D.C. 20037, USA
[3] Box 533/RM 1520, University of Pennsylvania, 3650 Chestnut Street, Philadelphia PA 19104, USA

SUMMARY

Snails and digeneans have been associated for at least 200 million years. Their inter-relationships over such a time-span must have been complex and varied. Few studies have attempted to explore these relationships in the light of knowledge of the phylogeny of both host and parasite groups. Here we focus on two important families of digeneans, the Schistosomatidae and the Paragonimidae, for which molecular phylogenies are available. We investigate the types of evolutionary relationships between host and parasite, operating at different phylogenetic depths, that might explain current host specificity and distributions of both associates. Both families of parasites utilise a number of highly diverged gastropod families, indicating that host extensions have featured in their histories. However, schistosomatids and paragonimids show different patterns of association with their snail hosts. As befits the apparently more ancient group, schistosomatids utilise snails from across a wide phylogenetic range within the Gastropoda. The genus *Schistosoma* itself has experienced one long-range host switch between pulmonates and caenogastropods. By contrast, paragonimids are restricted to two superfamilies of caenogastropods. Despite these differences, modern schistosomatid species appear to be more host specific than are paragonimids and host additions, at the level of host family, are far less common among species of schistosomatids than among paragonimids. Some species of *Paragonimus* exhibit remarkably low levels of host specificity, with different populations utilising snails of different families. Existing knowledge relating to the phenomenon will be presented in the context of phylogenies of schistosomatids, paragonimids, and their snail hosts. Discussion focuses on the usefulness of current theories of snail–digenean coevolution for interpreting these findings. In the past, much emphasis has been placed on the idea that digeneans engage in a one-to-one arms race with their snail host. We consider that phylogenetic tracking rather than an arms-race relationship might be a common alternative. Not being bound by the restrictions imposed by an arms race, some digeneans might be able to extend to new host species more easily than the literature suggests. Switches into related host taxa are most likely. However, ecologically equivalent but unrelated gastropod hosts may also be exploited. Given the right ecological setting, digeneans are able to switch across considerable phylogenetic distances. Examples from the Paragonimidae and Schistosomatidae are given.

Key words: Coevolution, DNA sequences, gastropods, historical biogeography, host-parasite relationships, *Paragonimus*, phylogeny, *Schistosoma*.

INTRODUCTION

In recent years, there have been considerable advances in understanding the systematics and phylogenies of both snail and parasite taxa involved in the evolution and deployment of two families of digeneans, the Schistosomatidae and Paragonimidae. These advances permit a critical evaluation of the extent to which these parasites have a shared evolutionary history with their snail hosts and of evolutionary concepts concerning host association and host switching.

Schistosomatids inhabit the blood vessels of birds and mammals, with a single genus and species known from crocodilians. Schistosomatids are found in all temperate and tropical regions of the world,

primarily associated with freshwater habitats. They are atypical digeneans in that they are dioecious and have only two hosts in their life cycles, a gastropod and a vertebrate. There are 13 genera in the family (Basch, 1991). Taxonomy of the type genus, *Schistosoma*, is reasonably settled but the same cannot be said of some other genera. The most recent general texts are by Rollinson & Simpson (1987) and Basch (1991). Paragonimids are lung flukes of mammals (including humans) and utilise fresh or (rarely) brackish water snails and crustaceans as first and second intermediate hosts respectively. The ~ 50 nominal species occur in Asia, Africa and the Americas, mostly in tropical or subtropical areas. More than half of all nominal species occur in East Asia, especially China. A few species are found in temperate regions of North America and Northern Asia. Only two genera are recognized, *Paragonimus* and *Euparagonimus*, but the taxonomy of the family remains confused (see Blair, Xu & Agatsuma (1999) for a discussion of the family).

* Corresponding author: D. Blair, School of Tropical Biology, James Cook University, Townsville, Qld 4811, Australia. Tel: +61 7 4781 4322. Fax: +61 7 4725 1570. E-mail: david.blair@jcu.edu.au

Parasitology (2001), **123**, S229–S243. © 2001 Cambridge University Press
DOI: 10.1017/S003118200100837X

A critical examination of molluscan/digenean evolutionary relationships requires knowledge of phylogenies for both hosts and parasites. Few papers have been specifically concerned with comparisons of such relationships. Rather, attention has been paid to phylogenetic analyses of either parasite or snail lineages. Earlier papers on evolutionary associations (Davis, 1980, 1992) focused on the genus *Schistosoma* with special emphasis on Asian species and their snail hosts. Davis, Spolsky & Zhang (1995) and Davis *et al.* (1999) subsequently 'mapped' the historical relationships between species of *Schistosoma* in Asia, and of *Paragonimus* worldwide, on the clades of relevant evolving snail lineages. A later molecular-based phylogenetic analysis of snail lineages transmitting *Paragonimus* in Asia (Wilke *et al.* 2000) resulted in changes in our understanding of some of these lineages and brought into question the long-held assumption of strict molluscan host specificity in tightly coevolved clades of parasites and snails.

Our study relies on accurate identification of snail hosts and parasite species. Most of our host identifications, especially for blood flukes, come from the literature. This is not without its difficulties. Mollusc species are easy to misidentify and the identity of cercariae emerging from a snail may have been proposed because of morphological resemblances to known species. Furthermore, some taxa of digeneans have been based, in part, on the molluscan host in which their cercariae were found. This may lead to excessive taxonomic splitting and can obscure cases of host switching or of continuing uses of multiple hosts.

Molluscan hosts of paragonimids are less well known than those of schistosomatids. Davis *et al.* (1994) made an in-depth annotated review of snail hosts with special emphasis on Asian species and Blair *et al.* (1999) listed known host species. For some species in the present analysis, host identity is inferred from knowledge of the molluscan fauna occurring in the habitats from which infected crustaceans were collected. Consequently, the paragonimid tree contains many question marks against host family names and these host assignments should be regarded as very tentative.

Here we map the molluscan hosts used by schistosomatids and paragonimids onto the most recently available molecular phylogenies of these digenean families, and we map the digeneans on a gastropod phylogeny based on both detailed comparative anatomy and molecular data. The gastropod tree (Fig. 1) is based on anatomical data (Ponder & Lindberg, 1997) with relationships among rissooidean families determined using DNA sequences (Davis *et al.* 1998; Wilke *et al.* 2000). The trees of schistosomatids and paragonimids presented here (Figs 2–4) are compiled from several sources and include some previously unpublished infor-

mation. The data used in each case were sequences of mitochondrial and/or nuclear genes. Approaches to tree construction and interpretation are many and varied. Here we simply present conservative trees, based on the many trees available to us, to provide a summary of relationships. The schistosomatid tree is based largely on Snyder & Loker (2000) with some additional information from Agatsuma (personal communication and Agatsuma *et al.* in press *a*, *b*). A detailed discussion of the paragonimid taxa sequenced, the trees, their interpretation and the data used in their construction, is in preparation.

Evolutionary terminology

Coevolution is the single term most often applied to describe the evolutionary association between digeneans and gastropods. It was coined by Mode (1958) to describe the genetics of virulence and host resistance among plants and their obligate pathogens. Strictly speaking, the term should not be used except to describe cases where the population genetic interactions of two or more species are such that a genetic change in one elicits a reciprocal genetic change in the other(s). Such mutual evolutionary interactions, in the form of an arms race, have been demonstrated a number of times between digeneans and their snail hosts (see Lively, this supplement), most particularly in cases where a single species of digenean has a high prevalence in the snail population. However, evolutionary interactions between digeneans and molluscs may not always be of this kind, a possibility generally overlooked. Snails possess an internal defence system that lacks immunological memory and seems to be general rather than able to target individual species of digeneans (Wright & Southgate, 1981; Adema & Loker, 1997; Adema, Hertel & Loker, 1999). Consequently, in the typical situation where the prevalence of infection with a single digenean species is extremely low, that particular species is unlikely to have a significant evolutionary impact on the snail. This is not to say that the overall suite of digeneans in a snail population does not elicit an evolutionary response, only that the snail mounts a general response to the different challenges represented by the various digeneans, each with its own evolutionary trajectory. Digeneans utilising the same gastropod species might themselves have conflicts of interest (Poulin, Steeper & Miller, 2000) that could tend to drive snail responses in different directions.

Contrary to the evolutionary indifference that the snail might display towards any single digenean species, the impact of evolutionary or population-genetic changes in the snail must have an enormous impact on each digenean species that utilises it. In other words, the evolutionary fate of the snail may not be influenced by any single digenean species (unless prevalences are atypically high), but that of the digenean is influenced by the snail. Digeneans must track snails through time but snails need not reciprocate. Use of presumptive terms such as coevolution and its near-synonyms (e.g. co-accommodation and co-adaptation) is inappropriate in such cases. Congruence between host and associate phylogenies can be explained by one of three general scenarios (Brooks & McLennan, 1991): allopatric cospeciation (their 'null' model), phylogenetic tracking and coevolutionary arms race. We consider that coevolution should, strictly, be used to describe only the arms-race scenario. The first two scenarios do not absolutely require *mutual* adaptive responses, a defining condition for coevolution, but may nevertheless lead to congruence of host and associate phylogenies. Strict coevolution is probably best detected at local scales in snail–digenean systems where present-day interactions can be studied in both field and laboratory: its presence cannot be inferred directly from patterns of association.

Other terms to consider are host extension and its components, host-switching and host-addition. Host-switching occurs when a lineage of digeneans establishes itself in, and evolves with, a previously unexploited snail lineage, abandoning the original host lineage completely or leaving a sister parasite lineage within it. Host addition occurs when a digenean adds a new taxon to its existing hosts, most likely in allopatry. Host-switching and host-addition are not mutually exclusive and may operate at different evolutionary scales. The former is more apparent above the taxonomic level of species in digeneans, the latter at the level of species or perhaps species-group. Host addition may be preliminary to host switching. Host extensions are most likely to involve host taxa that are closely related and have a considerable shared genetic heritage. For example, for a parasite normally utilising a pomatiopsid snail, host extension to another pomatiopsid is more likely than to snails of another family. Given the taxonomic uncertainties concerning many hosts and digeneans alike, extensions at shallow phylogenetic levels are likely to be difficult to detect. Consequently, in this paper, we probably over-emphasize the proportion of long-range switches or additions.

We can make a number of predictions about host addition. It involves extending the range of a digenean species by adding new snail hosts, almost certainly in allopatry. We predict that the first individuals of a species that are successful in parasitising a novel snail host produce few cercariae and may face other fitness penalties. Consequently, when novel hosts are added in sympatry with canonical hosts, the alleles permitting the addition are likely to be diluted out and hence selected against. In allopatry, this cannot happen and any parasites persisting in the novel host presumably increase their compatibility with it through time.

Host addition may induce parasite speciation through geographical isolation (allopatric speciation) and/or founder effects involving the parasite population. Note that allopatric speciation in such a case is not cospeciation (the 'null' model of Brooks & McLennan, 1991): the parasite need not be speciating in response to speciation by a host. The geographical dispersal of parasites that might lead to host addition is presumably mediated by vagile and migratory hosts such as birds. Parasites may thus be maintained in unrelated snail species at different points along a migratory route. Of course, migrating hosts could easily transport, to the same locality, individuals of a parasite species (which may also inhibit parasite speciation) that differ in snail compatibility. This might explain the discovery in the same pond of two forms of the echinostomatid digenean *Echinoparyphium recurvatum*, one infecting a lymnaeid snail and the other a valvatoid (McCarthy, 1990). Similarly, a single nominal species of *Typhlocoelum* (Digenea: Cyclocoelidae) has sympatric forms that each utilise one of three families of pulmonates (Scott, Rau & McClaughlin, 1982). *Echinoparyphium* and *Typhlocoelum* both occur in birds.

We predict that examples of host addition will best be recognized by comparing digeneans from different points along migratory routes, on opposite sides of biogeographic boundaries where snail faunas differ, and between marine and freshwater habitats. It might also be instructive to examine digenean taxa that include among their host taxa, snail families that are apparently permissive hosts, such as pomatiopsids (see below) and possibly also ancylids (Cable & Peters, 1986). Patterns of host addition by digenean species distributed around the world by human activities might also repay study. To confirm host addition, it will be necessary to show that the parasites occurring in different hosts and locations belong to the same species or species group, a task for which molecular genetic methods are well suited.

MOLLUSCAN PHYLOGENY AND RADIATION

Re-evaluation of molluscan phylogeny over the past several years has radically changed the traditional classifications of the past 60 years. The most important synthesis to date is that of Ponder & Lindberg (1997) providing the basis of the phylogeny presented here (Fig. 1).

There are over 100 000 species of molluscs of which some 60 % are gastropods that are distributed among three main lineages (Caenogastropoda, Valvatoidea and Euthyneura, the last two in the Heterobranchia). The 'prosobranch' caenogastropods arose first in the sea and then moved into freshwater and onto land. The early heterobranchs gave rise to the Valvatoidea and Euthyneura in the Carboniferous. The former lineage has retained some prosobranch features whereas the latter is more derived. The

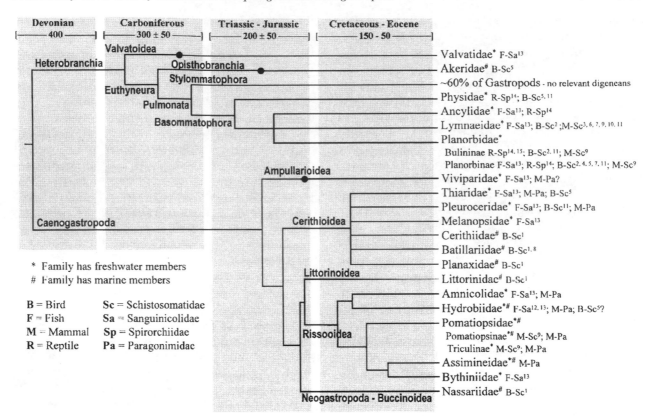

Fig. 1. Phylogeny, classification and approximate ages of the gastropod taxa relevant to this paper. Divergence dates within the Devonian and Carboniferous periods are uncertain, and these parts of the tree should be regarded only as an indication of order of divergence. Closed circles on branches leading to Valvatidae, Akeridae and Viviparidae indicate approximate times at which those families appeared. Note that our understanding of relationships among rissooidean families discussed here is likely to change in the near future (Wilke *et al.* 2000 and Davis, unpublished). In particular, hosts of *Paragonimus* in Central and South America that are usually regarded as members of the family Hydrobiidae should probably be referred to the Cochliopidae. Parasite families and their vertebrate host taxa are shown after each gastropod family name. Superscript numbers denote parasite genera as follows. Schistosomatidae 1, *Austrobilharzia*; 2, *Bilharziella*; 3, *Bivitellobilharzia* (not included in molecular tree); 4, *Dendritobilharzia*; 5, *Gigantobilharzia*; 6, *Heterobilharzia*; 7, *Orientobilharzia*; 8, *Ornithobilharzia*; 9, *Schistosoma*; 10, *Schistosomatium*; 11, *Trichobilharzia*. Sanguinicolidae 12, *Paracardicoloides*; 13, *Sanguinicola*. Spirorchiidae 14, *Spirorchis*; 15, *Enterohaematotrema*. Gastropod families with freshwater members are indicated by * and those with marine members by #.

major euthyneurous lineages are the marine Opisthobranchia (marked by a progressive and rapid loss of the shell) and the Pulmonata that evolved on land and in freshwater (gills replaced by a 'lung' and typically with a shell). The Pulmonata have two orders (or sub-orders). The Basommatophora are freshwater and the Stylommatophora are terrestrial. It is acknowledged that the Opisthobranchia and Pulmonata are paraphyletic or polyphyletic to some degree yet to be determined (Ponder & Lindberg, 1997), but the terms still serve the purpose of pointing out the main evolving clades of concern here.

The divergence of the Caenogastropoda and the Heterobranchia can be traced back to at least the Lower Carboniferous (325 to 360 MYa) (Kollmann & Yochelson, 1976). The heterobranchs were already well diversified by then suggesting an even earlier origin (Ponder & Lindberg, 1997), perhaps in the Devonian as shown in Fig. 1. All the major snail clades (caenogastropods, valvatoids, opisthobranchs and pulmonates) were present during, and impacted greatly by, the major Permian extinction (286–245 MYa). Consequently, we trace the flowering of the modern gastropod radiations in freshwater and land to the Mesozoic, especially the Jurassic and Cretaceous when sufficient fossil beds were laid down enabling some discernment of the distributions of snail families and their ecologies. The pulmonate Physidae possibly arose in the Carboniferous, but were well established in the fossil record along with the Lymnaeidae and Planorbidae (Bulininae) by the Jurassic (208–146 MYa) (Wenz, 1939). The rissooidean-hydrobioid radiations were well established by the late Triassic and early Jurassic (210–146 MYa) in what are now South Africa and India (Davis, 1979, 1980; Ponder, 1988). The Amnicolidae stem from the Cretaceous (Wenz, 1939). *Littorina* (Caenogastropoda: Littorinoidea) was found in the Triassic (245–208 MYa). The

Valvatoidea are found in the Carboniferous; the Akeridae, the Jurassic and the relevant cerithioideans, in the Cretaceous. Some brackish and freshwater families can only be traced to the early or mid-Tertiary of the Cenozoic Era (65–5 MYa), e.g. the rissooidean Assimineidae and Bithyniidae, but probably had earlier origins.

We can trace with some confidence the bio-geographic histories of many of the gastropod families important to our discussion. Davis (1979, 1980) documented the centre of origin of the family Pomatiopsidae in what are now southern Africa and India, with the break-up of Gondwana moving snail families to what are now South America, Australia, India, and presumably Antarctica. The Miocene collision of the Indian Plate with Asia created the Himalayan orogeny, which initiated an ecological revolution by rapidly creating and destroying fresh-water environments and accelerating the evolution of the major rivers of eastern Asia with all their attendant stream captures. Pomatiopsids, already differentiated into triculines and pomatiopsines, moved east and south along these evolving river systems. The amphibious *Oncomelania*, the only genus of the subfamily Pomatiopsinae in China and eastern Asia (Japan, Taiwan, Philippines, eastern Indonesia), tracked a relatively uniform ecology and formed a limited morphostatic radiation of two species, one of which, *O. hupensis*, comprises several subspecies. The genus also dispersed to North America where it gave rise to the sister genus *Pomatiopsis* (Davis, 1979, 1980, 1981). The sub-family Triculinae is found only in an arc from northeast India into northern Burma and China-southeast Asia. The rapid ecological changes during the late Miocene–Pliocene left behind a rich fossil record, especially of triculine taxa, in ancient lake beds in northern Burma. Triculines form a wholly aquatic group exhibiting today an amazing adaptive radiation of three tribes, over 20 genera and more than 120 species. Two tribes are remarkable adaptive radiations invading lakes, streams, and every conceivable niche in the Mekong River that experiences dramatic annual fluctuations in flow due to the monsoons.

The same generalised track as followed by the pomatiopsids, from India into northern Burma and China, and hence into southeast Asia, is observed in other relevant snail families, such as the Planorbidae and Thiaridae. The pulmonate family Planorbidae has two prominent subfamilies, the Planorbinae and Bulininae. The latter radiated in Africa as the genus *Bulinus* and in India as *Indoplanorbis* with subsequent introduction and dispersal in southeast Asia. The former subfamily, the Planorbinae, occurs in Africa and South America. The cerithioidean *Brotia* (Thiaridae), transmitting *Paragonimus* in eastern Asia, also occurs in a generalised track from India into southeast Asia (Malaysia to the Philippines).

ORIGINS OF SNAIL–DIGENEAN ASSOCIATIONS

The origins of digeneans and the steps leading to their host commitment to molluscs are unknown but often speculated upon. Gibson (1987) and Gibson & Bray (1994) proposed that proto-trematodes first parasitised molluscs (possibly bivalves) at least 400 MYa, later transferring to gastropods and, in one case, to scaphopods. The association with vertebrates, and hence the appearance of digeneans, probably started about 200 MYa when modern teleosts evolved. By this time the families of gastropod heterobranchs, superfamilies of caeno-gastropods and all higher taxa were well established. If this view is correct, and the digeneans arose only once from the flatworms associated with the gastro-pods of the time, then the early radiation of the Digenea must have been accompanied by much host-switching among snail families from the plesio-morphic host. It might also have been the case that host specificity with respect to the gastropod was less marked than at present (Gibson & Bray, 1994), in which case several molluscan taxa could have provided effective alternative hosts for each early digenean group.

PARASITE PHYLOGENIES AND HOST RELATIONSHIPS

The Schistosomatidae

Blood flukes belong to three related families and occur as adults in the blood vessels of fish (Sanguini-colidae), turtles (Spirorchiidae) and crocodilians, birds and mammals (Schistosomatidae). Members of all families occur in both marine and freshwater habitats. A phylogeny of the blood flukes (Fig. 2) provides a molecular-based road map of their evolution and permits a discussion of snail host associations. This discussion should start by trying to infer the ancestral host taxon for the schisto-somatids, a task that is far from easy.

As seen in Fig. 2, sanguinicolids and spirorchiids are outgroups for the Schistosomatidae. Blood flukes and the related strigeoids appear to be the earliest diverging digeneans (Cribb *et al.* 2001). Sanguini-colids, associated with fish, may represent the basal diverging lineage within the Digenea. Modern freshwater sanguinicolids use a wide range of snail families in all main lineages (11 families in 6 superfamilies – see Fig. 1). Hosts for the many marine sanguinicolids, excluding the aberrant *Aporocotyle*, are not known. Where known, species of *Aporocotyle* utilise polychaete annelids as the sole intermediate host (Køie, 1982)! Cercariae thought to be those of sanguinicolids have been found in marine bivalves (Smith, 1997). The sanguinicolids therefore provide no clues as to the plesiomorphic molluscan host for blood flukes. Indeed, it is possible that this

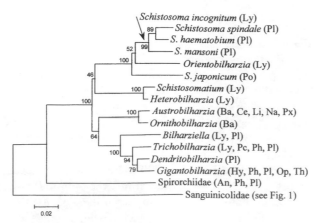

Fig. 2. Distance tree of relationships within the Schistosomatidae. This phylogeny is based on molecular data from the nuclear 28S ribosomal RNA gene and includes 10 of the 13 recognized genera of schistosomes (Rollinson & Southgate, 1987; Basch, 1991). A sequence for *Schistosoma spindale* (GenBank accession number Z46505) was added to the data of Snyder & Loker (2000). Sequences were aligned using ClustalW (Higgins, Thompson & Gibson, 1996). A distance matrix was constructed in MEGA2 (Kumar *et al.* 2000) using the Kimura 2-parameter model (with complete deletion of indels) and trees constructed using the minimum evolution method (1000 cycles of bootstrapping). Bootstrap values are shown in percentages. Molecular data from other gene regions indicate the internode (arrowed) on which *Schistosoma incognitum* should be joined to this tree (Agatsuma, personal communication and in press *b*). The tree is rooted by inclusion of the outgroup families Sanguinicolidae and Spirorchidae. Note that this tree differs somewhat from that published by Snyder & Loker (2000): The clades (*Heterobilharzia* + *Schistosomatium*) and (*Ornithobilharzia* + *Austrobilharzia*) have exchanged places and the root of the Schistosomatidae has been placed so as to include the former clade in a lineage with *Schistosoma* and *Orientobilharzia*. We regard these differences as due to the presence of two very short, adjacent internodes that produce instability in that portion of the tree. Where known, the family or higher taxon of snail hosts is indicated: An, Ancylidae; Ba, Batillariidae; Ce, Cerithiidae; Hy, Hydrobiide; Li, Littorinidae; Ly, Lymnaeidae; Na, Nassariidae; Op, Opisthobranchia; Pc, Pleuroceridae; Ph, Physidae; Pl, Planorbide; Po, Pomatiopsidae; Px, Planaxidae; Th, Thiaridae. For molluscan hosts of sanguinicolids, see Fig. 1 and the text.

family has retained the broad host specificity suggested for early digeneans by Gibson & Bray (1994). Such a breadth of host taxa does not occur in the more derived blood flukes (Spirorchiidae and Schistosomatidae). Perhaps by the time these had evolved, the bloodfluke lineage had already settled into tracking one or a few host lineages.

Spirorchiids in freshwater habitats utilise pulmonate snails. Hosts for this family in marine habitats, where many species occur, are unknown.

Marine pulmonates could possibly be involved, e.g. the Siphonariidae that are numerous and intertidal, or the Ellobiidae abounding in salt marshes, but data are lacking. Schistosomatids utilise pulmonates, caenogastropods and even opisthobranchs. The one schistosomatid species occurring in a freshwater crocodile (Platt *et al.* 1991) is presumably basal in the family. There is anecdotal evidence that the host for this schistosomatid is a pulmonate (Platt *et al.* 1991). We consider that pulmonates were the ancestral host taxon for schistosomatids, but acknowledge that the evidence is far from conclusive. Pulmonates are included among known host taxa for sanguinicolids and spirorchiids. They are also hosts (where known) for all schistosomatid genera except those in the clade *Austrobilharzia* + *Ornithobilharzia*, some species of *Gigantobilharzia*, one of *Trichobilharzia* (*T. corvi*) and for the *Schistosoma japonicum* and *S. sinensium* groups. The case of *Schistosoma* species in non-pulmonates is discussed in detail below and considered to be a consequence of a host-switch during the Tertiary. *Trichobilharzia corvi* belongs to a large, cosmopolitan genus in which all other members occur in pulmonates. The switch to a caenogastropod was presumably recent. In the remaining cases, the non-pulmonate hosts are marine. *Gigantobilharzia huttoni* is the only schistosomatid to utilise an opisthobranch. Again, this is presumably a relatively recent switch: most other members of the genus utilise pulmonates with two reports from caenogastropods. In the case of *Austrobilharzia* + *Ornithobilharzia*, given the rather basal position this clade occupies in the tree (Fig. 2) it is less easy to be certain that the snail hosts have not been retained from a plesiomorphic condition. However, we suspect that *Austrobilharzia* + *Ornithobilharzia* have switched to caenogastropod hosts relatively recently as a consequence of targeting sea birds. Host-switching associated with a change in ecology is discussed later.

We will now turn to the genus *Schistosoma*. Species within *Schistosoma* are usually placed into different groups according to snail host specificity and egg shape (Rollinson & Southgate, 1987). The *S. japonicum* group (defined by Davis & Greer, 1980) occurs in east and southeast Asia and utilises pomatiopsid snails (both Pomatiopsinae and Triculinae). Its members have round eggs with a recessed spine. The southeast Asian *S. sinensium* group has a rather more restricted range centred on southwest China and northern Thailand and occurs in triculine snails. It has elongated asymmetrical eggs with a lateral spine somewhat resembling those of the *S. mansoni* group.

The *S. mansoni* group, consisting of relatively few species, has elongated eggs with a lateral or subterminal spine. Two species, *S. mansoni* and *S. rodhaini*, are very closely related to each other (Després *et al.* 1992). The remaining two species occur in *Hippopotamus* and we are not convinced that they should be grouped with *S. mansoni* despite their egg shapes. Certainly, limited molecular data for one of these, *S. hippopotami*, suggest it lies basal to all African *Schistosoma* species (Després *et al.* 1995). The *S. mansoni* group is centred on Africa with introduction to the Americas (Després, Imbert-Establet & Monnerot, 1993). *Schistosoma mansoni* is associated with *Biomphalaria* of the Planorbidae, subfamily Planorbinae (see various papers in this supplement).

All remaining *Schistosoma* species (*S. haematobium* complex in Africa and the probably paraphyletic '*S. indicum*' group in India and southeast Asia) have elongated eggs of varying morphologies but always with a terminal spine of some sort. Most *Schistosoma* with terminal-spined eggs are associated with the Planorbidae, subfamily Bulininae. In Africa the bulinine snails are *Bulinus* while in India and southeast Asia the relevant genus is *Indoplanorbis*. The exception is *S. incognitum*, a member of the *S. indicum* group that utilises lymnaeids. In molecular trees (e.g. Fig. 2 and Agatsuma *et al.* in press *b*), this species lies basal to all African species (except possibly for *S. hippopotami*) and the remaining members of the *S. indicum* group. The four species of the *S. indicum* group have the same generalised track as *Indoplanorbis* and other snail taxa, from northeast India to Burma and dispersal into southeast Asia.

The genus *Orientobilharzia* (lymnaeid snail hosts) consists of about 4 species occurring in India, parts of southeast Asia, and in a broad band in the temperate zone from Turkey to northeastern China. Surprisingly, Snyder & Loker (2000) found that sequence data for *O. turkestanica* rendered *Schistosoma* paraphyletic in their molecular trees (Fig. 2). It may be that members of this genus should be absorbed into *Schistosoma*, or else that *Schistosoma* itself needs to be split. Certainly, molecular data (Barker & Blair, 1996; Snyder & Loker, 2000; Le *et al.* 2000) and cytogenetic evidence (Hirai *et al.* 2000) indicate marked divergence between the species using pomatiopsid snails and those using pulmonates. Baugh (1977) has suggested that *Orientobilharzia* may not itself be a natural taxon. Clearly, full appreciation of the history of *Schistosoma* requires further work to be done on the genus *Orientobilharzia*.

Interpretations of snail–parasite relationships require an understanding of the timing of events. For the species of *Schistosoma*, views on biogeographic origins and ages fall into two main camps differing on matters of timing: one invokes primarily vicariance, the other dispersal. The former hypothesis places the origin of *Schistosoma* in Gondwana (an area overlapping Gondwanan India, Africa and presumably Antarctica) during the Cretaceous with one or more lineages of the genus

subsequently introduced to Asia on the Indian Plate. The latter hypothesis has the genus originating somewhere in Asia, with a Miocene/Pliocene dispersal to Africa.

The Gondwanan-origin (vicariance) hypothesis was proposed and voluminously documented by Davis (1979, 1980, 1992). According to this, ancestral *Schistosoma* species were associated with pulmonate snails in Gondwana, especially in an area overlapping conjoined eastern Africa and India. As the supercontinent broke up, the ancestors of the Asian *Schistosoma* species rafted there on the Indian plate (which split from Africa at least 75 MYa in the late Cretaceous), leaving behind in Africa the ancestors of the modern African species. The collision between India and Asia during the Miocene caused the Himalayan orogeny and had a profound effect on the evolution of *Schistosoma* in the region. One clade, committed to the pomatiopsid snails which had accompanied them on the Indian plate, evolved down the huge river systems developing to the east and south and gave rise to species of both the *S. japonicum* and *S. sinensium* groups. Members of the *S. indicum* group retained their allegiance to pulmonates and remained in India or dispersed into southeast Asia, following the same pathway of introduction as the pomatiopsid snails.

The strengths of this argument are the congruence of the tectonic history, area cladograms and evolution of the relevant snail lineages. For example, the *S. haematobium* group has formed a considerable radiation in Africa in association with bulinine planorbids of the genus *Bulinus* while other terminal-spined schistosomes radiated in India in *Indoplanorbis* (the only other bulinine genus) with subsequent introduction to southeast Asia. These bulinine taxa are known to be an ancient lineage (fossils date to pre-Cretaceous, Wenz, 1939; Planorbinae and Bulininae in Africa and India in the mid to upper Cretaceous – Newton, 1920). A possible weakness of the argument is that, according to molecular trees (Fig. 2), two lineages (*S. japonicum* + *S. sinensium* and the lineage leading to *Orientobilharzia*) occur only in Asia and the basal member (*S. incognitum*) of the third lineage occurs in Asia today. (Some molecular evidence indicates that one African species, *S. hippopotami*, lies at the base of the clade including the African species (Després *et al.* 1995) and members of the *S. indicum* group (Agatsuma *et al.* in press *b*), but further data are required to confirm this.) If Gondwana is the ancestral home of the genus, then the three main clades must have existed before the break-up of Gondwana and two of them subsequently became extinct in Africa or were never in the African part of Gondwana. This also implies an origin for the genus in the mid-Cretaceous, followed by little morphological diversification subsequently. Other molecular data help demonstrate that *Schistosoma* is indeed

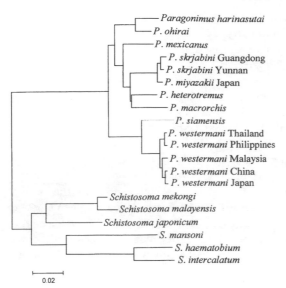

Fig. 3. Distance tree showing relationships of
Schistosoma and paragonimid species based on 372 bp of
the mitochondrial *cox*1 gene. Analyses in DAMBE (Xia,
2000) indicated that the sequences were not saturated
with transversions. Consequently, a transversion
distance matrix was constructed in MEGA2 (Kumar *et
al*. 2000) using the Kimura 2-parameter model and a
tree constructed using the minimum evolution method.
Divergence depths are far greater among members of
the genus *Schistosoma* than among paragonimids. Note
that some relationships among paragonimids differ from
those in Fig. 4.

very ancient. Our analyses of sequences of a 372 bp
fragment of DNA from the mitochondrial *cox*1 gene
from a number of schistosomatids and paragonimids
(Fig. 3) suggested that they were not saturated with
transversions (see caption to Fig. 3). The sequences
might therefore provide a reasonable phylogeny of
the two groups and a relative indication of divergence
times. Examination of the relative depths of nodes in
the tree (Fig. 3) implies that the divergence in
Schistosoma is much older than in *Paragonimus*. Of
course, caution must be exercised in interpreting
such a tree: rates of evolution might differ in the two
lineages. Nevertheless, given that *Paragonimus* has a
typical Gondwanan distribution, the implication is
that *Schistosoma* diversified well before the break-up
of Gondwana.

The dispersal hypothesis places the ancestral home
of *Schistosoma* in Asia (Snyder & Loker, 2000 and
reviewed in Morgan *et al.* in this supplement), with
subsequent dispersal into Africa in the Miocene/
Pliocene. The strength of this hypothesis is its
agreement with the topology of molecular trees
showing early-diverging taxa to be in Asia. There are
also some weaknesses. The relative depths of
branching of *Schistosoma* and *Paragonimus*, shown in
Fig. 3, have already been mentioned. The supporters
of the Asian-origin hypothesis did not specify an
exact region of origin. Wherever the genus arose, its

members would have had to disperse into both India
and Africa. The ramparts of mountains thrown up
by the collision of the Indian plate with Asia would
have made colonisation of India by existing Asian
freshwater fauna difficult. With regard to Africa, the
Tethys Sea separated this continent from Asia until
probably as recently as 5–10 MYa (Smith, Smith &
Funnell, 1994). The sea was most likely shallow and
narrow, but would nevertheless have been a barrier
to terrestrial species (but see Cox, 2000). If *Schistosoma* species did first enter Africa in the Miocene/
Pliocene, then they must have diversified extremely
rapidly on that continent. In particular, as mentioned
above, members of the *S. haematobium* group must
have radiated across Africa in a wide range of snail
species from the ancient bulinine lineage in a
relatively short space of time.

One more element of the debate about the
radiation of *Schistosoma* species and their hosts
needs to be touched on. *Schistosoma mansoni* now
occurs in both Africa and South America. Although
Davis (1979) originally proposed that this was a
consequence of ancient vicariance, recent molecular
evidence (Després *et al.* 1993) is more consistent
with a human introduction of the parasite from
Africa during the slave trade several centuries ago.
However, the story concerning the snail hosts cannot
be so simple. Like the schistosome, the planorbine
genus *Biomphalaria* occurs in both Africa and South
America. One South American species, *B. glabrata*,
is more similar, both morphologically and
genetically, to African species than to other South
American species (Pilsbry, 1911; Woodruff &
Mulvey, 1997; Campbell *et al.* 2000). Two classes of
explanation exist. One is that *Biomphalaria* species
were already diverse in Gondwana before the
separation of Africa and South America and the
observed relationships are therefore due to vicariance
(Davis, 1979, 1980). The other explanation is that
the genus diversified in the Americas and dispersed
to Africa, probably during the Pliocene (Woodruff &
Mulvey, 1997; Campbell *et al.* 2000). Fossils are
known from the Americas at least as far back as the
Paleocene (South America – Parodiz, 1969) and
Oligocene (North America – Pierce, 1993). According to Woodruff & Mulvey (1997), fossils of
Biomphalaria are not known before the mid-
Pleistocene in Africa. However, this overlooks
reports of relevant planorbid fossils from the Fayum
of Egypt in the upper Cretaceous (Newton, 1920).

Molecular data used to infer a recent dispersal of
Biomphalaria from South America to Africa (e.g.
Woodruff & Mulvey, 1997; Campbell *et al.* 2000)
also imply that divergences among South American
species are very recent, a scenario not consistent with
the existence of fossils over 50 MY old. This problem
is particularly apparent in the COI data reported by
Campbell *et al.* (2000). Based on an average rate of
change of COI genes, they suggested that the

ancestor of African *Biomphalaria* species dispersed to Africa not more than 3·6 MYa. Although they did not report estimated dates of divergence between *B. glabrata* and other South American species, their Fig. 2b implies that such divergences could hardly have been more than 5–10 MYa. This discrepancy needs to be investigated further, and additional South American taxa included in future studies.

Despite the case made for this by Woodruff & Mulvey (1997) and Campbell *et al.* (2000), it is hard to imagine freshwater pulmonates dispersing such great distances across an ocean. If snails were able to do so, then there should be evidence that other freshwater or terrestrial animals of similar size have also made the passage. No generalised track for such animals has a recent west to east dispersal over the ocean between South America and Africa (Croizat, 1958).

Knowledge of the origins of *Schistosoma* and its species is essential if we are to understand the association of *Schistosoma* and its snail hosts through time. The Gondwanan-origin hypothesis is more parsimonious in emphasising extensive phylogenetic congruence of parasites and snails, although there must have been a host switch at some stage between planorbid and lymnaeid hosts, and another between pulmonates and pomatiopsids. The Asian-origin hypothesis not only requires the host switches mentioned above, but also implies additional and extensive host-switches and host additions, albeit at relatively low taxonomic levels, as *Schistosoma* species entered Africa and diversified. Unless all hosts disperse together with their parasites, dispersalist scenarios such as this will always require more instances of host extension to explain them and will therefore seem less parsimonious.

The switch from pulmonates to pomatiopsids by the ancestor of the *S. japonicum* and *S. sinensium* groups is important. This seems to be a clear case of host switching and was a pronounced departure for pulmonate-dominated *Schistosoma*. This host switch is puzzling given the ubiquity of pulmonates. However, if the Gondwanan-origin hypothesis is correct, geological events might provide an explanation. As the Indian Plate moved towards Asia it passed over an active centre of volcanism that obliterated life over much of southern and western India and created the Deccan Traps (Davis, 1979). Such an event might have eliminated pulmonates locally. This, coupled with the ecological revolution caused by the rapid Himalayan orogeny that created new environments, could have promoted host-switching.

The *S. japonicum* species complex consists of three species in east and southeast Asia. The best known species, *S. japonicum*, occurs in *Oncomelania* over a large geographic area that includes China, Japan, Sulawesi and the Philippines. The remaining two species (*S. mekongi* and *S. malayensis*), almost identical to *S. japonicum* in morphology of all stages, including the egg, occur in triculines and have limited distributions. The divergence among these species must date to the Miocene separation of the snail subfamilies some 10 million years ago. Davis & Greer (1980) predicted, on the basis of snail systematics, distribution and timing of river formation and direction of evolution of the Pomatiopsidae in time and space, that the two triculine-transmitted schistosomes would be more closely related to each other than either is to *S. japonicum*. With the advent of molecular data for the schistosomes (Blair *et al.* 1997*b*) this has been confirmed. There is complete congruence between the area cladograms for river formation, snail phylogeny and *S. japonicum* group phylogeny based on molecular data. The *S. japonicum* group therefore seems to constitute a limited radiation in allopatry.

Schistosoma mekongi is transmitted in a small portion of the lower Mekong River by only one species of *Neotricula* (Tribe Pachydrobiinae) that is sympatric with numerous genera and species of the same tribe and the sister tribe Julieniini. *Schistosoma mekongi* will not develop in any other pomatiopsid snail (experimental infections reviewed in Davis, 1980, 1992). *Schistosoma malayensis* is found in rainforests in a small region of Malaysia where Davis & Greer (1980) discovered the snail host by deliberately looking in the correct ecological setting for the sister taxon to *S. mekongi* in a triculine snail. They found the snail and parasite and described *Robertsiella*, a sister genus to *Neotricula*. That there are only two species transmitted by triculines within the *S. japonicum* group may seem puzzling given the enormous radiation of the host subfamily. However, *S. mekongi* and *S. malayensis* were described as distinct species only within the last two decades and only after it was discovered that triculine snails transmit *Schistosoma* spp. (see Davis, Kitikoon & Temcharoen, 1976). We suspect that more species will be discovered as causative agents of human schistosomiasis in Asia are investigated more critically.

Associated with the Triculinae is an unusual schistosome, *Schistosoma sinensium*. This nominal species infects rodents and has eggs reminiscent of those of *S. mansoni*. Davis (1992) considered it to represent a complex of at least three species in Thailand and China that is a sister to the *S. japonicum* complex and evolved from an early ancestor on the Indian Plate. Molecular data from two populations of *S. sinensium* have shown these to be very distinct from one another and that they do indeed form a sister clade to the *S. japonicum* group (Agatsuma *et al.* in press *a*). Although little is yet known about this group, it appears to resemble the *S. japonicum* group in constituting a limited radiation in allopatry despite the vast radiation experienced by the host group. As suggested above for the triculine-borne members of the *S. japonicum* group, discovery of further taxa in the *S. sinensium* group is likely.

The Paragonimidae

Sister taxa for the paragonimids have not been identified with certainty. All paragonimids, where known, parasitise caenogastropods of the superfamilies Cerithioidea and Rissooidea (Fig. 1). Likely sister families, according to Odening (1974) are the Troglotrematidae (in which paragonimids are sometimes placed) and the Nanophyetidae. Snail hosts are known for only two species in the former family and are rissooidean amnicolids (genus *Bythinella*) in each case. Hosts for the latter family include *Semisulcospira*, *Juga*, *Oxytrema* (cerithioideans) and *Campeloma* (ampullarioidean). Species of *Paragonimus* and *Euparagonimus* fall into two clades in Fig. 4. One utilises cerithioidean snails, the other rissooideans. If nanophyetids and troglotrematids are valid sisters, then the earliest paragonimids could have utilised either cerithioidean or rissooidean snails.

One of the two clades contains *P. westermani* and the closely related *P. siamensis* as well as (tentatively) the genus *Euparagonimus* (Fig. 4). *Paragonimus westermani* uses cerithioidean snails and we have inferred that this is also the likely host group for *Euparagonimus cenocopiosus* (unpublished). Yaemput, Dekumyoy & Visiassuk (1994) proposed that the host for *P. siamensis* is the viviparid *Filopaludina martensi*, a member of another caeno-

gastropod superfamily, the Ampullarioidea. This is unlikely given the host relations of other members of the clade, but not impossible. The second main clade contains all remaining *Paragonimus* species, all of which utilise rissooidean snails.

One of the paragonimids in the east Asian radiation utilising pomatiopsids requires special mention. '*Paragonimus skrjabini*' in China must represent a complex of very closely related taxa at or below the level of species. Barring misidentification of cercariae, at least 33 nominal species of triculine and amnicolid snails have been reported as hosts (Davis *et al.* 1994; Wilke *et al.* 2000). Ecological factors and/or isolation have probably led to extensive allopatric radiations of the snails, especially the triculines, and Davis *et al.* (1999) suggested that 50 % of the relevant species have yet to be described. Genetic differences exist among geographically distinct populations of the parasite (Blair *et al.* unpublished), presumably mirroring the situation among the snail populations. We are probably seeing incipient allopatric speciation in the *P. skrjabini* group as a consequence of cospeciation with the diverging host snails.

We can now also identify an instance of host addition or switching in the *P. skrjabini* group. The defining situation was the discovery in Fujian, China, of putative *P. skrjabini* transmitted in microsympatry by two snail species, one a species of *Tricula* (Pomatiopsidae: Triculinae), the other a species of *Erhaia* (replaces *Pseudobythinella* as used in China, a junior homonym of an English fossil rissooidean taxon; see Davis & Kang, 1995), until recently classified in the Pomatiopsidae (e.g. Davis *et al.* 1999) but now, on the basis of molecular data, assigned to the Amnicolidae (Wilke *et al.* 2000). Species of *Erhaia* range from northern India into Yunnan, Hubei, Hunan and Fujian Provinces in China. They are similar in size (very small) and choice of habitat (small mountain streams) to species of *Tricula* and some other triculine genera. Molecular data involving DNA sequences of three genes have shown not only that *Erhaia* is an amnicolid but also that it is closely related to *Moria* in Japan (Wilke *et al.* 2000) that also transmits a species of *Paragonimus*. We are currently attempting to obtain molecular information on the *Paragonimus* species utilising these different hosts in Fujian. If two different sibling species of parasite are involved, then the field observations are best interpreted as a host switch. If a single form of *P. skrjabini* occurs in both snails, this is an example of host addition. In either case it is interesting that the host extension has been to an available rissooid-grade snail in the very limited ecological setting necessary for the transmission of *P. skrjabini*.

Amnicolid snails were able to colonise the southern parts of the Japanese archipelago while triculines did not. *Paragonimus miyazakii* is transmitted in Japan

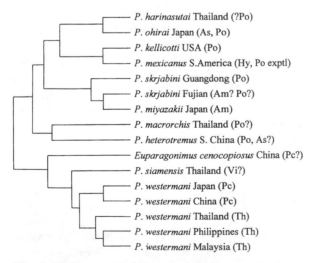

Fig. 4. Phylogeny (cladogram) of the Paragonimidae based on recent molecular studies by Blair and coworkers. The tree is a summary of data from two gene regions, the mitochondrial *cox*1 gene and the nuclear second ribosomal internal transcribed spacer. The tree is midpoint-rooted. Localities are indicated. Snail families are as follows: Am, Amnicolidae; As, Assimineidae; Hy, Hydrobiidae; Pc, Pleuroceridae; Po, Pomatiopsidae; Th, Thiaridae; Vi, Viviparidae. A question mark (?) indicates a tentative host assignment.

by the amnicolid genus *Moria*. Available molecular data are unambiguous in placing *P. miyazakii* as a close sister to *P. skrjabini* (see Blair *et al.* 1999). Furthermore, sequences from at least one form of nominal *P. skrjabini* from Fujian are more similar to those of *P. miyazakii* than they are to nominal *P. skrjabini* from southwestern China. Various scenarios leading to this situation can be imagined. For example, a Chinese population of '*P. skrjabini*' might have extended its range to Japan by capturing an amnicolid host there, diverged in allopatry to become *P. miyazakii*, and subsequently re-established in China in *Erhaia* in sympatry with local *P. skrjabini*. However, any acceptable interpretation must await more data.

Molluscan host relations in *Paragonimus westermani* (Fig. 4) probably provide another example of host addition. *Paragonimus westermani* populations in east Asia utilise pleurocerids and those in southeast Asia utilise thiarids. At first glance, these relations might be taken as an example of phylogenetic tracking with host taxa and parasite taxa associated since the origins of both. However, relative taxonomic levels of parasite (species or species group) and host (family) argue against this. Host-addition as defined earlier, followed by genetic divergence towards speciation in allopatry, appear to offer a better explanation. Molecular genetic studies show that populations of *P. westermani* utilising thiarid snails (that arrived in southeast Asia via a generalised track from Africa–India) and those utilising pleurocerids (coming from North America along a generalised track into Japan–Korea to Taiwan and southern China) are now as distinct from each other as are a number of other species in the genus (Blair *et al.* 1997a). Host addition is characterized by allopatric distributions of the host taxa. Thiarids and pleurocerids are both distributed widely but their range of overlap is limited to parts of southern and southwestern China. Other examples of probable host addition may be seen among paragonimids. For example, the widespread species *P. ohirai*, occurring in both fresh and brackish waters of coastal east Asia, uses several host species within the families Assimineidae and Pomatiopsidae. These host species do not co-occur in any locality where *P. ohirai* is found. Experimental infections can extend the snail host range of *P. ohirai* to include several pomatiopsids not known to be natural hosts, including an American species, *Pomatiopsis lapidaria* (reviewed in Blair, Xu & Agatsuma, 1999).

SYNTHESIS

Although phylogenies of schistosomatids and paragonimids appear to exhibit broad congruence with those of their snail hosts, there are differences between the two families of digeneans, notably in the frequency of host additions. For the Schisto-somatidae as a whole, phylogenetic tracking seems to explain the patterns of host usage (Fig. 2). There may have been switches among lymnaeids, physids and planorbids, often apparently at a shallow phylogenetic level with respect to parasites. It is understandable that switches might occur among related families such as these (all within the Basommatophora), which presumably have a high degree of genetic similarity. Nevertheless, there have also been longer-range switches on at least two occasions into marine gastropods, and a wide range of marine families have been exploited (Figs 1, 2), although generally only by a single species or genus in each case. The processes driving this are discussed below. Within *Schistosoma*, the situation is confused by uncertainty about the origins and evolution of the genus. Association with pulmonates is evident, but the extent of host switching implied will depend on which evolutionary model for the genus is accepted. One major host switch, between a pulmonate and caenogastropod has certainly occurred, as discussed earlier. Host addition beyond the level of a single snail genus is not apparent in schistosomatid species with the possible exception of members of *Austrobilharzia*.

In *Paragonimus*, various host switches at the level of family or superfamily can also be observed (Fig. 4). Basal in the family, there must have been a switch between a cerithioidean and a rissooidean host (assuming that both superfamilies were in existence before the evolution of *Paragonimus*). Thereafter, all paragonimids remained faithful to hosts of the same grade of organisation (cerithioideans or rissooideans) and did not exhibit switches over vast phylogenetic spans as schistosomatids did. One large clade remained faithful to pomatiopsids in Asia, but switched to another rissooidean family in Central/South America. (Rissooideans hosting Paragonimus species in Central and South America should probably be referred to the family Cochliopidae rather than to the Hydrobiidae as is customary: Davis unpublished.) Unlike the situation for single species or species complexes of schistosomatids, clear host additions by single paragonimid species or complexes are observed at the level of host family. Within the *P. skrjabini* group, amnicolid snails have been added to the usual pomatiopsid hosts. *Paragonimus ohirai*, for which pomatiopsids are presumably plesiomorphic hosts, can now utilise both pomatiopsids and assimineids within a small geographical area. Within the *P. westermani* group, two different cerithioidean families are used, indicating another addition event.

Another difference between *Paragonimus* and *Schistosoma* concerns the extent of radiations in a given geographic area. A relatively large number of closely related *Paragonimus* species occur in east Asia (Blair, Xu & Agatsuma, 1999). Many of these utilise pomatiopsid snails and some occur in sympatry where they may utilise the same snail species (yet to be confirmed). This might represent an adaptive radiation. However, we are mindful of the difficulties of confirming the 'adaptive' aspect of a radiation. The term 'evolutionary radiation' might be more appropriate (Futuyma, 1998). This pattern of distribution and host use is quite different from that seen in, for example, the *Schistosoma japonicum* group. The three known species in the *S. japonicum* group do not overlap geographically or in host use and probably represent a limited radiation in allopatry. If members of the *S. japonicum* group have tracked the pomatiopsids, many instances of parasite extinction, or of 'missing the boat' (Paterson & Gray, 1997), must be postulated: there are three schistosome species, each utilising one snail species, and at least 120 pomatiopsid species in east Asia, all but two belonging to the Triculinae. However, ecological factors, discussed below, might help explain the abundance of schistosome-free pomatiopsid species.

Given that host-extension is possible, why have schistosomatids and paragonimids 'chosen' the species they have? All else being equal, we should expect a parasite to involve a closely related host in a host switch/addition. Another important factor is ecological opportunity. For example, the *Paragonimus skrjabini* complex has added to its snail hosts the amnicolid *Erhaia*, a genus ecologically very similar to its widespread triculine hosts. Similarly, *Semisulcospira* (Pleuroceridae) and *Brotia* (Thiaridae), both hosts for *Paragonimus westermani* in different parts of Asia, are similar in their habitat preference and ecology. Ecological factors may also prevent certain phylogenetically-appropriate snails from acting as hosts. Successful transmission of a digenean requires a setting where the definitive host can become exposed to infection. For example, the reason why members of the enormous radiation of Stylommatophora are impervious to infection is, simply, that they are all terrestrial. Schistosomatid transmission requires an aquatic environment for miracidia and cercariae. The triculine radiations of the Pachydrobiini and Jullieniini fill every conceivable ecological niche in rivers or in deep lakes, niches unavailable to potential mammalian hosts. Thus, most triculines are ecologically removed from schistosome transmission.

We need to ask why so few schistosomatids and no paragonimids occur in strictly marine habitats. Of the approximately 22 families of freshwater snails worldwide (Davis, 1982; Wilke *et al.* 2000), schistosomatids and paragonimids both utilise about six. In the case of paragonimids, the families are scattered throughout the Cerithioidea and Rissooidea, both superfamilies that also contain many marine families. The phylogenetic range of schistosomatid hosts in freshwater is even greater, encompassing both heterobranchs and caenogastropods (Fig. 1). There

are therefore no absolute phyletic constraints preventing colonisation of marine gastropod taxa by these parasite families. The explanation for their restriction to freshwater habitats is more likely to be that the physiology of the parasites, or host specificity with regard to non-molluscan hosts, is responsible. When a switch to a marine host has been made, the host family exploited is often phyletically removed from freshwater host families. For example, *Paragonimus ohirai* utilises the family Pomatiopsidae in freshwater habitats, but populations cycling through brackish water habitats exploit assimineid snails. Where multiple marine families are used by a single schistosomatid genus, they can be unrelated to one another, suggesting multiple colonisations. For example, species of *Gigantobilharzia* generally occur in planorbids in freshwater but one, *G. huttoni*, utilises a marine opisthobranch and cercariae ascribed to this genus have been found in the brackish-water snail *Hydrobia stagnalis* (Rissooidea). (Note, however, that there is confusion concerning this name which probably refers to a member of the Hydrobiidae, but has also been used for a species within the Cochliopidae). Cercariae of *Austrobilharzia* have been reported from five families of marine snails in three superfamilies (Fig. 1). The marine species into which schistosomatids have switched generally occur in huge numbers in shallow and intertidal coastal areas where the avian definitive hosts are abundant. Ecology and opportunity are presumably linked here and phylogeny is of less importance in the selection of snail host.

It might be expected that a switch into a new host lineage and habitat would open up new opportunities for a parasite lineage and that a radiation would ensue. This does not seem to be the case with *Austrobilharzia* + *Ornithobilharzia* which today constitutes small group of about 10 recognized species (Basch, 1991). This is not a large number considering the opportunities that the marine habitat would seem to offer and the high densities of snails found in many inshore localities. However, the relevant bird hosts are highly vagile: their ability to traverse the globe might militate against parasite speciation. Even in freshwaters, where habitats are fragmented and hosts less vagile, extensive radiations are not seen in schistosomes. Neither the *S. japonicum* group nor the *S. sinensium* group has yielded more than a handful of known species despite their geographic ranges and the extensive radiations experienced within a similar time frame by the snail host groups.

One family of snails, the Pomatiopsidae (range also includes North America), is host to both schistosomatids and paragonimids. It may be that this family is particularly permissive as a host for digeneans and hence unusually prone to host extension. In particular, the genus *Oncomelania* appears to be a permissive host for many species of *Paragonimus* to which it is not normally exposed (reviewed in Blair, Xu & Agatsuma, 1999). Japanese *Oncomelania hupensis nosophora* is susceptible to experimental infection by two American species, *P. kellicotti* (normal host *Pomatiopsis lapidaria* also in the Pomatiopsidae) and *P. mexicanus* (normal host *Aroapyrgus* spp. in the Hydrobiidae *sensu lato*). Only a small proportion of snails yielded mature infections in the latter case. Asian species for which *O. hupensis nosophora* can act as an experimental host include *P. ohirai* (see above), *P. miyazakii* (normal host in the Amnicolidae) and *P. heterotremus* (normal hosts also in the Pomatiopsidae). The American species *P. lapidaria* is also permissive for *S. japonicum*. Further studies on the digenean faunas of pomatiopsids might provide valuable data for those interested in host specificity and host extension.

The absence of parasite species-specific internal defence mechanisms in snails may mean that digeneans are not bound by restrictions imposed by a coevolutionary arms race. Consequently, it may be easier than previously assumed for digeneans to switch to host taxa with convergently similar defence mechanisms (Adema & Loker, 1997), or indeed, to host taxa with similar ecologies but not necessarily similar phylogenies. Our thinking about the evolutionary interactions of digeneans and snails has been heavily influenced in the past by assumptions of strict coevolution. Many studies on coevolution have focused on vertebrates that have immune systems able to focus on individual species of pathogen. We predict that coevolutionary studies on invertebrates (and plants) and their parasites/associates will find many examples of phylogenetic tracking rather than strict coevolution. The extent of reciprocal evolutionary responses by the host against any particular associate will depend on the relative prevalence of that associate.

ACKNOWLEDEMENTS

A review of this depth would not have been possible without the benefit of research and resources at the TMRC. Shanghai supported by USA., NIH grant AI39461. We also wish to acknowledge information from the Host-Parasite Checklist maintained by the staff of the Parasitic Worms section, Natural History Museum, London. For unpublished information, we thank Dr T. Agatsuma, Kochi Medical School, Japan. Critical comments on drafts were provided by Dr E. S. Loker, Dr J. A. T. Morgan and Dr T. Platt.

REFERENCES

ADEMA, C. M., HERTEL, L. A. & LOKER, E. S. (1999). Evidence from two planorbid snails of a complex and dedicated response to digenean (echinostome) infection. *Parasitology* **119**, 395–404.

ADEMA, C. M. & LOKER, E. S. (1997). Specificity and immunobiology of larval digenean–snail associations. In *Advances in Trematode Biology* (ed. Fried, B. & Graczyk, T. K.), pp. 229–263. Baton Rouge & New York, CRC Press.

AGATSUMA, T., IWAGAMI, M., LIU, C. X., SAITOH, Y., KAWANAKA, M., UPATHAM, S., QIU, D. & HIGUCHI, T. (in press *a*). Molecular phylogenetic position of *S. sinensium* in the genus *Schistosoma*. *Journal of Helminthology*.

AGATSUMA, T., IWAGAMI, M., LIU, C. X., RAJAPAKSE, R. P. V. J., MONDAL, M. M. H., KITIKOON, V., AMBU, S., AGATSUMA, Y., BLAIR, D. & HIGUCHI, T. (in press *b*). Affinities between Asian non-human *Schistosoma* species, the *S. indicum* group, and the African human schistosomes. *Journal of Helminthology*.

BARKER, S. C. & BLAIR, D. (1996). Molecular phylogeny of *Schistosoma* species supports traditional groupings within the genus. *Journal of Parasitology* **82**, 292–298.

BASCH, P. F. (1991). *Schistosomes: Development, Reproduction and Host Relations*. New York and Oxford, Oxford University Press.

BAUGH, S. C. (1977). On the systematic position of the Thailand blood fluke *Orientobilharzia harinasutai* Kruatrachue, Bhaibulaya and Harinasuta, 1965. In *Excerta parasitológia en memoria del Doctor Eduardo Caballero y Caballero*, pp. 121–125. Mexico, Universidad Nacional Autónoma de México.

BLAIR, D., AGATSUMA, T., WATANOBE, T., OKAMOTO, M. & ITO, I. (1997*a*). Geographical genetic structure within the human lung fluke, *Paragonimus westermani*, detected from DNA sequences. *Parasitology* **115**, 411–417.

BLAIR, D., VAN HERWERDEN, L., HIRAI, H., TAGUCHI, T., HABE, S., HIRATA, M., LAI, K., UPATHAM, S. & AGATSUMA, T. (1997*b*). Relationships between *Schistosoma malayensis* and other Asian schistosome species deduced from DNA sequences. *Molecular and Biochemical Parasitology* **85**, 259–263.

BLAIR, D., WU, B., CHANG, Z. S., GONG, X., AGATSUMA, T., ZHANG, Y. N., CHEN, S. H., LIN, J. X., CHEN, M. G., WAIKAGUL, J., GUEVARA, A. G., FENG, Z. & DAVIS, G. M. (1999). A molecular perspective on the genera *Paragonimus* Braun, *Euparagonimus* Chen and *Pagumogonimus* Chen. *Journal of Helminthology* **73**, 295–300.

BLAIR, D., XU, Z. B. & AGATSUMA, T. (1999). *Paragonimus* and paragonimiasis. *Advances in Parasitology* **42**, 113–222.

BROOKS, D. R. & MCLENNAN, D. A. (1991). *Phylogeny, Ecology and Behavior: A Research Program in Comparative Biology*. Chicago & London, University of Chicago Press.

CABLE, R. M. & PETERS, L. E. (1986). The cercaria of *Allocreadium ictaluri* Pearse (Digenea: Allocreadiidae). *Journal of Parasitoloty* **72**, 369–371.

CAMPBELL, G., JONES, S. J., LOCKYER, A. E., HUGHES, S., BROWN, D., NOBEL, L. R. & ROLLINSON, D. (2000). Molecular evidence supports an African affinity of the Neotropical freshwater gastropod, *Biomphalaria glabra*, Say 1818, an intermediate host for *Schistosoma mansoni*. *Proceedings of the Royal Society of London B* **267**, 2351–2358.

COX, C. B. (2000). Plate tectonics, seaways and climate in the historical biogeography of mammals. *Memorias do Instituto Oswaldo Cruz, Rio de Janeiro* **95**, 509–516.

CRIBB, T. H., BRAY, R. A., LITTLEWOOD, D. T. J., PICHELIN, S. P. & HERNIOU, E. A. (2001). The Digenea. In *Interrelationships of the Platyhelminthes* (ed.

Littlewood, D. T. J. & Bray, R. A.), pp. 168–185. London, Taylor & Francis.

CROIZAT, L. (1958). *Pan-Biogeography. Volume 1. The New World.* Published by the author, Caracas, Venezuela.

DAVIS, G. M. (1979). The origin and evolution of the gastropod family Pomatiopsidae, with emphasis on the Mekong River Triculinae. *Monograph of the Academy of Natural Sciences of Philadelphia* **20**, 1–120.

DAVIS, G. M. (1980). Snail hosts of Asian *Schistosoma* infecting man: origin and coevolution. In *The Mekong Schistosome. (Malacological Review, Suppl. 2)* (ed. Bruce, J. and Sornmani, S.), pp. 195–238. Ann Arbor, Michigan.

DAVIS, G. M. (1981). Different modes of evolution and adaptive radiation in the Pomatiopsidae (Prosobranchia: Mesogastropoda). *Malacologia* **21**, 209–262.

DAVIS, G. M. (1982). Historical and ecological factors in the evolution, adaptive radiation and biogeography of freshwater mollusks. *American Zoologist* **22**, 375–395.

DAVIS, G. M. (1992). Evolution of prosobranch snails transmitting Asian *Schistosoma*; coevolution with *Schistosoma*: A review. *Progress in Clinical Parasitology* **3**, 145–204.

DAVIS, G. M., CHEN, C. E., KANG, Z. B. & LIU, Y. Y. (1994). Snails hosts of *Paragonimus* in Asia and Americas. *Chinese Journal of Parasitology and Parasitic Diseases* **12**, 279–284.

DAVIS, G. M. & GREER, G. (1980). A new genus and two new species of Triculinae (Gastropoda: Prosobranchia) and the transmission of a Malaysian mammalian *Schistosoma* sp. *Proceedings of the Academy of Natural Sciences of Philadelphia* **132**, 245–276.

DAVIS, G. M. & KANG, Z. B. (1995). Advances in the systematics of *Erhaia* (Gastropoda: Pomatiopsidae) from the People's Republic of China. *Proceedings of the Academy of Natural Sciences of Philadelphia* **146**, 391–427.

DAVIS, G. M., KITIKOON, V. & TEMCHAROEN, P. (1976). Monograph on "*Lithoglyphopsis*" *aperta*, the snail host of Mekong River schistosomiasis. *Malacologia* **15**, 241–287.

DAVIS, G. M., SPOLSKY, C. M. & ZHANG, Y. (1995). Malacology, biotechnology and snail-borne diseases. *Acta Medica Philippina* **31**, (Ser. 2) 45–60.

DAVIS, G. M., WILKE, T., SPOLSKY, C. M., QIU, C. P., QIU, D., XIA, M. Y., ZHANG, Y. & ROSENBERG, G. (1998). Cytochrome oxidase I-based phylogenetic relationships among the Pomatiopsidae, Hydrobiidae, Rissoidae, and Truncatellidae (Gastropoda: Caenogastropoda: Rissoacea). *Malacologia* **40**, 251–266.

DAVIS, G. M., WILKE, T., ZHANG, Y., XU, X., QIU, C. P., SPOLSKY, C. M., QIU, D., LI, Y. S., XIA, M. Y. & FENG, Z. (1999). Snail–*Schistosoma, Paragonimus* interactions in China: Population ecology, genetic diversity, coevolution and emerging diseases. *Malacologia* **41**, 355–377.

DESPRÉS, L., IMBERT-ESTABLET, D., COMBES, C. & BONHOMME, F. (1992). Molecular evidence linking hominid evolution to recent radiation of schistosomes (Platythelminthes: Trematoda). *Molecular Phylogenetics and Evolution* **1**, 295–304.

DESPRÉS, L., IMBERT-ESTABLET, D. & MONNEROT, M. (1993). Molecular characterization of mitochondrial DNA provides evidence for the recent introduction of *Schistosoma* into America. *Molecular and Biochemical Parasitology* **60**, 221–230.

DESPRÉS, L., KRUGER, F. J., IMBERT-ESTABLET, D. & ADAMSON, M. L. (1995). ITS2 ribosomal RNA indicates *Schistosoma hippopotami* is a distinct species. *International Journal for Parasitology* **25**, 1509–1514.

FUTUYMA, D. J. (1998). *Evolutionary Biology* (3rd Edition). Sunderland, Massachusetts, Sinauer Associates Inc.

GIBSON, D. I. (1987). Questions in digenean systematics and evolution. *Parasitology* **95**, 429–460.

GIBSON, D. I. & BRAY, R. A. (1994). The evolutionary expansion and host–parasite relationships of the Digenea. *International Journal for Parasitology* **24**, 1213–1226.

HIGGINS, D. G., THOMPSON, J. D. & GIBSON, T. J. (1996). Using CLUSTAL for multiple sequence alignments. *Methods in Enzymology* **266**, 383–402.

HIRAI, H., TAGUCHI, T., SAITOH, T., KAWANAKA, M., SUGIYAMA, H., HABE, S., OKAMOTO, M., HIRATA, M., SHIMADA, M., TIU, W. U., LAI, K., UPATHAM, E. S. & AGATSUMA, T. (2000). Chromosomal differentiation of the *Schistosoma japonicum* complex. *International Journal for Parasitology* **30**, 441–452.

KØIE, M. (1982). The redia, cercaria and early stages of *Aporocotyle simplex* Odhner, 1900 (Sanguinicolidae): a digenetic trematode which has a polychaete annelid as the only intermediate host. *Ophelia* **21**, 115–145.

KOLLMANN, H. A. & YOCHELSON, E. L. (1976). Survey of Paleozoic gastropods possibly belonging to the subclass Opisthobranchia. *Annalen des Naturhistorischen Museums in Wien* **80**, 207–220.

KUMAR, S., TAMURA, K., JACOBSEN, I. & NEI, M. (2000). MEGA2: Molecular Evolutionary Genetics Analysis, version 2.0. Pennsylvania and Arizona State Universities, University Park, Pennsylvania and Tempe, Arizona.

LE, T. H., BLAIR, D., AGATSUMA, T., HUMAIR, P.-F., CAMPBELL, N. J. H., IWAGAMI, M., LITTLEWOOD, D. T. J., PEACOCK, B., JOHNSTON, D. A., BARTLEY, J., ROLLINSON, D., HERNIOU, E. A., ZARLENGA, D. S. & McMANUS, D. P. (2000). Phylogenies inferred from mitochondrial gene orders–a cautionary tale from the parasitic flatworms. *Molecular Biology and Evolution* **17**, 1123–1125.

McCARTHY, A. M. (1990). Speciation of echinostomes: evidence for the existence of two sympatric sibling species in the complex *Echinoparyphium recurvatum* (von Linstow 1873) (Digenea: Echinostomatidae). *Parasitology* **101**, 35–42.

MODE, C. J. (1958). A mathematical model for the co-evolution of obligate parasites and their hosts. *Evolution* **12**, 158–165.

NEWTON, R. B. (1920). On some freshwater fossils from central South Africa. *Annals and Magazine of Natural History* **V** (9th Series), 241–249.

ODENING, K. (1974). Verwandschaft, System und Zyklo-Ontogenetische Besonderheiten der Trematoden. *Zoologische Jahrbücher* **101**, 343–396.

PARODIZ, J. J. (1969). The Tertiary non-marine Mollusca of South America. *Annals of the Carnegie Museum* **40**, 1–242.

PATERSON, A. M. & GRAY, R. D. (1997). Host-parasite co-speciation, host switching, and missing the boat. In *Host–Parasite Evolution: General Principles and Avian Models* (ed. Clayton, D. H. & Moore, J.), pp. 236–250. Oxford, Oxford University Press.

PIERCE, H. G. (1993). The non-marine mollusks of the late Oligocene–early Miocene Cabbage Patch Fauna of western Montana III. Aquatic mollusks and conclusions. *Journal of Paleontology* **67**, 980–993.

PILSBRY, H. A. (1911). Non-marine Mollusca of Patagonia. In *Reports of the Princeton University Expeditions to Patagonia, 1896–1899* (ed. Scott, W. B.) 3(2), part 5, pp. 513–633. Princeton, New Jersey.

PLATT, T. R., BLAIR, D., PURDIE, J. & MELVILLE, L. (1991). *Griphobilharzia amoena* n. gen., n. sp. (Digenea: Schistosomatidae), a parasite of the freshwater crocodile *Crocodylus johnstoni* (Reptilia: Crocodylia) from Australia, with the erection of a new subfamily, Griphobilharziinae. *Journal of Parasitology* **77**, 65–68.

PONDER, W. H. (1988). The truncatelloidean (= rissoacean) radiation – a preliminary phylogeny. In *Prosobranch Phylogeny* (*Malacological Review* (*Suppl.* 4) (ed. Ponder, W. F.), 129–164. Ann Arbor, Michigan.

PONDER, W. H. & LINDBERG, D. (1997). Towards a phylogeny of gastropod molluscs: an analysis using morphological characters. *Zoological Journal of the Linnean Society* **119**, 83–265.

POULIN, R., STEEPER, M. J. & MILLER, A. A. (2000). Non-random patterns of host use by the different parasite species exploiting a cockle population. *Parasitology* **121**, 289–295.

ROLLINSON, D. & SIMPSON, A. J. G. (eds.). (1987). *The Biology of Schistosomes from Genes to Latrines*. London, Academic Press.

ROLLINSON, D. & SOUTHGATE, V. R. (1987). The genus *Schistosoma*: a taxonomic appraisal. In *The Biology of Schistosomes from Genes to Latrines* (ed. Rollinson, D. & Simpson, A. J. G.), pp. 1–49. London, Academic Press.

SCOTT, M. E., RAU, M. E. & MCCLAUGHLIN, J. D. (1982). A comparison of aspects of the biology of two subspecies of *Typhlocoelum cucumerinum* (Digenea: Cyclocoelidae) in three families of snails (Physidae, Lymnaeidae and Planorbidae). *International Journal for Parasitology* **12**, 123–133.

SMITH, A. G., SMITH, D. G. & FUNNELL, B. M. (1994). *Atlas of Mesozoic and Cenozoic Coastlines*. Cambridge, Cambridge University Press.

SMITH, J. W. (1997). The blood flukes (Digenea: Sanguinicolidae and Spirorchidae) of cold-blooded vertebrates: Part 1. A review of the literature published since 1971 and bibliography. *Helminthological Abstracts* **66**, 255–294.

SNYDER, S. D. & LOKER, E. S. (2000). Evolutionary relationships among the Schistosomatidae (Platyhelminthes: Digenea) and an Asian origin for *Schistosoma*. *Journal of Parasitology* **86**, 283–288.

WENZ, W. (1939). Gastropoda: *Handbuch der Palaeozoologie*. 1 (Lief) 1–7, pp. 1–1639. Berlin-Zehlendorf, Verlag von Gebruder Borntrager.

WILKE, T., DAVIS, G. M., GONG, X. & LIU, H. X. (2000). *Erhaia* (Gastropoda: Rissooidea): phylogenetic relationships and the question of *Paragonimus* coevolution in Asia. *American Journal of Tropical Medicine and Hygiene* **62**, 453–549.

WOODRUFF, D. S. & MULVEY, M. (1997). Neotropical schistosomiasis: African affinities of the host snail *Biomphalaria glabrata* (Gastropoda: Planorbidae). *Biological Journal of the Linnean Society* **60**, 505–516.

WRIGHT, C. A. & SOUTHGATE, V. R. (1981). Coevolution of digeneans and molluscs, with special reference to schistosomes and their intermediate hosts. Pages 191–205. In *The Evolving Biosphere (Chance, Change and Challenge)* (ed. Forey, P. L.), pp. 191–205. Cambridge, Cambridge University Press.

XIA, X. (2000). *Data Analysis in Molecular Biology and Evolution*. Boston/Dordrecht/London, Kluwer Academic Publishers.

YAEMPUT, S., DEKUMYOY, P. & VISIASSUK, K. (1994). The natural first intermediate host of *Paragonimus siamensis* (Miyazaki and Wykoff, 1965) in Thailand. *Southeast Asian Journal of Tropical Medicine and Public Health* **25**, 284–290.

Interactions between intermediate snail hosts of the genus *Bulinus* and schistosomes of the *Schistosoma haematobium* group

D. ROLLINSON*, J. R. STOTHARD *and* V. R. SOUTHGATE

Wolfson Wellcome Biomedical Laboratories, Department of Zoology, The Natural History Museum, Cromwell Road, London SW7 5BD

SUMMARY

Within each of the four species groups of *Bulinus* there are species that act as intermediate hosts for one or more of the seven species of schistosomes in the *Schistosoma haematobium* group, which includes the important human pathogens *S. haematobium* and *S. intercalatum*. *Bulinus* species have an extensive distribution throughout much of Africa and some surrounding islands including Madagascar, parts of the Middle East and the Mediterranean region. Considerable variation in intermediate host specificity can be found and differences in compatibility between snail and parasite can be observed over small geographical areas. Molecular studies for detection of genetic variation and the discrimination of *Bulinus* species are reviewed and two novel assays, allele-specific amplification (ASA) and SNaPshot®, are introduced and shown to be of value for detecting nucleotide changes in characterized genes such as cytochrome oxidase 1. The value and complexity of compatibility studies is illustrated by case studies of *S. haematobium* transmission. In Senegal, where *B. globosus*, *B. umbilicatus*, *B. truncatus* and *B. senegalensis* may act as intermediate hosts, distinct differences have been observed in the infectivity of different isolates of *S. haematobium*. In Zanzibar, molecular characterization studies to discriminate between *B. globosus* and *B. nasutus* have been essential to elucidate the roles of snails in transmission. *B. globosus* is an intermediate host on Unguja and Pemba. Further studies are required to establish the intermediate hosts in the coastal areas of East Africa. Biological factors central to the transmission of schistosomes, including cercarial emergence rhythms and interactions with other parasites and abiotic factors including temperature, rainfall, water velocity, desiccation and salinity are shown to impact on the intermediate host-parasite relationship.

Key words: *Bulinus*, *Schistosoma haematobium* group, compatibility, transmission, molecular characterization.

INTRODUCTION

The interactions observed between snails of the genus *Bulinus* and parasites within the *Schistosoma haematobium* group provide an excellent example of how host-parasite compatibility may change over space and time. The co-evolution of these host-parasite relationships is reflected by the present day geographical distribution of human and animal schistosomiasis caused by *S. haematobium* and related parasites in Africa and adjacent regions. Observations on parasites and snails isolated from different parts of their ranges reveal that the interplay between them has resulted in the emergence of many different intermediate host specificities. This applies not only on a large geographical scale but may be observed at a local level, with parasites being generally more infective to sympatric hosts than to allopatric hosts of the same species.

Perhaps to enhance the genetic diversity required for selection processes to act upon, both snails and parasites have reproductive strategies that are striking in their complexity. *Bulinus* are freshwater planorbids and, for the most part, are true hermaphrodites being able to outcross as both male and female. They can also self-fertilise and some species preferentially appear to do so, whereas others predominantly outcross. One of the four *Bulinus* species groups exists in different ploidy states, from diploids, tetraploids, hexaploids to octoploids, and some populations consist of aphallic individuals that are unable to cross-fertilise as males. Most digenean parasites are hermaphrodite but schistosomes are unusual in possessing separate sexes. Male and female worms pair in the vasculature system of the definitive host, and intriguing mating strategies have evolved that may result in mating competition and even hybridization between different species in the *S. haematobium* group. In the snail host, schistosomes multiply prodigiously and asexually placing considerable demands on the snail's resources. Place this complicated scenario of multiplication, gene flow and development against a continually changing environmental backdrop, with temperature and water levels oscillating through the seasons, together with various ecological parameters including predators and other parasites and it becomes apparent that multiple factors are involved in shaping the interactions and evolution of the host-parasite relationship.

Currently some 37 species of *Bulinus* are recognized (Brown, 1994), which have been divided into

* E-mail: D. Rollinson@nhm.ac.uk

Parasitology (2001), **123**, S245–S260. © 2001 Cambridge University Press
DOI: 10.1017/S0031182001008046

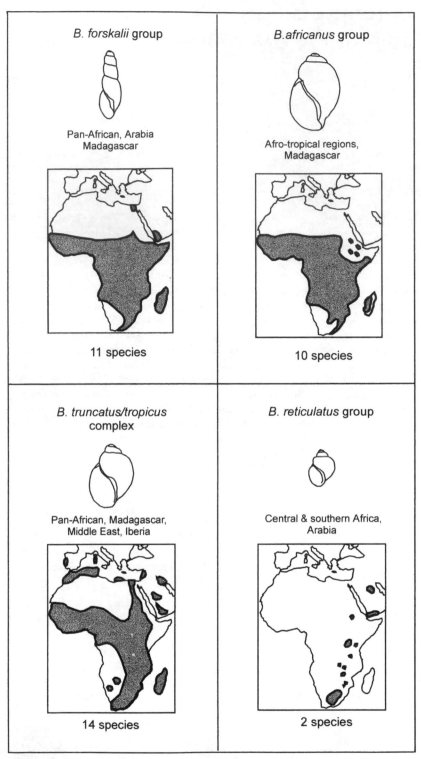

Fig. 1. Sketch map of the general distribution of species groups of *Bulinus* in Africa, the Middle East and in the Mediterranean Basin area based upon the maps of Brown (1994). A total of 37 species is presently recognized which are partitioned into four species groups. A shell, representing the characteristic adult morphology of a typical species within each of the species groups is presented and scaled accordingly; the shell of an adult rarely exceeds 2·5 cm in length and is usually between 0·5–2 cm. Within each of the species groups, there are snails capable of natural transmission of human schistosomes.

four species groups. The *B. forskalii* group contains 11 species with slender shells and usually high spires and is practically pan-African in distribution with species occurring on some of the surrounding islands and the Arabian peninsula. The *B. africanus* group

has 10 species confined to the Afrotropical region. The *B. truncatus/tropicus* complex, which contains polyploid species, is again pan-African with 14 representatives extending into the Middle East, Mediterranean islands and the Iberian Peninsula;

the two species included in the *B. reticulatus* group both have restricted distributions (Fig. 1). The species groups are for the most part well differentiated and early enzyme studies revealed large genetic distances within the genus (Biocca *et al.* 1979). The proposal was made to divide the genus into three genera but this classification has not been generally accepted and was considered to have serious disadvantages (Brown, 1981, 1994). Interestingly, substantial nucleotide variation of the internal transcribed spacer (ITSI) of the ribosomal RNA gene (rRNA) also indicated large divergence between representatives of the species groups (Stothard, Hughes & Rollinson, 1996).

Within each of the species group of *Bulinus* there are species that act as intermediate hosts for one or more of the seven species of schistosomes in the *S. haematobium* group in part or all of their geographical range (Rollinson & Southgate, 1987). Two of these species, *S. haematobium* and *S. intercalatum*, are primarily parasites of man. *S. haematobium*, responsible for urinary schistosomiasis, is found in 53 countries in the Middle East and Africa including the islands of Madagascar and Mauritius, the majority of cases being located in the Afrotropical region. *S. intercalatum* causes a form of intestinal schistosomiasis and has a much more restricted distribution primarily in Equatorial Guinea, Zaire, Gabon, São Tomé, Nigeria and Cameroon. Recent studies suggest that the two recognized biological strains of *S. intercalatum*, which display quite different intermediate host specificities, may well be distinct species (Pagès *et al.* 2001). The closely related *S. bovis*, *S. mattheei* and *S. curassoni* can be found in sheep and cattle, whereas *S. margrebowiei* and *S. leiperi* are more commonly found in antelopes. Many molecular studies point to the close relationship of the *S. haematobium* group species (see Rollinson *et al.* 1997*a*) which is further emphasised by observations on natural hybridization between species infecting the same definitive host and the ease at which both pairing and mating take place in experimental studies (see Southgate, Jourdane & Tchuem Tchuenté, 1998). The ability of different species within the *S. haematobium* group to develop in the same species of *Bulinus* in sympatric situations and infect the same definitive host (e.g. *S. haematobium* and *S. mattheei* in *B. globosus* in Zimbabwe and South Africa) creates conditions where hybridization occurs, resulting in gene introgression with the resulting epidemiological consequences (Jourdane & Southgate, 1992).

Unravelling the interplay between snails and schistosomes throughout their range has been central to the understanding of the geographical distribution and transmission of schistosomiasis. In a paper published in 1974 concerning experiments with hybrid schistosomes and observations on the inheritance of infectivity, Wright posed the fundamental question "Snail susceptibility or trematode infectivity?". He showed that the progeny of a cross between *S. mattheei* males X *S. intercalatum* females were equally infective to *B. globosus* and *B. scalaris*, whereas the parental forms were restricted in the case of *S. mattheei* to *B. globosus* and *S. intercalatum* to *B. scalaris* (Wright, 1974). Now we are fully aware that compatibility comprises both infectivity of the parasite and susceptibility of the snail but there is still much to learn concerning genetically-based variation among and within populations of both snail and schistosomes and the genes that influence the host parasite relationship (Woolhouse & Webster, 2000).

Aided by an increasing armoury of molecular and morphological techniques to identify and describe the genetic makeup of parasites and hosts, significant advances in our understanding of the snails responsible for transmission and the epidemiology of *S. haematobium* group schistosomes have been made (see Rollinson *et al.* 1998; Davies *et al.* 1999; Jones *et al.* 1997, 1999). In this paper the identification and discrimination of *Bulinus* species based on techniques of molecular characterization are briefly considered and new methods of detecting known nucleotide changes are introduced. Snails responsible for the transmission of *S. haematobium* are reviewed and examples from our recent studies in both east and west Africa are used to illustrate the importance of compatibility studies in understanding the changing epidemiology of urinary schistosomiasis. Environmental factors important in the transmission of schistosomes are considered and, although the *Bulinus/Schistosoma* model is not as amenable to laboratory study as others, the current state of knowledge concerning the internal defence mechanisms of *Bulinus* is highlighted as this is important in our understanding of snail parasite interactions.

MOLECULAR CHARACTERIZATION OF *BULINUS*

Reliable methods for the differentiation and identification of *Bulinus* species have been sought in order to determine those species and strains playing a major role in schistosomiasis transmission. Early descriptions of taxa relied upon analysis of morphological variation, including observations on shell shape and size, on the radula and on the reproductive system. Knowledge of a particular snail species was often confined to collections of snails from a few localities in one geographical area. These data sets were added to by determination of chromosome number and the use of biochemical characters. Enzyme electrophoresis has been used to supplement such studies and has proved useful for identification, elucidation of relationships between taxa and for studying aspects of reproductive biology and population structure (e.g. Biocca *et al.* 1979; Jelnes, 1986;

Njiokou *et al.* 1993; Rollinson & Southgate, 1979; Rollinson & Wright, 1984). More recently, studies on ribosomal RNA genes (rRNA), random amplified polymorphic DNA (RAPDs) and the mitochondrial gene cytochrome oxidase 1 (CO1) have added further insight into the relationships between species and provided new markers for species differentiation as described below.

DNA probes and microsatellites

Early explorations of DNA methods for identification of *Bulinus* were made by Strahan, Kane & Rollinson (1991) and Rollinson & Kane (1991). Shotgun-cloned genomic DNA from *B. cernicus* was used in both dot and Southern blot methodologies to screen recombinant DNA inserts for 'species' specificity. A representative selection of *Bulinus* species was used to investigate the probes potential for cross hybridisation. Four recombinant clones were characterized; two clones appeared population specific while the remainder were either non-specific or showed some species group resolution (Strahan *et al.* 1991). Rollinson & Kane (1991) investigated the use of restriction fragment length polymorphism (RFLP) analysis and Southern blotting to identify variation within the ribosomal internal transcribed spacer (ITS) between species of *Bulinus*. Clear differences in the sizes of restriction fragments between species of *Bulinus* representing the four species groups were observed when DNA was digested with either *Bam*HI or *Bgl*II and hybridized to the schistosome derived probe pSM889. This probe is a 4·4 kb fragment that encompasses part of the small rRNA, internal transcribed spacer and large rRNA gene. As it contains highly conserved regions it shows homology with a wide variety of organisms and will readily bind to restricted fragments of snail DNA. No differences were observed between samples of *B. tropicus* and *B. truncatus* but intraspecific variation was observed between samples of *B. forskalii* from São Tomé and Angola (Rollinson & Kane, 1991). This work suggested that sequence variation in the spacer regions may be useful for discrimination as was later shown by Stothard *et al.* (1996).

Population genetic analysis of *B. globosus* and *B. truncatus* using human minisatellite sequences in a multi-locus profiling approach was introduced by Jarne *et al.* (1990, 1992). More detailed population studies have been made possible by the development of microsatellites for *Bulinus*. Microsatellites are useful tools for investigating selfing rates in highly inbred species such as selfing snails. Enzyme studies by Njiokou (1993) on the tetraploid *B. truncatus* showed little variation in this predominantly selfing species, but Viard *et al.* (1996) demonstrated higher levels of polymorphism using microsatellites. Low genetic variation was also detected in *B. forskalii*

using allozymes (Mimpfoundi & Greer, 1990), but Gow *et al.* (in press) have recently isolated and characterized eleven polymorphic microsatellites in *B. forskalii*, which will allow further elucidation of the population structure and mating system.

RAPD assays

Langand *et al.* (1993) were the first to apply RAPD-PCR to populations of *Bulinus*. RAPD profiles could differentiate samples of *B. globosus* and *B. umbilicatus*, the number of shared or co-migrating fragments between profiles was 33%. Amplified fragments within a profile varied across populations of *B. forskalii* from West Africa. In a slightly more formal use of RAPD profiles for identification purposes using congruence between different sets of primers and electrophoretic methods, Stothard & Rollinson (1996) showed that in principle, RAPDs would not be ultimately reliable within all groups of the genus. Phenetic analysis of RAPD data showed that distance estimates between species were often non-additive, indicative of non-metric data, resulting in unstable, conflicting dendrograms. Whilst RAPDs could clearly show differences between species, it was unlikely that there would be species-diagnostic profiles for all species (Stothard & Rollinson, 1996). It was concluded that the nucleotide divergence within the genus went beyond the phylogenetic scope of the technique which is limited to samples of less than 10% divergence across actual or potential RAPD priming sites [see simulations of Clark & Lanigan (1993) and Stothard (1997)].

Nevertheless, on a regional level or within a species group of *Bulinus*, RAPDs have been useful especially when used in conjunction with other methods of identification. The levels of heterogeneity detected with RAPDs vary considerably within the species groups of *Bulinus*. For example, there is substantial diversity within and between populations of *B. globosus* (Stothard *et al.* 1997; Davies *et al.* 1999), a member of the *B. africanus* group. However, in a recent survey of *B. truncatus* (*B. truncatus/tropicus*) complex species from Sudan nearly all snails inspected had identical RAPD profiles (J. R. Stothard, unpublished data). RAPDs have been used in local comparisons of *B. africanus* group species in Zanzibar (Stothard *et al.* 1997) and around Kisumu, western Kenya (Raahauge & Kristensen, 2000).

Genetic population structures of both *S. haematobium* and *B. globosus* from the Zimbabwean highveld were compared using RAPDs by Davies *et al.* (1999). The observations reflected the dispersal mechanisms of snails and schistosomes with the genetic distance of snail populations correlating to position on river systems. In contrast *S. haematobium* from different sites within a river system were not differentiated, although populations from different river systems appeared to be, suggesting that the definitive host plays an important role in dispersal of the parasite.

One potential drawback with RAPD analysis is the problem of homology of amplified fragments such that co-migrating fragments may be of identical size but of totally different sequence (Jones *et al.* 1999). After inspection of RAPD profiles from *B. forskalii* group snails, Jones *et al.* (1997) selected RAPD fragments to which further PCR fragments were designed. These new primer sets allowed amplification of the characterized regions using conventional PCR with two primers of specific sequence. This new assay was easier to interpret as only single amplified fragments of known sizes were amplified. The procedure named RAPD-SCARs, Sequence Characterized Amplified Regions provides a useful method for differentiating *B. forskalii* and related taxa.

The ribosomal internal transcribed spacer (ITS)

Parallel with the developments utilizing PCR with random primers, universal primers have been of value for studies on *Bulinus*. Using primers developed by Kane & Rollinson (1994) to amplify the ribosomal internal transcribed spacer (ITS) from schistosomes, amplification products from all *Bulinus* tested were of approximately 1·3 kbp (Stothard *et al.* 1996). Direct digestion of these products with a battery of restriction enzymes produced RFLP profiles that could be easily visualized by agarose electrophoresis and ethidium bromide staining. Substantial variation between species groups was recorded both by inspection of the RFLP profiles themselves and by DNA sequencing of the ITS1, an approximately 500 bp sub-region within the ITS. The levels of divergence questioned the systematic placement of all such species within a single genus (Stothard *et al.* 1996). One restriction assay proved particularly useful. Digestion of the ITS with *Rsa*I clearly differentiated *B. truncatus* from *B. tropicus*. Separation of these two species often requires detailed inspection of radulae preparations or chromosomal squashes from egg mass material, *B. truncatus* is tetraploid whereas *B. tropicus* is diploid. PCR-RFLP analysis of the ITS has been used by Sène & Southgate (1998) for *B. truncatus* from Mali and Senegal. More recently a RFLP assay proved to be useful for differentiation of *B. forskalii* group snails from Mafia Island, Tanzania using digestion with *Sac*I (Stothard, Loxton & Rollinson, unpublished).

Nucleotide divergence within the *B. africanus* group was detectable with PCR-RFLP of the ITS and originally RFLP patterns appeared to be stable, offering potential for identification purposes for nearly all species within this group (Stothard *et al.* 1996). Further sampling of *B. africanus* group snails

from Zanzibar revealed greater variation suggesting the occurrence of multiple ITS types within an individual snail (Stothard & Rollinson, 1997*a*). Whilst the profiles could differentiate *B. globosus* and *B. nasutus*, interpretation of RFLP patterns had to be conducted with care as the multiple or paralogous ITS types could potentially confound comparisons. Raahauge & Kristensen (2000) similarly found multiple ITS types within *B. africanus* group snails from western Kenya. This intragenomic heterogeneity was thought to be localized within the ITS2 region. Further PCR primers were designed to amplify only the ITS1 and restriction digestion of this product differentiated *B. globosus* from *B. nasutus* and *B. africanus*. It failed, however, to detect variation between *B. nasutus* and *B. africanus* (Raahauge & Kristensen, 2000).

Cytochrome Oxidase 1

Stothard & Rollinson (1997*b*) amplified a 450 bp region within the COI and determined by DNA sequencing that 33 positions within the sequence were variable between *B. globosus* and *B. nasutus*. Moreover, using double digestion with *Rsa*I and *Alu*I, RFLP patterns were produced that could differentiate *B. globosus* from *B. nasutus*. In addition, a species-specific restriction enzyme site *Ssp*I was identified such that only *B. globosus* sequences would be cut whilst those from *B. nasutus* or *B. africanus* remained intact (Stothard & Rollinson, 1997*b*). A simple restriction test such as this had certain advantages. Firstly, as the COI was amplified in all snails there was no need for further experimental controls during PCR and secondly, positive controls to check for restriction digestion could be performed with either *Alu*I or *Rsa*I or both. The assay appeared to be validated on three *B. africanus* group species originating from Zanzibar, Kenya, Zimbabwe and South Africa. However, as further snail populations were inspected from further afield, e.g. *B. globosus* from Niger, it become clear that this *Ssp*I site, whilst present in some individuals, was not universally shared (Rollinson & Hughes, unpublished data). Restriction enzymes are highly specific, hence their value decreases with increased sampling of genetic diversity and the detection of point mutations.

Point mutation detection: ASA and SNaPshot™

Molecular assays may have advantages and disadvantages when compared against each other (see Table 1 of Jones *et al.* 1999). New techniques under development may be valuable for devising routine procedures for the detection of single nucleotide changes in known sequences. One possibility is the use of allele-specific amplification (ASA). During the PCR amplification, if there is a base pair mismatch at the 3′ end of the PCR primer, the

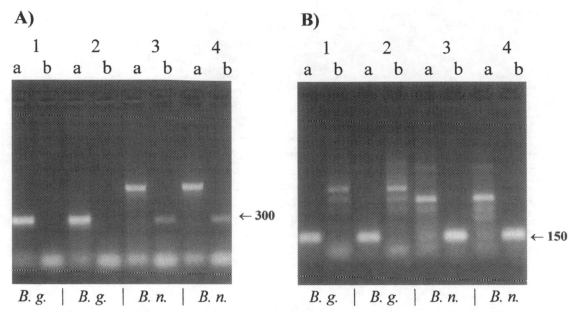

Fig. 2. Allele Specific Amplification (ASA) assay using a gel-based detection method for discrimination of *Bulinus* using point mutations within the CO1. Fig. 2A. ASA primers specific for either *B. globosus* (lanes a) or *B. nasutus* (lanes b) are used separately upon genomic DNA preparations from 4 single snails. An amplification product of specific size (300 bp) is targetted (Tracks 1 and 2 *B. globosus*, Tracks 3 and 4 *B. nasutus*). Presence or absence of the 300 bp product allows differentiation of the two taxa but in lanes 3a and 4a, a spurious amplification product is also produced. Fig. 2B. ASA primers specific for either *B. globosus* (lanes a) or *B. nasutus* (lanes b) are used separately upon genomic DNA preparations from 4 single snails and an amplification product of specific size (150 bp) is now targetted (Tracks 1 and 2 *B. globosus*, Track 3 & 4 *B. nasutus*). Whilst the target 150 bp amplification product is specific to each taxon with each ASA primer-pair, further spurious amplification products are seen e.g. lanes 1b and 2b and 3a and 4a.

enzyme Taq will not extend this oligonucleotide (Ugozzoli & Wallace, 1991). If the DNA variants/ mutations are known, PCR primers can therefore be designed to match only one allele or variant. Providing that this allele or variant is specific to a species, ASA can provide a useful taxonomic assay but it is limited to the extent of sequence data available for candidate loci.

A good candidate locus is the COI, where there are numerous sequence differences between species but only a few are detectable with restriction enzymes. After inspection of several DNA point mutations that differentiated *B. nasutus* and *B. globosus*, four variant positions were selected to which ASA primers were designed to match either species. Initially non-gel based methods were used by incorporation of ethidium bromide into the PCR reaction itself, however, false positives were obtained because of primer-dimer formation and (or) the amplification of non-specific bands. After agarose electrophoresis, non-specific bands or primers-dimers could be discernible from the desired amplification product as its exact size was known (Fig. 2). Whilst ASA is highly discriminatory, it still is affected by additional mutations near the tested nucleotide that might also cause mis-priming and reaction failure. More importantly, the assay is biased towards false negatives if less than four primers are used to type each position. For example,

specific primers for the nucleotide A would fail for all non-A nucleotides therefore mutations with C, G or T are not discernible unless additional ASA primers are used, or the region is later sequenced.

SNaPshot™ (Applied BioSystems Inc., U.K.) has recently been marketed as a fluorescent-based, primer extension assay and offers an alternative approach. In brief, after the gene target is amplified by PCR, a short oligonucleotide primer, or probe, is used to abut next to the variant position to be typed (Fig. 3). The SNaPshot primer can be of the same sequence as an ASA primer except that the 3′ terminal nucleotide is omitted. The SNaPshot primer is then used in a cycle sequencing reaction that has only dideoxynucleotides terminators with fluorescent labels and AmpliTaq DNA polymerase. In essence, a sequencing reaction is performed for the variable, single base. Upon completion of this reaction, the SNaPshot primer now carries a fluorescent label complementary to the nucleotide typed. This labelled primer is then separated on a standard denaturing polyacrylamide gel and analysed with GeneScan (analysis software on an automated DNA sequencer, e.g. ABI PRISM 377). The whole procedure is very quick, less than a working day and has the potential to be multiplexed where several variant positions, not necessarily from the same gene target, are genetically typed.

Preliminary results using SnaPshot™ typing of

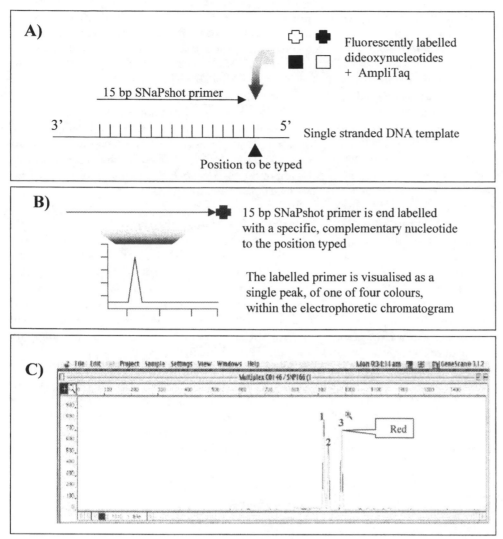

Fig. 3. SNaPshot™ offers a new primer extension based assay to type mutations at specific locations within an amplification product. Fig. 3A. A purified amplification product undergoes the addition of a 3ʹ fluorescent label complementary to the variant position to be typed in a SNaPshot™ reaction using a 15 bp primer. Fig. 3B. The SNaPshot™ primer is now fluorescently labelled and subjected to denaturing electrophoresis on an ABI 377 automated DNA sequencer. This primer can be visualised as a single peak within the electrophoretic chromatogram and is of a specific colour to the incorporated dideoxynucleotide. Fig. 3C. A typical SNaPshot™ reaction as visualised using GeneScan software. In this instance, three reactions are sequentially loaded at 2 and 4 minute intervals. Multiplexing of several reactions within a single lane can considerably reduce costs. Up to 96 individuals can be typed individually for several point mutations in several target amplification products and analysed within a single gel.

variant nucleotides within the COI have shown the potential of this method as mutations can be quickly recognized and typed as well as detection of 'mixed templates' or heterozygotes (Fig. 3). Further optimization and use of multiplexing might make this methodology ideal for routine identification.

SNAIL HOSTS OF *S. HAEMATOBIUM*

Species and distributions of the intermediate hosts of *S. haematobium* have been summarized by Brown (1994). Modifications to the list of species would include the addition of *B. truncatus* for Jordan (Arbaji *et al.* 1998) and Senegal (Southgate *et al.* 2000). Intermediate hosts can be found in all of the species groups but most commonly in the *B.*

africanus group and the *B. truncatus/tropicus* complex. The existence of strains of *S. haematobium* has been recognized for some time based primarily on compatibility with different species of *Bulinus*. The main divisions seem to lie between strains adapted more closely to *B. truncatus*, to *B. globosus* and to members of the *B. forskalii* group. Many experiments have reported the differences in parasite infectivity with *S. haematobium* from *B. truncatus* failing to develop in *B. africanus* group snails and *vice versa* (McCullough, 1959; Wright & Knowles, 1972; Chu, Kpo & Klumpp, 1978). A population study of the compatibility between *S. haematobium* and its potential snail hosts in Niger, exposing populations of snails from the same focus (sympatric) and populations from other foci (allopatric), emphasise the range of the 'compatibility polymorphism' within the *S. haematobium–Bulinus* system (Vera *et al.* 1990).

In some endemic areas the situation is simple with only one species of *Bulinus* being implicated in transmission. For example, in areas of North Africa including Algeria, Egypt, Libya, Morocco and Tunisia, *B. truncatus* is the only intermediate host, in Mauritius *B. cernicus* is the sole host: although variation in parasite compatibility may occur over small areas it is possible that only *B. globosus* acts as a host in Zambia, Zimbabwe and Zanzibar (Manning, Woolhouse & Ndamba, 1995; Mukaratirwa *et al.* 1996). Elsewhere more than one potential snail host may occur and transmission may be through more than one species. For example Fryer & Probert (1988) showed that in north-eastern Nigeria *S. haematobium* derived from *B. truncatus* and from *B. globosus* could also develop in *B. senegalensis*. The following two case studies illustrate how changes in compatibility can occur over a small geographical range.

Bulinus *and* Schistosoma *spp. in Senegal*

In Senegal there have been dramatic increases in the prevalence and intensity of human infection with *S. mansoni* and to a lesser extent *S. haematobium* (see Picquet *et al.* 1996; Southgate, 1997 and Sturrock *et al.* this supplement) following the construction of the Diama Dam on the Senegal River and the Manantali Dam on the Bafang River, a tributary of the Senegal River, in Mali. In the Lower and Middle Valleys, *S. haematobium* is transmitted primarily by *B. globosus* and *B. senegalensis* respectively (Vercruysse *et al.* 1994) and yet in the Upper Valley in Mali the main host appears to be *B. truncatus* (Rollinson *et al.* 1997b).

Interestingly, *B. truncatus* from the Lower and Middle Valleys were shown experimentally to be susceptible to *S. haematobium* isolated from the urines of children in Tenegue, Office du Niger, Mali where *B. truncatus* is the intermediate host. However they were not susceptible to a parasite isolated from the Lower Valley where natural transmission is normally associated with *B. globosus*. The authors drew attention to the possible appearance of a *B. truncatus*-borne parasite in the Middle and Lower Valleys (Rollinson *et al.* 1997b).

Southgate *et al.* (2000) carried out further snail infection experiments with *S. haematobium* from various sites in Senegal using laboratory-bred and wild-caught snails. They were able to show that isolates of *S. haematobium* from the Middle Valley do show some compatibility with *B. truncatus* although the experimental snails were from Mali and not Senegal. However, M. Sène (unpublished obser-

vations) has recently found *B. truncatus* from the Middle Valley naturally infected with *S. haematobium*. Therefore it seems that the situation in the Lower and Middle Valley of the SRB may be changing in that there is the likelihood that the currently most widespread bulinid snail, *B. truncatus*, may become more important in the epidemiology of urinary schistosomiasis in the Senegal River Basin (SRB) in addition to *B. senegalensis* and *B. globosus*. In the Tambacounda region of Senegal *B. umbilicatus* also acts as a host for *S. haematobium* and *S. curassoni* (Rollinson *et al.* 1998).

When other schistosome species are taken into account it can be seen that the situation is more complex with some snails potentially acting as hosts for more than one schistosome parasite. For example, *S. curassoni* from Senegal is incompatible with *B. truncatus*, marginally compatible with *B. senegalensis* and compatible with *B. umbilicatus*, whereas *S. bovis* is compatible with all three species (Southgate *et al.* 1985a). Therefore, in the SRB it is necessary to be able to differentiate cercariae of *S. bovis* and *S. haematobium* shed from naturally infected *B. truncatus*. *S. bovis* has a natural wide intermediate host range and is compatible experimentally with species in all four *Bulinus* species groups (for review see Moné, Mouahid & Morand, 2000).

Bulinus *and* S. haematobium *transmission in Zanzibar*

In East Africa hosts for *S. haematobium* are found in the *B. africanus* group. Four species of the *B. africanus* group are currently recognized in Kenya and Tanzania: *B. globosus*, *B. nasutus*, *B. africanus* and *B. ugandae* (Brown, 1994) and all except *B. ugandae* have been implicated in transmission. Two of the taxa, *B. globosus* and *B. nasutus*, are difficult to identify, as morphometric variation within each species appears to form a continuum and in the north eastern region of Tanzania specimens from either taxa almost appear conspecific (Mandahl-Barth, 1957). To resolve the taxonomic dilemma between *B. globosus* and *B. nasutus* a detailed study of the snails was conducted in Zanzibar with individual snails being characterized by three methods: shell morphometry, enzyme analysis and molecular analysis with RAPDs.

Zanzibar incorporates two islands Unguja, more commonly known as Zanzibar island, and the more northerly island of Pemba. Schistosomiasis is major public health problem on both Pemba and Unguja and these islands have been the focus of a WHO-funded disease control programme through chemotherapy (Savioli *et al.* 1989). Both islands lie close to the coastal region of mainland Tanzania where earlier Rollinson & Southgate (1979) found three enzyme systems which appeared to discriminate *B. africanus* group species.

On Zanzibar it became clear that the two species could be differentiated by isoenzyme analysis and DNA typing (Stothard *et al.* 1997; Stothard & Rollinson, 1997 *a, b*). Detailed surveys suggest that the distributions of *B. globosus* and *B. nasutus* are allopatric although *B. forskalii* may be found in association with both species. On Unguja, *B. globosus* is restricted to northern areas whereas *B. nasutus* is confined mainly to the South. On Pemba, *B. globosus* is widespread whereas *B. nasutus* is associated with the eastern border of the central region (Stothard *et al.* 1997). With clear species markers it was possible to assess the roles played by the snails in transmission of *S. haematobium*. *B. nasutus* was refractory to experimental infection and no evidence of natural infection has been observed. In contrast, *B. globosus* has been found naturally infected and showed a high infection rate on experimental challenge with miracidia hatched directly from infected urines. In the north of Unguja the high prevalence of infection in school children fits with the known distribution of *B. globosus* (Stothard & Rollinson, 1997*b*: Stothard *et al.* 2000).

It must now be clarified as to whether *B. nasutus* is playing a role in transmission on the African mainland especially in the coastal areas of Kenya and Tanzania (Webbe & Msangi, 1958; Sturrock, 1965; Pringle *et al.* 1971). This will be possible if morphological identifications are supplemented with enzyme and DNA markers. Also observations on naturally infected snails will need to take into account that *B. africanus* group snails may also transmit *S. bovis*, hence cercariae emerging from naturally infected snails will need to be identified. If *S. haematobium* from coastal areas is shown to be transmitted by *B. nasutus* it will be of interest to determine whether the parasite is compatible with *B. nasutus* from Zanzibar.

OBSERVATIONS ON TRANSMISSION

Chronobiology

There is considerable interspecific diversity in cercarial emergence rhythms within the genus *Schistosoma*, with emergence peaks occurring at different times of the nycthemere (a 24 h photocycle), ranging between 08·00 and 09·00 h for the earliest, to 22·00 h and 23·00 h for the latest (Combes *et al.* 1994). The rhythmic emergence of cercariae is synchronized by exogenous factors to which the cercariae respond, and photoperiod is the most important factor. Furthermore, generally there is a clear correlation between the maximum shedding period of cercarial and water contact patterns of the definitive host, thus enhancing the chances of infection of the definitive host and continuation of the parasite's life cycle. The cercarial emergence rhythms limit their temporal dispersion by concentrating them in the period of the nycthemere when the chances of meeting the host are greatest. Thus, cercarial rhythms may be considered as an adaptive behaviour increasing the probabilities of a meeting between parasite and host (Combes *et al.* 1994). Intraspecific variation in cercarial emergence rhythms has been well documented in cases where the parasite has adapted to different hosts which themselves possess dissimilar water contact patterns. Indeed, experimental cross-mating experiments between different species and populations with different chronobiological phenotypes demonstrate that the rhythms are genetically controlled (Théron, 1989; Théron & Combes, 1988).

There is less information available regarding intraspecific variation in cercarial emergence patterns with the same definitive host. A study by N'Goran *et al.* (1997) on nine isolates of *S. haematobium* that utilized either *B. truncatus* or *B. globosus* from the Ivory Coast as intermediate hosts demonstrated the existence of intraspecific variation in cercarial emergence rhythms. A statistical analysis on the data demonstrated that the nine isolates fell into 3 groups which corresponded to 3 different climatic and vegetal zones: (i) sub-Sudanian zone with aquatic habitats in the wooded savannah; (ii) transitory zone between savannah and forest and (iii) Guinea climatic zone where transmission sites are located in the dense forest. Illumination is the primary factor influencing cercarial shedding, and the cercarial emergence of *S. haematobium* shows rapid reaction to variation in light intensity (Raymond & Probert, 1987). N'Goran *et al.* (1997) postulated that *S. haematobium* cercariae from the forest foci in Ivory Coast may be more sensitive to light than those from the savannah habitat, thus enabling the cercariae from the forest foci to maintain the shadow response, that is, the stimulation of cercarial emergence and increased activity of cercariae caused by human activity forming a shadow. Thus, the response to a shadow will increase the chances of infection. N'Goran *et al.* (1997) considered the differential sensitivity to light to be an ecological adaptation to maintaining the shadow response.

Studies by Pitchford *et al.* (1969) examined the influence of changing seasons on the chronobiology of cercarial emergence of a number of schistosome species, including *S. haematobium*, *S. mattheei* and *S. bovis*. These authors noted interspecific differences and seasonal differences in shedding patterns: for example, *S. mattheei* and *S. bovis* shed between 06·00 h and 08·00 h, whereas *S. haematobium* shed between 11·00 h and 15·00 h. Temperature was also found to influence the numbers of cercariae released, generally more cercariae were shed at higher temperatures, fewer at lower temperatures. *S. mattheei* was found to be also a nocturnal shedder, in addition to the early morning shed, at lower temperatures. *S. curassoni*, a parasite of domestic stock, has a shedding pattern which peaks in the early morning (Mouchet *et al.* 1992), whereas *S. intercalatum*, a parasite of man, peaks at the middle of the day (Pagès & Théron, 1990). However, *S. margrebowiei* is unusual in that it possesses an ultradian shedding rhythm with two distinct emergence peaks, at dawn and at dusk. The two peaks of emergence pattern coincide with visits to watering places by antelopes and waterbuck (Raymond & Probert, 1991).

Other parasites can influence compatibility

A polyploid series occurs within the *B. truncatus/tropicus* group, with diploid, tetraploid, hexaploid and octoploid species. The diploid species are found in the Afrotropical region and, although there is some overlap between the tetraploid and diploid species, the tetraploid species have a more northern distribution. Hexaploid and octoploid species are associated with streams at high altitude in the Ethiopian highlands. *B. tropicus*, a diploid species, was generally considered to be incompatible with schistosome parasites until *Schistosoma margrebowiei* had been shown to be naturally transmitted by *B. tropicus* in Lochinvar National Park, Zambia (Southgate *et al.* 1985*b*). Around the same time Southgate *et al.* (1985*c*) reported that of 112 *B. tropicus* collected from the Mau Escarpment, Kenya, 10 were infected with *S. bovis* and an amphistome parasite. Exposure of laboratory-bred *B. tropicus* from the Mau Escarpment, Kenya to *S. bovis* proved negative, thus suggesting that the presence of amphistome infection may have a suppressive effect on the immune system of the snail, thereby allowing *S. bovis* to develop. Subsequent infection studies in the laboratory demonstrated that it is possible to infect *B. tropicus* with *S. bovis*, but only if the snails had been exposed to miracidia of an amphistome, *Calicophoron microbothrium* (Southgate *et al.* 1989). Thus the laboratory studies supported the field studies that a primary infection of another trematode is necessary for *B. tropicus* to be susceptible to *S. bovis*. Studies by Loker, Bayne & Yui (1986) on a different host parasite relationship showed that when cytotoxic *Biomphalaria glabrata* haemocytes were co-incubated with *S. mansoni* sporocysts and *E. paraensi* rediae, the echinostome depressed the cytotoxic capacity of the haemocytes for the larval schistosomes. Thus, it is possible that larval amphistomes secrete interfering factor(s) that somehow alter the behaviour of the *Bulinus* haemocytes, preventing them from recognizing and encapsulating the young *S. bovis* sporocysts.

Internal defence mechanisms in Bulinus

The recognition of non-self by the snail is suspected to be mediated by lectins or other humoral factors which potentially function as opsonins or recognition agents that promote antigen clearance or encapsulation by haemocytes. Functional differences of haemocytes between various snail species are known; for example, peroxidases can be lacking (Adema, Harris & Van Deutekom-Mulder, 1992), and there appears to be at least two morphologically distinct types (Preston & Southgate, 1994).

With regard to the immuno-biology of *Bulinus*, a different set of molecules are thought to operate from those in *Biomphalaria*. For example, in *Biomphalaria glabrata* there appears to be two families of lectins, which can be differentially induced. These are broadly placed within two groups: G1M and G2M – carbohydrate binding proteins of 150–220 kDa and 75–130 kDa, respectively (Couch, Hertel & Loker, 1990). These molecules are also able to bind mammalian erythrocytes, hence have agglutinin properties, and are able to differentiate non-self epitopes. Agglutinins have been shown to be present in two species of *Bulinus*, *B. nasutus*, and *B. truncatus*, though surprisingly, of the other species of snail tested, none was detected.

Harris (1990) studied the binding of these agglutinins in *Bulinus* and found they were not inhibited by either monosaccharides or disaccharides, hence were unlikely to be lectins: the agglutinating activity was later traced to a glycoprotein (Harris, 1990). In the reduced form on SDS-PAGE gels this protein was shown to be approximately 135 kDa and inducible upon infection with miracidia. It was also shown to bind to invading miracidia but not to mother sporocysts, daughter sporocysts or to emerging cercariae. It is speculated that this molecule is involved in humoral surveillance (Harris, 1990). Proteins of similar M_r to the haemagglutinin of *B. nasutus* were shown to be present in a different species of bulinid tested, and Cleveland mapping of the M_r 135 kDa proteins revealed identical polypeptide patterns thus suggesting that the primary structure of these proteins is similar (Harris, Preston & Southgate, 1993). The glycoprotein is thought to have similar primary structure and could be categorized into two groups: agglutinin or non-agglutinin. Further subgroups could be recognized immunologically: polyclonal antisera were obtained from BALB/c mice immunized with the 135 kDa protein from *B. truncatus* and these antibodies were used as probes in Western blots against 135 kDa polypeptides from several other *Bulinus* species (Harris *et al.* 1993). The antibody cross-reactivity was confined to species within the *B. truncatus/tropicus* complex. Despite there being some primary sequence homology between species groups, the protein epitopes recognized by mice were not shared across species groups of *Bulinus* (Harris *et al.* 1993). Since this protein is glycosylated, it may potentially allow considerable variation in the shape of this molecule and may be of considerable importance in immuno-surveillance (Preston & Southgate, 1994).

ABIOTIC FACTORS AFFECTING TRANSMISSION

Numerous abiotic factors (see below) are important in how they affect the distribution patterns of snails, and influence the life cycles of snails, population dynamics and hence patterns of transmission. One of the remarkable features of the biology of *Bulinus* is the high rate of increase in populations which is related to the ability to reproduce by both cross-fertilization (outcrossing) and self-fertilization (selfing) and responding to environmental stimuli (e.g. temperature, rainfall etc.).

For example, the life cycle of *B. globosus* has been described by many authors (O'Keefe, 1985*a*, *b*; Marti, 1986; Woolhouse & Chandiwana, 1989, 1990*a*, *b*). Young snails commence egg-laying at an age of 5–7 weeks and a height of 6–7 mm and continue breeding while growing to a height of 12–17 mm. Eggs develop immediately and hatch about one week after being laid.

Temperature

Temperature is considered as the most important abiotic factor influencing the distribution of snails in lentic environments. Snails may be killed by temperatures above or below lethal limits. However, the assessment of the effects of temperature in the field is difficult because there are considerable temperature gradients within a water body and snails can seek microhabitats where temperature is most favourable. Temperature influences snail distribution through its effects on reproduction and growth of juveniles, as well as on the survival of adults.

For cohorts of *B. globosus* reared in reservoirs on the Kenyan coast, r (the intrinsic increase in a population) was inversely related to increasing mean water-temperature above 25 °C (O'Keefe, 1985*a*). Nevertheless, *B. globosus* seems better adapted to high temperatures than *Biomphalaria pfeifferi* which is absent from the coastal area of Kenya. Mark recapture data show the recruitment rate of *B. pfeifferi* into populations in Zimbabwe to increase over the range 13–24 °C; therefore, under the cooler climatic conditions of southern Africa the breeding season is short. Although optimal temperature of *B. globosus* in the laboratory lies at about 25 °C, whereas for *B. pfeifferi* it is between 20–27 °C, it seems as though *B. globosus* is more tolerant of higher temperatures. For example, in a cohort of *B. globosus* egg production was delayed at 18 °C until snails were about 23 weeks old, more than twice the age of first oviposition at 25 °C (Shiff, 1964 *a*). In cohorts maintained at different temperatures, 18, 22·5, 25

and 27 °C, the most rapid increase of growth was observed at 25 °C (Shiff, 1964*b*). Laboratory investigations have demonstrated that although the temperatures at which snails attained an optimal r were similar, the range at which r remained high varied for different species. The marked peak in r attained by *B. globosus* at 25 °C and 27 °C probably enables it to take advantage of harsh environmental conditions found in temporary water bodies, for example, multiplying rapidly at the optimal temperature range. These examples illustrate how temperature plays a vital role in the biology of pulmonate snails by influencing growth and reproduction.

Rainfall and water velocity

Rainfall has a marked effect on snail populations by causing population fluctuations through drought and flooding, and influencing rates of oviposition and survival. For example, Woolhouse & Chandiwana (1990*a*, *b*) investigated the population dynamics of *B. globosus* in Zimbabwe and reported that year-to-year fluctuations were correlated with the effects of sudden spates, which washed away the snails from some sites and deposited them in others. Gradual increases in water level can encourage breeding probably through reducing the population density thereby removing density-dependent inhibition of reproduction. O'Keefe (1985*a*, *b*) noted peaks of egg production in reservoirs on the coast of Kenya following rainfall in the cooler months.

Water velocity is determined by geomorphic factors, especially the resistance offered by different types of bedrock: it is recognized as the most important factor in lotic environments determining the distribution of intermediate hosts. Measuring currents in the field experienced by snails is difficult because of variations in time and space. Appleton (1975) estimated tolerance ranges with upper limits of 0·3 m s^{-1} for *Bulinus*, but where the bedrock is hard a river may contain many boulders providing refuges and protection for snails in currents above 0·3 m s^{-1}. In small rivers and streams sudden spates following heavy rainfall will cause major fluctuations in population density and may be an important factor in the dispersal of snails. In such habitats, the rainy season will coincide with the reduction in snail population and transmission. Southgate *et al.* (1994) carried out a survey of *B. forskalii* on the island of São Tomé and noted that the snails were confined to the north east of the island where water velocity was considerably less than elsewhere on the island. In order to investigate the limiting factor(s) of the distribution of *B. forskalii* on São Tomé, a survey was conducted at 4 sites along the Agua Serra, taking physical measurements and searching for freshwater snails, and again snails were only found in one locality, Bombom, where the water velocity was much reduced. A one year study at two weekly intervals from 5 different sites showed that the snail populations generally increased during the dry period of June, July and August, and during the wet season snails were completely absent from two localities for two months and one month, respectively and were much reduced in the other three localities, with the obvious implications for reduced rates of transmission (Southgate et al. 1994).

Desiccation

Desiccation is a catastrophe for a snail population and it is the major restraint on the number of species that live in seasonal water bodies. There is a distinction between aestivation (prolonged survival by dormant snails out of water) and diapause (spontaneous climbing out of water and entering a state of dormancy). A third type of behaviour has been recognized, water-quitting. Diapause and water quitting are of interest in relation to the avoidance by snails of molluscicide. Aestivation is important because some species of snail, for example, B. senegalensis and B. nasutus, are able to survive without water for between 5–8 months. This ability enables these species to inhabit ephemeral water bodies which become transmission foci only during the rainy season. There is a general correlation between success in aestivation and a high capacity for increase (r), enabling rapid repopulation of a habitat when water returns. Field observations on pre-aestivation behaviour by Bulinus indicate two different strategies. B. nasutus and B. senegalensis aestivated around the margins of temporary pools, whereas B. globosus and B. truncatus aestivated towards the bottom of drying out pools. Aestivation around the margins of the pools prevents snails emerging when revived by isolated showers; in fact, the snails only emerge when pools are well filled. This strategy is advantageous preventing the depletion of the emerging population. It is known that some snails are able to carry over an infection after a period of aestivation (Webbe, 1962).

Salinity

Salinity (also known as total dissolved chemical content or total electrolytes) is commonly estimated as electrical conductivity. The concentration of dissolved salts limits the distribution of pulmonate snails when it is unusually high or low. For example, in West Africa the intrusion of sea water into rivers in the dry season is known to limit the distribution of intermediate hosts. There is a progressive elimination of gastropod species at salinities of as little 1‰. Donnelly, Appleton & Schutte (1983) demonstrated that the fecundity and survival of B. africanus was adversely affected with salinities of 1‰ with the most significant reductions occurring between 3·5‰ and 4·5‰. The severe outbreak of S. mansoni and S.

haematobium in the Senegal river basin is correlated with reductions in salinity due to the construction of a barrage at Diama, approximately 40 km from the sea, on the Senegal River thus preventing the intrusion of sea water into the Senegal River during the dry season (Southgate, 1997).

Recent studies on Zanzibar Island (Unguja) demonstrated a clear distinction in the distribution of B. globosus and B. nasutus on the island and this appears to be correlated to the sharp division of the geological zones of the island and water conductivity (Stothard et al. 2000). B. globosus, which is highly compatible with S. haematobium is found only in the north of the island on clayey soils with subordinate limestone, whereas B. nasutus is found primarily in the south and is confined to fossiliferous limestone/ marly sand substratum. There is some overlap of the two species in the central region, but no occurrences of sympatric populations. B. nasutus on Zanzibar Island is refractory to S. haematobium.

CONCLUDING REMARKS

The identification and characterization of Bulinus spp. is sometimes fraught with difficulties, especially in those cases where there is considerable overlap of characters. The development of molecular techniques has added significantly to the armoury of morphological, cytological and biochemical techniques that have been used in the identification and characterization of species and populations of Bulinus, and has facilitated a more detailed elucidation of the intermediate host parasite relationship. More specifically studies on ribosomal genes (rRNA), random amplified polymorphic DNA (RAPDs) and mitochondrial gene cytochrome oxidases 1 (CO1) have been used to determine insights into relationships between species and provide markers for species differentiation. Morphometric variation poses problems in distinguishing B. globosus from B. nasutus on Zanzibar Island and Pemba, but these have been solved by the use of enzymes and DNA typing. More detailed population studies have been made possible by the development of microsatellites. RAPDs-PCR have been shown to be useful in certain situations for diagnosis, but it is thought unlikely that this approach will lead to a species specific diagnosis for all species of Bulinus. A further refinement, RAPD-SCARs, sequence amplification of characterized regions, has enabled amplification of characterized regions using PCR with two pairs of specific sequence. Additional techniques are being developed and investigated to refine diagnosis such as the allele specific amplification (ASA) and SNaPshot™ typing of variant nucleotide within CO1 where mutations can be quickly recognized and scored. It seems likely molecular techniques will continue to develop in the area of diagnosis. The epidemiology and Bulinus/S. haematobium group relationships are complex. For example, some species of *Bulinus* within the 4 species groups may act as an intermediate host for *S. haematobium* but strains of *S. haematobium* show considerable variation in levels of compatibility or indeed incompatibility with different species of *Bulinus*. It is important that this intermediate host parasite specificity is recognized to gain a further understanding of the intricacies of epidemiology. The intermediate host response has both humoral and cellular components, and Preston & Southgate (1994) suggested that a glycosylated protein found in *Bulinus* spp. may be of importance in immunosurveillance. It is interesting that the 'normal' intermediate host response to an incompatible schistosome may be modified by earlier infection with a different parasite. The epidemiological implications of such interactions and modification of host response have yet to be fully elucidated. Examples are given to demonstrate how environmental factors such as temperature, rainfall, water velocity and desiccation influence the distribution of *Bulinus* spp. and how different species adapt in varying ways to environmental changes. The influence of abiotic factors on the distribution of *Bulinus* spp. inevitably has a bearing on the distribution of the parasites. The situation in the SRB is given as an example where environmental changes (primarily reductions in levels of salinity) have given rise to marked increases in the spread, prevalence and intensity of *S. mansoni* and to a lesser extent *S. haematobium* (Picquet *et al.* 1996). Recent field and laboratory data support the view that changes are occurring in the intermediate host–parasite relationships between *Bulinus* spp. and *S. haematobium* in the SRB possibly through the introduction and establishment of different strains of *S. haematobium*: such changes emphasise the dynamic nature of the intermediate host-parasite relationship. With the ever-increasing mobility of the definitive host (man) and environmental changes induced by increases in the human population it is not surprising that new intermediate host parasite relationships are being initiated resulting in changes in epidemiology.

ACKNOWLEDGEMENTS

Russell Stothard gratefully acknowledges funding from The Wellcome Trust. We thank Julia Bartley for her excellent technical assistance with the SNaPshot™ and use of the ABI 377 automated DNA sequencer.

REFERENCES

ADEMA, C. M., HARRIS, R. A. & VAN DEUTEKOM-MULDER, E. C. (1992). A comparative study of hemocytes from six different snails: morphology and functional aspects. *Journal of Invertebrate Pathology* **59**, 24–32.

APPLETON, C. C. (1975). The influence of stream geology on the distribution of the bilharzia host snails

Biomphalaria pfeifferi and *Bulinus* (*Physopsis*) sp. *Annals of Tropical Medicine and Parasitology* **69**, 241–255.

ARBAJI, A., AMR, Z., ABBAS, A., AL-ORAN, R., AL-KHARABSHEH, S. & AL-MELHIM, W. (1998). New sites of *Bulinus truncatus* and indigenous cases of urinary schistosomiasis in Jordan. *Parasite – Journal de la Societé Française de Parasitologie* **5**, 379–382.

BIOCCA, E., BULLINI, L., CHABAUD, A., NASCETTI, G., ORECCHIA, P. & PAGGI, L. (1979). Subdivisions su base morfologica e genetica del genere *Bulinus* in tre generi: *Bulinus* Muller, *Physopsis* Krauss e *Mandahlbarthia* gen.nov. *Rendiconti della Classe di Scienza fisiche, matematiche e naturali, Accademia Nazionale dei Lincei, ser. 8* **66**, 276–282.

BROWN, D. S. (1981). Generic nomenclature of freshwater snails commonly classified in the genus *Bulinus* (Mollusca: Basommatophora). *Journal of Natural History* **15**, 909–915.

BROWN, D. S. (1994). *Freshwater snails of Africa and their Medical Importance*, London. Taylor & Francis, pp. 609

CHU, K. Y., KPO, H. K. & KLUMPP, R. K. (1978). Mixing of *Schistosoma haematobium* strains in Ghana. *Bulletin of the World Health Organisation* **59**, 549–554.

CLARK, A. G. & LANIGAN, M. S. (1993). Prospects for estimating nucleotide divergence with RAPDs. *Molecular Biology and Evolution* **10**, 1096–1111.

COMBES, C., FOURNIER, A., MONÉ, H. & THÉRON, A. (1994). Behaviours in trematode cercariae that enhance parasite transmission patterns and processes. *Parasitology* **109**, S3–S13.

COUCH, L., HERTEL, L. A. & LOKER, E. S. (1990). Humoral response of the snail *Biomphalaria glabrata* to trematode infection observations on a circulating hemagglutinin. *Journal of Experimental Zoology* **255**, 340–349.

DAVIES, C. M., WEBSTER, J. P., KRUGER, O., MUNATSI, A., NDAMBA, J. & WOOLHOUSE, M. E. J. (1999). Host-parasite population genetics: a cross-sectional comparison of *Bulinus globosus* and *Schistosoma haematobium*. *Parasitology* **119**, 295–302.

DONNELLY, F. A., APPLETON, C. C. & SCHUTTE, C. H. J. (1983). The influence of salinity on certain aspects of the biology of *Bulinus* (*Physopsis*) *africanus*. *International Journal of Parasitology* **13**, 539–545.

FRYER, S. E. & PROBERT, A. J. (1988). The cercarial output from three Nigerian bulinids infected with two strains of *Schistosoma haematobium*. *Journal of Helminthology* **62**, 133–140.

HARRIS, R. A. (1990). *Haemolymph Proteins and the Snail Immune Response*. Ph.D. thesis, University of London, pp. 292.

HARRIS, R. A., PRESTON, T. M. & SOUTHGATE, V. R. (1993). Purification of an agglutinin from the haemolymph of the snail *Bulinus nasutus* and demonstration of related proteins in other *Bulinus* spp. *Parasitology* **106**, 127–135.

GOW, J. L., NOBLE, L. R., ROLLINSON, D. & JONES, C. S. (2001). Polymorphic microsatellites in the African freshwater snail *Bulinus forskalii* (Gastropoda, Pulmonata). *Molecular Ecology Notes* (in press).

JARNE, P., DELAY, B., BELLEC, C., ROIZES, G. & CUNY, G.

(1990). DNA fingerprinting in schistosome-vector snails. *Biochemical Genetics* **28**, 577–583.

JARNE, P., DELAY, B., BELLECE, C., ROIZES, G. & CUNY, G. (1992). Analysis of mating systems in the schistosome-vector hermaphrodite snail *Bulinus globosus* by DNA fingerprinting. *Heredity* **68**, 141–146.

JELNES, J. E. (1986). Experimental taxonomy of *Bulinus* (Gastropoda: Planorbidae): the West and North African species reconsidered, based upon an electrophoretic study of several enzymes per individual. *Zoological Journal of the Linnean Society* **87**, 1–26.

JONES, C. S., NOBLE, L. R., LOCKYER, A. E., BROWN, D. S. & ROLLINSON, D. (1997). Species-specific primers discriminate intermediate hosts of schistosomes: unambiguous PCR diagnosis of *Bulinus forskalii* group taxa (Gastropoda: Planorbidae). *Molecular Ecology* **3**, 172–179.

JONES, C. S., NOBLE, L. R., OUMA, J., KARIUKI, H. C., MIMPFOUNDI, R., BROWN, D. S. & ROLLINSON, D. (1999). Molecular identification of schistosome intermediate hosts: case studies of *Bulinus forskalii* group species (Gastropoda: Planorbidae) from West and East Africa. *Biological Journal of the Linnean Society* **68**, 215–240.

JOURDANE, J. & SOUTHGATE, V. R. (1992). Genetic changes and sexual interactions between species of the genus *Schistosoma*. *Research and Reviews in Parasitology* **52**, 21–26.

KANE, R. A. & ROLLINSON, D. (1994). Repetitive sequences in the ribosomal DNA internal transcribed spacer of *Schistosoma haematobium*, *Schistosoma intercalatum* and *Schistosoma mattheei*. *Molecular and Biochemical Parasitology* **63**, 153–156.

LANGAND, J., BARRAL, V., DELAY, B. & JOURDANE, J. (1993). Detection of genetic diversity within snail intermediate hosts of the genus *Bulinus* by using Random Amplified Polymorphic DNA markers (RAPDs). *Acta Tropica* **55**, 205–215.

LOKER, E. S., BAYNE, C. J. & YUI, M. A. (1986). *Echinostoma paraensei* hemocytes of *Biomphalaria glabrata* as targets of echinostome mediated interference with host snail resistance to *Schistosoma mansoni*. *Experimental Parasitology* **62**, 149–152.

MANDAHL-BARTH, G. (1957). Intermediate hosts of *Schistosoma*. African *Biomphalaria* and *Bulinus*: 2. *Bulinus*. *Bulletin of the World Health Organisation* **17**, 1–65.

MANNING, S. D., WOOLHOUSE, M. E. J. & NDAMBA, J. (1995). Geographic compatibility of the freshwater snail *Bulinus globosus* and schistosomes from the Zimbabwean highveld. *International Journal for Parasitology* **25**, 37–42.

MARTI, H. P. (1986). Field observations on the dynamics of *Bulinus globosus*, the intermediate host of *Schistosoma haematobium* in the Ifakara area, Tanzania. *Acta Tropica* **42**, 171–187.

McCULLOUGH, F. S. (1959). The susceptibility and resistance of *Bulinus* (*P.*) *globosus* and *B.* (*B.*) *truncatus rohlfsi* to two strains of *S. haematobium* in Ghana. *Bulletin of the World Health Organisation* **20**, 75–85.

MIMPFOUNDI, R. & GREER, G. J. (1990). Allozyme variation among populations of *Bulinus forskalii* (Ehrenberg,

1831) (Gastropoda: Planorbidae) in Cameroon. *Journal of Molluscan Studies* **56**, 363–371.

MONÉ, M., MOUAHID, G. & MORAND, S. (1999). The distribution of *Schistosoma bovis* Sonsino, 1876 in relation to intermediate host-parasite relationships. *Advances in Parasitology* **44**, 99–138.

MOUCHET, F., THÉRON, A., BREMOND, P., SELLIN, E. & SELLIN, B. (1992). Pattern of cercarial emergence of *Schistosoma curassoni* from Niger and comparison with 3 sympatric species of schistosomes. *Journal of Parasitology* **78**, 61–63.

MUKARATIRWA, S., SEIGISMUND, H. T., KRISTENSEN, T. K. & CHANDIWANA, S. K. (1996). Genetic structure and parasite compatibility of *Bulinus globosus* (Gastropoda: Planorbidae) from two areas of different endemicity of *Schistosoma haematobium* in Zimbabwe. *International Journal for Parasitology* **26**, 269–280.

N'GORAN, E., BRÉMOND, P., SELLIN, E., SELLIN, B. & THÉRON, A. (1997). Intraspecific diversity of *Schistosoma haematobium* in West Africa: Chronobiology of cercarial emergence. *Acta Tropica* **66**, 35–44.

NJIOKOU, F., BELLEC, C., BERREBI, P., DELAY, B. & JARNE, P. (1993). Do self-fertilization and genetic drift promote a very low genetic variability in the allotetraploid *Bulinus truncatus* (Gastropoda: Planorbidae) populations? *Genetical Research Cambridge* **62**, 89–100.

O'KEEFE, J. H. (1985a). Population biology of the freshwater snail *Bulinus globosus* on the Kenya coast. 1. Population fluctuations in relation to climate. *Journal of Applied Ecology* **22**, 73–84.

O'KEEFE, J. H. (1985b). Population biology of the freshwater snail *Bulinus globosus* on the Kenya coast. 2. Feeding and density effects on population parameters. *Journal of Applied Ecology* **22**, 85–90.

PAGÈS, J. R., DURAND, P., SOUTHGATE, V. R., TCHUEM TCHUENTÉ, L. A. & JOURDANE, J. (2001). Molecular arguments for splitting *Schistosoma intercalatum* into two distinct species. *Parasitology Research* **87**, 57–62.

PAGÈS, J. R. & THÉRON, A. (1990). *Schistosoma intercalatum* from Cameroon and Zaire – chronobiological differentiation of cercarial emergence. *Journal of Parasitology* **76**, 743–745.

PICQUET, M., ERNOULD, J. C., VERCRUYSSE, J., SOUTHGATE, V. R., MBAYA, A., SAMBOU, B., NIANG, M. & ROLLINSON, D. (1996). The epidemiology of human schistosomiasis in the Senegal River Basin. *Transactions of the Royal Society of Tropical Medicine and Hygiene* **90**, 340–346.

PITCHFORD, R. J., MEYLING, A. H., MEYHLING, J. & DU TOIT, J. F. (1969). Cercarial shedding patterns of various schistosome species under outdoor conditions in the Transvaal. *Annals of Tropical Medicine and Parasitology* **63**, 359–371.

PRINGLE, G., OTIENO, L. H. & CHIMTAWI, M. B. (1971). Notes on the morphology, susceptibility to *S. haematobium* and genetic relationships of *Bulinus* (*P.*) *globosus* and *B.* (*P.*) *nasutus* from north-eastern Tanzania. *Annals of Tropical Medicine and Parasitology* **65**, 211–219.

PRESTON, T. M. & SOUTHGATE, V. R. (1994). The species specificity of *Bulinus–Schistosoma* interactions. *Parasitology Today* **10**, 69–73.

RAAHAUGE, P. & KRISTENSEN, T. K. (2000). A comparison of *Bulinus africanus* group species (Planorbidae; Gastropoda) by use of the internal transcribed spacer 1 region combined by morphological and anatomical characters. *Acta Tropica* 75, 85–94.

RAYMOND, K. & PROBERT, A. J. (1987). The effect of light and darkness on the production of cercariae of *Schistosoma haematobium* from *Bulinus globosus*. *Journal of Helminthology* 61, 291–296.

RAYMOND, K. & PROBERT, A. J. (1991). The daily cercarial emission rhythm of *Schistosoma margrebowiei* with particular reference to dark period stimuli. *Journal of Helminthology* 65, 159–168.

ROLLINSON, D., DE CLERCQ, D., SACKO, M., TRAORÉ, M., SÈNE, M., SOUTHGATE, V. R. & VERCRUYSSE, J. (1997b). Observations on compatibility between *Bulinus truncatus* and *Schistosoma haematobium* in the Senegal River Basin. *Annals of Tropical Medicine and Parasitology* 91, 371–378.

ROLLINSON, D. & KANE, R. A. (1991). Restriction enzyme analysis of DNA from species of *Bulinus* (Basommatophora: Planorbidae) using a cloned ribosomal gene probe. *Journal of Molluscan Studies* 57, 93–98.

ROLLINSON, D., KAUKAS, A., JOHNSTON, D. A., SIMPSON, A. J. G. & TANAKA, M. (1997a). Some molecular insights into schistosome evolution. *International Journal for Parasitology* 27, 11–28.

ROLLINSON, D. & SOUTHGATE, V. R. (1979). Enzyme analyses of *Bulinus africanus* group snails (Mollusca: Planorbidae) from Tanzania. *Transactions of the Royal Society of Tropical Medicine and Hygiene* 73, 667–672.

ROLLINSON, D. & SOUTHGATE, V. R. (1987). The genus *Schistosoma*: a taxonomic appraisal. In *The Biology of Schistosomes: From Genes to Latrines* (ed. Rollinson, D. & Simpson, A. J. G.), pp. 347–378. London, Academic Press.

ROLLINSON, D., STOTHARD, J. R., JONES, C. S., LOCKYER, A. E., PEREIRA DE SOUZA, C. & NOBLE, L. R. (1998). Molecular characterisation of intermediate snail hosts and search for resistance genes. *Memorias do Instituto Oswaldo Cruz* 93, 111–116.

ROLLINSON, D., VERCRUYSSE, J., SOUTHGATE, V. R., MOORE, P. J., ROSS, G. C., WALKER, T. K. & KNOWLES, R. J. (1987). Observations on human and animal schistosomiasis in Senegal. In *Helminth Zoonoses* (ed. Geerts, S., Kumar, V. & Brandt, J.), pp. 119–131. Dordrecht, The Netherlands, Martinus Nijhoff Publishers.

ROLLINSON, D. & WRIGHT, C. A. (1984). Population studies on *Bulinus cernicus* from Mauritius. *Malacologia* 25, 447–463.

SAVIOLI, L., DIXON, H., KISUMKU, U. M. & MOTT, K. E. (1989). Control of morbidity due to *S. haematobium* on Pemba island: programme organization and management. *Tropical Medicine and Parasitology* 40, 189–194.

SÈNE, M. & SOUTHGATE, V. R. (1998). Comparison of *Bulinus truncatus* from Mali and Senegal, West Africa, by restriction fragment length polymorphism (RFLP) analysis of the ribosomal internal transcribed spacer DNA (ITS). In *Proceedings of Workshop on Medical Malacology in Africa*. Harare, Zimbabwe, pp. 231–236 (ed. Madsen, H., Appleton, C. C. & Chimbari, M.), Danish Bilharziasis Laboratory, Charlottenlund.

SHIFF, C. J. (1964a). Studies on *Bulinus* (*Physopsis*) *globosus* in Rhodesia. 1. The influence of temperature on the intrinsic rate of natural increase. *Annals of Tropical Medicine and Parasitology* 58, 94–105.

SHIFF, C. J. (1964b). Studies on *Bulinus* (*Physopsis*) *globosus* in Rhodesia. 3. Bionomics of a natural population existing in a temporary habit. *Annals of Tropical Medicine and Parasitology* 58, 240–255.

SOUTHGATE, V. R. (1997). Schistosomiasis in the Senegal River Basin: before and after the construction of dams at Diama, Senegal and Manantali, Mali and future prospects. *Journal of Helminthology* 71, 125–132.

SOUTHGATE, V. R., BROWN, D. S., ROLLINSON, D., ROSS, G. C. & KNOWLES, R. J. (1985c). *Bulinus tropicus* from Central Kenya acting as a host for *Schistosoma bovis*. *Zeitschrift für Parasitenkunde* 71, 61–69.

SOUTHGATE, V. R., BROWN, D. S., WARLOW, A., KNOWLES, R. J. & JONES, A. (1989). The influence of *Calicophoron microbothrium* on the susceptibility of *Bulinus tropicus* to *Schistosoma bovis*. *Parasitology Research* 75, 381–391.

SOUTHGATE, V. R., DE CLERCQ, D., SÈNE, M., ROLLINSON, D., LY, A. & VERCRUYSSE, J. (2000). Observations on the compatibility between *Bulinus* spp. and *Schistosoma haematobium* in the Senegal River basin. *Annals of Tropical Medicine and Parasitology* 94, 157–164.

SOUTHGATE, V. R., HOWARD, G., ROLLINSON, D., BROWN, D. S., ROSS, G. C. & KNOWLES, R. J. (1985b). *Bulinus tropicus* a natural intermediate host for *Schistosoma margrebowiei*. *Journal of Helminthology* 59, 153–155.

SOUTHGATE, V. R., JOURDANE, J. & TCHUEM TCHUENTÉ, L. A. (1998). Recent studies on the reproductive biology of the schistosomes and their relevance to speciation in the Digenea. *International Journal for Parasitology* 28, 1159–1172.

SOUTHGATE, V. R., ROLLINSON, D., KAUKAS, A., ALMEDA, J., CASTRO, F., SOARES, E. & CORACHAN, M. (1994). Schistosomiasis in the Republic of São Tomé and Principe: characterisation of *Schistosoma intercalatum*. *Transactions of the Royal Society of Tropical Medicine and Hygiene* 88, 479–486.

SOUTHGATE, V. R., ROLLINSON, D., ROSS, G. C., KNOWLES, R. J. & VERCRUYSSE, J. (1985a). On *Schistosoma curassoni*, *S. haematobium* and *S. bovis* from Senegal: development in *Mesocricetus auratus*, compatibility with species of *Bulinus* and enzymes. *Journal of Natural History* 19, 1249–1267.

STOTHARD, J. R. (1997). Phylogenetic inference with RAPDs: some observations involving computer simulation with viral genomes. *Journal of Heredity* 88, 222–228.

STOTHARD, J. R., HUGHES, S. & ROLLINSON, D. (1996). Variation within the ribosomal internal transcribed spacer (ITS) from freshwater snails of the genus *Bulinus*. *Acta Tropica* 61, 19–29.

STOTHARD, J. R., LOXTON, N., ROLLINSON, D., MGENI, A. F., KHAMIS, S., AMERI, H., RAMSAN, M. & SAVIOLI, L. (2000). The transmission status of *Bulinus* on Zanzibar Island (Unguja), with implications for control of urinary schistosomiasis. *Annals of Tropical Medicine and Parasitology* 94, 87–94.

STOTHARD, J. R., MGENI, A. F., ALAWI, K. S., SAVIOLI, L. & ROLLINSON, D. (1997). Observations on shell morphology, enzymes and Random Amplified

Polymorphic DNA (RAPD) in *Bulinus africanus* group snails (Gastropoda: Planorbidae) in Zanzibar. *Journal of Molluscan Studies* **63**, 489–503.

STOTHARD, J. R. & ROLLINSON, D. (1996). An evaluation of random amplified polymorphic DNA for identification and phylogeny of freshwater snails of the genus *Bulinus* (Gastropoda: Planorbidae). *Journal of Molluscan Studies* **62**, 165–176.

STOTHARD, J. R. & ROLLINSON, D. (1997a). Molecular characterisation of *Bulinus globosus* and *B. nasutus* on Zanzibar, and an investigation of their roles in the epidemiology of *Schistosoma haematobium*. *Transactions of the Royal Society of Tropical Medicine and Hygiene* **91**, 353–357.

STOTHARD, J. R. & ROLLINSON, D. (1997b). Partial DNA sequences from the mitochondrial cytochrome oxidase subunit I (COI) gene can differentiate the intermediate snail hosts *Bulinus globosus* and *B. nasutus* (Gastropoda: Planorbidae). *Journal of Natural History* **31**, 727–737.

STRAHAN, K., KANE, R. A. & ROLLINSON, D. (1991). Development of cloned DNA probes for identification of snail intermediate hosts within the genus *Bulinus*. *Acta Tropica* **48**, 117–126.

STURROCK, R. F. (1965). The development of irrigation and its influence on the transmission of bilharziasis in Tanganyika. *Bulletin of the World Health Organisation* **32**, 225–236.

THÉRON, A. (1989). Hybrids between *Schistosoma mansoni* and *Schistosoma rodhaini* characterization by cercarial emergence rhythms. *Parasitology* **99**, 225–228.

THÉRON, A. & COMBES, C. (1988). Genetic-analysis of cercarial emergence rhythms of *Schistosoma mansoni*. *Behavioural Genetics* **18**, 201–209.

UGOZZOLI, L. & WALLACE, R. B. (1991). Allele-specific polymerase chain reaction. *Methods: A Companion to Methods in Enzymology* **2**, 42–48.

VERA, C., JOURDANE, J., SELLIN, B. & COMBES, C. (1990). Genetic variability in the compatibility between *Schistosoma haematobium* and its potential vectors in Niger. Epidemiological implications. *Tropical Medical Parasitology* **41**, 143–148.

VERCRUYSSE, J., DE CLERCQ, D., SACKO, M., TRAORÉ, M., SOUTHGATE, V. R., ROLLINSON, D., DE BONT, J., MUNGOMBA, L. & NIANG, M. (1994). Studies on transmission and schistosome interactions in Senegal, Mali and Zambia. *Tropical and Geographical Medicine* **46**, 220–226.

VIARD, F., BRÉMOND, P., LABBO, R., JUSTY, F., DELAY, B. & JARNE, P. (1996). Microsatellites and the genetics of highly selfing populations in the freshwater snail *Bulinus truncatus*. *Genetics* **142**, 1237–1247.

WEBBE, G. (1962). The transmission of *Schistosoma haematobium* in an area of lake province, Tanganyika. *Bulletin of the World Health Organisation* **27**, 59–85.

WEBBE, G. & MSANGI, A. S. (1958). Observations on three species of *Bulinus* on the East coast of Africa. *Annals of Tropical Medicine and Parasitology* **52**, 302–314.

WOOLHOUSE, M. E. J. & CHANDIWANA, S. K. (1989). Spatial and temporal heterogeneity in the population dynamics of *Bulinus globosus* and *Biomphalaria pfeifferi* and in the epidemiology of their infection with schistosomes. *Parasitology* **98**, 21–34.

WOOLHOUSE, M. E. J. & CHANDIWANA, S. K. (1990a). Population biology of the freshwater snail *Bulinus globosus* in the Zimbabwe highveld. *Journal of Applied Ecology* **27**, 41–59.

WOOLHOUSE, M. E. J. & CHANDIWANA, S. K. (1990b). Population dynamics model for *Bulinus globosus*, intermediate host for *Schistosoma haematobium*, in river habitats. *Acta Tropica* **47**, 151–160.

WOOLHOUSE, M. E. J. & WEBSTER, J. P. (2000). In search of the Red Queen. *Parasitology Today* **16**, 506–508.

WRIGHT, C. A. (1974). Snail susceptibility or trematode infectivity? *Journal of Natural History* **8**, 545–548.

WRIGHT, C. A. & KNOWLES, R. J. (1972). Studies on *Schistosoma haematobium* in the laboratory. 3. Strains from Iran, Mauritius and Ghana. *Transactions of the Royal Society of Tropical Medicine and Hygiene* **66**, 108–118.

Bulinus species on Madagascar: molecular evolution, genetic markers and compatibility with *Schistosoma haematobium*

J. R. STOTHARD[1]*, P. BRÉMOND[2], L. ANDRIAMARO[2], B. SELLIN[2],
E. SELLIN[2] *and* D. ROLLINSON[1]

[1] *Wolfson Wellcome Biomedical Laboratories, Department of Zoology, The Natural History Museum, Cromwell Road, London, SW7 5BD*
[2] *Equipe Schistosomoses I.R.D., BP 434, 101 Antananarivo, Madagascar*

SUMMARY

Of the four species of *Bulinus* found on Madagascar, three species: *B. obtusispira*, *B. liratus* and *B. bavayi* are endemic while the fourth, *B. forskalii*, is probably a recent introduction from the African mainland. The evolutionary relationships of these species with *Bulinus* species from Africa were studied by phylogenetic analysis of DNA sequence variation at two mitochondrial loci: cytochrome oxidase subunit I (COI) and large ribosomal subunit (LSU) or 16S. The observed levels of nucleotide divergence within *Bulinus* were substantial but may underestimate the true levels as there was evidence of 'saturation' of transitional substitutions at both loci. A putative secondary structure model for the sequenced segment of the 16S was developed. Subsequent phylogenetic analysis using transversional changes only for both loci, showed that there were contrasting levels of divergence within the four species groups. *B. obtusispira* was consistently placed within the *B. africanus* group, appearing ancestral to this group and was closest to the basal node within *Bulinus*. Together with *B. bavayi*, the two species appear to have been isolated on Madagascar for a long time, contrasting with both *B. liratus* and *B. forskalii* that appear more recent colonisers; however, estimate of exact times of divergence is problematic. A PCR-RFLP assay was developed to enable identification and discrimination of *B. obtusispira* and *B. liratus* using discriminatory variation within the COI. To enable population genetic analysis within *B. obtusispira*, microsatellite markers were developed using an enrichment method and 8 primer pairs are reported. Laboratory infection experiments using Madasgacan *S. haematobium* from the Mahabo area showed that certain populations of *B. obtusispira*, *B. liratus* and *B. bavayi* were compatible.

Key words: Madagascar, *Bulinus*, *Schistosoma haematobium*, mitochondrial DNA, phlylogeny, microsatellites.

INTRODUCTION

Madagascar is the fourth largest island in the world and is located off the Mozambique coast of Africa (Fig. 1). The human population of the Republic of Madagascar numbers over 16 million (Anon, 1999) and the Malagasy are of mainly Malay-Indonesian origin (Preston-Mafham, 1991). Schistosomiasis is endemic on Madagascar with estimates of 2 million infections of *Schistosoma mansoni* and 0·5 million of *S. haematobium* (Ollivier, Brutus & Cot, 1999). With one or two exceptions, the two forms of the disease follow a strong east-west division. Intestinal schistosomiasis is found in eastern and central regions, where it is also spreading (Ollivier *et al.* 1999). Urinary schistosomiasis is in central and western regions but is absent in the east (Doumenge *et al.* 1987).

The pulmonate snail family Planorbidae is widely distributed throughout the world with origins thought to be within the Triassic (Baker, 1945). The subfamily Bulininae contains two genera, *Bulinus*

and *Indoplanorbis*. The genera are conchologically different, physoid and discoid respectively, but are both associated with the transmission of schistosomes (Rollinson & Southgate, 1987). Several of the 37 species of *Bulinus* act as intermediate hosts for parasites of the *S. haematobium* group throughout Africa, Madagascar and adjacent regions (Brown, 1994). Whilst *Indoplanorbis* is absent from the African mainland save for recent introductions (Brown, 1994), *I. exustus* is an intermediate host for members of the *S. indicum* group throughout India and south-east Asia (Rollinson & Southgate, 1987).

Many elements of the fauna of Madagascar are strongly associated with endemicity (Jolly, Oberle & Albignac, 1984). The paleogeographic history of Madagascar is firmly established based upon sedimentological, structural and paleomagnetic data (Rabinowitz, Coffin & Falkey 1983; Piqué, 1999). The island was once part of Gondwana (Fig. 1, inset A) and separated from Africa at a present position close to Kenya/Tanzania (Piqué, 1999). The conjoined Madagascar and India subsequently drifted southwards towards the island's present position around 120 MYA (Storey *et al.* 1995). Madagascar finally separated from India at a time when dis-

* Address for correspondence: e-mail: j.r.stothard@
nhm.ac.uk

Parasitology (2001), **123**, S261–S275.   © 2001 Cambridge University Press
DOI: 10.1017/S003118200100806X

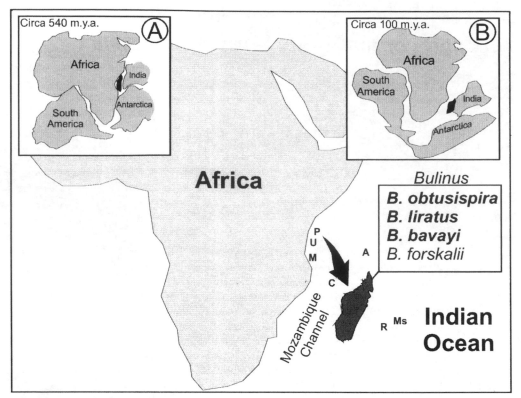

Fig. 1. Sketch map of the geography and geological evolution of Madagascar with present species of *Bulinus*. Insert A – depiction of Madagascar's placement in the land mass of Gondwana during the Protozeroic taken from Piqué (1999). Insert B – depiction of continental break up of Gondwana land masses during the Cretaceous taken from Smith *et al.* (1994). The three species of *Bulinus* in bold are endemic to Madagascar, *B. obtusispira* (*B. africanus* group), *B. bavayi* (*B. forskalii* group) and *B. liratus* (*B. truncatus/tropicus* complex). Key of Indian Ocean Islands: P–Pemba, U–Unguja (Zanzibar Island), M–Mafia, C–Comoros, A–Aldabra, R–Réunion, Ms–Mauritius.

integration of virtually all of the individual Gondwanan land masses had begun, circa 88 MYA (Mildenhall, 1980), Fig. 1, inset B. At that time, although physically isolated, dispersal pathways for many organisms via Antarctica to South America were thought to remain up to the late Eocene, circa 40 MYA (Raven, 1979).

Tristram (1863) provided the first account of freshwater snails on Madagascar. Since then there have been several taxonomic revisions, see Starmühlner (1969) & Wright (1971a). While Starmühlner (1969) recognized only two species of *Bulinus*, *B. liratus* and *B. mariei*, with the latter now treated as synonymous with *B. forskalii* (Brown, 1994), four species are accepted today. The four species have affinities with three of the four species groups of *Bulinus* (Fig. 1). Two species, *B. obtusispira* and *B. liratus* are restricted to Madagascar, whereas *B. bavayi* is also found on Aldabra (Wright, 1971a), see Fig. 1. In contrast, *B. forskalii* has a pan-African distribution.

B. obtusispira is placed within the *B. africanus* group due to immunological and biochemical assays (Wright, 1971a, b; Brown & Wright, 1978; Wright & Rollinson, 1979; Jelnes, 1984). Conchologically and anatomically it can be confused with members of the *B. truncatus/tropicus* complex in which *B. liratus* is placed (Wright, 1971b). Its specific status was

reconsidered when a collection of *B. liratus*, highly compatible with *S. haematobium*, was encountered (Brygoo & Moreau, 1966). Wright (1971b) concluded that *B. obtusispira* was a 'relic' species of *Bulinus*, perhaps paralleling the unusual evolution of other endemic Madagascan groups, and proposed that ancestral *B. obtusispira* became isolated on Madagascar at a time when the *B. africanus* group had only recently diverged from other bulinids.

Wright (1971b) noted that while *B. obtusispira* was highly compatible with *B. forskalii* group-borne *S. haematobium* and compatible with *B. truncatus*-borne *S. haematobium*, it was curiously incompatible with *B. africanus* group-borne *S. haematobium*. The sporocyst development of Madagascan *S. haematobium* in *B. obtusispira* has some unusual features (Jourdane, 1983). The transmission status of *B. liratus* was less clear. Wright (1971a) concluded that like other diploids of the *B. truncatus/tropicus* complex it did not prove to be susceptible to any strain of *S. haematobium* to which it had been exposed. In the Mangoky region, however, Degrémont (1973), who also used immuno-diffusion tests for snail identifications similar to those of Wright (1971a), found naturally infected *B. liratus*. By laboratory infections, Degrémont (1973) was also able to demonstrate compatibility with local *S. haematobium*. Similarly the transmission status of *B.*

Table 1. Details of *Bulinus* populations used for COI* and 16S sequences

Marker	Species/Country	Population (NHM no.)	GenBank no.
COI	*B. obtusispira*/Madagascar	Andohaviana (2000)	AF369588
	B. obtusispira/Madagascar	Ambilobe (1998)	AF369589
	B. liratus/Madagascar	Befeno (1966)	AF369590
	B. liratus/Madagascar	Ranohira (1967)	AF369591
	B. forskalii/Madagascar	Vohémar (2092)	AF369592
	B. bavayi/Madagascar	Ranohira (1967)	AF369593
	B. bavayi/Madagascar	Nosibe (2093)	AF369594
	B. globosus/Zanzibar Is.	Kisongoni (2019)	AF369595
	B. globosus/Niger	Liboré (3003)	AF369596
	B. globosus/Ivory Coast	Kan Gare (3002)	AF369597
	B. nasutus/Mafia Is.	Site M15 (2013)	AF369598
	B. umbilicatus/Niger	Baban Tabkin (3001)	AF369599
	B. umbilicatus/Niger	Bouboute (3000)	AF369600
	B. africanus/South Africa	Pietermaritzburg (2094)	AF369601
	B. tropicus/Kenya	Lanet (1253)	AF369602
	B. tropicus/Zambia	Kalumba (1638)	AF369603
	B. truncatus/Malawi	Nyemba (1087)	AF369604
	B. truncatus/Sudan	Keriab (2095)	AF369605
	B. truncatus/Sudan	Abu-Ushar (2096)	AF369606
	B. truncatus/Sudan	Gamooeya (2097)	AF369607
	B. truncatus/Sudan	Faki Hashem (2098)	AF369608
	B. truncatus/Sudan	Managil (2099)	AF369609
	B. forskalii/Mafia Is.	Site M3 (2012)	AF369610
	B. forskalii/Pemba Is.	Kinowe (2012)	AF369611
	B. sp./Mafia Is.	Kilindoni (2014)	AF369612
	B. cernicus/Mauritius	Vallée Pitot (1689)	AF369613
	B. wrighti/France	Lab. stock (1697)	AF369614
	B. wrighti/Oman	Baushar (1698)	AF369615
16S	*B. obtusispira*/Madagascar	Andohaviana (2000)	AY029542
	B. liratus/Madagascar	Befeno (1966)	AY029543
	B. bavayi/Madagascar	Ranohira (1967)	AY029544
	B. forskalii/Madagascar	Vohémar (2092)	AY029545
	B. globosus/Zanzibar Is.	Kisongoni (2019)	AY029546
	B. nasutus/Zanzibar Is.	Kibonde (2019)	AY029547
	B. truncatus/Burundi	Cyohoha (1087)	AY029548
	B. truncatus/Malawi	Nyemba (1087)	AY029549
	B. forskalii/Mafia Is.	Site M3 (2012)	AY029550
	B. sp./Mafia Is.	Kilindoni (2014)	AY029551
	B. wrighti/Oman	Baushar (1698)	AY029552

* Other COI sequences were obtained from Stothard & Rollinson (1997 *a*).

bavayi remained uncertain; whilst Degrémont (1973) was unable to show compatibility with local *S. haematobium*, Wright (1971 *a*), using isolates of *S. haematobium* from Mauritius, found *B. bavayi* to be compatible.

The studies in this paper stem from a collaborative project concerned with the distribution and transmission status of *Bulinus* species on Madgascar. This paper will focus upon the molecular taxonomy and evolution of the *Bulinus* species encountered and on laboratory compatibility studies. Phylogenetic analysis of DNA sequences from mitochondrial cytochrome oxidase subunit I (COI) was chosen since this gene target has been useful in interspecific comparisons within *Bulinus* (Stothard & Rollinson, 1997 *a*; Jones *et al.* 1999) and other gastropods (Davis *et al.* 1998; Campbell *et al.* 2000). In addition, variation in the mitochondrial ribosomal large subunit (16S) was analysed and a secondary structure

model for this region in *Bulinus* is proposed. To compare the genetic structures of populations of *B. obtusispira* across the island, recent progress in isolation of microsatellite marker loci is reported.

MATERIALS AND METHODS

Snail identification

Identification of snails was by conchological examination and inspection of isoenzyme profiles for four enzyme systems: glucose phosphate isomerase (GPI), malate dehydrogenase (MDH), acid phosphatase (ACP) and phosphoglucomutase (PGM) following separation by iso-electric focusing (Wright & Rollinson, 1979). Morphological descriptions, enzyme profiles and distribution of snail species on Madagascar will be described later by Brémond *et al.* (unpublished).

Amplification COI, 16S and phylogenetic analysis

The COI was amplified according to methods detailed in Stothard & Rollinson (1997a). In total a 340 bp data set covering the region of the COI was analysed. From the 35 sequences generated in this study, 28 have been deposited in GenBank and are detailed in Table 1 to which other, existing COI data were added (Stothard & Rollinson, 1997a). The sequences were aligned by eye since there were no insertions or deletions.

A sub-region of the 16S was amplified using the conserved primers 16Sar-L [5'-cgcctgtttatcaaaaacat] and 16Sbr-H [5'-ccggtctgaactcagatcacgt] detailed by Thollesson (1999) and Remigio & Blair (1997). A PCR product of approximately 450 bp was produced, extracted from an agarose gel using QIAEX II (Qiagen, Germany) according to manufacturer's instructions and then sequenced using ABI PRISM BigDye cycle sequencing with dye terminators (ABI, U.K.). The 11 sequences of the 16S generated in this study have been deposited in GenBank and are detailed in Table 1. With the addition of 16S data from two outgroup taxa, *Biomphalaria pfeifferi* and *Lymnaea stagnalis* (Remigio & Blair, 1997), the sequence data were aligned using Clustal-V (Higgins, Bleasby & Fuchs, 1992) using default parameters. This alignment was subsequently modified with the use of secondary structure models that are available electronically from "http://www.rna.icmb.utexas.edu". Stem and loop regions were identified and sequence nomenclature followed that proposed by Kjer (1995). The secondary structure, however, within the region between L10–L11 was not easily identified by direct comparison to other taxa and an investigation of the intra-strand folding of this region was conducted with the program RNAdraw 1·0 (Matzura & Wennborg, 1996). A sliding window analysis over the 16S alignment, window size – 7 bp, was performed to identify the most variable regions in the alignment.

Basic nucleotide statistics (nucleotide composition, transition (Ts) and transversion (Tv) frequencies, number of variable and parsimony informative sites) were calculated for the COI and 16S in MEGA (Kumar, Tamura & Nei, 1993). Analysis of the COI and 16S data, treated separately, was conducted using distance methods with programs PHYLIP (Felsenstein, 1993) and MEGA (Kumar *et al.* 1993), maximum parsimony with PAUP* (Swofford, 1999), and maximum likelihood with PAUP* (Swofford, 1999) and PUZZLE 3·1 (Strimmer & Von Haeseler, 1997). The resultant phylograms were compared between methods and assessed for congruence. Tests for phylogenetic signal such as skewness of observed tree length to that of randomly generated trees and Tajima's test for neutrality were also conducted (Tajima, 1989).

Microsatellite isolation

Twenty individual *B. obtusispira* from Miandrivazo (site 20, Table 4) were selected and total genomic DNA was extracted following protocols of Stothard, Hughes & Rollinson (1996) for each snail and then pooled, adjusting for DNA concentrations, to form a genomic stock. This genomic stock was digested with *Mbo*I and microsatellites were isolated from this preparation using a single-strand-capture (SSC) enrichment method as described by Hammond *et al.* (1998). Mirocosatellite hybridization probes of either $(ca)_{18}$ and $(ct)_{18}$ sequence that were 5' biotin labelled were used for SSC. Post-enrichment preparations were then blunt-end cloned into *E. coli* (Stratagene, U.K.). Clones containing repetitive regions were identified after hybridisation to microsatellite probes. Briefly, bacterial colonies were transferred onto nylon membranes followed by cell lysis then hybridisation to 5' poly-digoxigenin-labelled $(ca)_{18}$ and $(ct)_{18}$ probes. Microsatellite containing clones were identified after autoradiography using chemiluminescence detection methods according to the manufacturer's instructions (Roche Molecular Biochemicals, U.K.). Those clones with positive inserts after hybridisation were subsequently amplified by PCR using plasmid primers, purified and sequenced. Cloned inserts found to contain microsatellite motifs, with sufficient flanking sequence to anchor PCR primers, were selected to which PCR primers sets were designed. A single clone containing a minisatellite motif 'AATCATGCTC' was also found to which primers were designed. Primers sets were tested and optimised by varying either $MgCl_2$ or annealing temperature of the PCR upon the original genomic stock from Miandrivazo and DNA preparations from individual snails from the same population.

Snail infections and parasite isolation

Snail infection experiments were conducted from December 1996 to July 1999 as laboratory snail material became available. All snails used for infections originated from laboratory colonies that ranged from first to eighth generation stocks. The colonies had been established from field collected snails. The snail colonies were identified by morphological examination and, where possible, isoenzyme and(or) DNA analysis. The majority of exposed snails were either of second or third generation. Snails were individually exposed to *S. haematobium* collected from the Mahabo area (the villages: Antsakoavaky, Soaserana, Andohaviana, Soafoasa) using approximately five miracidia per snail. Schistosome eggs were directly obtained from infected urines that were taken, after the patient's informed consent, as part of routine epidemiological surveillance of urinary schistosomiasis in Mahabo

A)

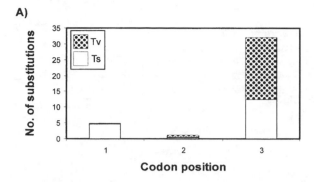

B)

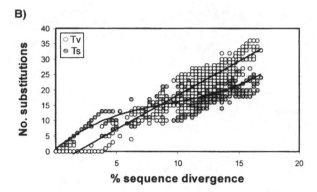

C)

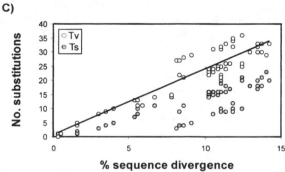

Fig. 2. Plots of transitions (Ts) and transversions (Tv) for *Bulinus* with respect to codon position and % sequence divergence (p-distance) show (non)linear relationships. Fig. 2A. Variation of Ts and Tv at 1st, 2nd and 3rd codon positions in the COI. Fig. 2B. Plot of Ts and Tv against % sequence divergence for the COI shows nonlinear relationship for Ts and linear relationship Tv. Fig. 2C. Plot of Ts and Tv against % sequence divergence for the 16S shows a linear relationship for Tv, and that there are less Ts than Tv substitutions.

area. Patients found infected with schistosomes were provided with a curative dose of praziquantel. The schistosome eggs were concentrated by sedimentation in saline, eggs were hatched after exposure to freshwater. After 28 days post-challenge, snails were visually inspected for shedding of schistosome cercariae once a week until 50 days after exposure.

RESULTS

COI

The 40 aligned COI sequences from *Bulinus* were found to be A:T rich (70·5%), with nucleotide

compositions of: T (42%), C (11·9%), A (28·5%) and G (17·6%). Of the 340 nucleotide positions, 218 (64%) were found to be invariant. A total of 122 (36%) sites varied within *Bulinus*; 104 were parsimony informative while the remainder (18) were singletons. Using the *Drosophilia* mitochondrion codon table, 17 (15%) positions of the 113 amino acids across *Bulinus* would be variable. Tajima's test of neutrality showed that substitutions did not deviate significantly from that expected under neutral evolution (D = −0·21, P > 0·1).

The number of Transitions (Ts) and Transversions (Tv) substitutions was plotted against codon position (Fig. 2A). Within the aligned data set there were on average ± one standard deviation, 4·8 ± 2·3 substitutions in the 1st position, 1 ± 0·8 in the 2nd and 32·1 ± 12 in the 3rd. The Ts 'v' Tv ratio differed considerably at each of the codon positions: 1st = 23, 2nd = 0·67 & 3rd = 0·64. The total number of Ts and Tv substitutions was plotted against % sequence divergence (p-distance × 100) to investigate (non)linear relationships (Fig. 2B). Up to 5% sequence divergence the number of Ts substitutions is in excess to that of Tv, however, at values > 5%, there is a greater number of Tv than Ts substitutions. A linear regression line could be fitted for Tv substitutions (y = 0·81x−3·5) with a high coefficient of determination (R^2 = 0·9). The trend of Ts against % sequence divergence does not follow a linear relationship, with a 3rd order polynomial (y = $0·0005x^3 − 0·04x^2 + 1·05x + 0·84$) the most predictive but with a low coefficient of determination (R^2 = 0·64).

Inspection of Ts 'v' Tv plots suggests that 'multiple hits' or 'saturation' of transitions was occurring at the 3rd codon position. Subsequent phylogenetic analysis either omitted the 3rd position entirely or analysed Tv substitutions alone since a linear relationship could be demonstrated. Phylogenetic signal could be shown for Tv substitutions by comparison with the observed tree length to those of 1000 random trees. The most justifiable analysis for this data is presented, Fig. 3, and was conducted as follows: calculation of Kimura-2-Parameter (K-2-P) distance only considering Tv substitutions, neighbour-joining clustering algorithm to infer the topology of the phylogram with bootstrap support of each node (1000 replicates). The phylogram's distances were linearised to enable comparison of the relative order of bifurcations. Where bootstrap values were below 65% the nodes were collapsed into a polytomy. Similar phylograms recovered with maximum parsimony and maximum likelihood, using HYK85 substitution model, with the exceptions that the relationships of *B. umbilicatus* and *B. africanus* within the *B. africanus* group were sometimes reversed. In consideration of Kimura-2-Parameter distance using Tv substitutions alone, there are contrasting levels of sequence divergence

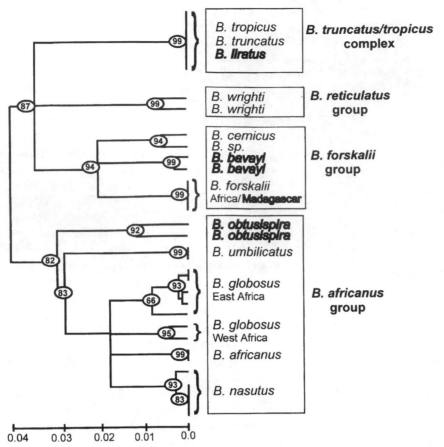

Fig. 3. A phylogram of relationships within *Bulinus* inferred from the COI calculated from distance data taken from Tv substitutions alone. Bootstrap values are placed on each node, the distances on the neighbor-joining phylogram were linearised, topology unchanged, with respect to relative time and the tree is unrooted. [If both Ts and Tv were considered many of the interspecific relationships that were previously resolvable became polytomies. With the exception that relationships within the *B. truncatus/tropicus* complex could now be determined]. Madagascan species are in bold typeface.

Table 2. *Nucleotide divergence (K-2-P) between species groups of* Bulinus. *Above diagonal, Tv & Ts considered, below diagonal Tv alone, with standard deviation (estimated by bootstrapping)*

	B. forskalii group	*B. africanus* group	*B. reticulatus* group	*B. truncatus/tropicus* complex
B. forskalii group		0·157 ± 0·019	0·150 ± 0·020	0·136 ± 0·018
B. africanus group	0·090 ± 0·013		0·167 ± 0·021	0·139 ± 0·018
B. reticulatus group	0·075 ± 0·013	0·096 ± 0·015		0·124 ± 0·018
B. truncatus/tropicus complex	0·076 ± 0·014	0·080 ± 0·013	0·064 ± 0·014	

within each of the species groups of *Bulinus* that were surveyed. In a descending order, *B. africanus* group (0·041), *B. forskalii* group (0·031), *B. reticulatus* group (0·015) and *B. truncatus/tropicus* complex (0·0). Table 2 shows the Kimura-2-parameter distance between the species groups calculated for all substitutions (above diagonal) and Tv alone (below diagonal).

The Madagascan species are clearly assigned to the species groups of *Bulinus*. Madagascan *B. forskalii* is almost identical in sequence to that sampled from mainland Africa, differing by only two substitutions across the alignment. Similarly, *B. liratus* is placed

with other *B. truncatus/tropicus* species and, as there were no Tv substitutions found, these species cannot be differentiated. If, however, Ts substitutions were considered then relationships reanalysed, the species could be differentiated giving the following relationships ((*B. tropicus* [Kenya & Zambia], (*B. truncatus* [Malawi], *B. liratus*)), (*B. truncatus* [Sudan])) with a mean nucleotide divergence within these species of 2·3%. In contrast the divergence of *B. bavayi* and *B. obtusispira* to other members of the same species group was much greater. However, whilst *B. bavayi* is placed within the *B. forskalii* group, the relationships of this species with other members were

A)

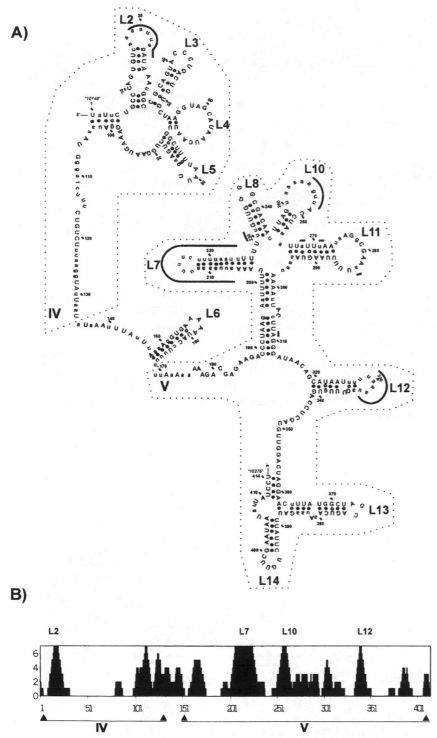

B)

Fig. 4. Putative secondary structure of the sequenced 16S region for *Bulinus* and sliding window analysis to detect heterogeneity across this region. Fig. 4A. Secondary structure for *B. liratus* that approximately covers 10748–10276 bp of the mitochondrial genome of *Katharina tunicata* (MIKTTU098). Nucleotides in caps denote invariant positions within *Bulinus*, nucleotide in lower case denote positions that vary within *Bulinus*. The broken line arcs refer to domains IV and V, the unbroken line arcs denote regions of the 16S that varied within *Bulinus* attributable to insertions/deletions. Fig. 4A. Sliding window analysis reveals heterogeneity across regions of the 16S, the stem and loop regions of L2, L7, L10 and L12 are particularly variable.

not resolvable i.e. a deep polytomy was encountered. *B. obtusispira* was consistently placed within the *B. africanus* group and appears to be ancestral to other members. Moreover it is the species lineage placed closest to the basal node within the phylogram.

16S

The length of the homologous region of the 16S between species of *Bulinus* varied considerably, by 32 bp, with a mean of approximately 400 bp. *B.*

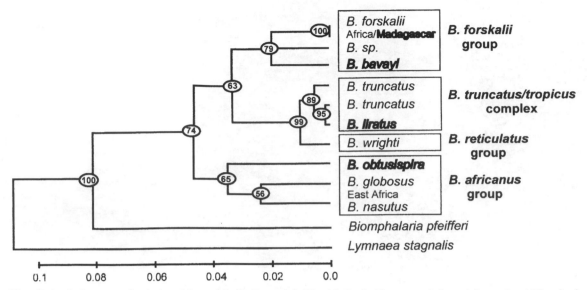

Fig. 5. A phylogram of relationships with *Bulinus* and *Biomphalaria/Lymnaea* inferred from the 16S calculated from distance data from Tv substitutions alone. Bootstrap values are placed on each node, the distances on the neighbour-joining phylogram were linearised, topology unchanged, with respect to relative time and the tree is rooted to the outgroups. Madagascan species are in bold typeface.

Table 3. *Microsatellite primer pair sets for* B. obtusispira *with predicted size and type of repeat motif*

Locus	Expected size (bp)	Type & Repeat	Primers
7S.22*	190	Broken $(GT)_3...(GT)_{13}$	F 5'-AGACCTAGGTCATGGTGAAA R 5'-CCCTTGCGATCCAATATAC
7S.25	171	Broken $(CA)_4...(CA)_7$	F 5'-CACTCACACTCACAAACAC R 5'-AGAGGGTGTATATTGTGGTGA
7S.26	243	Broken $(GT)_7...(GT)_{10}$	F 5'-TCAAGAAATTAATACACACATGA R 5'-ATTTTCCCATTGCAAATTGC
7S.29	190	Broken $(CT)_{10}...(CT)_3$	F 5'-TGTATTCATAATGTGCATGC R 5'-TTTCATCAAACACATTTTACG
7S.33	247	Pure $(GT)_4$	F 5'-AGAACCATAGGAATTTTCCC R 5'-GGTGGGCTTAGTACAGCTTA
7S.34*	264	Broken $(CA)_{14}...(CA)_8$	F 5'-GTCATTGGAATGAGGGATAC R 5'-GGCTTTTATAGTCTCCCTTG
7S.35	325	Broken, compound $(GT)_3 ... (CA)_4$	F 5'-ACCAATCAAAACACGAATCT R 5'-GGTTGGTGGATCACATACTC
7S.36*	140	Minisatellite $(AATCATGCTC)_5$	F 5'-ATCGATCTTTGTTTTGGTGA R 5'-CCAAGGGGTGATAACCAAC

* size variation detected in initial primer screen.

truncatus from Senegal possessed the longest sequence (419 bp) while *B. obtusispira* possessed the shortest (387 bp). The 16S sequences from *Bulinus* could be aligned over 419 positions and the nucleotide composition was A:T rich (71·1%) with individual nucleotide compositions of T (34·7%), C (11·9%), A (36·4%) and G (17%). Since the 16S does not code for a polypeptide, an analysis of substitutions at each codon position is not possible; however, an analysis across the total number of variable sites (112) was possible and is presented in Fig. 2C. The number Tv was in excess of Ts, a linear relationship of Tv against % sequence divergence could be shown (y = 2·34x + 0·77) with an acceptable coefficient of determination (R^2 = 0·85).

Across the alignment (419 bp), within *Bulinus*, there were 112 variable positions in terms of both insertions/deletions and substitutions. Of the substitutional changes, 72 were parsimony informative and 39 were singletons.

A putative secondary structure model for this part of the 16S is presented (Fig. 4). Further sequence data are required to confirm complementary base pairings over region IV 107–134 bp, region V 169–183 bp and 350–357 bp as these are presumed to pair with bases outside this sequenced region. A sliding window analysis of *Bulinus* sequences across the sequence reveals considerable heterogeneity (Fig. 4B). L2, L7, L10 and L12 are particularly variable where there were numerous insertions or deletions

Table 4. Transmission status of laboratory populations of *Bulinus*

Site	Locality	Species	GPS Latitude	GPS Longitude	Exposed (n)	Survival (%)	Infected (%)
1	Vohémar	*B. forskalii* group	13·25·518 S	49·57·316 E	107	49·0	0·0
2	Ambilobe	*B. obtusispira*	13·12·001 S	49·09·832 E	214	57·0	30·3
3	Bealanana 2	*B. liratus*	14·33·456 S	48·43·875 E	297	65·7	0·0
4	Ambatosia 2	*B. liratus*	14·39·933 S	48·39·611 E	158	67·1	0·0
5	Marotandrano #	*B. liratus*	16·10·849 S	48·50·821 E	240	62·5	78·0
6	Route Brieville	*B. liratus*	17·38·455 S	47·45·077 E	25	56·0	0·0
7	Bemokotra	*B. obtusispira*	17·03·423 S	46·41·955 E	249	55·4	5·8
8	Andranomena *	*B. liratus*	18·50·800 S	47·33·500 E	7	71·4	0·0
9	Anosizato-Ouest	*B. forskalii* group	18·56·000 S	47·30·200 E	22	45·5	0·0
10	1800m alt	*B. liratus*	18·53·320 S	46·41·864 E	12	58·3	0·0
11	Fonoraty	*B. forskalii* group	18·56·017 S	46·16·483 E	48	37·5	0·0
12	Manandona #	*B. liratus*	18·53·320 S	46·41·864 E	193	55·4	7·5
13	Sakay	*B. liratus*	19·00·310 S	46·26·990 E	27	48·1	0·0
14	Belanitra	*B. liratus*	19·36·580 S	46·26·393 E	142	69·7	1·0
15	Ambohimanga	*B. liratus*	19·51·073 S	46·55·000 E	46	52·2	0·0
16	Ireninoro	*B. liratus*	19·51·423 S	46·54·833 E	50	74·0	2·7
17	Andoharano	*B. liratus*	19·55·691 S	46·55·806 E	49	49·0	0·0
18	Bemasoandro	*B. liratus*	19·55·069 S	46·55·379 E	62	43·5	0·0
19	Miarinatsimo	*B. liratus*	19·55·197 S	46·55·834 E	12	41·7	0·0
20	Miandrivazo	*B. obtusispira*	18·56·983 S	45·13·583 E	449	46·5	31·1
21	Tsitelo*	*B. forskalii* group	20·15·800 S	44·22·000 E	48	60·4	0·0
22	Andranovorogise	*B. forskalii* group	20·20·495 S	44·33·300 E	89	14·6	0·0
23	Nosibe	*B. forskalii* group	20·15·770 S	44·45·486 E	60	53·3	0·0
24	Manombo	*B. obtusispira*	20·14·308 S	44·47·825 E	437	69·3	26·7
25	Anjamahitsy	*B. forskalii* group	20·17·998 S	44·43·006 E	12	58·3	0·0
26	Tanandava I*	*B. forskalii* group	20·22·200 S	44·41·000 E	36	11·1	0·0
27	Analamitsivalana	*B. forskalii* group	20·19·125 S	44·41·403 E	115	54·8	0·0
28	Tsinjorano-toby	*B. forskalii* group	20·20·520 S	44·35·117 E	42	47·6	0·0
29	Mahabo	*B. forskalii* group	20·22·235 S	44·40·259 E	6	66·7	0·0
30	Bemokoty	*B. forskalii* group	20·21·715 S	44·41·069 E	131	35·9	0·0
31	Soaserana	*B. forskalii* group	20·21·199 S	44·40·877 E	221	23·1	5·9
32	Ankilimida*	*B. forskalii* group	20·21·600 S	44·43·700 E	66	30·3	0·0
33	Ampanihy #	*B. forskalii* group	20·24·196 S	44·43·790 E	19	21·1	50·0
34	Andohaviana	*B. obtusispira*	20·28·980 S	44·38·101 E	178	28·1	82·0
35	Andriamilato	*B. liratus*	21·39·862 S	47·02·774 E	50	66·0	3·0
36	Tsaramandroso	*B. forskalii* group	21·23·361 S	48·01·757 E	16	56·3	0·0
37	Samangoky*	*B. obtusispira*	21·43·000 S	43·45·600 E	145	39·3	0·0
38	Ankiliabo #	*B. liratus*	22·46·326 S	43·35·802 E	74	85·1	9·5
39	Ankililoaka #	*B. liratus*	22·46·684 S	43·36·759 E	252	75·0	2·1
40	Sakaraha	*B. forskalii* group	22·54·926 S	44·32·110 E	96	39·6	0·0
41	Sakaraha	*B. liratus*	22·54·926 S	44·32·110 E	92	85·9	0·0
42	Ranohira #	*B. liratus*	22·34·130 S	45·25·982 E	150	50·7	6·6
43	Ranohira	*B. forskalii* group	22·34·130 S	45·25·982 E	124	36·3	26·7
44	Besely #	*B. liratus*	23·29·436 S	44·31·246 E	53	66·0	20·0
45	Kasaria #	*B. liratus*	24·12·384 S	45·39·843 E	413	64·4	8·3
46	Ampanihy (Sakatovo)	*B. liratus*	24·39·925 S	44·42·780 E	102	68·6	7·1
47	Andrahora	*B. liratus*	24·20·342 S	45·37·631 E	18	83·3	0·0
48	Kariera	*B. liratus*	25·00·039 S	46·36·568 E	135	70·4	35·8
49	Ambolofasy	*B. liratus*	25·02·092 S	46·44·786 E	210	65·7	11·6
50	Ampasy	*B. liratus*	25·02·181 S	46·44·404 E	86	37·2	21·9
51	Befeno #	*B. liratus*	25·01·737 S	46·46·379 E	129	79·8	12·6
52	Sonjorano	*B. liratus*	22·34·220 S	45·52·690 E	17	35·3	0·0
53	Analavoka	*B. liratus*	22·32·760 S	46·29·500 E	18	100·0	0·0

* approximate locations from FTM maps.
\# snails from these locations were also challenged with two isolates of *S. haematobium* from Senegal and Zambia.

(Fig. 4A & 4B). *B. obtusispira* was exceptional in that the L7 region is completely truncated. To enable a direct phylogenetic comparison of the 16S and COI, phylogenetic analysis of the 16S was conducted in the same manner as that for COI. A K-2-P distance only considering Tv substitutions was calculated but gamma corrected ($\gamma = 2$) to account for heterogeneity between sites. Neighbour-joining algorithm was used to infer the topology of the phylogram with bootstrap support of each node (1000 replicates).

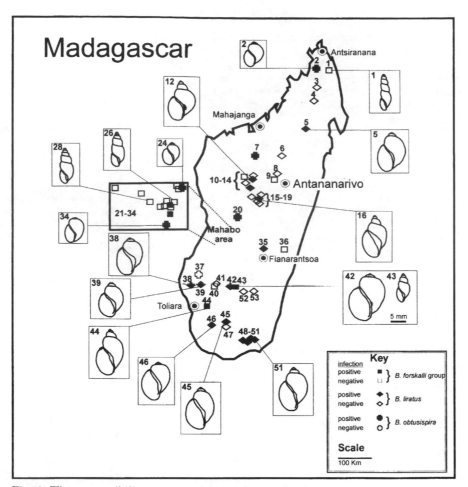

Fig. 6. The compatibility status and approximate distribution of *Bulinus* on Madagascar determined after laboratory challenge to progeny of field-collected populations. Outline shell drawings were made of representatives from some populations/species.

The phylogram distances were linearised to enable comparison of the relative order of bifurcations and two outgroup taxa are also considered (Fig. 5).

The four species groups of *Bulinus* are again obvious and there are contrasting levels of sequence divergence within species groups but there is resolution between the relationships of *B. forskalii* group, *B. truncatus/tropicus* complex and *B. reticulatus* group. There is minimal divergence between Madagascan and African *B. forskalii*, only a single substitutional change was detected. Slightly greater divergence was detected between *B. liratus* and *B. truncatus*, approximately 1·5 % when both Ts and Tv were considered. The relationships within the *B. forskalii* group remain unresolveable. *B. obtusispira* is clearly placed within the *B. africanus* group and appears ancestral within this group and a taxon placed towards the basal node of *Bulinus*.

Microsatellites

A total of 150 transformed, recombinant colonies were picked and probed for microsatellite motifs. Approximately 15 % of transformed colonies were positive by hybridisation. Twenty-two inserts were

sequenced in both +ve and −ve directions. From the 20 sequences that could be obtained, 11 primer pairs were designed and tested against genomic DNA from *B. obtusispira*. Upon the basis of an amplification product of approximately the correct size in an agarose gel, 8 primer pairs appeared successful while 3 primers pairs gave non-specific amplification smears and could not be optimised to yield specific amplification products. Size variation, ±50 bp, of the amplification products within the genomic pool was detectable for 3 of the primer pairs. The successful primer pairs and their microsatellite motifs are detailed in Table 3.

Compatibility with S. haematobium

In total the compatibility status of 53 populations of *Bulinus* has been determined: 29 *B. liratus*, 6 *B. obtusispira* and 18 *B. forskalii* group (the distinction between *B. bavayi* and *B. forskalii* was not made), Table 4. Percentage infection was calculated from the proportion of surviving snails found to be shedding cercariae. Where possible, outline drawing of representative shells were made. The distribution of the populations whose progeny were experimentally infected is shown (Fig. 6). Fifteen *B. liratus* populations were compatible with *S. haematobium*, percent infection of survivors ranged from 2–35 % but with one population, Marotandrano (site 5), with an infection rate of 78 %. *B. liratus* from Befeno and Ranohira were used for DNA analysis (see above). With the exception of Samangoky (site 37), all of the *B. obtusispira* populations were compatible 6–82 %. *B. obtusispira* from Andohaviana (site 34) and Ambilobe (site 2) were used for DNA analysis (see above). The majority of the *B. forskalii* group populations were incompatible. Three populations, however, were able to act as hosts: Ampanihy (site 33)–50 %, Ranohira (site 43)–27 % and Soaserana (site 31)–6 %. Snails from Ranohira were identified as *B. bavayi* and were used for DNA analysis (see above).

DISCUSSION

Molecular evolution

Phylogenetic analysis of the COI and 16S has clarified that *B. obtusispira* is a member of the *B. africanus* group. It appears that this lineage could also be one of the earliest members of the group. Moreover it is placed closest to the basal node of *Bulinus*. Wright's (1971 *a, b*) supposition that the *B. obtusispira* was a 'relic' species of *Bulinus* that became isolated on Madagascar at a time when the *B. africanus* group had only recently diverged from other bulinids is supported. The topology of the phylograms indicates that *Bulinus* could have colonised Madagascar on at least four separate occasions, with a relative temporal order of proto-*B. obtusispira* first, -*B. bavayi* second, -*B. liratus* third then -*B. forskalii* fourth (Fig. 5). The levels of sequence divergence between Madagascan and African taxa can also be ranked in the reverse of this order, minimal divergence of *B. forskalii* to considerable divergence of *B. obtusispira*.

Whilst the relative temporal placement of nodes can be inferred, one potential problem, however, is placing these events within a specific time-scale, as there are no extrinsic data to calibrate nodes with actual times, neither within *Bulinus*, nor those between *Bulinus* and *Biomphalaria* or *Lymnaea*. If we apply an approximate general mitochondrial molecular clock rate of 2–4 % Myr^{-1} (million years) utilised by Campbell *et al.* (2000) for COI sequences from *Biomphalaria*, then divergence between *B. obtusispira* and other *B. africanus* taxa could have been at the end of the Miocence/beginning of the Pliocene, circa 7 Myr^{-1}. This would be inferred if both transitions and transversions are considered [K-2-P, 16S $\sim$ 10 %, COI $\sim$ 14 %]. Presumably shortly before this time, *Bulinus* would have undergone a radiation of species such that the present day dispersal of populations does not reflect the long-term patterns of vicariances. Even if the general molecular clock is true the observed nucleotide

distances between *Bulinus* may be wrong and a gross underestimate, since Ts introduce unacceptable errors to the data, so the time frame of the Miocence/Pliocene is too early.

Inspection of the non-linear relationships of transitions against percent sequence divergence is indicative of strong 'saturation' or 'multiple hits'. Transitional saturation has been associated with long diveregence times (Moritz, Dowling & Brown, 1987). Despite application of a correction factor, the K-2-P distance, the calculated divergence will still, however, be an underestimate of the true value. If this value is incorrect by approximately one order of magnitude then a vicariance theory incorporating Gondwana break-up may better explain the distribution of *B. obtusispira*. It would also point towards an early invasion of *B. bavayi* to Madagascar from nearby Africa with much later invasions of proto-*B. liratus* and *B. forskalii*. There are perhaps two other observations that favour the latter interpretation. Firstly, comparisons between populations of *B. truncatus/tropicus* species followed the classical Ts:Tv ratio $\gg$ 1, whereas for interspecific comparisons within other species groups, this ratio often inverted i.e. < 1 (Fig. 2B). Secondly, when transversions alone were considered, there was good resolution between *Bulinus* and the outgroup taxa *Biomphalaria* and *Lymnaea*. If both Ts and Tv are considered, the molecular data implies the most recent common ancestor of *Bulinus* and *Biomphalaria/Lymnaea* to have been during the early Miocene, circa 20 Myr^{-1} which would be clearly at odds with the fossil record (Baker, 1945) that places the ancestral lymnaeid stock in the Permian, circa 250 Myr^{-1} (Remigio & Blair, 1997). Sampling of 16S variation within *Indoplanorbis* will help to date the relationships within *Bulinus* as the division of these two taxa might also be linked to continental break-up during the Cretaceous. For example, *Indoplanorbis* might have arrived to present-day Asia via the northward migrating Indian land-mass.

Molecular markers

Previous molecular and biochemical evidence has pointed towards *Bulinus* as a divergent group of snails that do not neatly fit within the variation shown by a genus (Biocca *et al.* 1979; Jelnes, 1987; Stothard *et al.* 1996). Morphologically, however, the snails are perhaps not differentiated enough to justify a generic revision (Brown, 1981). The lack of clear cut morphological characters also hampers specific identifications. For example, whilst *B. globosus* and *B. nasutus* on Zanzibar are clearly differentiated upon molecular characters (Stothard *et al.* 1997; Stothard & Rollinson, 1997 *a, b*) and phylogenetic analysis (Stothard & Rollinson, 1997 *a*), conchologically they can be confused since there appears to be overlapping variation of the shell (Stothard *et al.* 1997). Even placement of species, e.g. *B. obtusispira* and *B. umbilicatus*, to a species group of *Bulinus* is problematic if morphological characters alone are considered. The problematic morphology calls for molecular characters for identification and for assays to be developed that can easily differentiate taxa e.g. *B. forskalii* group (Jones *et al.* 1999) or *B. africanus* group (Rollinson, Stothard & Southgate, this supplement).

Better methods of identification and discrimination of *B. obtusispira* and *B. liratus* are required especially where populations exhibit atypical morphology or there is only access to alcohol-preserved specimens that would preclude isoenzyme analysis. By taking advantage of the variable nucleotides within COI a PCR-RFLP assay can be developed by selecting restriction enzymes that cut specifically at sites that differentiate the two taxa. For example, upon *Hpa*II digestion of the COI, two fragments of approximately 240 and 210 bp are produced for *B. obtusispira*, while *B. liratus* amplification products would remain intact since there are no cutting sites; conversely, *Rsa*I digestion would produce two fragments of approximately 350 and 100 bp for *B. liratus*, while *B. obtusispira* amplification products would remain intact. As both *Hpa*II and *Rsa*I can be combined within the same digestion reaction, double digestions would be a powerful, simple method to differentiate each taxa at these enzyme cutting sites.

To enable calculation of Hardy-Weinberg equilibrium and (or) gene flow between populations of snails, nuclear markers that vary within populations are required. Microsatellites have several advantages as markers for detailed genetic analysis (Schlötterer & Pemberton, 1994) and have been used to examine mating systems and genetic structure of *Bulinus truncatus* (Jarne *et al.* 1994; Viard & Jarne, 1997). Microsatellite primers designed on one species, however, generally do not cross-amplify microsatellites in other species. To initiate such a population genetics study for a previously uncharacterized species, PCR primers have to be designed afresh (Jarne & Lagoda, 1996). This stage of development is recognized to be time-consuming. Even after isolation and primer design there can be significant loss of primer sets since they may fail to amplify the locus from genomic DNA, null alleles are encountered or the loci proves to be invariant. The 8 primer sets designed to *B. obtusispira* pass the first requirements, discrete amplification product(s) were obtained. For two of the primer sets, size variation was also apparent. Further analysis of populations of *B. obtusispira* is ongoing, as well as the use of denaturing polyacrylamide gels for better fragment sizing, to determine if all primers sets will be useful for detailed population genetic analysis. Using these markers, it would be interesting to assess the genetic diversity of *B. obtusispira* populations across Madagascar and the effect that parasite pressure has upon them.

Compatibility studies

After laboratory infections, three species: *B. obtusispira*, *B. liratus* and *B. bavayi* have been shown to be able to act as intermediate hosts for *S. haematobium* from the Mahabo area. However, our ongoing malacological surveys of *Bulinus* have so far failed to find evidence of natural infections of human schistosomes; this contrasts with *Biomphalaria* where natural infections are often encountered. Degrémont (1973) found natural infections in both *B. obtusispira* and in a single population of *B. liratus*. From laboratory infection studies, compatibility of both species could be shown while *B. bavayi* appeared refractory (Degrémont, 1973). Wright (1971*a*), however, was able to show that *B. bavayi* was compatible with a *B. cernicus*-strain of *S. haematobium* from Mauritius and with a strain from South Africa (Wright & Knowles, 1972).

Given the specific characteristics of the Malagasy fauna, Degrémont (1973) concluded that it would be dangerous to systematically eliminate *B. liratus* from the epidemiology of urinary schistosomiasis on Madagascar only on the pretext that diploid *B. truncatus/tropicus* complex species do not transmit human schistosomes in Africa. The COI data suggest that *B. liratus* is closely related but different from African *B. tropicus* and despite being diploid, shares a slightly greater affinity to *B. truncatus* from Malawi than other species examined. Expanding upon Degrémont's findings, there appear to be populations of *B. liratus* on Madagascar that have some compatibility with *S. haematobium*. To examine this compatibility further, nine populations (sites 5, 12, 33, 38, 39, 43, 44, 45 and 51) were exposed to two *B. globosus*-borne strains of *S. haematobium* from Senegal and Zambia. No compatibility could be shown with either of these strains despite twice repeating exposure with 5 miracidia per snail (J. R. Stothard, unpublished observations). It would be interesting to challenge *B. liratus* against other isolates of African *S. haematobium*.

The potentially broad compatibility of Madagascan *S. haematobium* with three of the species groups of *Bulinus* is rather unusual as normally strains of *S. haematobium* that develop in *B. africanus* group snails are not compatible with *B. truncatus/tropicus* complex snails and *vice versa* (see Rollinson, Stothard & Southgate, this supplement). The development of Madagascar *S. haematobium* in *B. obtusispira* is unusual since many secondary sporocysts remain in the snail foot, an adaptation to enhance production of cercariae (Jourdane, 1983). It will be of interest to make molecular comparisons of parasite isolates from Madagascar and mainland Africa as it is not yet known from where, or when, the parasite was introduced. This could be as recently as during the last Millennium. Madagascar was only 'discovered' in the 15th Century by the Portuguese, although other human settlements predate. The first humans arrived, after leaving their Malaysian-Indonesian homeland, somewhere between 500 BC and 500 AD (Preston-Mafham, 1991).

The presence of compatible species of *Bulinus* precedes that of human movements to the island so that once *S. haematobium* was introduced, natural transmission could be sustained. The distribution of *S. mansoni* on Madagascar is spreading (Ollivier *et al.* 1999) but it is not clear when its intermediate snail host *Biomphalaria pfeifferi* first arrived. Other schistosome intermediate hosts have either failed to colonise the island or are yet to be introduced; despite the influx of people from the Far East infected with *S. japonicum*, this form of the disease has not established since there are no suitable intermediate snail hosts (Brygoo, 1972). The snail compatibility studies suggest that *S. haematobium* with different intermediate host specificities may have been introduced to Madagascar and (or) that Madagascan *Bulinus* in the absence of parasite pressure lost, or never possessed, the internal defence systems that evolved in African *Bulinus*. Further studies on natural transmission of *S. haematobium* and defence mechanisms of *Bulinus* will help clarify the intermediate host range on Madagascar.

ACKNOWLEDGEMENTS

JRS gratefully acknowledges funding from The Wellcome Trust. We thank the staff of Institut Pasteur, Antananarivo and I.R.D. for assistance with malacological surveys and maintenance of snail laboratory cultures. We thank Mike Anderson and Viv Tuffney for snail maintenance and Julia Bartley for running the ABI automated sequencer at the NHM. The study benefited from computer facilities at the Human Genome Mapping project "www.hgmp.mrc.ac.uk".

REFERENCES

ANON. (1999). Country file: Madagascar. *Geographical* **71**, 40.

BAKER, F. C. (1945). *The Molluscan Family Planorbidae*. Urbana, University of Illinois Press, p. 530.

BIOCCA, F., BULLINI, L., CHABAUD, A., NASCETTI, G., ORECCHIA, P. & PAGGI, L. (1979). Subdivisions su base morfologica e genetica del genre *Bulinus* in tre generi: *Bulinus* Müller, *Physopsis* Kraus e *Mandahlnarthia* gen. Nov. Rendiconti della classe di Scicenze Fisiche. Mateematiche e Naturali. *Accademia Nazionale del Lincei* **66**, 276–282.

BROWN, D. S. (1981). Generic nomenclature of freshwater snails commonly classified within the genus *Bulinus* (Gastropoda: Basommatophora). *Journal of Natural History* **15**, 909–915.

BROWN, D. S. (1994). *Freshwater Snails of Africa and their Medical Importance*, 2nd Edition. London, Taylor & Francis, p. 609.

BROWN, D. S. & WRIGHT, C. A. (1978). A new species of *Bulinus* from temporary freshwater pools in Kenya. *Journal of Natural History* **12**, 217–229.

BRYGOO, E. R. (1972). Human diseases and their relationship to the environment. In *Biogeography and Ecology in Madagascar*, pp. 703–725. (ed. Battistini, R. & Richard-Vindard, G.) The Hague, Dr. W. Junk B. V., Publishers.

BRYGOO, E. R. & MOREAU, J. P. (1966). *Bulinus obtusispira* (Smith, 1882), hôte intermédiarire de la Bilharziose à la *Schistosoma haematobium* dans le nord-ouest de Madagascar. *Bulletin de la Société de Pathologie Exotique* **59**, 835–839.

CAMPBELL, G., JONES, C. S., LOCKYER, A. E., HUGHES, S., BROWN, D., NOBLE, L. R. & ROLLINSON, D. (2000). Molecular evidence supports an African affinity of the neotropical freshwater gastropod, *Biomphalaria glabrata*, Say 1818, an intermediate host for *Schistosoma mansoni*. *Proceedings of the Royal Society of London, B* **267**, 2352–2358.

DAVIS, G. M., WILKE, T., SPOLSKY, C., QIU, C.-P., QIU, D.-C., XIA, M.-Y., ZHANG, Y. & ROSENBERG, G. (1998). Cytochrome oxidase I-based phylogenetic relationships among the Pomatiopsidae, Hydrobiidae, Rissoidae and Truncatellidae (Gastropoda: Caenogastropoda: Rissoacea). *Malacologia* **40**, 251–266.

DEGRÉMONT, A. E. (1973). *Mangoky Project: Campaign against Schistosomiasis in the Lower-Mangoky (Madagascar)*. Basle, Swiss Tropical Institute, pp. 261.

DOUMENGE, J. P., MOTT, K. E., CHEUNG, C., VILLENAVE, D., CHAPUIS, O., PERRIN, M. F. & REAUD-THOMAS, G. (1987). *Atlas de la Répartition Mondiale des Schistosomiases/Atlas of the Global Distribution of Schistosomiasis*. Talcnce, CEGET-CNRS; Genève, OMS/WHO; Talence, PUB, pp. 400.

FELSENSTEIN, J. (1993). *PHYLIP – Phylogeny Inference Package*, Version 3.5. Seattle, Washington: Department of Genetics, University of Washington.

HAMMOND, R. L., SACCHERI, I. J., CIOFI, C., COOTE, T., FUNK, S. M., MCMILLAN, W. O., BAYES, M. K., TAYLOR, E. & BRUFORD, M. W. (1998). Isolation of Microsatellite Markers in Animals. In *Molecular Tools for Screening Biodiversity*, pp. 279–296. (ed. Karp, A., Isaace, P. & Ingram, D.) London, Chapman & Hall.

HIGGINS, D. G., BLEASBY, A. J. & FUCHS, R. (1992). CLUSTAL-V: improved software for multiple sequence alignment. *Computer Applications in the Biological Sciences* **8**, 189–191.

JARNE, P. & LAGODA, P. J. L. (1996). Microsatellites, from molecules to populations and back. *Trends in Ecology and Evolution* **11**, 424–429.

JARNE, P., VIARD, F., DELAY, B. & CUNY, G. (1994). Variable microsatellites in the highly selfing snail *Bulinus truncatus* (Basommatophora: Planorbidae). *Molecular Ecology* **3**, 527–528.

JELNES, J. E. (1984). Taxonomie expérimental de *Bulinus*. 6. Possibilité d'utiliser des charactères enzymatiques pour la distinction entre des espèces *Bulinus liratus* et *B. obtusispira* à Madagascar. *Archives de l'Institut Pasteur de Madagascar* **51**, 89–96.

JELNES, J. E. (1987). Enzymelektroforese avendt til belysning af afrikanske og amerikanske bilharziosesnegles systematik. *Privatety published* (in Danish with English summary).

JOLLY, A., OBERLE, P. H. & ALBIGNAC, R. (1984). Introduction. In *Madagascar* (eds. Jolly, A., Oberle, P. H. & Albignac, R.) Oxford, Pergamon Press Ltd., p. 239.

JONES, C. S., NOBLE, L. R., OUMA, J., KARIUKI, H. C., MIMPFOUNDI, R., BROWN, D. S. & ROLLINSON, D. (1999). Molecular identification of schistosome intermediate hosts: case studies of *Bulinus forskalii* group species (Gastropoda: Planorbidae) from central and East Africa. *Biological Journal of the Linnean Society* 68, 215–240.

JOURDANE, J. (1983). Mise en évidence d'un processus original de la reproduction asexuée chez *Schistosoma haematobium*. *Comptes Rendus de l'Académie des Sciences, Paris* 296, 419–424.

KJER, K. M. (1995). Use of rRNA secondary structure in phylogenetic studies to identify homologous positions: an example of alignment and data presentation from the frogs. *Molecular Phylogenetics and Evolution* 4, 314–330.

KUMAR, S., TAMURA, K. & NEI, M. (1993). *MEGA: Molecular Evolutionary Genetics Analysis, Ver. 1.1.* University Park, PA, The Pennsylvania State Univ.

MATZURA, O. & WENNBORG, A. (1996). RNAdraw: an integrated program for RNA secondary structure calculation and analysis under 32-bit Microsoft Windows. *Computer Applications in the Biological Sciences* 12, 247–249.

MILDENHALL, D. C. (1980). New Zealand Late Cretaceous and Cenozoic plant biogeography: a contribution. *Palaeogeography Palaeoclimatology Palaeoecology* 31, 197–233.

MORITZ, C., DOWLING, T. E. & BROWN, W. M. (1987). Evolution of animal mitochondrial DNA: relationships for population biology and systematics. *Annual Review of Ecology and Systematics* 18, 269–292.

OLLIVIER, G., BRUTUS, L. & COT, M. (1999). Schistosomiasis due to *Schistosoma mansoni* in Madagascar: spread and focal patterns. *Bulletin de la Société de Pathologie Exotique* 92, 99–103.

PIQUÉ, A. (1999). The geological evolution of Madagascar: an introduction. *Journal of African Earth Sciences* 28, 919–930.

PRESTON-MAFHAM, K. (1991). *Madagascar: A Natural History*. Oxford, Facts on File, p. 224.

RABINOWITZ, P. D., COFFIN, M. F. & FALVEY, D. (1983). The separation of Madagascar and Africa. *Science* 220, 67–69.

RAVEN, P. H. (1979). Plate tectonics and Southern Hemisphere biogeography. In *Tropical Botany* (ed. Larsen, K.), pp. 1–24. London, Academic Press Ltd.

REMIGIO, E. A. & BLAIR, D. (1997). Molecular systematics of the freshwater snail family Lymnaeidae (Pulmonata, Basommatophora) utilising mitochondrial ribosomal DNA sequences. *Journal of Molluscan Studies* 63, 173–185.

ROLLINSON, D. & SOUTHGATE, V. R. (1987). The genus *Schistosoma*: a taxonomic appraisal. In *The Biology of Schistosomes: Genes to Latrines* (eds. Rollinson, D. &

Simpson, A. J. G.), pp. 1–49. London, Academic Press Ltd.

ROLLINSON, D., STOTHARD, J. R. & SOUTHGATE, V. R. (2001). Interactions of *Bulinus* and schistosomes of the *Schistosoma haematobium* group. *Parasitology*, in press.

SCHLÖTTERER, C. & PEMBERTON, J. (1994). The use of microsatellites for genetic analysis of natural populations. In *Molecular Ecology and Evolution: Approaches and Applications.* (eds. Schierwater, B., Streit, B., Wagner, G. P. & DeSalle, R.) Basel, Birkhäuser Verlag.

SMITH, A. G., SMITH, D. G. & FUNNELL, B. M. (1994). *Atlas of Mesozoic and Cenozoic Coastlines*. Cambridge, Cambridge University Press.

STARMÜHLNER, F. (1969). Die gastropoden der Madagassischen binnengewässer. *Malacologia* 8, 1–434.

STOREY, M., MAHONEY, J. J., SAUNDERS, A. D., DUNCAN, R. A., KELLEY, S. P. & COFFIN, M. F. (1995). Timing of hot spot-related volcanism and the break-up of Madagascar and India. *Science* 267, 852–855.

STOTHARD, J. R., HUGHES, S. & ROLLINSON, D. (1996). Variation within the Internal Transcribed Spacer (ITS) of ribosomal DNA genes of intermediate snails hosts within the genus *Bulinus* (Gastropoda: Planorbidae). *Acta Tropica* 61, 19–29.

STOTHARD, J. R., MGENI, A. F., ALAWI, K. S., SAVIOLI, L. & ROLLINSON, D. (1997). Observations on shell morphology, enzymes and random amplified polymorphic DNA (RAPD) in *Bulinus africanus* group snails (Gastropoda: Planorbidae) in Zanzibar. *Journal of Molluscan Studies* 63, 489–503.

STOTHARD, J. R. & ROLLINSON, D. (1997a). Partial DNA sequences from the mitochondrial cytochrome oxidase subunit I (COI) gene can differentiate the intermediate snail hosts *Bulinus globosus* and *B. nasutus* (Gastropoda: Planorbidae). *Journal of Natural History* 31, 727–737.

STOTHARD, J. R. & ROLLINSON, D. (1997b). Molecular characterisation of *Bulinus globosus* and *B. nasutus* on Zanzibar, and an investigation of their roles in the epidemiology of *Schistosoma haematobium*. *Transactions of the Royal Society of Tropical Medicine and Hygiene* 91, 353–357.

STRIMMER, K. & VON HAESELER, A. (1997). *Puzzle. Maximum Likelihood Analysis for Nucleotide and Amino Acid Alignments*. Munchen, Germany, Zoologisches Institut.

SWOFFORD, D. L. (1999). *PAUP*4.0. Phylogenetic Analysis Using Parsimony (*and other methods)*. Sunderland, MA, Sinauer.

TAJIMA, F. (1989). Statistical method for testing the neutral mutation hypothesis by DNA polymorphism. *Genetics* 123, 585–595.

THOLLESSON, M. (1999). Phylogenetic analysis of dorid nudibranchs (Gastropoda: Doridacea) using the mitochondrial 16S rRNA gene. *Journal of Molluscan Studies* 65, 335–353.

TRISTRAM, H. B. (1863). Note on some freshwater shells sent from Madagascar by J. Caldwell, Esq. *Proceeding of the London Zoological Society*, 60–61.

VIARD, F. C. & JARNE, P. (1997). Selfing, sexual polymorphism and microsatellites in the hermaphroditic freshwater snail *Bulinus truncatus*. *Proceedings of the Royal Society London B* **264**, 39–44.

WRIGHT, C. A. (1971*a*). *Bulinus* on Aldabra and the subfamily Bulininae in the Indian Ocean area. *Philosophical Transactions of the Royal Society London B* **260**, 299–313.

WRIGHT, C. A (1971*b*). *Flukes and Snails.* Science of Biology Series, No. 4. London, George Allen and Unwin Ltd, p. 168.

WRIGHT, C. A. & KNOWLES, R. J. (1972). Studies on *Schistosoma haematobium* in the laboratory III. Strains from Iran, Mauritius and Ghana. *Transactions of the Royal Society of Tropical Medicine and Hygiene* **66**, 108–118.

WRIGHT, C. A. & ROLLINSON, D. (1979). Analysis of enzymes in the *Bulinus africanus* group (Mollusca: Planorbidae) by isoelectric focusing. *Journal of Natural History* **13**, 263–273.

Molecular evolution of freshwater snail intermediate hosts within the *Bulinus forskalii* group

C. S. JONES[1]*, D. ROLLINSON[2], R. MIMPFOUNDI[3], J. OUMA[4], H. C. KARIUKI[4] and L. R. NOBLE[1]

[1] *Zoology Department, Aberdeen University, Tillydrone Avenue, Aberdeen, AB24 2TZ, UK*
[2] *Wolfson Wellcome Biomedical Laboratories, Zoology Department, The Natural History Museum, Cromwell Road, London, SW7 5BD*
[3] *Faculty of Science, BP812, Université de Yaoundé, Cameroun*
[4] *Ministry of Health, Division of Vector Borne Diseases, P.O. Box 20750, Nairobi, Kenya*

SUMMARY

Freshwater snails of the *Bulinus forskalii* group are one of four *Bulinus* species complexes responsible for the transmission of schistosomes in Africa and adjacent regions. The species status of these conchologically variable and widely distributed planorbids remains unclear, and parasite compatibility varies considerably amongst the eleven taxa defined, making unambiguous identification and differentiation important prerequisites for determining their distributions and evolutionary relationships. Random Amplified Polymorphic DNA (RAPD) analyses were used to investigate relationships between taxa, with particular emphasis on Central and West African representatives. RAPD-derived phylogenies were compared with those from other independent molecular markers, including partial sequences of mitochondrial cytochrome oxidase subunit I (COI) gene, and the nuclear ribosomal RNA internal transcribed spacer 1 region (ITS1). The phylogenetic reconstructions from the three approaches were essentially congruent, in that all methods of analysis gave unstable tree topologies or largely unresolved branches. There were large sequence divergence estimates between species, with few characters useful for determining relationships between species and limited within species differentiation. Nuclear and mtDNA sequence data from Central and East African representatives of the pan-African *B. forskalii* showed little evidence of geographical structuring. Despite the unresolved structure within the phylogenies, specimens from the same species clustered together indicating that all methods were capable of differentiating taxa but could not establish the inter-specific relationships with confidence. The limited genetic variation displayed by *B. forskalii*, and the evolution and speciose nature of the group, are discussed in the context of the increasingly arid climate of the late Miocene and early Pliocene of Africa.

Key words: *Bulinus forskalii*, mitochondrial DNA, nuclear ribosomal DNA, Randomly Amplified Polymorphic DNA, phylogeny, evolution.

INTRODUCTION

African freshwater snails of the *Bulinus forskalii* group serve as intermediate hosts of *Schistosoma* species, blood flukes responsible for human and animal schistosomiasis. Taxa of this group are amongst the most conchologically variable and widely distributed planorbids, with parasite compatibility varying considerably between taxa (Brown, 1994). It is essential to identify species reliably for a better understanding of the role of individual taxa in disease transmission. Defining the boundaries of species variation and reconstructing the phylogeny of freshwater snails of the *B. forskalii* group has been problematic as there is often ecophenotypic conchological variation between sites (Mandahl-Barth, 1957). Such localized variation is often exaggerated by the high incidence of inbreeding and self-fertilization, promoted by restricted dispersal between small, isolated water bodies. Although poorly

represented in the fossil record (Van Damme, 1984) to date, eleven *B. forskalii* group species have been defined, with varying precision, nine according to morphological criteria (Brown, 1994).

Although the group is widely distributed throughout Africa only *B. forskalii* (Ehrenberg, 1831) has a truely pan-African distribution and is a host for *S. intercalatum*, the causative agent of one form of intestinal schistosomiasis; its range overlaps several other *B. forskalii* group taxa, which have comparatively restricted distributions. For instance, *B. senegalensis* Muller, 1781, a host for *S. haematobium* which causes urinary schistosomiasis, is found mainly in the sub-Saharan belt throughout Senegal, Gambia, Mauritania, Chad, Nigeria, Cameroon and Niger (Mimpfoundi & Slootweg, 1991). Where the distributions of *B. forskalii* and *B. senegalensis* are sympatric, unambiguous species identification from shells alone can be difficult (Goll, 1981; Betterton, Fryer & Wright, 1983), as is typical of some other *B. forskalii* taxa comparisons. Conversely, *B. camerunensis*, confined to two crater lake localities, Barombi Kotto (the type locality) and Lake Debundsha, in South West Cameroon, although conchologically

* Corresponding author: Zoology Department, Aberdeen University, Tillydrone Avenue, AB24 2TZ, UK. Tel: 01224 272403. Fax: 01224 272396. E-mail: c.s.jones@abdn.ac.uk

Parasitology (2001), **123**, S277–S292. © 2001 Cambridge University Press
DOI: 10.1017/S0031182001008381

distinct, is biochemically and genetically indistinct from *B. forskalii* (Mimpfoundi & Greer, 1989; Jones *et al.* 1997, 1999). A major problem limiting our understanding of the evolutionary ecology and host-parasite relationships within the *B. forskalii* group is that most surveys have necessarily been localized in their extent (Greer *et al.* 1990; Mimpfoundi & Greer, 1989, 1990), and although useful, for broader evolutionary questions and epidemiological surveys further data are required in the context of the group's pan-African distribution. The present study evaluates the application of RAPD analyses to generate polymorphic markers for both species discrimination and phylogenetic analysis of taxa within *B. forskalii* group from West and Central Africa.

To avoid major errors, it has been suggested that the application of RAPD analyses for phylogenetics be restricted to closely related taxa (Bowditch *et al.* 1993; Stothard & Rollinson, 1996), or even sibling species (Van de Zande & Bijlsma, 1995). Although *B. forskalii* group taxa share morphological characters, at present there are few pointers to their genetic relationships. With such limited information it is impossible to determine if these species are in fact closely or distantly related. To strengthen the analysis several RAPD fragments apparently shared between taxa were investigated by Southern blotting and hybridization. Phylogenies derived from RAPD analyses were compared with other independent molecular markers; partial sequences of mitochondrial DNA (mtDNA) cytochrome oxidase I (COI) gene and the nuclear ribosomal RNA internal transcribed spacer 1 region (ITS1).

MATERIALS AND METHODS

Samples

Study material (Table 1) comprised 6 species of the *B. forskalii* group taxa including *B. forskalii* (Ehrenberg 1831), *B. scalaris* (Dunker, 1845), *B. senegalensis* Muller, 1781, *B. camerunensis* Mandahl-Barth, 1957, *B. crystallinus* (Morelet, 1868), and *B. cernicus* (Morelet, 1867). Material was field-collected from Cameroon and Kenya or obtained from the Natural History Museum collections held in the Biomedical Parasitology Division. Field-collected specimens were picked from emergent aquatic vegetation using long-handled nets, relaxed for approximately 30 min by the addition of a drop of menthol-saturated ethanol to the collecting vial, and placed in 100% ethanol for long-term storage. Parasite-free first generation laboratory stocks of two *B. forskalii* populations were included for comparison with field collections for RAPD data sets (asterisked in Table 1).

B. forskalii group taxa were identified on the basis of shell characters (Brown, 1994) and distal genitalia morphology (Mandahl-Barth, 1957). *B. senegalensis*

can usually be differentiated conchologically from extreme forms of *B. forskalii* by the complete absence of a carina, allowing initial identification based on the shell alone. Specimens with intermediate shell morphologies were identified by comparison of their RAPD profiles with those of specimens which had unequivocal shell morphology and confirmed using RAPD-derived species-specific primers (Jones *et al.* 1997).

DNA extraction and RAPD analyses

Tissue preserved in 95% ethanol was first vacuum dried and total genomic DNA extracted according to Vernon, Jones & Noble (1995). General RAPD and electrophoresis conditions are detailed in Jones, Okamura & Noble (1994). Positive controls used DNA samples with known banding patterns and negative controls omitted template DNA. Forty primers from kits Y, R & F from Operon Technologies, Du Pont, USA and three decamer primers from Okamura, Jones & Noble (1993) were initially screened for RAPD analysis of *Bulinus* snails. Twenty primers were chosen to characterize, in detail, 27 *B. forskalii* populations and 2 other *Bulinus* species. Primers chosen include three arbitrary decamers (Primer 02, 10 and 12) from Okamura *et al.* (1993), and the remaining seventeen were Operon primers (OPY14–18, OPR8–13 and OPF4–9). Several repeat amplifications were always performed for each primer to ensure profile reproducibility.

Negatives of the gel photos were examined on a light box and each RAPD fragment between 350 and 1400 base pairs (bp) was scored for presence (1) or absence (0) to produce a binary matrix for all taxa and primers examined. To be scored as present the fragments had to be strongly fluorescent and reproducible; all samples were coded and gels scored blind with no knowledge of the identity of each sample in each lane to avoid any scoring bias.

RAPDs are considered unsuitable for phylogeny inference using parsimony methods as the shared absence of a character in this instance cannot be considered informative (Backeljau *et al.* 1995). Hence a purely phenetic approach was adopted. Pairwise similarities (S) were calculated between taxa using Dice's coefficient (Jackson, Somers & Harvey, 1989). This similarity index was chosen because it is based upon sharing of present bands alone which is more robust than those which weight shared presence alleles equally with shared nulls (Rossetto, Weaver & Dixon, 1995). The similarity coefficients (S) were converted to distance (distance, $P = 1 - S$; percent sequence divergence is $P \times 100$; Swofford & Olsen, 1990).

Shared sequence identity and hybridization

Shared sequence identity between common bands between species was investigated by Southern analy-

Table 1. Snails used in this study and the collection localities

Species	Population no. and locality	Latitude, longitude or NHM/Aberdeen† accession no.	Markers		
			RAPD	mtDNA	ITS1
Bulinus forskalii species					
B. senegalensis (type locality)	1. Podor, Senegal	1745	✓	✓	—
B. senegalensis	2. Diator, Senegal	1746	✓	—	—
B. senegalensis	3. Nianga, near Podor, Senegal	1749	✓	—	—
B. senegalensis	4. Lambata, near Mora, extreme north Cameroon	N11.02.017 E14.09.533	✓	—	✓
B. senegalensis	5. 3 km from Mora, extreme north Cameroon	N11.03.767 E14.09.319	✓	✓	—
B. senegalensis	6. Guidiguis, past Lara, extreme north Cameroon	N10.08.333 E14.41.622	✓	—	—
B. senegalensis	7. Boro, Mali	1901	✓	✓	—
B. forskalii	8. Mbakhana, Senegal	1750	✓	—	—
B. forskalii	9. Mbeyssus, Senegal	1751	✓	—	—
B. forskalii	10. Lampsar, Senegal	1754	✓	—	—
B. forskalii	11. Maklingay, Far North Cameroon	N10.52.059 E14.14.179	✓	✓	✓
B. forskalii	12. Udkia, Mokolo, Far North Cameroon	N10.42.216 E13.50.285	✓	—	—
B. forskalii	13. Yagoua, Maya Daray river, Far North Cameroon	CY9/93†	✓	—	—
B. forskalii	14. Sangmelina, South Cameroon	N02.56.515 E11.58.953	✓	✓	—
B. forskalii	15. River Mabanga, Kumba, SW Cameroon	N04.38.161 E09.28.349	✓	✓	✓
B. forskalii	16. Mabanga water, Kumba, SW Cameroon	N04.37.997 E09.25.608	✓	✓	—
B. forskalii	17. Loum, Southwest Cameroon	N04.42.411 E09.44.517	✓	✓	—
B. forskalii	18. Ebebda village, Sanaga River, S Cameroon	N04.21.845 E11.16.130	✓	✓	—
*B. forskalii**	19. Ebebda village, Sanaga River, S Cameroon, F1	1770	✓	✓	—
B. forskalii	20. Bafia, Ritob stream, South Cameroon	N04.45.001 E11.13.676	✓	✓	—
*B. forskalii**	21. Bafia, Ritob stream, South Cameroon, F1	1769	✓	—	—
B. forskalii	22. Nigeria	1019	✓	—	—
B. forskalii	23. Sao Tome	1625	✓	✓	✓
B. camerunensis (type locality)	24. Barombi Kotto, Southwest Cameroon	N04.27.949 E09.15.435	✓	✓	✓
B. forskalii mixed	25. Toukou, Far North Cameroon	N10.24.695 E15.14.836	✓	—	—
B. senegalensis/*B. crystallinus*	26. Angola	247	✓	✓	—
B. cernicus	27. Mauritius	1687	✓	✓	✓
B. scalaris	28. Kwanziuv stream, Nairobi, Kenya	KN2/95†	—	✓	✓
B. scalaris	29. Murrum, Kamagaga area, Kisumu, Kenya	S00.05.781 E34.57.832	—	✓	✓
B. scalaris/*B. forskalii* mixed	30. Murrum, Kamagaga area, Kisumu, Kenya	S00.05.008 E34.57.846	—	✓	✓
B. forskalii	31. Masongaleni, Kibwenzi, Nairobi, Kenya,	1900	—	✓	✓
Other *Bulinus* species					
B. truncatus (Outgroup)	32. Barombi Kotto, Southwest Cameroon	N04.27.949 E09.15.435	✓	✓	✓
B. truncatus (Outgroup)	33. Senegal	2060	—	—	✓
B. wrighti	34. Oman	1698	✓	—	—

sis of RAPD gels using DNA isolated from these bands as probes. Amplified DNA fragments were separated by agarose gel electrophoresis and alkali blotted onto Hybond N+ nylon membrane (Amersham International, UK). Fragments to be used as probes were excised from the gel prior to Southern blotting and purified directly using Gene clean (Bio 101, Inc.). The purified DNA fragments were kept at −70 °C until required. These fragments were radiolabelled by either nick translation (fragments less than 400 bp; Rigby et al. 1977) or random priming (fragments greater than 400 bp; Feinberg & Vogelstein, 1983), with [α-^{32}P]dCTP (3000 Ci mmol^{-1}, Amersham International). Standard methods for Southern blots followed Sambrook, Fritsch, & Maniatis (1989). Blots were prehybridized in 1% (w/v) BSA, 1 mM EDTA, 0·5 M Na$_2$HPO$_4$ pH 7·2, 7% (w/v) SDS, according to Church & Gilbert (1984), at 65 °C for 4–6 h and hybridized overnight at 65 °C, by the addition of the labelled probe. After hybridization, blots were washed to a high stringency by briefly rinsing in 400 mM Na$_2$HPO$_4$ pH 7·2, at room temperature, and incubated twice in each of the following solutions of 400 mM, 100 mM, 40 mM, 10 mM and 1 mM Na$_2$HPO$_4$ pH 7·2, each with 0·1% SDS, for 15 min at 65 °C. Autoradiography was carried out for using X-ray film (Hyperfilm, Amersham) for 24–72 h with no screens.

PCR amplification and sequencing

Primer sequences used to amplify a 450 bp section of the mtDNA COI gene were: forward 5′ TTTTT-TGGGCATCCTGAGGTTTAT 3′ and reverse 5′ TAAAGAAAGAACATAATGAAAATG 3′, following Bowles, Blair & McManus (1992). A 569 bp fragment of the ITS1 was amplified using the primers ETTS2 (Kane & Rollinson, 1994) and ETTS16 (Stothard, Hughes & Rollinson, 1996). PCR cycling conditions for mtDNA COI and ITS1 followed Stothard & Rollinson (1997) and Stothard et al. (1996), respectively. PCR products were purified using Qiaquick PCR™ purification columns (Qiagen Inc.) and sequenced on both strands using the SequiTherm EXCEL™ II DNA sequencing kit-LC (Cambio, UK) and run on a Li-COR Long-ReadIR automatic sequencer (MWG Biotech, UK). Sequences were edited using e-seq™ (version 1.0), 5′ and 3′ ends aligned using Align IR (version 1.2) and multiple alignments performed in Clustal X (Thompson et al. 1997).

Phylogenetic analyses

Phylogenetic reconstruction was estimated using Fitch–Margoliash criteria that utilize a least squares optimization method (Fitch & Margoliash, 1967) for all data sets. Fitch-generated distance trees were achieved using PHYLIP 3.5 (Felsenstein, 1993), with options set to prohibit negative branch lengths, jumble input order and for global tree rearrangements. As the rate of transitional (Ts) nucleotide substitution is often higher than transversional (Tv) substitution, especially for mtDNA, the Kimura 2-Parameter (K2P) distance model was applied to the sequencing data sets; this model incorporates transitional bias and computes the numbers of transitional and transversional nucleotide substitutions per site and their variances.

Maximum likelihood (ML) methods (quartet puzzling; Strimmer & Von Haeseler, 1996), and maximum parsimony (Swofford, 1996) were additionally employed for the sequence data sets. All characters (nucleotide sites) were weighted equally. The heuristic search option of PAUP 4.0 v 3.c (Phylogenetic Analyses Using Parsimony; Swofford, 1996) was used with gaps (indels) scored as missing data and coded separately, with each gap representing one character with the states either present (1) or absent (0). Gaps which spanned more than one nucleotide position were scored as a single character only. Nodal branch support for distance and parsimony analyses was assessed using 1000 bootstrap replications (Felsenstein, 1985) for the sequencing data sets only. Computation of maximum likelihood trees using PUZZLE 3.1 was achieved without incorporating a molecular clock, but with substitution rates using the HYK model (which allows transitions and transversions to occur at different rates and allows base frequencies to vary as well) following Hasegawa, Kishino & Yano (1985), and the model of rate heterogeneity set at a uniform rate. Alternative models of substitution (e.g. Tamura-Nei) and different rates of heterogeneity (e.g. gamma distribution) were implemented, but had no impact upon the main branch topology of the tree or support values. The adequacy of the phylogenetic information content of the sequence data sets were assessed by the proportion of unresolved maximum likelihood quartets reported by PUZZLE 3.1. The program MEGA (Kumar, Tamura & Nei, 1993) was used to calculate nucleotide statistics such as nucleotide composition and number of Ts and Tvs.

Where possible representatives of the same ingroup taxa were used for all analyses; samples of B. scalaris were not available when the RAPD analyses commenced, and despite several attempts, B. crystallinus samples failed to give good quality sequence data with the ITS1 primers. Outgroup taxa from Gastropoda, Pulmonata for the sequencing data sets included Albinaria caerulea and Bulinus truncatus (from the B. truncatus/tropicus complex). Further outgroups tested due to the availability of the additional sequences included Cepaea nemoralis for the mtDNA COI and Biomphalaria pfeifferi for the ITS1, data sets respectively. B. truncatus and B.

wrighti were included as outgroups for the RAPD analyses.

RESULTS

Shared RAPD sequence identity and scoring

To test the assumption that co-migrating RAPD bands from different taxa can be scored as shared characters for phylogenetic analyses, selected fragments were examined for shared sequence identity. A sample of between 2 and 4 different fragments from each of 3 selected primers were investigated by Southern analysis of RAPD gels, using isolated 'shared' fragments as radiolabelled probes. Bands with high sequence homology to the probes were identified by washing the blots at high stringency. In nine out of ten probings homology of the fragments between taxa was as predicted; fragments shared between taxa and of equal fluorescence were shown to be homologous (Table 2). For example, probe 2/4 (1200 bp product from primer 2) showed similar sized fragments of equal fluorescence from *B. forskalii* and *B. camerunensis* to be homologous. Weakly amplifying fragments, with respect to the rest of the profile, of approximately the same size were not homologous. For example, a 330 bp fragment from *B. senegalensis* (Senegal) was highly fluorescent, whereas a similarly sized fragment was less fluorescent in *B. forskalii* and *B. camerunensis*. When this fragment was excised and used as a probe (2/2), it was not homologous to *B. forskalii* or *B. camerunensis* but was specific to *B. senegalensis* from both the type locality (Senegal), North Cameroon and Mali. In all cases, with one exception, hybridization signal of the probe and the fluorescent intensity of the original band on ethidium stained agarose gels corresponded. A single case of homoplasy was detected between fragments of the same size and intensity. Homology was detected as predicted for fragment 12/1 between *B. forskalii*, *B. senegalensis* and *B. camerunensis* but not in *B. truncatus*; this may be explained by their relative positions on the gel as the first three species were run in adjacent lanes and the latter was on the gel edge.

140 RAPD bands in total were included in the RAPD data set scored using the criteria of fragments of the same size and equal fluorescent intensity assumed to be homologous, from 20 primers, with an average of 7 bands each, from 27 *B. forskalii* group populations and two non-*B. forskalii* species as outgroups, *B. truncatus* and *B. wrighti*, respectively. Each primer generated characteristic fragment profiles for each species, comprising mainly of common species-diagnostic monomorphic bands, with few intraspecific polymorphic bands and with few similarities between taxa (Fig. 1). Mean percent sequence divergence estimates within and between each of 5 *B. forskalii* group taxa and 2 outgroup taxa reflect these observations from RAPD profiles, ranging from 0·0 to 15·8 % for within species, 9·9–88·4 % for between ingroup taxa and 71·8–90·3 % between ingroup and outgroup taxa comparisons (Table 3).

RAPD phylogeny

Analysis of the pairwise genetic distance matrix using the Fitch–Margoliash least squares criterion is shown in Fig. 2. This tree was rooted with *B. wrighti* and shows two main branches, with *B. senegalensis* branching earlier than *B. forskalii-cernicus-crystallinus* clade. Unexpectedly, *B. truncatus*, a non-*B. forskalii* group species falls within the latter ingroup, the *B. forskalii* clade. Within the main *B. forskalii* clade, *B. cernicus* branches before *B. crystallinus*, followed by the remaining *B. forskalii* specimens. Additionally, although identified as a separate species on morphology, *B. camerunensis* clearly clusters with specimens identified as true *B. forskalii* and is most closely related to the nearest *B. forskalii* population (locality 15, 16 from Table 1). Conversely, the molecular data clusters what has been described tentatively as an extreme geographical variant of *B. forskalii* from Sâo Tomé on shell morphology, with *B. crystallinus* from Angola. Within the *B. forskalii* branch no clear geographical structuring of samples was evident, with samples from Senegal falling amongst Cameroonian specimens. While geographically separate localities of *B. senegalensis* could be differentiated, clustering on separate lineages on the main *B. senegalensis* branch.

However, the tree topology is unstable with some significant changes occurring, for instance when the tree is rooted with *B. truncatus* (data not shown), causing the affinities of *B. wrighti*, *B. cernicus* and *B. crystallinus* to swap to the *B. senegalensis* clade.

Molecular sequence phylogenies

Phylogenetic reconstruction used 21 nucleotide sequences obtained from 6 species of *B. forskalii* group taxa, *B. truncatus* and 2 outgroups, for 350 aligned bases of the mtDNA COI gene. Comparison of *Bulinus* COI sequences showed that they were A:T rich (68·9 %) (nucleotide compositions: A = 27·8 %, C = 12·3 %, G = 18·8 %, T = 41·1 %), there were no major insertions or deletions and that 266 (76 %) of the 350 nucleotide sites were invariant. Of the 84 variable sites, there were 8 (9·5 %) at the first, none at the second and 76 (90·5 %) at the third codon position. This pattern of variation, most substitutions accumulating at the third codon position, is typical of sequences under strong functional constraints. 74 of the variable sites were phylogenetically informative. Seven of the eight variable first codon positions, including all of the phylogenetically informative ones, occurred in leucine codons and

Table 2. Shared sequence identity of RAPD bands and species-specificity by hybridization analysis

Primer used	Fragment name	Size (in bp)	Species the fragment was derived from	Prediction; reason for probing[a]	Outcome
10	10/5	800	B. senegalensis	NH; similar sized fragment weakly amplified in B. forskalii	NH; specific to B. senegalensis
10	10/7	380	B. senegalensis	NH; weakly amplifying fragments in B. forskalii & B. camerunensis	NH; fragment specific to B. senegalensis
10	10/10	940	B. forskalii	H; similar sized fragment in B. camerunensis	H; fragment homologous in B. camerunensis & B. forskalii
2	2/2	330	B. senegalensis	NH; weakly amplifying fragments in B. forskalii & B. camerunensis	NH; fragment specific to B. senegalensis
2	2/3	1400	B. senegalensis	H; similar sized fragment in B. forskalii & B. camerunensis	H; fragment homologous in B. forskalii, B. senegalensis & B. camerunensis
2	2/4	1200	B. camerunensis	H; similar sized fragment in B. forskalii	H; fragment homologous from B. camerunensis & B. forskalii
2	2/5	1100	B. camerunensis	H; similar sized fragment in B. forskalii	H; fragment specific to B. camerunensis
12	12/1	750	B. forskalii	H; similar sized fragment in B. senegalensis, B. camerunensis & B. truncatus	NH/X[b]; fragment homologous in B. forskalii, B. senegalensis & B. camerunensis but not B. truncatus
12	12/2	420	B. senegalensis	H; similar sized fragment in B. forskalii & B. camerunensis	H; fragment homologous in B. senegalensis, B. forskalii & B. camerunensis

[a] H – To test shared sequence identity of similar sized fragments with similar fluorescent intensity from other taxa compared, predicting probe hybridization; NH – To test shared sequence identity of strongly amplifying with weakly amplifying fragments, predicting no probe hybridization; [b] NH/X – mixed outcome of probing indicating fragments of shared sequence identity in some taxa but not all.

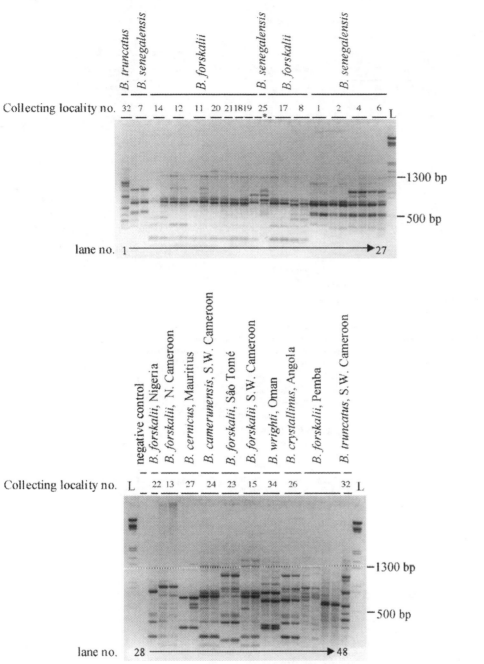

Fig. 1. Amplification products from RAPD primer OPF06 (Operon Technologies, Du Pont, USA) illustrating representative profiles from *B. forskalii* group taxa and outgroups. Lane designations (numbers in brackets refer to collecting localities listed in Table 1): 1. *B. truncatus*, S.W. Cameroon (32); 2–3. *B. senegalensis*, Mali (7); 4–5. *B. forskalii*, S. Cameroon (14); 6–7. *B. forskalii*, N. Cameroon (12); 8–9. *B. forskalii*, N. Cameroon (11); 10. *B. forskalii*, S. Cameroon (20); 11. *B. forskalii*, S. Cameroon (21); 12. *B. forskalii*, S. Cameroon (18); 13. *B. forskalii*, S. Cameroon (19); 14. *B. forskalii*, S. Cameroon (25)*; 15. *B. senegalensis*, S. Cameroon (25)*; 16–17. *B. forskalii*, S.W. Cameroon (17); 18–19. *B. forskalii*, Senegal (8); 20–23. *B. senegalensis*, Senegal (1, 2); 24–27. *B. senegalensis*, *B. forskalii*, N. Cameroon (4, 6); 28. negative control; 29. *B. forskalii*, Nigeria (22); 30–31. *B. forskalii*, N. Cameroon (13); 32–33. *B. cernicus*, Mauritius (27); 34–35. *B. camerunensis*, S.W. Cameroon (24); 36–37. *B. forskalii*, São Tomé (23); 38–39. *B. forskalii*, S.W. Cameroon (15); 40–41. *B. wrighti*, Oman (34); 42–43. *B. crystallinus*, Angola (26); 44–45. *B. forskalii*, Pemba (NHM accession no. 1828); 46–47. *B. forskalii*, Pemba (NHM accession No. 1826); 48. *B. truncatus*, S.W. Cameroon (32). L. molecular weight ladder in base pairs (1KB ladder, BRL–Gibco, UK). Taxa depicted by asterisks indicate two distinct species from the same collecting locality.

were silent substitutions. The remaining first codon position change occurred in *B. truncatus*, changing a valine to an isoleucine.

The transition/transversion ratio across all codon positions was 1·74, indicating that there is a tran-

sitional bias, typical of animal mitochondrial DNA (Wakeley, 1996). The number of Ts and Tv substitutions across all codon positions was plotted against percent sequence divergence (*P*-distance ×100) to investigate whether Ts substitutions

Table 3. *Estimates of RAPD mean percent sequence divergence (P distance × 100) within and between taxa (ranges given in brackets for within species comparisons)*

	B. senegalensis	*B. forskalii*	*B. camerunensis*	*B. cernicus*	*B. crystallinus*	*B. truncatus*	*B. wrighti*
B. senegalensis	7·47 (1·40–15·8)	88·17	88·45	74·53	82·32	89·46	85·43
B. forskalii		8·02 (1·30–14·3)	7·96	75·00	69·53	71·69	87·95
B. camerunensis			0·00	74·78	69·07	71·83	87·87
B. carnicus				3·12 (0·00–3·12)	64·04	85·71	82·76
B. crystallinus					0·00	77·99	90·32
B. truncatus						0·00	89·47
B. wrighti							0·00

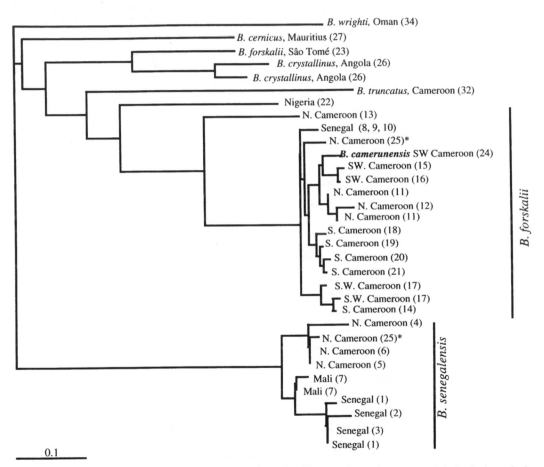

Fig. 2. Fitch-generated distance phylogram using the Kimura 2-parameter model depicting phylogenetic relationships among *B. forskalii* group species from Cameroon using the RAPD data. The scale denotes pairwise nucleotide substitution. Taxa depicted by asterisks indicate two separate species from the same collecting locality. *B. camerunensis* (in bold) falls within the *B. forskalii* clade. Numbers in brackets refer to collecting localities listed in Table 1.

become saturated with increasing sequence divergence (Fig. 3). The number of Ts substitutions is generally in excess of Tv across almost the entire sequence divergence range (up to 12%), significant linear regressions could be fitted to both Ts ($y = 1.653x + 3.396$; $R^2 = 0.898$, $P = 0.0001$) and Tv ($y = 1.776x - 3.625$; $R^2 = 0.891$, $P = 0.0001$) substitutions, suggesting that the data set was not significantly affected by 'saturation' of transitions. Sub-

sequent phylogenetic analyses used distance estimates calculated using the K2P model, which incorporates transitional bias.

COI data set displayed largely congruent topologies for each phylogenetic methods employed. Parsimony analysis yielded 6 equally parsimonious trees (length = 294 steps, consistency index, CI = 0·701) and the 50% majority-rule consensus tree showed three unresolved branches: *B. truncatus, B.*

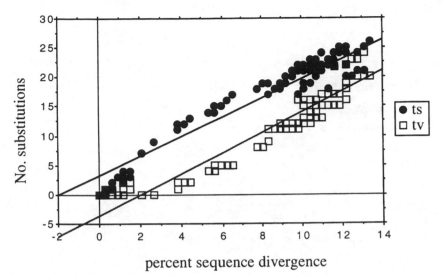

Fig. 3. Plot of transitions (Ts; shaded circles) and transversions (Tv; open boxes) against percent sequence divergence for the partial mtDNA COI sequences from *Bulinus*, depicting linear relationships.

senegalensis and a *B. forskalii-cernicus-scalaris-crystallinus* clade. Maximum likelihood analyses shows that *B. truncatus* and *B. senegalensis* branch earlier than an unresolved, weakly supported trichotomy (quartet puzzling value of 66) of *B. forskalii*, *B. crystallinus* with *B. forskalii* from Sâo Tomé and finally *B. cernicus* with *B. scalaris* (Fig. 4a). The low frequency of unresolved ML quartets (1·3 %) indicated a good phylogenetic signal in this data set. Distance methods resolve the latter main branch into 2 sister groups with *B. cernicus* clustering with *B. scalaris* and in the second group *B. crystallinus* clustering with *B. forskalii* from Sâo Tomé which branch earlier than the rest of the *B. forskalii* taxa. In all cases, *B. camerunensis* clusters tightly *B. forskalii*, with nodal bootstrap support values of 77 and 78 for parsimony and distance methods, respectively, and a quartet puzzling value of 100.

The ITS1 data matrix analysed here contained 15 sequences, representing 5 *B. forskalii* group species, *B. truncatus* and 2 outgroup taxa with 493 aligned bases. ITS1 sequences from 3 *B. forskalii* samples representing different geographical localities in Cameroon and *B. camerunensis* from South West Cameroon, proved identical and replicates were removed from the data set to speed up analyses. *Bulinus* ITS1 sequences were G:C rich (59·4 %) (nucleotide compositions: A = 19·5 %, C = 30·4 %, G = 29 %, T = 21·2 %), characterized by minor insertions and deletions, with mainly 1–4 gaps, but up to 8 were recorded for some taxa (*B. scalaris*). A total of 40 binary gap characters were scored, and added to the data set, with 28 being phylogenetically informative. 178 (36 %) of the 493 nucleotide characters were constant. Of the remaining variable sites, 183 were phylogenetically informative. Phylogenetic reconstruction using the ITS1 region is largely congruent with the mtDNA COI results. Parsimony analysis gave 2 equally parsimonious

trees with 464 steps and a high CI value (0·916) suggesting a robust phylogeny, although 10 % of puzzling quartets were unresolved in the ML analyses. For all algorithms, two main branches were detected, one clade with *B. truncatus* and the other with the remaining taxa split into an unresolved trichotomy of *B. senegalensis*, *B. cernicus-scalaris* and *B. forskalii* taxa, with *B. forskalii* from Sâo Tomé branching earlier (Fig. 4b).

Mean pairwise percent sequence divergences for both the mtDNA COI and ITS1, are given in Table 4. Sequence divergence estimates for mtDNA COI indicate that the majority of *B. forskalii* group species are as distant from each other as the non-*B. forskalii* group taxa, *B. truncatus*. Values ranged from 3·9–14·8 % between *B. forskalii* ingroup taxa, 8·7–13·8 % for between the ingroup *B. forskalii* taxa and *B. truncatus* and 16·8–27·9 % between ingroup and outgroup taxa. Sequence divergence within *B. forskalii* ranged from 0·00 to 6·75 % (mean = 4·24 %). No geographical structuring of *B. forskalii* with COI sequences was evident using all phylogenetic approaches, with samples from Kenya, East Africa and Cameroon, Central Africa clustering together (Fig. 4a); sequence divergence between these 2 localities range from 0·89–6·75 % (mean = 4·47 %). Sequence divergence within *B. senegalensis* ranged from 0·59–2·68 % (mean = 1·78 %) with taxa separating according to major geographical locality.

The ITS1 sequence divergence estimates show that while ingroup taxa appear equally distant from each other, ranging from 0·44–9·43 %, there are larger differences between the ingroup *B. forskalii* taxa and *B. truncatus* (27·1–30·9 %) and between ingroup and outgroup taxa (39·4–57·8 %). No intraspecific sequence divergence was detected in *B. forskalii* from Cameroon, with only negligible variation between Kenyan and Cameroonian samples (0·44 %).

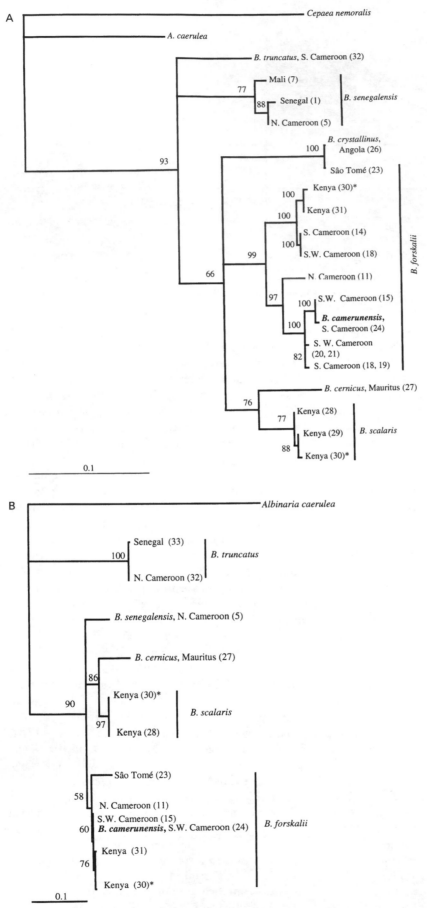

Fig. 4. Maximum likelihood neighbour-joining tree using quartet puzzling (Strimmer & Haeseler, 1996) depicting phylogenetic relationships among *B. forskalii* group species with sequences from (a) with partial mtDNA COI

DISCUSSION

Phylogenetic reconstructions of the *B. forskalii* group species, based on RAPD analyses, partial mtDNA gene sequences (COI) and nuclear region sequences (ITS1 rDNA) were essentially congruent. With few exceptions all methods of analysis gave largely unresolved branches, with relatively large sequence divergence estimates between species and limited within species differentiation. The ITS1 and COI of pan-African *B. forskalii*, with representatives from Central and East Africa, showed little evidence of geographical structuring. Despite the unresolved or unstable nature of the phylogenies for all data sets, specimens from the same taxon clustered together, indicating that all methods were capable of differentiating species.

RAPDs – homology and phylogeny

The speed, simplicity and arbitrary sequence of primers has made RAPDs (Williams *et al.* 1990; Welsh & McClelland, 1990) one of the most frequently used molecular tools for identification and differentiation of many genetically anonymous groups (Rollinson & Stothard, 1994), including planorbid snails (Langand *et al.* 1993; Stothard & Rollinson, 1997; Jones *et al.* 1997; Kristensen, Yousif & Raahauge, 1999). Preliminary analyses of *B. forskalii* group snails (Jones *et al.* 1999), and data from this study, have demonstrated the utility of RAPDs for differentiation of morphologically similar species such as *B. forskalii* and *B. senegalensis*.

However, for estimation of nucleotide divergence between closely related taxa Clarke & Lanigan (1993) suggested RAPDs must conform strictly to a number of criteria, the most important of which is arguably verification of band homology. Co-migrating RAPD PCR products of similar molecular weight amplified from different species of the same genus may represent different sequences. Our limited Southern blot analysis of probes derived from excised RAPD fragments demonstrated that 90% of co-migrating bands in different taxa were of shared sequence identity, suggestive of homology, providing that fragments of the same size and fluorescence were compared from closely related species, a finding concordant with other studies (Smith *et al.* 1994; Rieseberg, 1996). However, unless loci are mapped to genomic locations it is impossible to distinguish whether these are paralogous or orthologous copies.

Phenetic analysis of RAPD data produced an apparently resolved tree, with *B. senegalensis* branching separately and earlier than the remaining *B. forskalii* group taxa. However, the tree topology was unstable, with affiliations of some taxa changing with the outgroup. This suggests it may be difficult to determine the relationships of *B. forskalii* group taxa using RAPDs. *B. truncatus* (a non-*B. forskalii* group species) was unexpectedly found clustered within the ingroup. This anomalous finding may be partially attributable to the large nucleotide divergence estimates between species, extending beyond the phylogenetic scope of RAPDs, concordant with the findings of Stothard & Rollinson (1997), who evaluated their use in phylogenetic analysis of *Bulinus* species groups. A reliable phylogenetic signal with RAPD markers is attainable if there is less than 10% sequence divergence between taxa, otherwise the variance in the estimate of divergence becomes too large to be informative (Clark & Lanigan, 1993).

Sequence phylogenies

COI is known to exhibit a relatively slow rate of evolution for a mtDNA gene and is subject to considerable selective constraints, the rate of amino acid substitutions being very low, even between taxa belonging to different phyla (Simon *et al.* 1994). Consequently, this marker exhibits taxonomically useful levels of variation for between species comparisons, including *Bulinus* (Stothard & Rollinson, 1997). Similarly, Porter & Collins (1991) noted that rDNA ITS sequences are probably free to accumulate substitutions at rates at least as high as those recorded for silent sites in protein coding genes. ITS sequences have often been shown to be phylogenetically informative at the species level (Jeandroz, Roy & Bousquet, 1997) including snail intermediate hosts (Stothard *et al.* 1996). Campbell *et al.* (2000) used COI and ITS1 gene sequences to reconstruct the phylogenetic relationships of species of the genus *Biomphalaria*, intermediate hosts of *S. mansoni* and both data sets gave concordant, resolved phylogenies, revealing that the Neotropical species *B. glabrata* had an African affinity.

In this study phylogenetic analyses of COI, ITS1 and RAPDs produce approximately concordant trees which are unresolved at the main branches. Despite this, relationships of some *B. forskalii* group members showed congruencies across all data sets, for all algorithms used. *B. forskalii* and *B. senegalensis* form

sequences and (b) nuclear rDNA ITS1 region. Numbers at the branch nodes indicates percentage quartet puzzling support values for 1000 puzzling steps. The scale denotes pairwise nucleotide substitution. Taxa depicted by asterisks indicate two separate species from the same collecting locality. *B. camerunensis* (in bold) falls within the *B. forskalii* clade. Numbers in brackets refer to collecting localities listed in Table 1. Sequences are deposited under GenBank accession numbers AF369729–AF369747, and AY–30345–AY030354, for the mtDNA COI and ITS1 data sets, respectively.

Table 4. Estimates of mtDNA COI (above diagonal) and nuclear rDNA ITS1 region (below diagonal) percent sequence divergence with the two-parameter model of Kimura (1980)

	B. senegalensis	B. forskalii	B. camerunensis	B. cernicus	B. scalaris	B. crystallinus	B. truncatus	A. caerulea
B. senegalensis	—	14·58	14·48	12·59	13·63	14·83	12·72	27·03
B. forskalii	5·15	—	3·96	11·34	10·04	11·57	13·80	22·65
B. camerunensis	5·15	0·44	—	11·18	10·77	11·14	13·47	26·21
B. cernicus	9·43	8·16	8·16	—	8·39	13·67	12·98	27·98
B. scalaris	6·25	4·11	4·11	6·40	—	13·23	13·15	26·12
B. crystallinus	n/a	n/a	n/a	n/a	n/a	—	8·70	17·71
B. truncatus	29·74	27·12	26·89	30·82	28·70	n/a	—	16·79
B. pfeifferi	41·66	39·44	39·44	43·19	38·20	n/a	44·59	—
A. caerulea	53·89	51·13	51·13	57·88	53·60	n/a	57·60	—

distinct branches, demonstrating that these morphologically similar species are clearly differentiated by all molecular approaches. Conversely, cluster analyses place B. camerunensis with specimens unequivocally identified as B. forskalii, confirming the taxonomic status of B. camerunensis is debatable, supporting the observation that although conchologically distinct and compatible with a different parasite, it possesses no allozyme alleles additional to those of B. forskalii (Mimpfoundi & Greer, 1989). In particular its affinity with B. forskalii from nearby Kumba (the closest B. forskalii population to the B. camerunensis site) suggests that B. camerunensis has probably evolved in situ from local B. forskalii which became isolated in the crater lake. A further unexpected outcome of molecular analysis was the clustering of a conchologically extreme geographical variant of B. forskalii from Sâo Tomé (Brown, 1991), with Angolan B. crystallinus. Given the historical links between these two areas such movement of B. crystallinus seems plausible. It is important to test the affiliations between these taxa because B. forskalii from Sâo Tomé is an intermediate host for S. intercalatum while B. crystallinus is suspected to transmit S. haematobium in Angola (Wright, 1963) and maybe a host of S. intercalatum in Gabon (Jelnes & Highton, 1984). Finally, B. cernicus from the island of Mauritius forms another separate group and where included in the analyses, clusters with East African B. scalaris.

Divergence estimates

Owing to the paucity of the fossil record, the evolutionary history of B. forskalii must be inferred from extant species (Brown, 1994). Although grouped by similar morphological traits, there are few indicators of relationships between the B. forskalii group species. However, some indications may be achieved by applying a general time frame using the average rate of mtDNA divergence across the entire molecule; for most organisms 2–4 % per million years (Wilson, Ochman & Prager, 1987). A linear relationship of transitions against percent sequence divergence for the COI suggests that the data set was not significantly affected by 'saturation' or 'multiple hits', which could have caused an underestimate of sequence divergence. An average COI sequence divergence between the sympatric species B. senegalensis and B. forskalii is 14·58 %, indicating an approximate time-scale of 3·6–7·2 million years. Estimated sequence divergence between Central (Cameroon) and East African (Kenya) B. forskalii is 4·47 % dating an approximate divergence to 1·1–2·2 million years. Following Desprès et al. (1992) a time frame specific to the COI region sequenced in Bulinus can be calibrated with sequences homologous to this region from rat and mouse with a known divergence of 9–12 Mya (Jaeger,

Tong & Denys, 1986). Calculated at 1·8–2·4 % per million years this encompasses the lower mtDNA sequence divergence estimates assumed for the general mtDNA clock. However, these should be considered conservative estimates as mollusc mtDNA may evolve more rapidly (Hoeh *et al.* 1996) than vertebrates, suggesting that the use of upper range of the general mtDNA clock may be more appropriate.

Evolution of the B. forskalii *group*

As a group these taxa show little intra-specific variation, but are separated by large sequence divergences which produce unresolved branches on phylogenetic trees. Several scenarios involving taxon-wide catastrophic population crashes, a likely possibility for tropical freshwater snails, could be invoked to explain both range expansion of certain individual taxa and their genesis. East Africa began to cool and become progressively drier some 10 Mya, during the Miocene, leading to the eventual drying out of the lake basins and the extinction of many species (Van Damme, 1984). The palaeoclimate of the Late Miocene–Early Pliocene, about 4–8 Mya, would have provided a mosaic of extreme and temporally unstable habitats for freshwater snails (Roberts *et al.* 1993). This period correlates with the suggested divergence time of *B. forskalii* and *B. senegalensis* and several other *B. forskalii* group taxa suggesting that genesis of the species complex may have been concomitant with environmental change.

Environmental change can promote rapid genesis of novel self-compatible taxa, with little substantive genetic change or adaptive divergence (Gottlieb, 1978; Carson, 1971), often from single propagules. Sudden and severe environmental stress, of the kind probably experienced by freshwater snails during the Miocene, can lead to catastrophic selection of a very few robust individuals to produce derived taxa (Lewis, 1963; Davies, 1993). That the breeding system of pulmonate molluscs, unusual amongst higher invertebrates in their high incidence of self-compatibility, can lead to speciation via this route is reflected in the many taxonomic complexes of both terrestrial (Noble & Jones, 1996) and freshwater representatives, such as the *B. forskalii* group.

While phyletic gradualism inevitably produces a legacy of distinct intermediate forms, arising from common ancestral stock, rapid and abrupt saltational speciation constructs a derivative genome in a few generations, containing a limited repertoire of parental allelic variation and linkages (Noble & Jones, 1996). Molecular analysis of phyletic gradualism produces a well differentiated phylogenetic tree with most branches resolved reflecting the gradual differentiation which has taken place in contrast to the unresolved branches of genetically similar rapidly derived taxa. Some derived taxa (Gottlieb, 1978) may fortuitously isolate complexes of adaptive genes and so persist, eventually expanding their range, but most are ephemeral. The extreme ecological and physiological tolerance of many of the less widespread *B. forskalii* taxa is consistent with this scenario (Brown, 1994).

Anomalous transmission of Schistosoma in the B. forskalii *group*

Only the 'terminal spined' schistosomes use species of the genus *Bulinus* as intermediate hosts in Africa and the Middle East (Rollinson & Southgate, 1987; Rollinson *et al.* 1997). This group contains 11 species, 2 infecting man, *S. haematobium* and *S. intercalatum* both transmitted by species of the *B. forskalii* group and taxa from other *Bulinus* groups. Though *B. forskalii* has a wide distribution throughout much of tropical Africa its transmission of *S. intercalatum* is restricted to Cameroon, São Tomé and Gabon. While *B. senegalensis*, with a more restricted, mainly Sahelian distribution from Senegambia through to Nigeria and Cameroon, transmits *S. haematobium* throughout much of its range. A natural hybridization event between *S. haematobium* and *S. intercalatum* has been documented from Loum, Southwest Cameroon (Southgate, Van Wijk & Wright, 1976; Tchuem Tchuente *et al.* 1997). *S. haematobium* has completely replaced *S. intercalatum* where originally only *S. intercalatum* was present, the shift occurring progressively over 30 years. Hybrid parasites exhibit heterosis, are more fecund and can infect both *B. truncatus* and *B. forskalii*, the local snail intermediate hosts.

This hybridization event emphasises the low fitness of *S. intercalatum* and affords an explanation of the restricted distribution of this parasite, despite the widespread distribution of the intermediate (*B. forskalii*) and definitive hosts (Tchuem Tchuente *et al.* 1996). The situation in Barombi Kotto, a crater lake in Southwest Cameroon, may reflect just such a parasite hybridization event, where *B. camerunensis*, effectively a geographical variant of *B. forskalii*, anomalously transmits *S. haematobium* and not the expected *S. intercalatum*. Comparison of nuclear and mtDNA sequences from schistosomes sampled from Loum and Barombi Kotto would substantiate their origin and afford an explanation for the anomalous transmission of *S. haematobium* in *B. forskalii*.

Within-species variation

Sequence analyses of COI and ITS1 from East and Central African *B. forskalii* show little differentiation within and between localities in Kenya and Cameroon, with no geographic structuring evident, suggesting no clear isolation by distance effect.

Further, RAPD analyses from a more limited survey of West and Central Africa also show limited within-*B. forskalii* differentiation and no evidence of geographical structuring. Enzyme electrophoresis had previously indicated rather low genetic diversity in *B. forskalii* (Jelnes, 1980, 1986; Mimpfoundi & Greer, 1989, 1990), although some alleles showed a geographical pattern in Cameroon (Mimpfoundi & Slootweg, 1991). Further genetic analyses (allozymes – Mimpfoundi & Greer, 1990; microsatellites – Gow *et al.* in press) of wild-caught *B. forskalii* population samples suggests this taxon may be facultatively self-fertile under natural conditions. The lack of pan-African differentiation could be a consequence of the homogenizing effects of considerable gene flow, but this seems unlikely considering the snail's breeding system. Alternatively there could have been little genetic diversity available to the taxon upon its genesis, or selection for a generalist genotype capable of coping with climatic exigencies, might lead to a loss of genetic diversity (Mimpfoundi & Greer, 1990; Noble & Jones, 1996).

Rapid range expansion from a segment of the *B. forskalii* range would similarly result in a widely distributed, genetically depauperate taxon, as would the converse, severe and prolonged bottlenecking followed by range expansion. The very limited differentiation of ITS1 sequences from Cameroon and Kenya, combined with lack of geographical structuring of COI and RAPD data, suggest a rapid and recent origin is the most likely explanation for our observations of these populations.

Resolution of the evolutionary history of the *B. forskalii* complex shows its genesis is linked with the story of climatic change during the Late Tertiary of Africa. The breeding system of these snails, by permitting the rapid establishment of evolutionary novelties, allows the group to exploit marginal habitats. Establishment of novel gene complexes, protected by self-fertilization and experiencing only limited gene flow, clearly provides the most adaptive evolutionary strategy to a constantly changing abiotic or biotic environment. The less widespread *B. forskalii* taxa may represent successful colonization of marginal environments, whereas *B. forskalii* itself has expanded rapidly, coming to dominate more equable habitats.

However, the rapid genesis of taxa within this complex has led to incomplete phylogenetic reconstruction, demanding further sequence analysis and examination of additional taxa, including East African representatives such as *B. barthi* and *B. browni*, for a fuller resolution of the *B. forskalii* group.

ACKNOWLEDGEMENTS

CSJ was initially supported by a Wellcome Trust Biodiversity Fellowship and is currently funded by a BBSRC Advanced Fellowship (grant number 1/AF09056).

REFERENCES

BACKELJAU, T., DEBRUYN, L., DEWOLF, H., JORDAENS, K., VANDONGEN, S., VERHAGEN, R. & WINNEPENNINCKX, B. (1995). Random Amplified Polymorphic DNA (RAPD) and parsimony methods. *Cladistics* **11**, 119–130.

BETTERTON, C., FRYER, S. E. & WRIGHT, C. A. (1983). *Bulinus senegalensis* (Mollusca: Planorbidae) in northern Nigeria. *Annals of Tropical Medicine and Parasitology* **77**, 143–149.

BOWDITCH, B. M., ALBRIGHT, D. G., WILLIAMS, J. G. K. & BRAUN, M. J. (1993). Use of randomlly amplified polymorphic DNA markers in comparative genome studies. In *Methods in Enzymology, Volume 224. Molecular Evolution: Producing the Biochemical Data* (eds. Zimmer, E. A., White, T. J., Cann, R. L. & Wilson, A. C.), pp. 294–309. San Diego, CA: Academic Press.

BOWLES, J., BLAIR, D. & MCMANUS, D. (1992). Genetic variants within the genus *Echinococcus* identified by mitochondrial DNA sequencing. *Molecular and Biochemical Parasitology* **54**, 165–174.

BROWN, D. S. (1991). Freshwater snails of Sâo Tomé, with special reference to *Bulinus forskalii* (Ehrenberg), host of *Schistosoma intercalatum*. *Hydrobiologia* **209**, 141–153.

BROWN, D. S. (1994). *Freshwater Snails of Africa and their Medical Importance*. 2nd edn. London: Taylor & Francis Ltd.

CAMPBELL, G., JONES, C. S., LOCKYER, A. E., HUGHES, S., BROWN, D., NOBLE, L. R. & ROLLINSON, D. (2000). Molecular evidence supports an African affinity of the Neotropical freshwater gastropod, *Biomphalaria glabrata* (Say, 1818), an intermediate host for *Schistosoma mansoni*. *Proceeding of the Royal Society of London, B* **267**, 2351–2358.

CARSON, H. L. (1971). Speciation and the founder principle. *Stadler Symposium* **3**, 51–70.

CHURCH, G. M. & GILBERT, W. (1984). Genomic sequencing. *Proceedings of the National Academy of Sciences, USA* **81**, 1991.

CLARK, A. G. & LANIGAN, C. M. S. (1983). Prospects for estimating nucleotide divergence with RAPDs. *Molecular Biology and Evolution* **10**, 1096–1111.

DAVIS, M. S. (1993). Rapid speciation in plant populations. In *Evolutionary Patterns and Processes* (eds. Lees, D. R. and Edwards, D.), pp. 171–188. Linnean Society Symposium Series 14, Linnean Society of London. London: Academic Press.

DESPRÈS, L., IMBERT-ESTABLET, D., COMBES, C. & BONHOMME, F. (1992). Molecular phylogeny linking hominoid evolution to recent radiation of schistosomes (Platyhelminthes: Trematoda). *Molecular Phylogenetics and Evolution* **1**, 295–304.

FEINBERG, A. P. & VOGELSTEIN, B. (1983). A technique for radiolabeling DNA restriction endonuclease fragments to high specific activity. *Analytical Biochemistry* **132**, 6–13.

FELSENSTEIN, J. (1985). Confidence limits on phylogenies: an approach using the bootstrap. *Evolution* **39**, 783–791.

FELSENSTEIN, J. (1993). *PHYLIP (Phylogeny Inference*

Package) version 3.5c. Seattle, WA: University of Washington.

FITCH, W. M. & MARGOLIASH, E. (1967). Construction of phylogenetic trees. *Science* **155**, 279–284.

GOLL, P. H. (1981). Mixed populations of *Bulinus senegalensis* (Muller) and *B. forskalii* (Ehrenberg) in the Gambia. *Transactions of the Royal Society of Tropical Medicine and Hygiene* **75**, 576–578.

GOTTLIEB, L. D. (1978). Biochemical consequences of speciation of plants. In *Molecular Evolution* (ed. Ayala, F. J.), pp. 123–140. Sunderland, Massachusetts: Sinauer Associates.

GOW, J. L., NOBLE, L. R., ROLLINSON, D. & JONES, C. S. (in press). Polymorphic microsatellites in the African freshwater snail, *Bulinus forskalii* (Gastropoda, Pulmonata). *Molecular Ecology.*

GREER, G. J., MIMPFOUNDI, R., MALEK, E. A., JOKY, A., NGONSEU, E. & RATARD, R. C. (1990). Human schistosomiasis in Cameroon. II. Distribution of the snail hosts. *American Journal of Tropical Medicine and Hygiene* **42**, 573–580.

HASEGAWA, M., KISHINO, H. & YANO, K. (1985). Dating of the human-ape splitting by a molecular clock of mitochondrial DNA. *Journal of Molecular Evolution* **22**, 160–174.

HOEH, W. R., STEWART, D. T., SUTHERLAND, B. W. & ZOUROS, E. (1996). Cytochrome *c* oxidase sequence comparisons suggest an unusually high rate of mitochondrial DNA evolution in *Mytilus* (Mollusca: Bivalvia). *Molecular Biology and Evolution* **13**, 418–421.

JACKSON, D. A., SOMERS, K. M. & HARVEY, H. H. (1989). Similarity coefficients: measures of co-occurrence and association or simply measures of occurrence? *American Naturalist* **133**, 436–453.

JAEGER, J. J., TONG, H. & DENYS, C. (1986). The age of *Mus-Rattus* divergence: paleontological data compared with the molecular clock. *Comptes Rendus de l'Académie des Sciences Series II Paris* **302**, 917–922.

JEANDROZ, S., ROY, A. & BOUSQUET, J. (1997). Phylogeny and phylogeography of the circumpolar genus *Fraxinus* (Oleaceae) based on internal transcribed spacer sequences of nuclear ribosomal RNA. *Molecular Phylogenetics and Evolution* **7**, 241–251.

JELNES, J. E. (1980). Experimental taxonomy on *Bulinus* (Gastropoda: Planorbidae). III. Electrophoretic observations on *Bulinus forskalii, B. browni, B. barthi* and *B. scalaris* from East Africa, with additional electrophoretic data on the subgenus *Bulinus sensu strictu* from other parts of Africa. *Steenstrupia* **6**, 177–193.

JELNES, J. E. (1986). Experimental taxonomy of *Bulinus*: the West and North African species reconsidered, based upon an electrophoretic study of several enzymes per individual. *Zoological Journal of the Linnean Society* **87**, 1–26.

JELNES, J. E. & HIGHTON, R. B. (1984). *Bulinus crystallinus* (Morelet, 1868) acting as intermediate host for *Schistosoma intercalatum* Fisher 1934 in Gabon. *Transactions of the Royal Society of Tropical Medicine and Hygiene* **78**, 412.

JONES, C. S., OKAMURA, B. & NOBLE, L. R. (1994). Parent and larval RAPD fingerprints reveal outcrossing in freshwater bryozoans. *Molecular Ecology* **3**, 172–179.

JONES, C. S., NOBLE, L. R., LOCKYER, A. E., BROWN, D. S. & ROLLINSON, D. (1997). Species-specific primers discriminate intermediate hosts of schistosomes: unambiguous PCR diagnosis of *Bulinus forskalii* group taxa (Gastropoda: Planorbidae). *Molecular Ecology* **6**, 843–849.

JONES, C. S., NOBLE, L. R., OUMA, J., KARIUKI, H. C., MIMPFOUNDI, R., BROWN, D. & ROLLINSON, D. (1999). Molecular identification of schistosome intermediate hosts: case studies of *Bulinus forskalii* group species (Gastropoda: Planorbidae) from Central and East Africa. *Biological Journal of the Linnean Society* **68**, 215–220.

KANE, R. A. & ROLLINSON, D. (1994). Repetitive sequences in the ribosomal DNA internal transcribed spacer of *Schistosoma haematobium, Schistosoma intercalatum,* and *Schistosoma mattheei. Molecular and Biochemical Parasitology* **63**, 153–156.

KRISTENSEN, T. K., YOUSIF, F. & RAAHAUGE, P. (1999). Molecular characterisation of *Biomphalaria* spp. in Egypt. *Journal of Molluscan Studies* **65**, 133–136.

KUMAR, S., TAMURA, K. & NEI, M. (1993). *MEGA: Molecular Evolutionary Genetics Analysis, Version 1.1.* University Park, PA: The Pennysylvania State University.

LANGAND, J., BARRAL, V., DELAY, B. & JOURDANE, J. (1993). Detection of genetic diversity within snail intermediate hosts of the genus *Bulinus* by using Random Amplified Polymorphic DNA markers (RAPDs). *Acta Tropica* **55**, 205–215.

LEWIS, H. (1963). Speciation in flowering plants. *Science* **152**, 167–171.

MANDAHL-BARTH, G. (1957). Intermediate hosts of *Schistosoma. Bulletin of the World Health Organization* **16**, 1103–1163; **17**, 1–65.

MIMPFOUNDI, R. & GREER, G. (1989). Allozyme comparisons among species of the *Bulinus forskalii* group (Mollusca: Planorbidae) in Cameroon. *Journal of Molluscan Studies* **55**, 405–410.

MIMPFOUNDI, R. & GREER, G. (1990). Allozyme variation among populations of *Bulinus forskalii* (Ehrenberg, 1831) in Cameroon. *Journal of Molluscan Studies* **56**, 363–371.

MIMPFOUNDI, R. & SLOOTWEG, R. (1991). Further observations on the distribution of *Bulinus senegalensis* Muller in Cameroon. *Journal of Molluscan Studies* **57**, 487–489.

NOBLE, L. R. & JONES, C. S. (1996). A molecular and ecological investigation of the large arionid slugs of N-W Europe: the potential for new pests. In *The Ecology of Agricultural Pests: Biochemical Approaches. Systematics Association Special Volume Series* (eds. Symondson, W. O. C. & Liddell, J. E.), pp. 93–131. Chapman & Hall: London.

OKAMURA, B., JONES, C. S. & NOBLE, L. R. (1993). Randomly amplified polymorphic DNA analysis of clonal population structure and geographic variation in a freshwater bryozoan. *Proceedings of the Royal Society of London B* **253**, 147–155.

PORTER, C. H. & COLLINS, F. H. (1991). Species-diagnostic difference in a ribosomal DNA internal transcribed spacer from the sibling species *Anopheles freeborni* and *Anopheles hermsi* (Diptera; Culicidae). *American Journal of Tropical Medicine and Hygiene* **45**, 271–279.

RIESEBERG, L. (1996). Homology among RAPD fragments in interspecific comparisons. *Molecular Ecology* 5, 99–105.

RIGBY, P. W. J., DIECKMANN, M., RHODES, C. & BERG, P. (1977). Labeling deoxy-ribonucleic acid to high specific activity by nick translation with DNA polymerase. *Journal of Molecular Biology* 113, 237.

ROBERTS, N., TAIEB, M., BARKER, P., DAMNATI, B., ICOLE, M. & WILLIAMSON, D. (1993). Timing of the Younger Dryas event in East Africa from lake-level changes. *Nature* 366, 146–148.

ROLLINSON, D., KAUKAS, A., JOHNSTON, D. A., SIMPSON, A. J. G. & TANAKA, M. (1997). Some molecular insights into schistosome evolution. *International Journal for Parasitology* 27, 11–28.

ROLLINSON, D. & SOUTHGATE, V. R. (1987). The genus *Schistosoma*: a taxonomic appraisal. In *The Biology of Schistosomes: From Genes to Latrines* (eds. Rollinson, D. & Simpson, Λ. J. G.), pp. 1–49. London: Academic Press.

ROLLINSON, D. & STOTHARD, J. R. (1994). Identification of pests and pathogens by random amplification of polymorphic DNA (RAPDs). In *Identification and Characterization of Pest Organisms* (ed. Hawksworth, D. L.), pp. 447–459. St. Albans, UK: CAB International.

ROSSETTO, M., WEAVER, P. K. & DIXON, K. W. (1995). Use of RAPD analysis in devising conservation strategies for the rare and endangered *Grevillea scapigera* (Proteaceae). *Molecular Ecology* 4, 321–329.

SAMBROOK, J., FRITSCH, E. F. & MANIATIS, T. (1989). *Molecular Cloning, A Laboratory Manual*. 2nd Edition. New York: Cold Spring Harbor Laboratory Press.

SIMON, C., FRATI, F., BECKENBACH, A., CRESPI, B., LIU, H. & FLOOK, P. (1994). Evolution, weighting, and phylogenetic utility of mitochondrial gene sequences and a compilation of conserved polymerase chain reaction primers. *Annals of the Entomological Society of America* 87, 651–701.

SMITH, J. L., SCOTT-CRAIG, J. S., LEADBETTER, J. R., BUSH, G. L., ROBERTS, D. L. & FULBRIGHT, D. W. (1994). Characterization of random amplified polymorphic DNA (RAPD) products from *Xanthomonas campestris* and some comments on the use of RAPD products in phylogenetic analysis. *Molecular Phylogenetics and Evolution* 3, 135–145.

SOUTHGATE, V. R., VAN WIJK, H. B. & WRIGHT, C. A. (1976). Schistosomiasis at Loum, Cameroon: *Schistosoma haematobium, Schistosoma intercalatum* and their natural hybrid. *Zeitschrift für Parasitenkunde* 56, 183–193.

STOTHARD, J. R., HUGHES, S. & ROLLINSON, D. (1996). Variation within the ribosomal DNA internal transcribed spacer (ITS) of intermediate snail hosts within the genus *Bulinus* (Gastropoda: Planorbidae). *Acta Tropica* 61, 19–29.

STOTHARD, J. R. & ROLLINSON, D. (1996). An evaluation of Randomly Amplified Polymorphic DNA (RAPD) for identification and phylogeny within the freshwater snails of the genus *Bulinus* (Gastropoda: Planorbidae). *Journal of Molluscan Studies* 62, 165–176.

STOTHARD, J. R. & ROLLINSON, D. (1997). Partial sequence from cytochrome oxidase subunit (COI) can differentiate *Bulinus globosus* and *B. nasutus*. *Journal of Natural History* 31, 727–737.

STRIMMER, K. & VON HAESELER, A. (1996). Quartet puzzling: a quartet maximum likelihood method for reconstructing tree topologies. *Molecular Biology and Evolution* 13, 964–969.

SWOFFORD, D. L. (1996). *PAUP*: Phylogenetic Analysis Using Parsimony (and Other Methods), version 4.3c*. Sunderland, Massachusetts, Sinauer Association

SWOFFORD, D. L. & OLSEN, G. L. (1990). Phylogeny reconstruction. In *Molecular Systematics* (eds. Hillis, D. M. & Moritz, C.), pp. 411–501. Sunderland, Massachusetts: Sinauer Associates, Inc.

TCHUEM TCHUENTE, L. A., MORAND, S., IMBERT-ESTABLET, D., DELAY, B. & JOURDANE, J. (1996). Competitive exclusion in human schistosomes: the restricted distribution of *Schistosoma intercalatum*. *Parasitology* 113, 129–136.

TCHUEM TCHUENTE, L. A., SOUTHGATE, V. R., NJIOKOU, F., NJINE, T., KOUEMENI, L. E. & JOURDANE, J. (1997). The evolution of schistosomiasis at Loum, Cameroon: replacement of *Schistosoma intercalatum* by *S. haematobium* through introgressive hybridization. *Transactions of the Royal Society of Tropical Medicine and Hygiene* 91, 664–665.

THOMPSON, J. D., GIBSON, T. J., PLEWNIAK, F., JEANMOUGIN, F. & HIGGINS, D. G. (1997). The CLUSTAL X windows interface: flexible strategies for multiple sequence alignment aided by quality analysis tools. *Nucleic Acids Research* 25, 4876–4882.

VAN DAMME, D. (1984). *The Freshwater Mollusca of Northern Africa. Distribution, Biogeography and Palaeoecology*. Dordrecht, The Netherlands: W. Junk.

VAN DE ZANDE, L. & BIJLSMA, R. (1995). Limitations of the RAPD technique in phylogeny reconstruction in *Drosophila*. *Journal of Evolutionary Biology* 8, 645–656.

VERNON, J. G., JONES, C. S. & NOBLE, L. R. (1995). Random polymorphic DNA (RAPD) markers reveal cross-fertilization in *Biomphalaria glabrata* (Pulmonata, Basommatophora). *Journal of Molluscan Studies* 61, 455–465.

WAKELEY, J. (1996). The excess of transitions among nucleotide substitutions: new methods of estimating the transitions bias underscore its significance. *Trends in Ecology and Evolution* 11, 158–163.

WELSH, J. & MCCLELLAND, M. I. (1990). Fingerprinting genomes using PCR with arbitrary primers. *Nucleic Acids Research* 18, 7213–7218.

WILLIAMS, J. G. K., KUBELIK, A. R., LIVAK, K. J., RAFALSKI, J. A. & TINGEY, S. V. (1990). DNA polymorphisms amplified by arbitrary primers are useful as genetic markers. *Nucleic Acids Research* 18, 6531–6535.

WILSON, A. C., OCHMAN, H. & PRAGER, E. M. (1987). Molecular timescale for evolution. *Trends in Genetics* 3, 241–247.

WRIGHT, C. A. (1963). The freshwater gastropod Mollusca of Angola. *Bulletin of the British Museum (Natural History), Zoology* 10, 449–528.

Subject Index

Index compiled by Dr Laurence Errington

Page numbers refer to the first page of each article in which entries are cited.

Printed in the United States
By Bookmasters